SpringerWienNewYork

# CISM COURSES AND LECTURES



The series presents lecture notes, monographs, edited works and proceedings in the field of Mechanics, Engineering, Computer Science and Applied Mathematics.
Purpose of the series is to make known in the international scientific and technical community results obtained in some of the activities organized by CISM, the International Centre for Mechanical Sciences.

INTERNATIONAL CENTRE FOR MECHANICAL SCIENCES

COURSES AND LECTURES - No. 534

# MECHANICAL BEHAVIOUR OF SOILS UNDER ENVIRONMENTALLY INDUCED CYCLIC LOADS

EDITED BY

CLAUDIO DI PRISCO
POLITECNICO DI MILANO, ITALY

DAVID MUIR WOOD
UNIVERSITY OF BRISTOL, UK

SpringerWienNewYork

This volume contains 437 illustrations



SPIN 86017959

All contributions have been typeset by the authors.

ISBN 978-3-7091-1067-6 SpringerWienNewYork

# *PREFACE*

*This volume originates from the lectures delivered during a CISM course held in Udine in Summer 2009. The editors organised that doctoral school on the problem of ratcheting and cyclic loading of soils with the aim of collecting the experience and knowledge on this subject that the geotechnical community has been accumulating over the last five decades. The book is thus aimed at doctoral students, researchers and geotechnical engineers who are interested in adopting the so-called displacement based design methods for geotechnical systems. Displacement based design is now encouraged by all current national and international regulations, where transient loads are not negligible in comparison with permanent actions.*

*The original suggestion that there was a need to gather together, systematically, the huge amount of work that had already been carried out on this topic, to organise the course, and to publish this volume came from Professor Michele Jamiolkowsi. He also proposed the title of the course and the names of the coordinators. For all these reasons, for his scientific support and for his esteem, the Editors want warmly to acknowledge the contribution made by Michele, who has as usual demonstrated his enthusiasm for the promotion of scientific activities related to geotechnical engineering and has always been aware of the current needs of geotechnical designers.*

*The course aimed to provide a comprehensive description of the mechanical response of soils (granular and cohesive materials) under cyclic loading, with the final goal of providing the geotechnical engineer with the theoretical and analytical tools necessary (i) for the evaluation of settlements developing with time under cyclic, environmentally induced, loads (such as wave motion, wind actions, water table level variation) and their consequences for the serviceability and durability of structures such as the shallow or deep foundations used in offshore engineering, caisson breakwaters, ballast and airport pavements and also (ii) to interpret monitoring data, obtained from both natural and artificial slopes and earth embankments, for the purposes of risk assessment and mitigation.*

*In fact, geotechnical designers currently often face many difficulties in evaluating the mechanical effects of non-monotonic loadings because (i) many aspects of the mechanical response of soils under very large numbers of cycles of loading are not yet well known, (ii) constitutive models capable of reproducing the observed response are often sophisticated and appear excessively complex for the end user, and (iii) the computational time required for numerical solution of boundary value problems in which large numbers of cycles of loading are applied is often substantial. In order to provide some rational route through the relevant research findings, this volume follows a path which begins with micromechanical observations, passes through the analysis of laboratory experimental test results on single element soil specimens, and arrives at the analysis of boundary value problems. One of the challenges facing anyone embarking on the modelling or analysis of geotechnical problems is the particle-continuum duality of soils. As an example, the accumulation of irreversible strains through the application of many cycles of very small amplitude of loading results essentially from the fact that in general the representative elementary volume of soil for the purposes of continuum analysis is itself a highly redundant micro-structure composed of millions of grains/clusters of soil particles.*

*The accumulation of irreversible strains, as in general redundant structures, is essentially the result of either non-symmetric loading conditions or non-symmetric geometries. The asymmetry of the mechanical response of soil roughly coincides with the anisotropy of the structure or fabric of the soil - the arrangement of particles and the contacts between particles. As soils are deformed, the probability of the evolution of the internal fabric is closely related to the stability of local contacts between the grains. The non-linearities of the mechanical response of soils are the result at the mesoscale of the large displacements (large in relation to the particle size) occurring at the microscale (grain sliding and grain rotation and rearrangement of stress chains). The evolution of fabric is dominated by the current microstructure and by the applied mechanical perturbation. There may be a progressive stabilisation of strains under cyclic loading towards a so-called plastic adaptation state which can be interpreted as a reduction in the probability of further fabric evolution, whereas, at the same time, an increase in the strain rate (cyclic fatigue) is usually associated with a progressive destabilisation which might result either from evolution of the nature of the contacts between particles (such as the breakdown of bonding in cemented sands or other structured soils) or from progressive damage of the grains themselves (as in carbonate sands, lightweight aggregates, where the particles themselves have a brittle mechanical response, breaking into many pieces when the stress conditions within the particle become too high). The problem of qualitatively and quantitatively assessing the progressive accumulation of strains - and the phenomena of shake-down, ratcheting, and progressive failure - in soils under quasi-static or more rapid cyclic loading is presented through discussion of experimental, theoretical, numerical and computational findings. Drained and undrained deformations of cohesive and frictional materials are considered. The presence of fluids within the pores of the soil generates coupling effects which can be interpreted as a further degree of redundancy for the material microstructure. Different classes of constitutive modelling theories are described including standard non-associated elasto-plastic constitutive models with anisotropic strain hardening rules, generalised plasticity, bounding surface plasticity and viscoplasticity. The response of these different classes of model to cyclic loading is explored.*

*Three chapters are devoted to the challenges of numerical implementation of constitutive models into finite element codes under static and dynamic conditions. Two chapters are used to explore the concept of macroelement modelling, which can provide a very useful engineering tool, built around the definition of appropriate generalised work-conjugate forces and displacements, which can be used for the solution of problems of soil-structure interaction in a simplified manner Such an approach is well supported by programmes of experimental observation and is particularly appropriate for cyclic loading since, by considerably reducing the number of unknowns to be determined, it seriously reduces the computational times and costs. Finally, some engineering problems related to offshore structures (including the soil-structure interaction of deep water risers placed on the sea bottom under the action of waves, and the progressive damage of the soil-pile interface under cyclic environmental loads) are presented in order to demonstrate the practical impact of the phenomena of cyclic loading on geotechnical design at prototype scale.. There is inevitably a great range of approaches to the problems of cyclic loading implicit in the various chapters. This heterogeneity (analysis, numerical modelling, physical modelling, laboratory testing, design, ...) will encourage the reader to approach the same problem from different and sometimes orthogonal viewpoints and to recognise the challenges that remain in the characterisation of the mechanical behaviour of soils under environmentally induced cyclic loads.*

*The editors, even on behalf of all the authors, want to warmly acknowledge Miss Dalila Vescovi for her precious help in revising the texts and editing all the papers.*

*Claudio di Prisco, Milano*
*David Muir Wood, Dundee*

## CONTENTS

Behaviour of Granular Soils Under Environmentally Induced Cyclic Loads
*by T. Wichtmann and T. Triantafyllidis* ................ 1

Constitutive Modelling
*by D. Muir Wood* ................ 137

Creep Versus Transient Loading Effects in Geotechnical Problems
*by C. di Prisco* ................ 227

Mathematical Models for Transient, Dynamic and Cyclic Problems in Geotechnical Engineering
*by M. Pastor, P. Mira and J.A. Fernández Merodo* ...... 263

Discretization Techniques for Transient, Dynamic and Cyclic Problems in Geotechnical Engineering: First Order Hyperbolic Partial Differential Equations
*by M. Pastor* ................ 291

Discretization Techniques for Transient, Dynamic and Cyclic Problems in Geotechnical Engineering: Second Order Equation
*by M. Pastor, J.A. Fernández Merodo, P. Mira, S. López Querol, I. Herreros and M. Mabssout* ................ 329

Cyclic Mechanical Response of Rigid Bodies Interacting With Sand Strata
*by C. di Prisco* ................ 363

Macroelement Modelling
*by D. Muir Wood* ................ 399

Offshore Design Approaches and Model Tests for Sub-Failure Cyclic Loading of Foundations
*by M. F. Randolph* ................ 441

Cyclic Interface Shearing in Sand and Cemented Soils and Application to Axial Response of Piles
*by M. F. Randolph* .......................................... 481

Evaluation of the Remoulded Shear Strength of Offshore Clays and Application to Pipeline-Soil and Riser-Soil Interaction
*by M. F. Randolph* .......................................... 529

# Behaviour of Granular Soils Under Environmentally Induced Cyclic Loads

Torsten Wichtmann * and Theodor Triantafyllidis *

* Institute for Soil Mechanics and Rock Mechanics, Karlsruhe Institute of Technology (KIT), Karlsruhe, Germany

## 1 Introduction

Cyclic loading and its impacts are of practical relevance for many problems in geotechnical engineering. Some examples of non-endogenous nature are illustrated in Fig. 1. A cyclic loading may be caused by traffic (high-speed trains, magnetic levitation trains), industrial sources (crane rails, machine foundations), wind and waves (on-shore and off-shore wind power plants, coastal structures) or repeated filling and emptying processes (watergates, tanks and silos). Furthermore, construction processes (e.g. vibro-installation of sheet piles) and mechanical compaction (e.g. vibratory compaction) impose cyclic loads into the soil. A cyclic loading of the soil may be also caused by endogenous sources. Earthquake events due to a slip between adjacent tectonic plates lead to a propagation of shear waves. The shear waves induce a cyclic shearing of the soil. The cyclic loading of the soil can lead to an accumulation of permanent deformations or to a possible liquefaction due to a build-up of excess pore water pressure.

If the repeated loading involves a large number of cycles ($N > 10^3$) with a relative small strain amplitude ($\varepsilon^{\mathrm{ampl}} < 10^{-3}$) one speaks of a *high-* or *polycyclic* loading. In many cases the cyclic loading is multiaxial, for example due to traffic loading (see the changes of stress components due to a passing wheel illustrated in Fig. 2a) or in the case of offshore wind power plants due to the different directions and frequencies of the wind and wave loading. Furthermore, the time history of the loading may be irregular, that means the amplitude changes from cycle to cycle. As an example, the irregular time history of shear stress during an earthquake is shown schematically in Fig. 2b.

It depends on the boundary conditions if a cyclic loading leads to an accumulation of residual strains and/or to changes in the average stress. Closed stress loops result in not perfectly closed strain loops or vice versa

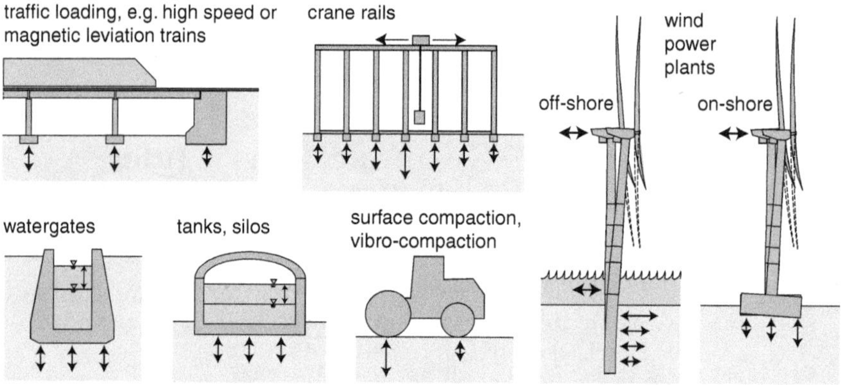


**Figure 1.** Sources of cyclic loading of soils

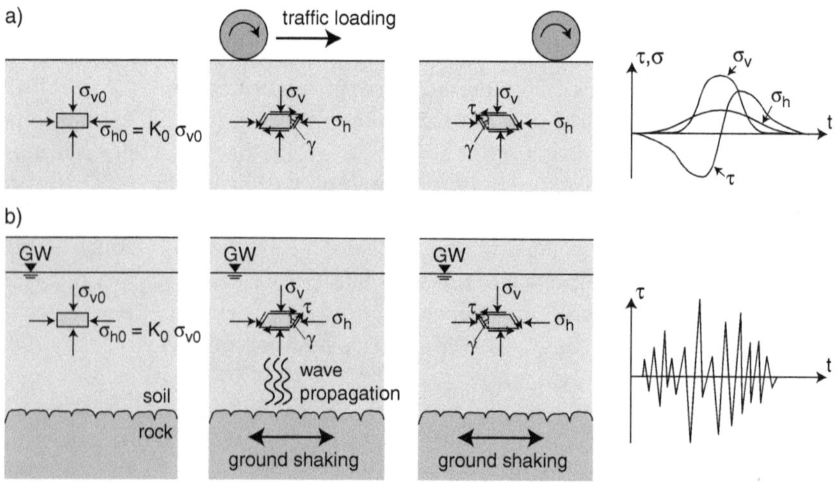


**Figure 2.** Cyclic stresses in a soil element a) due to a wheel passing on the ground surface and b) due to an earthquake loading

(Fig. 3a,b). A simultaneous accumulation of stress and strain in mixed boundary value problems is also possible (Fig. 3c). The accumulation of strain leads to residual settlements of foundations under cyclic loading (Fig. 4). The magnitude of these settlements depends on the loading (average load, load amplitude) and on the current state of the subsoil (void ratio, cyclic preloading). Even small amplitudes can significantly contribute to permanent settlements if the number of cycles is high. Some structures subjected to a cyclic loading are extremely sensitive to differential settlements, which must be kept small in order to ensure the operational or serviceability requirements. For example, offshore wind turbines loose their serviceability if a certain amount of tilting ($2° - 5°$) of the structure is surpassed. High-speed trains also require low tolerances concerning differential settlements of the railway tracks. In statically indeterminate structures the differential settlements may cause changes of internal forces leading to an acceleration of deterioration processes in the structure. Vice versa, a change in the reaction forces leads to a different rate of settlement accumulation. In all these cases an accurate prediction of the residual deformations for the life-time of a structure, that means for several decades of operation is necessary. A suitable method for such a prediction is discussed in detail in Section 3.

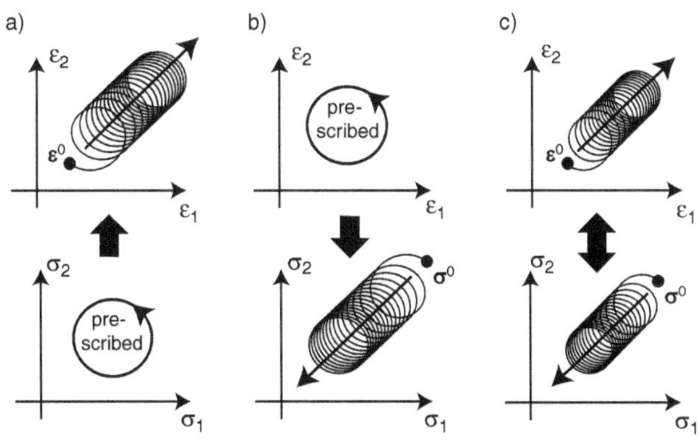


**Figure 3.** Accumulation of stress or strain, illustrated for the two-dimensional case

In water-saturated soils under either partly drained or undrained conditions the pore water pressure may accumulate with increasing number of load cycles. This is due to the fact that the compaction of the soil usually occurring under drained conditions is prevented. The effective stress, and with it both the shear strength and the stiffness, decrease. The decreased

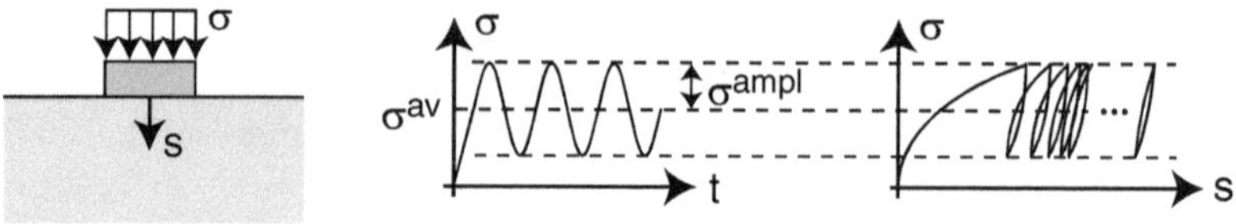


**Figure 4.** Accumulation of the settlement of a foundation due to cyclic loading

shear strength may lead to a loss of the bearing capacity of a foundation. In the extreme case the pore water pressure may equalize the total stress, the effective stress becomes nil and the shear strength vanishes. The soil then behaves like a suspension and this phenomenon is called "liquefaction" (a more detailed discussion of this aspect is in the following Section 4).

In the context of foundations subjected to cyclic loading, one may distinguish between the *short-term* and the *long-term* behaviour. Studies of the short-term behaviour deal with the deformations of the structure and the subsoil within a few cycles (e.g. examination of the dynamic characteristics of a system). In the majority of such studies a linear elastic response is assumed, neglecting any possible changes in the soil state variables during the event. A non-linear behaviour of the soil can be considered by performing implicit finite element (FE) calculations. In long-term studies the accumulation of settlements or the changes in the soil-structure interaction are of main concern.

If the load cycles are applied at a low amplitude and with a low frequency $f = \omega/(2\pi)$, the inertial forces are negligible and the cyclic loading is considered to be *quasi-static.* If the frequency is large, inertial forces are relevant and the loading is *dynamic.* A harmonic excitation with the displacement u $= \mathrm{u}^{\mathrm{ampl}} \cos(\omega t)$ can be considered as quasi-static, if $\mathrm{u}^{\mathrm{ampl}}\omega^2$ is significantly smaller than the acceleration of gravity $g$. Often the amplitude-dependence is ignored and the transition from a quasi-static to a dynamic loading is considered to be at a frequency of $f \approx 5$ Hz.

After a short description of available test devices in Section 2, the subsequent sections concentrate on three important aspects of cyclic loading of soils. Section 3 deals with the accumulation of strains due to a drained polycyclic loading. The main influencing parameters are discussed within the framework of a high-cycle accumulation (HCA) model which can be used for predictions of long-term deformations of foundations. Section 4 discusses the main influencing parameters with respect to the accumulation of pore water pressure due to an undrained cyclic loading. A brief summary of available methods for an estimation of the liquefaction risk in situ due to an earthquake loading is also given. Finally, Section 5 deals with the most important influencing parameters concerning the secant elastic stiffness (small-strain stiffness) which is usually used in elastic calculations of the short-term behaviour. Sections 3 to 5 are mainly dedicated to non-cohesive soils. Findings from the literature are compared to test results of the authors.

## 2 Test Devices

The behaviour of either soil or foundations under repeated loading can be studied by performing different types of tests:

- element tests in the laboratory
- small-scale model tests
- model tests with increased gravitation (centrifuge model tests)
- large-scale model tests
- in-situ tests and measurements at real buildings (observational method)

The following sections focus on soil behaviour measured in element tests performed in the laboratory. Different types of test devices are available. They are illustrated schematically in Fig. 5:

a) axisymmetric triaxial tests on cylindrical specimens
b) true triaxial tests on cubical specimens
c) torsional shear tests on hollow cylinder specimens
d) simple shear tests
e) direct shear tests
f) shaking table tests
g) resonant column tests
h) measurement of wave velocities by means of piezoelectric elements

In most test devices both, a control of the stresses or loads induced at the specimen boundaries or a control of the boundary displacements are possible. The different types of tests are briefly illustrated for a stress control in the following.

In most cyclic triaxial tests on cylindrical specimens (Fig. 5a), the axial stress $\sigma_1$ is cyclically varied, while the lateral stress $\sigma_2 = \sigma_3$ is kept constant. Therefore, only uniaxial cycles with an inclination of the stress path of 1:3 in the $p$-$q$-plane are tested (where $p = (\sigma_1 + 2\sigma_3)/3$ and $q = \sigma_1 - \sigma_3$ are Roscoe's stress invariants). When $\sigma_3$ is oscillating in phase with $\sigma_1$, different inclinations of the stress cycles in the $p$-$q$-plane can be studied. A phase shift between $\sigma_1(t)$ and $\sigma_3(t)$ results into elliptically shaped stress loops. A scheme of the cyclic triaxial device used by the authors is shown schematically in Fig. 6a.

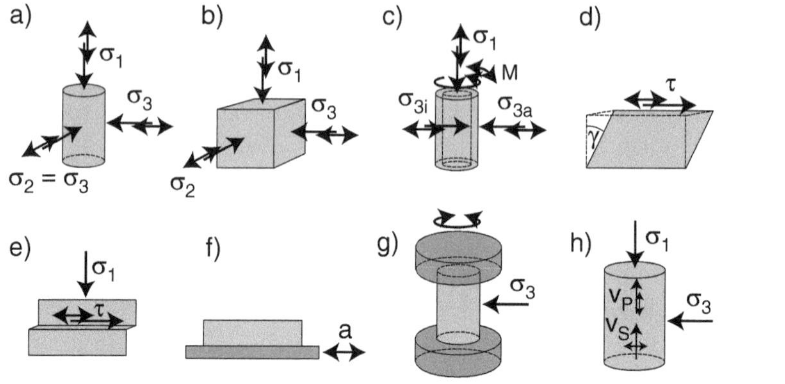


**Figure 5.** Different types of laboratory tests for studies of soil behaviour under cyclic loading

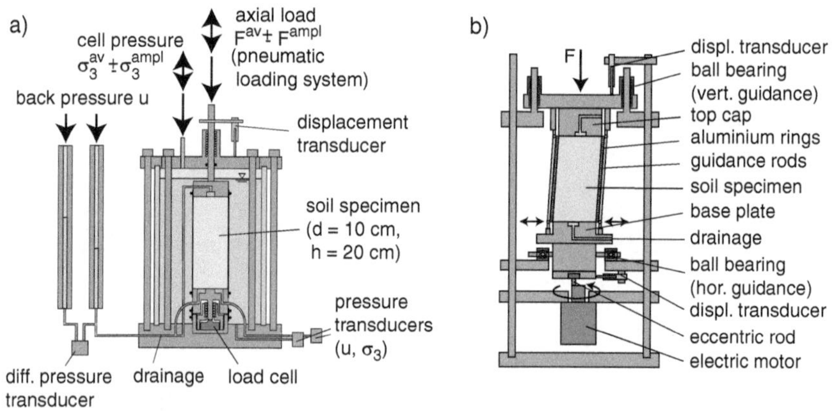


**Figure 6.** a) Cyclic triaxial device and b) cyclic multidimensional simple shear (CMDSS) device at Ruhr-University Bochum

The cycles in the axisymmetric triaxial test are at most two-dimensional. Three-dimensional cycles can be tested in a true triaxial test (Fig. 5b) with a cyclic variation in all three principal stresses. During torsional shear tests on hollow cylinder specimens (Fig. 5c) the cyclic variation of the inner pressure $\sigma_{3i}$, the outer pressure $\sigma_{3a}$, the axial stress $\sigma_1$ and the torsional moment $M$ results into four-dimensional stress loops.

In a simple shear test (Fig. 5d) either a shear stress or a displacement is imposed on the upper boundary and the lateral boundaries are forced to a linear displacement over the specimen height. The lateral stresses are seldom measured. The problem of inhomogeneous stress and strain fields within a specimen has been addressed for example by Budhu (1984) and Budhu and Britto (1987). Some modified simple shear devices have been used to apply a multiaxial cyclic shearing (see the CMDSS test device used by the authors shown schematically in Fig. 6b).

The applicability of a direct shear test (Fig. 5e) for studying the material behaviour under cyclic loading is limited. Such tests can be used to examine changes in the granulometry within a shear band during cyclic loading (Helm et al., 2000) or for studying the contact behaviour under cyclic loading (Malkus, 2000).

Shaking table tests (Fig. 5f) are sometimes applied in liquefaction studies. In those tests a water-saturated sand layer is subjected to an earthquake loading. In some studies several shaking tables were mounted along orthogonal directions onto each other in order to apply a multidimensional cyclic loading.

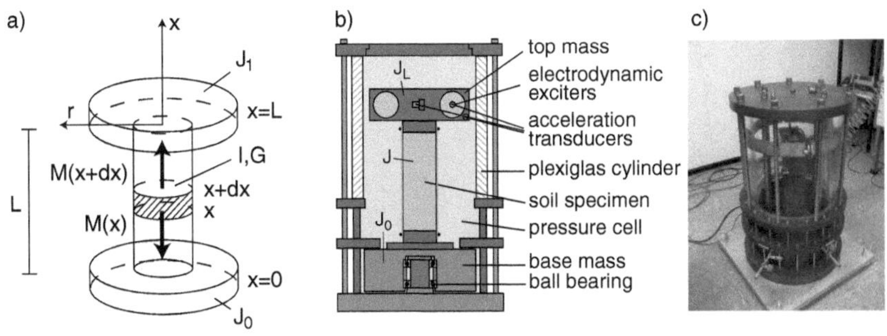


**Figure 7.** a) Principle, b) scheme and c) photo of the resonant column (RC) device at Ruhr-University Bochum

Resonant Column (RC) tests (Fig. 5g) are applied for the determination of the small-strain shear modulus. The system of the RC test consists of a cylindrical specimen, the base mass and the top mass (Fig. 7a). The RC devices are further distinguished depending on the bearing of these masses or the excitation system. The RC device used by the authors (Fig. 7b,c) is of the "free-free" type, that is both the top and the base mass are fully rotatable. The system of the specimen and the end masses is dynamically excited by a torsional moment. The frequency of excitation is varied in order to determine the resonant frequency $f_R$. The secant shear stiffness $G$ is determined from $f_R$:

$$G = \left(\frac{2\,\pi\,h\,f_R}{a}\right)^2 \varrho \tag{1}$$

Therein $h$ is the height of the specimen, $\varrho$ is the soil density and $a$ is a parameter depending on the mass moments of inertia $J$, $J_0$ and $J_L$ of the specimen and the end masses (Fig. 7a, for details see Wichtmann and Triantafyllidis (2009a)). Material damping can be obtained using different methods. Observing the decay of free vibration it can be calculated from

$$D = \frac{1}{2\pi}\,\Delta_1 \tag{2}$$

wherein $\Delta_1$ is the inclination of a log-linear plot of the strain amplitude versus the number of cycles (Fig. 8a). Alternatively, the damping ratio $D$ can be computed as it follows from the area $\Delta W$ of the Lissajous figure (Fig. 8b):

$$D = \frac{1}{4\pi}\,\frac{\Delta W}{W_e} \tag{3}$$

with $W_e$ being the elastic energy of the specimen when the maximum shear strain is applied (Wichtmann and Triantafyllidis, 2009a). Using the half power method (Fig. 8c) the damping ratio is calculated from

$$D = \frac{f_2 - f_1}{f_R} \tag{4}$$

wherein $f_1$ and $f_2$ are the frequencies at an amplitude which is $1/\sqrt{2}$ times smaller than that at the resonant frequency $f_R$.

Piezoelectric elements may be used to measure the shear (S-) wave velocity $v_S$ and the compression (P-) wave velocity $v_P$ in soil samples (e.g. in the triaxial test). The waves generated by piezoelectric elements usually propagate with strain amplitudes less than $10^{-6}$. The triaxial device used by the authors is schematically illustrated in Fig. 9a. The travel time of the wave in the specimen is determined by comparing the transmitted with the received signal (Fig. 9b). Since the distance between the two piezoelectric

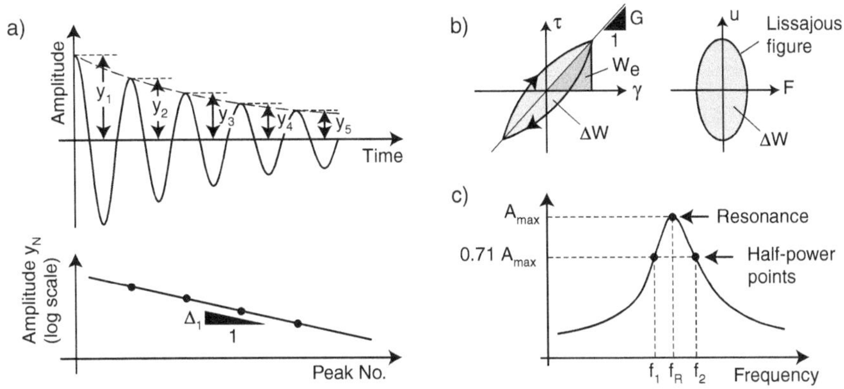


**Figure 8.** Different methods to determine material damping in a RC test: a) from the decay of free vibration, b) from the area of the Lissajous figure and c) using the half power method

elements on the base and the top cap is known (the distance of the tips is used in the case of bender elements) the wave velocities can be calculated. The small-strain shear modulus $G_{\max}$ and the small-strain constrained elastic modulus $M_{\max}$ can be obtained from

$$G_{\max} = \varrho\,(v_S)^2 \qquad \text{and} \qquad M_{\max} = \varrho\,(v_P)^2. \tag{5}$$

Five independent wave velocities can be determined in a cylindrical soil sample if piezoelectric elements for measurements in the horizontal direction are available (Fig. 9c).

Fig. 10 gives a useful overview about the range of strain amplitudes usually tested in the different devices.

# 3 Accumulation of Strain Due to a Drained Polycyclic Loading

## 3.1 Strategies for FE Calculations

To compute irreversible displacements induced by cyclic loading by means of finite element codes, two different calculation strategies can be used. One is called *pure implicit* and it is illustrated for a shallow foundation in Fig. 11a. Each cycle is calculated with many increments using a conventional constitutive model formulated in terms of the rates of stress ($\dot{\boldsymbol{\sigma}}$) and strain ($\dot{\boldsymbol{\varepsilon}}$). Elastoplastic multi-surface models (e.g. Mróz et al., 1978; Chaboche, 1994)

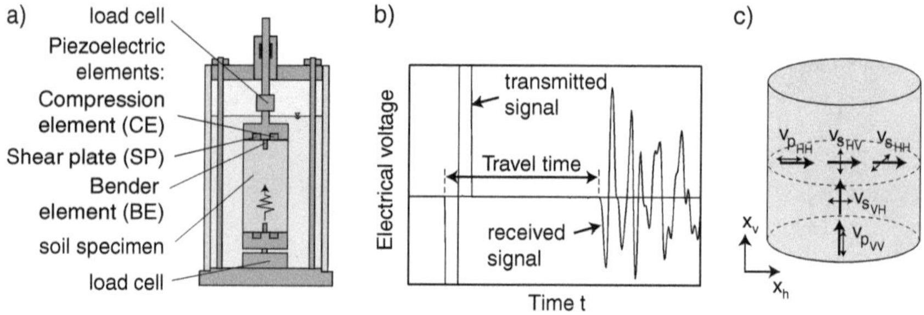


**Figure 9.** Measurement of wave velocities in triaxial specimens: a) test device at Ruhr-University Bochum, b) typical signals of a P-wave measurement, c) five independent wave velocities measured in the horizontal or vertical direction of a triaxial sample

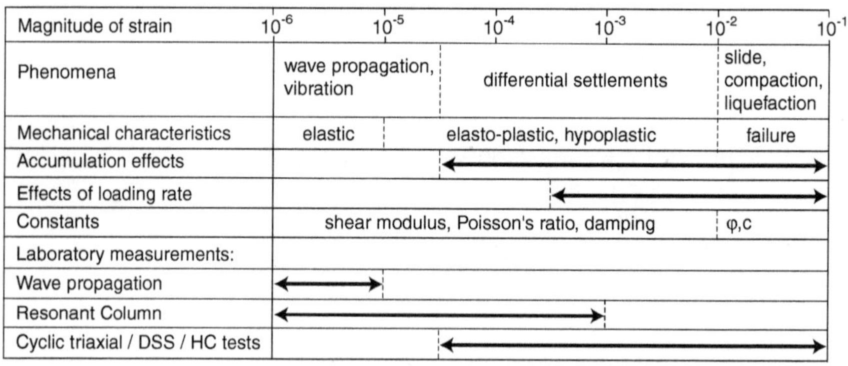


**Figure 10.** Typical ranges of shear strain amplitude $\gamma^{\mathrm{ampl}}$ for different types of tests (modified after Ishihara, 1995)

or the hypoplastic model with intergranular strain (e.g. von Wolffersdorff, 1996; Niemunis and Herle, 1997) may be used for this purpose. The pure implicit strategy is suitable only for small numbers of cycles ($N < 50$). For larger numbers of cycles the numerical error becomes excessive (Niemunis et al., 2005b), let alone the large calculation effort.

A calculation strategy suitable for larger numbers of cycles combines *implicit* and *explicit* parts (Fig. 11b). Only a few cycles are calculated implicitly with many strain increments. Larger packages of cycles between are calculated explicitly. The explicit parts of the calculation require a constitutive formulation which takes packages of cycles $\Delta N$ as input and which predicts the permanent strain due to these packages directly, without tracing the oscillating strain path during the individual cycles. Such a constitutive formulation is called "high-cycle accumulation (HCA) model". The external load is kept constant during the explicit parts of the calculation. Therefore, the accumulation of permanent strain due to cyclic loading is treated similarly as a creep deformation in viscoplastic models.

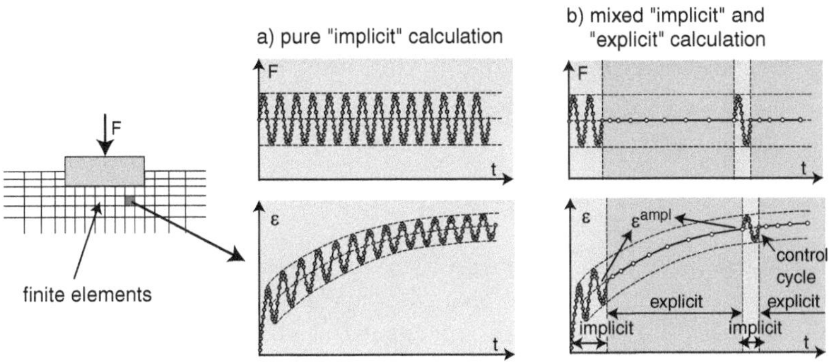


**Figure 11.** FE calculation of the settlements of a shallow foundation under cyclic loading: a) Pure implicit versus b) combined implicit and explicit calculation

An important input parameter of the HCA model proposed by Niemunis et al. (2005b) (Section 3.2) is the strain amplitude $\varepsilon^{\text{ampl}}$. In order to determine the spatial field of the strain amplitude the implicit parts of the calculation are necessary. The first cycle may be irregular since the deformations in the first cycle can significantly differ from those in the subsequent cycles. The second cycle is more representative for the elastic portion of deformation during the subsequent regular cycles. Therefore, the strain amplitude is determined from the second cycle. For that purpose the strain path during the cycle is recorded in each integration point of the FE model.

The strain amplitude is determined from the strain path according to the procedure described by Niemunis et al. (2005b). During the explicit parts of the calculation the strain amplitude is assumed to be constant. During cyclic loading the spatial field of the strain amplitude may change due to compaction or a re-distribution of stress. Therefore, the explicit calculation should be interrupted after a certain number of cycles (e.g. at $N = 10$, 100, 1000, etc.) and a so-called control cycle should be calculated implicitly in order to update the strain amplitude (Fig. 11b). The necessary number of control cycles depends on the boundary value problem and on the material under consideration.

Deficits (lack of generality, missing influencing parameters, 1D formulation) of older HCA models proposed in the literature (Marr and Christian, 1981; Diyaljee and Raymond, 1982; Bouckovalas et al., 1984; Sawicki and Świdziński, 1987, 1989; Kaggwa et al., 1991; Gotschol, 2002) have been discussed by Wichtmann (2005). The HCA model more recently proposed by Abdelkrim et al. (2006) for ballast materials needs an experimental validation based on cyclic laboratory tests.

### 3.2 HCA Model Proposed by Niemunis et al. (2005b)

The basic constitutive equation of the HCA model proposed by Niemunis et al. (2005b) reads

$$\dot{\boldsymbol{\sigma}} = \mathsf{E} : (\dot{\boldsymbol{\varepsilon}} - \dot{\boldsymbol{\varepsilon}}^{\text{acc}} - \dot{\boldsymbol{\varepsilon}}^{\text{pl}}) \tag{6}$$

with the rate $\dot{\boldsymbol{\sigma}}$ of the effective stress $\boldsymbol{\sigma}$ (compression positive), the strain rate $\dot{\boldsymbol{\varepsilon}}$ (compression positive), the given accumulation rate $\dot{\boldsymbol{\varepsilon}}^{\text{acc}}$, a plastic strain rate $\dot{\boldsymbol{\varepsilon}}^{\text{pl}}$ (for stress paths touching the yield surface, see Niemunis et al., 2005b) and the pressure-dependent elastic stiffness $\mathsf{E}$. In the context of HCA models "rate" means a derivative with respect to the number of cycles $N$, that means $\dot{\sqcup} = \partial \sqcup / \partial N$. Depending on the boundary conditions, the model predicts an accumulation of strain ($\dot{\boldsymbol{\varepsilon}} \neq \mathbf{0}$, pseudo-creep) and/or a change of stress ($\dot{\boldsymbol{\sigma}} \neq \mathbf{0}$, pseudo-relaxation).

For $\dot{\boldsymbol{\varepsilon}}^{\text{acc}}$ in Eq. (6) the following expression is used:

$$\dot{\boldsymbol{\varepsilon}}^{\text{acc}} = \dot{\varepsilon}^{\text{acc}} \, \mathbf{m} \tag{7}$$

with the *direction* of strain accumulation $\mathbf{m} = \dot{\boldsymbol{\varepsilon}}^{\text{acc}} / \|\dot{\boldsymbol{\varepsilon}}^{\text{acc}}\|$ (flow rule, unit tensor) and the *intensity* of strain accumulation $\dot{\varepsilon}^{\text{acc}} = \|\dot{\boldsymbol{\varepsilon}}^{\text{acc}}\|$. The flow rule of the modified Cam clay (MCC) model

$$\mathbf{m} = \left[ \frac{1}{3} \left( p - \frac{q^2}{M^2 p} \right) \mathbf{1} + \frac{3}{M^2} \boldsymbol{\sigma}^* \right]^{\rightarrow} \tag{8}$$

**Table 1.** Summary of the functions, material constants and reference quantities of the HCA model

| Influencing parameter | Function | Mat. const. | Reference quantities |
|---|---|---|---|
| Strain amplitude | $f_{\text{ampl}} = \min \left\{ \left( \frac{\varepsilon^{\text{ampl}}}{\varepsilon^{\text{ampl}}_{\text{ref}}} \right)^2 ; 100 \right\}$ | | $\varepsilon^{\text{ampl}}_{\text{ref}} = 10^{-4}$ |
| Cyclic preloading | $\dot{f}_N = \dot{f}_N^A + \dot{f}_N^B$ <br> $\dot{f}_N^A = C_{N1} C_{N2} \exp \left[ -\frac{g^A}{C_{N1} f_{\text{ampl}}} \right]$ <br> $\dot{f}_N^B = C_{N1} C_{N3}$ | $C_{N1}$ <br> $C_{N2}$ <br> $C_{N3}$ | |
| Mean pressure | $f_p = \exp \left[ -C_p \left( \frac{p^{\text{av}}}{p_{\text{ref}}} - 1 \right) \right]$ | $C_p$ | $p_{\text{ref}} = 100$ kPa |
| Stress ratio | $f_Y = \exp \left( C_Y \ \bar{Y}^{\text{av}} \right)$ | $C_Y$ | |
| Void ratio | $f_e = \frac{(C_e - e)^2}{1+e} \ \frac{1+e_{\text{ref}}}{(C_e - e_{\text{ref}})^2}$ | $C_e$ | $e_{\text{ref}} = e_{\max}$ |

is used for $\mathbf{m}$ since it approximates well the ratios $\dot{\varepsilon}_v^{\text{acc}} / \dot{\varepsilon}_q^{\text{acc}}$ measured in drained cyclic triaxial tests with $\dot{\varepsilon}_v = \dot{\varepsilon}_1 + 2\dot{\varepsilon}_3$ and $\dot{\varepsilon}_q = 2/3(\dot{\varepsilon}_1 - \dot{\varepsilon}_3)$ being the rates of volumetric or deviatoric strain accumulation, respectively. The superposed arrow in Eq. (8) denotes Euclidean normalization. The superposed star $\sqcup^*$ means the deviatoric part of $\sqcup$ and $p$,$q$ are Roscoe's invariants. For triaxial extension ($\eta = q/p < 0$) a small modification $M = F \ M_c$ is used with

$$F = \begin{cases} 1 + M_e/3 & \text{for } \ \eta \leq M_e \\ 1 + \eta/3 & \text{for } \ M_e < \eta < 0 \\ 1 & \text{for } \ \eta \geq 0 \end{cases} \tag{9}$$

wherein

$$M_c = \frac{6 \sin \varphi_c}{3 - \sin \varphi_c} \quad \text{and} \quad M_e = -\frac{6 \sin \varphi_c}{3 + \sin \varphi_c}. \tag{10}$$

$\varphi_c$ is the critical friction angle.

The intensity of strain accumulation $\dot{\varepsilon}^{\text{acc}}$ in Eq. (7) is instead calculated as a product of six functions:

$$\dot{\varepsilon}^{\text{acc}} = f_{\text{ampl}} \ \dot{f}_N \ f_e \ f_p \ f_Y \ f_\pi \tag{11}$$

Each function (see Tab. 1) considers the influence of a different parameter. The function $f_{\text{ampl}}$ describes the proportionality between $\dot{\varepsilon}^{\text{acc}}$ and the square of the strain amplitude $(\varepsilon^{\text{ampl}})^2$. It is valid up to strain amplitudes $\varepsilon^{\text{ampl}} \approx 10^{-3}$. For larger strain amplitudes, the accumulation rate was observed to be almost independent of $\varepsilon^{\text{ampl}}$ (Wichtmann, 2005). Therefore $f_{\text{ampl}} = 100$ is specified as an upper boundary in Tab. 1. The HCA model is primarily devoted to small strain amplitudes $\varepsilon^{\text{ampl}} < 10^{-3}$.

The model incorporates a tensorial definition of the amplitude for multidimensional strain loops (Niemunis, 2003; Niemunis et al., 2005b). Its scalar value $\varepsilon^{\text{ampl}}$ is determined from multiple projections of the original six-dimensional strain loop. For the three-dimensional case these projections are shown in Fig. 12. In the case of uniaxial cycles the novel amplitude definition is equal to the conventional one (i.e. $\sqcup^{\text{ampl}} = \frac{1}{2}(\sqcup^{\max} - \sqcup^{\min})$). The treatment of more complex strain loops (Fig. 13) has been discussed by Niemunis et al. (2007a).

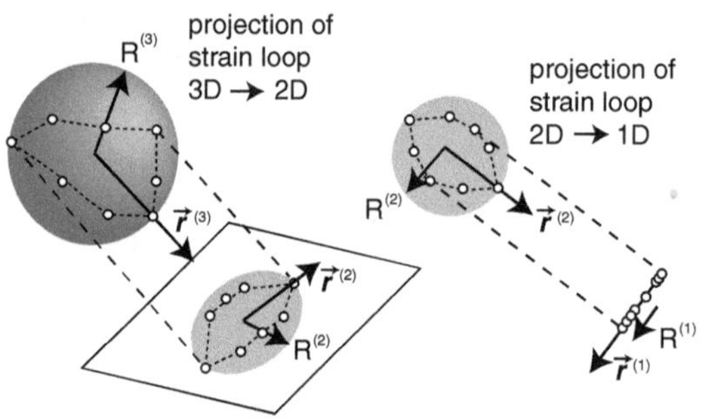


**Figure 12.** Definition of a multiaxial amplitude: Multiple projections of the 6-D strain path (Niemunis et al., 2005b)

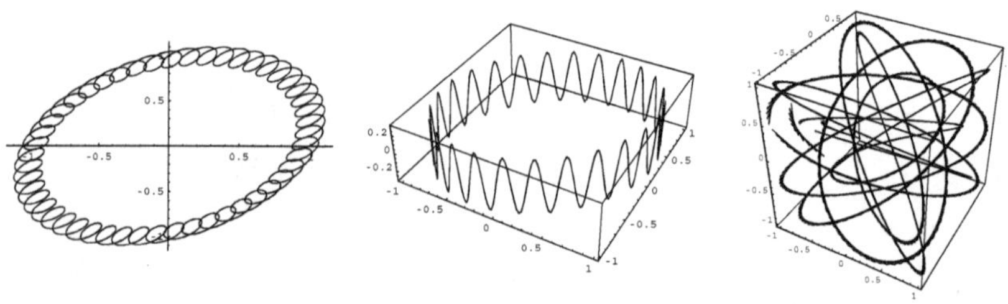

**Figure 13.** Strain loops generated by superposition of harmonic oscillations (Niemunis et al., 2007a)

The increase in $\dot{\varepsilon}^{\text{acc}}$ with decreasing average mean pressure $p^{\text{av}}$ at $\eta^{\text{av}}$

= constant and with increasing average stress ratio $\eta^{\mathrm{av}} = q^{\mathrm{av}}/p^{\mathrm{av}}$ at $p^{\mathrm{av}}$ = constant is captured by the functions $f_p$ and $f_Y$, respectively. In the function $f_Y$ the stress ratio is described by the normalized quantity $\bar{Y}^{\mathrm{av}}$ instead of $\eta^{\mathrm{av}}$, using the function $Y$ of Matsuoka and Nakai (1982):

$$\bar{Y} = \frac{Y-9}{Y_c-9} \quad \text{with} \quad Y_c = \frac{9-\sin^2\varphi_c}{1-\sin^2\varphi_c} \quad \text{and} \tag{12}$$

$$Y = -\frac{I_1 I_2}{I_3} = \frac{27(3+\eta)}{(3+2\eta)(3-\eta)} \tag{13}$$

The $I_i$ in Eq. (13) are the basic invariants of the effective stress $\boldsymbol{\sigma}$.

The function $f_e$ in Eq. (11) describes the increase in the rate $\dot{\varepsilon}^{\mathrm{acc}}$ with increasing void ratio $e$. In the cyclic triaxial tests the curves $\varepsilon^{\mathrm{acc}}(N)$ of the residual strain versus the number of cycles were found to run proportional to the function $f_N$:

$$f_N = C_{N1}\left[\ln(1+C_{N2}N) + C_{N3}N\right]. \tag{14}$$

It consists of a logarithmic and a linear portion. The derivative with respect to $N$ is

$$\dot{f}_N = \underbrace{\frac{C_{N1}C_{N2}}{1+C_{N2}N}}_{\dot{f}_N^A} + \underbrace{C_{N1}C_{N3}}_{\dot{f}_N^B} \tag{15}$$

It can be splitted into a $N$-dependent portion $\dot{f}_N^A$ and a constant portion $\dot{f}_N^B$. However, the number of cycles $N$ alone is not a suitable state variable for the quantification of cyclic preloading (historiotropy) since it contains no information about the intensity of the cycles in the past. For that reason, the preloading (historiotropic) variable $g^A$ was introduced into the HCA model. It counts the cycles weighting them with their amplitude

$$g^A = \int f_{\mathrm{ampl}}\, \dot{f}_N^A \,\mathrm{dN} \tag{16}$$

Only the $N$-dependent portion of $\dot{f}_N$ is considered for $g^A$. The function $\dot{f}_N^A$ was re-formulated using $g^A$ instead of $N$ (Tab. 1). The HCA model with $g^A$ is able of predicting correctly the accumulation of strains due to packages of cycles with different amplitudes applied in different sequences. The model approximately obeys Miner's rule (Miner, 1945) known from fatigue mechanics of metals, that means the sequence of the packages of cycles is of no importance, which is in accordance with the experimental results presented in Section 3.3.

The factor $f_\pi$ considers that a change of the polarization of the cycles, that means a change in the direction of the cyclic loading, leads to an increase in the rate of accumulation. A detailed description of $f_\pi$ has been provided by Niemunis et al. (2005b). For cyclic triaxial tests with $\sigma_3 =$ constant $f_\pi = 1$ holds.

For axisymmetric element tests it is convenient to rewrite Eq. (6) with Roscoe's invariants:

$$\begin{bmatrix} \dot{p} \\ \dot{q} \end{bmatrix} = \begin{bmatrix} K & 0 \\ 0 & 3G \end{bmatrix} \begin{bmatrix} \dot{\varepsilon}_v - \dot{\varepsilon}^{\text{acc}}\, m_v \\ \dot{\varepsilon}_q - \dot{\varepsilon}^{\text{acc}}\, m_q \end{bmatrix} \tag{17}$$

Omitting $\dot{\varepsilon}^{\text{pl}}$ in Eq. (6) is legitimate for homogeneous stress fields. The bulk modulus $K = \frac{E}{3(1-2\nu)}$ and the shear modulus $G = \frac{E}{2(1+\nu)}$ are expected to be pressure-dependent. The volumetric ($m_v$) and the deviatoric ($m_q$) portions of the flow rule are:

$$\begin{bmatrix} m_v \\ m_q \end{bmatrix} = \frac{1}{\sqrt{\frac{1}{3}\left(p - \frac{q^2}{M^2 p}\right)^2 + 6\left(\frac{q}{M^2}\right)^2}} \begin{bmatrix} p - \frac{q^2}{M^2 p} \\ 2\frac{q}{M^2} \end{bmatrix} \tag{18}$$

The determination of a set of material constants has been explained in detail by Wichtmann et al. (2010a). Wichtmann et al. (2009) proposed a simplified calibration procedure in which the parameters required in the HCA model can be estimated based on granulometric or index properties.

### 3.3 Parameters Influencing the Rate of Strain Accumulation

**Direction of Accumulation** Luong (1982) was probably the first one who observed, that it depends on the average stress if a sand shows a contractive or a dilative behaviour under cyclic loading. He applied packages of 20 cycles at different average deviatoric stresses $q^{\text{av}}$ (Fig. 14). Fig. 14b shows the measured $q$-$\varepsilon_v$-loops. Below a certain value of $q^{\text{av}}$ the material behaviour was contractive while it was dilative at larger average deviatoric stresses. Luong defined a borderline in the $p$-$q$-plane (the so-called "characteristic threshold (CT) line") separating the contractive and the dilative material behaviour. This borderline was found to be independent of the soil density.

A second important study on the direction of accumulation was conducted by Chang and Whitman (1988). In a series of cyclic triaxial tests on medium coarse to coarse sand they varied the average stress ratio $\eta^{\text{av}}$ while the average mean pressure $p^{\text{av}}$ was kept constant. Four tests were performed

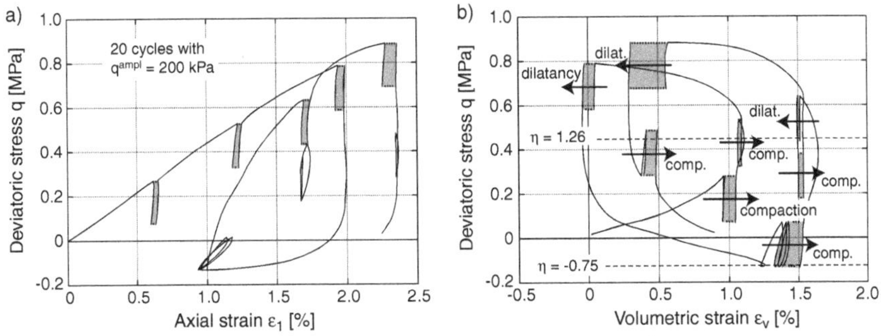


**Figure 14.** Contractive or dilative behaviour of sand under cyclic loading in dependence of the average stress: a) $q$-$\varepsilon_1$-loops and b) $q$-$\varepsilon_v$-loops measured by Luong (1982)

on loose and four other ones on dense samples. In Fig. 15a the residual volumetric strain after 100 cycles is shown as a function of the average stress ratio $\eta^{\text{av}} = q^{\text{av}}/p^{\text{av}}$. Independently of the density of the sand, the rate of volumetric strain accumulation $\dot{\varepsilon}_v^{\text{acc}}$ vanished at an average stress ratio $\eta^{\text{av}}$ equal to the critical one $M_c(\varphi_c)$. For $\eta^{\text{av}} < M_c(\varphi_c)$ a compaction and for $\eta^{\text{av}} > M_c(\varphi_c)$ a dilative material behaviour was observed. Chang and Whitman (1988) assumed the CT-line of Luong (1982) to be identical with the critical state line. They observed an increase in the ratio $\gamma^{\text{acc}}/\varepsilon_v^{\text{acc}}$ with increasing $\eta^{\text{av}}$ (Fig. 15b). For different sands they found a good approximation of the measured ratios $\varepsilon_v^{\text{acc}}/\gamma^{\text{acc}}$ by the flow rule of the modified Cam Clay model $\omega = ({M_c}^2 - (\eta^{\text{av}})^2)/(2\eta^{\text{av}})$. An influence of the average mean pressure $p^{\text{av}}$ and of the amplitude ratio $\zeta = q^{\text{ampl}}/p^{\text{av}}$ on the direction of strain accumulation could not be detected. Also the influence of the number of cycles was reported to be negligible. However, the maximum number of cycles in the tests of Chang and Whitman (1988) was rather small ($N_{\text{max}} \leq$ 1,050).

Our own test results also demonstrate that the direction of accumulation is mainly governed by the average stress ratio $\eta^{\text{av}} = q^{\text{av}}/p^{\text{av}}$. Fig. 16a presents the $\varepsilon_q^{\text{acc}}$-$\varepsilon_v^{\text{acc}}$-strain paths in cyclic triaxial tests with different average stress ratios $\eta^{\text{av}}$ (Wichtmann et al., 2006a). The average mean pressure was $p^{\text{av}} = 200$ kPa in all tests and all specimens were medium dense. At an isotropic average stress ($\eta^{\text{av}} = 0$), a pure volumetric accumulation (compaction) takes place, that means the rate of deviatoric strain accumulation $\dot{\varepsilon}_q^{\text{acc}}$ is almost zero. The deviatoric portion of accumulation increases with increasing absolute value of the average stress ratio

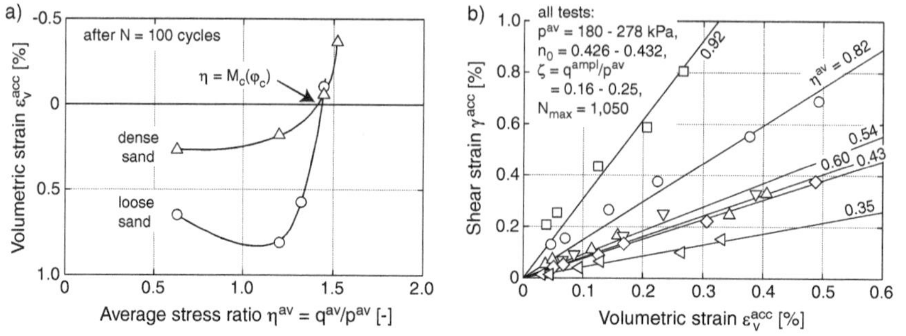


**Figure 15.** Tests of Chang and Whitman (1988): a) residual volumetric strain $\varepsilon_v^{\mathrm{acc}}$ as a function of the average stress ratio $\eta^{\mathrm{av}}$, b) residual shear strain $\gamma^{\mathrm{acc}}$ as a function of $\varepsilon_v^{\mathrm{acc}}$ for different values of $\eta^{\mathrm{av}}$

$|\eta^{\mathrm{av}}| = |q^{\mathrm{av}}/p^{\mathrm{av}}|$. At an average stress lying on the critical state line in compression ($\eta^{\mathrm{av}} = M_c(\varphi_c) = 1.25$) or in extension ($\eta^{\mathrm{av}} = M_e(\varphi_c) = -0.88$), a pure deviatoric strain accumulation is observed, that means the rate of volumetric strain accumulation vanishes ($\dot{\varepsilon}_v^{\mathrm{acc}} = 0$). While average stress ratios in the range $M_e(\varphi_c) < \eta^{\mathrm{av}} < M_c(\varphi_c)$ lead to compaction, a dilative behaviour is observed in the overcritical regime (see e.g. the test with $\eta^{\mathrm{av}} = 1.375$). Fig. 16b presents the measured directions of accumulation for all tested average stresses $\boldsymbol{\sigma}^{\mathrm{av}}$ as unit vectors in the $p$-$q$-plane. Each vector starts in the point of the average stress ($p^{\mathrm{av}}$,$q^{\mathrm{av}}$) and is inclined by $1/\omega = \varepsilon_q^{\mathrm{acc}}/\varepsilon_v^{\mathrm{acc}}$ towards the $p$-axis. The different grayscales correspond to different numbers of cycles. A slight increase in the volumetric portion of accumulation with the number of cycles becomes visible as a rotation of the vectors towards the positive $p$- or $\varepsilon_v^{\mathrm{acc}}$-axis. Possible causes are discussed by Wichtmann et al. (2006a). The direction of strain accumulation was found to be almost independent of the average mean pressure $p^{\mathrm{av}}$, of the strain loop (span, shape, polarization, Fig. 17), the void ratio and the loading frequency (Wichtmann et al., 2006a). It was also not affected by a static preloading.

The cyclic flow rule $\mathbf{m}$ of the HCA model (Eq. (8)) approximates well the measured ratios $\dot{\varepsilon}_v^{\mathrm{acc}}/\dot{\varepsilon}_q^{\mathrm{acc}}$ for various $\eta^{\mathrm{av}}$-values. $\mathbf{m}$ could be confirmed for air-pluviated specimens of eight quartz sands with different mean grain sizes in the range 0.15 mm $\leq d_{50} \leq$ 4.4 mm and with different coefficients of uniformity $C_u = d_{60}/d_{10}$ between 1.3 and 4.5 (Wichtmann et al., 2009). Fig. 18 shows the data for six of these sands. The corresponding grain size distribution curves are given in Fig. 33a.

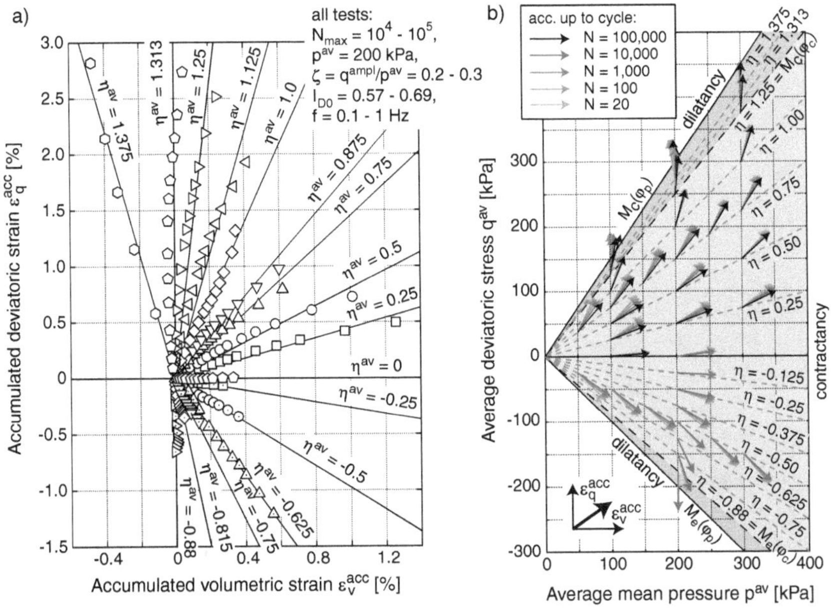


**Figure 16.** a) $\varepsilon_q^{\mathrm{acc}}$-$\varepsilon_v^{\mathrm{acc}}$-strain paths in cyclic triaxial tests with different average stress ratios $\eta^{\mathrm{av}}$, b) Direction of accumulation shown as unit vectors in the $p$-$q$-plane (Wichtmann et al., 2006a)

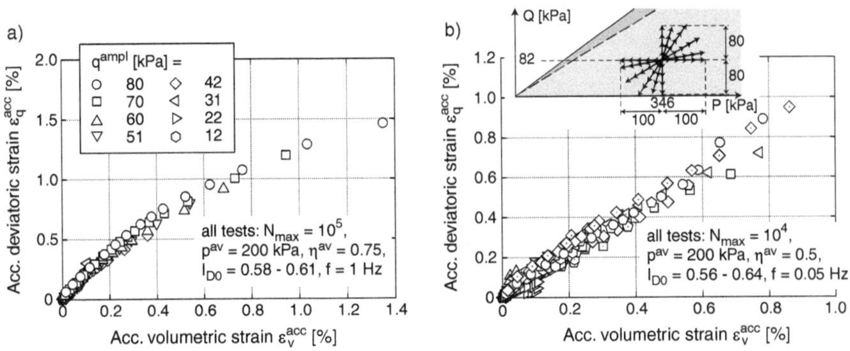


**Figure 17.** No influence of a) the stress amplitude and b) the polarization of the cycles on the direction of strain accumulation, cyclic triaxial tests of Wichtmann et al. (2006a)

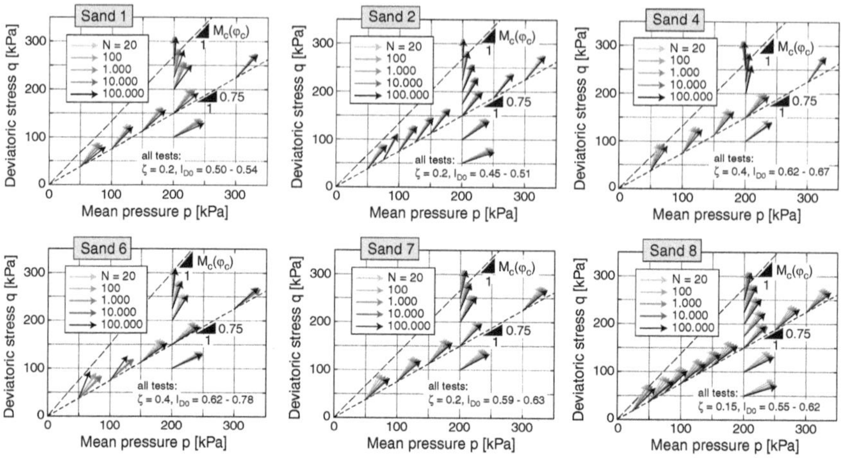


**Figure 18.** Direction of strain accumulation for six different sands tested by Wichtmann et al. (2009) (see the grain size distribution curves in Fig. 33a)

**Intensity of Accumulation** It is well-known that the intensity of accumulation increases with increasing stress or strain amplitude. This is evident for instance from the cyclic simple shear test results of Sawicki and Świdziński (1987, 1989) reported in Fig. 19a or from the cyclic triaxial test results of Marr and Christian (1981) of Fig. 19b. Similar results were obtained by Youd (1972) and Silver and Seed (1971a,b).

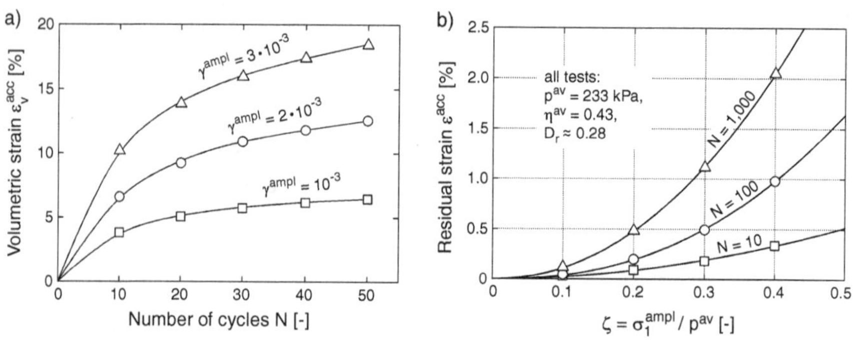


**Figure 19.** Increase of the residual strain with increasing strain or stress amplitude a) in simple shear tests of Sawicki and Świdziński (1987, 1989) and b) in cyclic triaxial tests of Marr and Christian (1981)

Also the tests of Wichtmann et al. (2005a) showed an increase in the rate $\dot{\varepsilon}^{\rm acc}$ of strain accumulation with increasing stress amplitude (Fig. 20a, $\varepsilon = \sqrt{(\varepsilon_1)^2 + 2(\varepsilon_3)^2}$). If the residual strain $\varepsilon^{\rm acc}$ is plotted versus the square of the strain amplitude $(\bar{\varepsilon}^{\rm ampl})^2$ linear curves are obtained independently of the number of cycles (Fig. 20b). Since in the stress-controlled tests the strain amplitude $\varepsilon^{\rm ampl}$ varies slightly with $N$, a mean value of the strain amplitude was used on the abscissa in Fig. 20b (here and in the following a bar over a quantity denotes that a mean value over $N$ is used, i.e. $\bar{\sqcup} = \frac{1}{N}\int_0^N \sqcup \, dN$). The division of $\varepsilon^{\rm acc}$ by the void ratio function $\bar{f}_e$ of the HCA model on the ordinate considers slightly different initial void ratios $e_0$ and different compaction rates $\dot{e}$. The proportionality between $\dot{\varepsilon}^{\rm acc}$ and the square of the strain amplitude $(\varepsilon^{\rm ampl})^2$ has been described by the function $f_{\rm ampl}$ (Tab. 1) in the HCA model and holds up to $\varepsilon^{\rm ampl} = 10^{-3}$ (Wichtmann, 2005). It was confirmed for air-pluviated specimens of the eight different sands shown in Fig. 33a. Fig. 21 shows data for six of these sands (Wichtmann et al., 2009). The linear curves trough the origin correspond with $\dot{\varepsilon}^{\rm acc} \sim (\varepsilon^{\rm ampl})^2$.

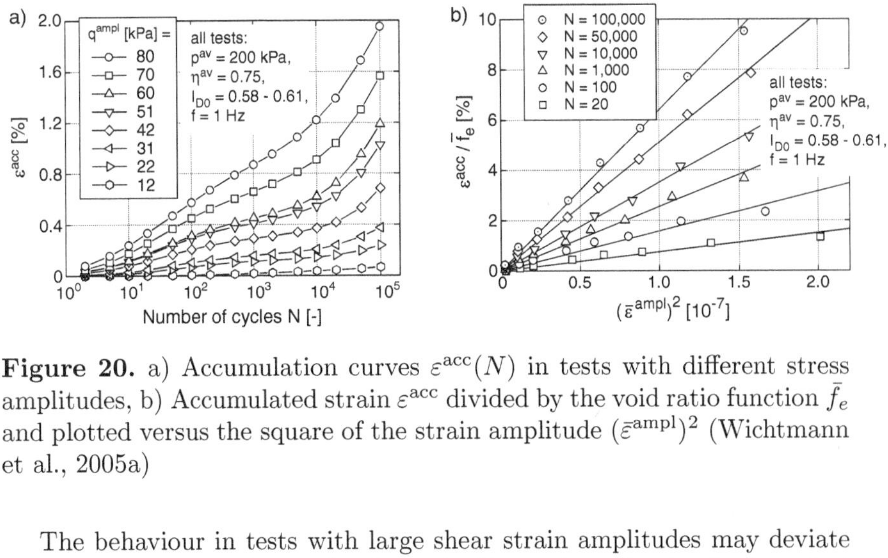


**Figure 20.** a) Accumulation curves $\varepsilon^{\rm acc}(N)$ in tests with different stress amplitudes, b) Accumulated strain $\varepsilon^{\rm acc}$ divided by the void ratio function $\bar{f}_e$ and plotted versus the square of the strain amplitude $(\bar{\varepsilon}^{\rm ampl})^2$ (Wichtmann et al., 2005a)

The behaviour in tests with large shear strain amplitudes may deviate from that described above. A schematic illustration of a simple shear test result is drawn in Fig. 22a. While during the first cycles exclusively densification takes place, an alternating contractive and dilative behaviour during a cycle is observed at larger $N$-values (Gudehus, 2000; Pradhan et al., 1989). In that case the frequency of volumetric deformation is doubled compared with the applied shear loading.

Fig. 22b presents the results of cyclic triaxial tests with a simultaneous

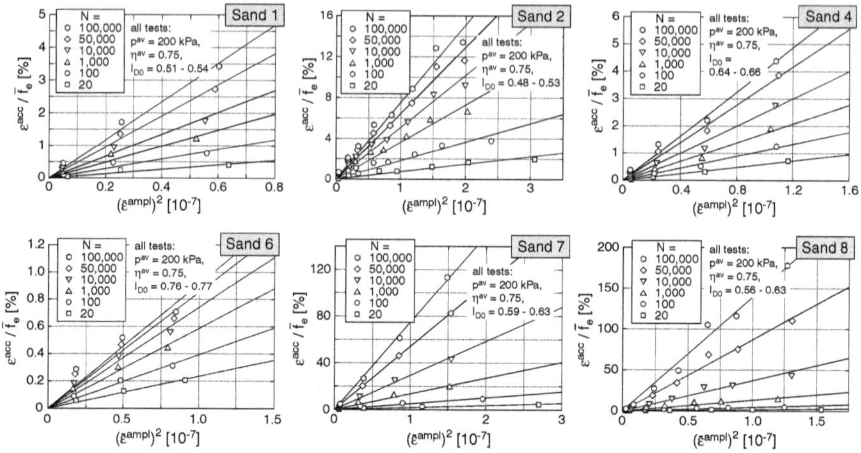


**Figure 21.** Accumulated strain $\varepsilon^{acc}$ divided by the void ratio function $\bar{f}_e$ and plotted versus the square of the strain amplitude $(\bar{\varepsilon}^{ampl})^2$ for six of the sands tested by Wichtmann et al. (2009)

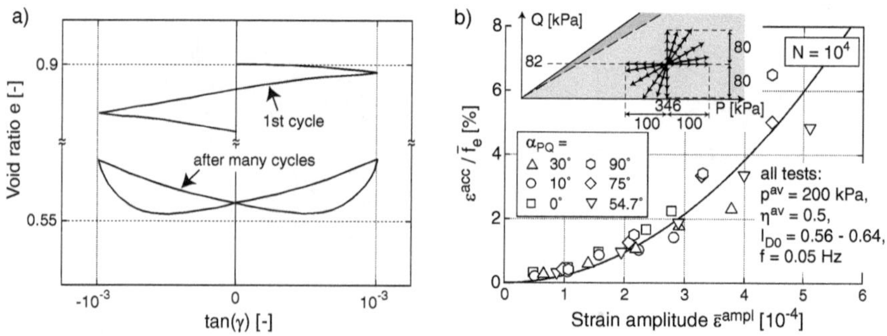


**Figure 22.** a) Doubling of frequency of the curves $e(t)$ after many cycles of simple shear with large amplitudes (Gudehus, 2000), b) No influence of the polarization of the cycles on the accumulation of residual strain (Wichtmann et al., 2007a)

oscillation of the axial and the lateral stress (Wichtmann et al., 2007a). Different polarizations of the cycles in the $P$-$Q$-plane were tested (described by the angle $\alpha_{PQ} = \arctan(Q/P)$ using the isomorphic stress invariants $P = \sqrt{3}p$ and $Q = \sqrt{2/3}q$). For each polarization tests with different amplitudes were performed. In Fig. 22b the residual strain after 10,000 cycles is plotted versus a mean value $\bar{\varepsilon}^{\mathrm{ampl}}$ of the strain amplitude. For a given $\bar{\varepsilon}^{\mathrm{ampl}}$ the residual strain does not significantly depend on the polarization of the cycles (as long as the polarization does not change, see below).

The shaking table tests of Pyke et al. (1975) (Fig. 23a) and the multidimensional simple shear tests of Wichtmann et al. (2007a) (Fig. 23b) demonstrate the importance of the shape of the strain cycles. Circular strain cycles produce almost twice larger accumulation rates than one-dimensional cycles with the same maximum span. In the HCA model (Section 3.2) this is captured by the tensorial amplitude definition.

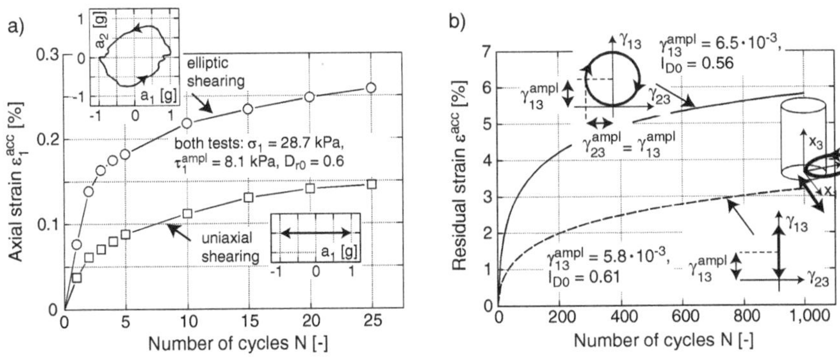


**Figure 23.** Large influence of the shape of the cycles observed a) in shaking table tests of Pyke et al. (1975) and b) in multidimensional simple shear tests of Wichtmann et al. (2007a)

It is well known that the rate of strain accumulation increases with decreasing soil density, that means with increasing void ratio. This becomes clear from the cyclic triaxial tests of Marr and Christian (1981) (Fig. 24a) and also from our own test results (Wichtmann et al., 2005a) (Fig. 24b). In Fig. 24b the residual strain $\varepsilon^{\mathrm{acc}}$ after different $N$-values has been divided by the amplitude function $f_{\mathrm{ampl}}$ of the HCA model in order to consider slightly different strain amplitudes and was plotted versus a mean value of the void ratio $\bar{e}$. In the HCA model the increase of $\dot{\varepsilon}^{\mathrm{acc}}$ with increasing void ratio is described by the function $f_e$ (Tab. 1).

In simple shear tests performed by Silver and Seed (1971a,b), Youd

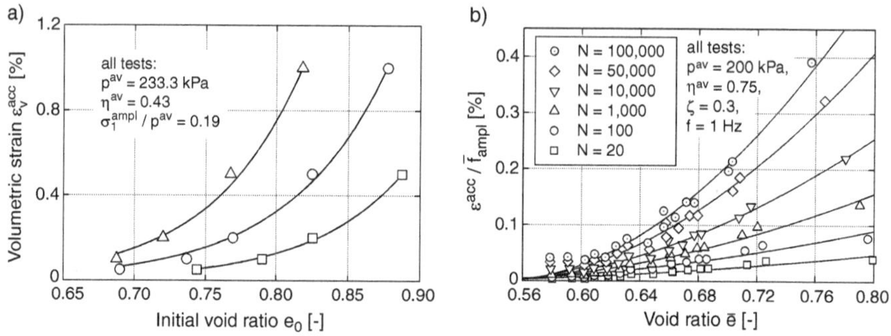


**Figure 24.** Increase of the rate of strain accumulation with increasing void ratio: Cyclic triaxial tests of a) Marr and Christian (1981) and b) Wichtmann et al. (2005a)

(1972) and Sawicki and Świdziński (1987, 1989) no influence of the vertical stress on the accumulation of residual strain was observed (Fig. 25a). In contrast to these studies, the more recent simple shear tests of Duku et al. (2008) showed a significant decrease of the accumulation rate with increasing vertical stress. This in good accordance with our results from triaxial tests (Fig. 25b). The decrease of the intensity of accumulation with increasing average mean pressure $p^{\mathrm{av}}$ becomes more pronounced with increasing number of cycles. It is described by the function $f_p$ in the HCA model (Tab. 1). The $\sigma_1$-independence observed in the early simple shear test studies may be due to the low number of cycles applied or due to the large strain amplitudes.

The data from cyclic triaxial tests with different average stress ratios $\eta^{\mathrm{av}}$ are plotted in Fig. 26 (Wichtmann et al., 2005a). The normalized stress ratio $\bar{Y}^{\mathrm{av}}$ has been used on the $x$-axis. The accumulation rate increases strongly with increasing average stress ratio. The function $f_Y$ of the HCA model (Tab. 1) describes this observation.

Most researchers found no influence of the loading frequency on the rate of strain accumulation. As an example results from drained simple shear tests of Youd (1972) with loading frequencies in the range 0.2 Hz $\leq f \leq$ 1.9 Hz are presented in Fig. 27a. Shenton (1978) also found no influence of the loading frequency in drained cyclic triaxial tests on ballast with 0.1 Hz $\leq f \leq$ 30 Hz. Our own test results are given in Fig. 27b. A similar accumulation of strain was observed for loading frequencies between 0.05 and 2 Hz. Therefore, the loading frequency has not to be considered as an influencing parameter in the HCA model.

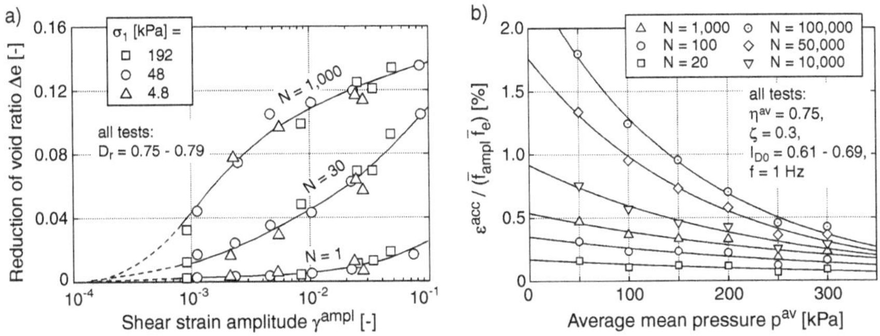


**Figure 25.** a) No influence of the axial stress in simple shear tests of Youd (1972), b) Large decrease of the rate of strain accumulation with increasing average mean pressure in cyclic triaxial tests of Wichtmann et al. (2005a)

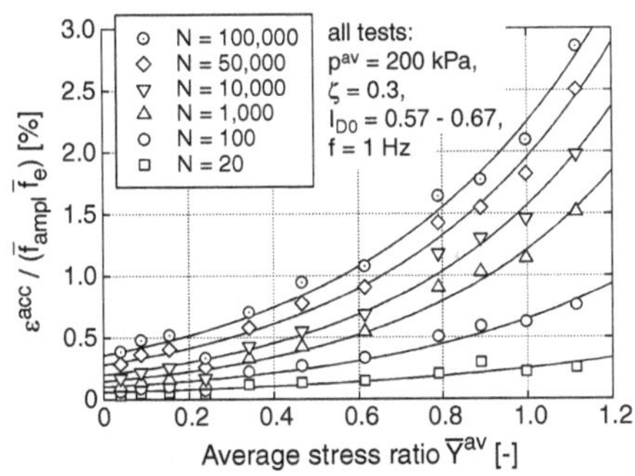


**Figure 26.** Increase of the rate of strain accumulation with increasing average stress ratio, cyclic triaxial tests of Wichtmann et al. (2005a)

The effect of changes of the polarization of the cycles was studied by Wichtmann et al. (2007a) in multi-dimensional simple shear tests. 1,000 cycles with a certain polarization were followed by 4,000 cycles with a perpendicular polarization. Fig. 28a reveals that the sudden change of the polarization causes a temporary increase of the accumulation rate. A strong increase of the rate of strain accumulation due to polarization changes was also observed by Yamada and Ishihara (1982) in true triaxial tests (Fig. 28b). After consolidation under an isotropic stress they applied four cycles. Keeping $p$ constant, in the first cycle the vertical stress was increased until a certain octahedral shear stress $\tau_{\text{oct}}$ was reached. After that, the direction of loading was rotated by a certain angle $\theta$ in the octahedral plane and the second cycle

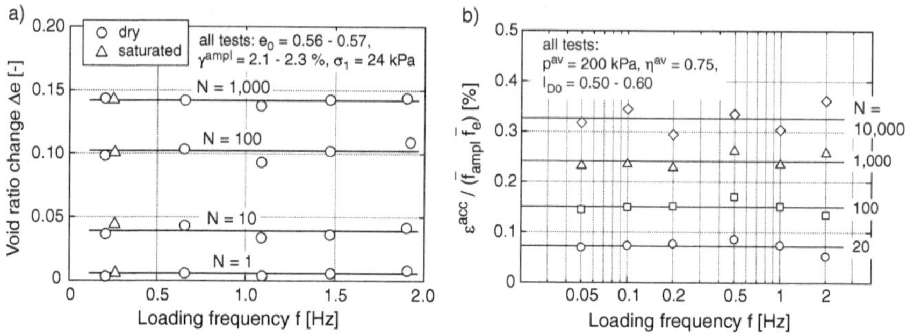


**Figure 27.** No influence of the loading frequency a) in simple shear tests of Youd (1972) and b) in cyclic triaxial tests of Wichtmann et al. (2005a)

was applied with the same amplitude in this new direction. The third and the fourth cycles had the same polarizations as the first and the second one but the amplitude was larger. The residual strain increased significantly with increasing value of $\theta$. In the HCA model the effect of polarization changes is described by the function $f_\pi$ (Wichtmann et al., 2007a).

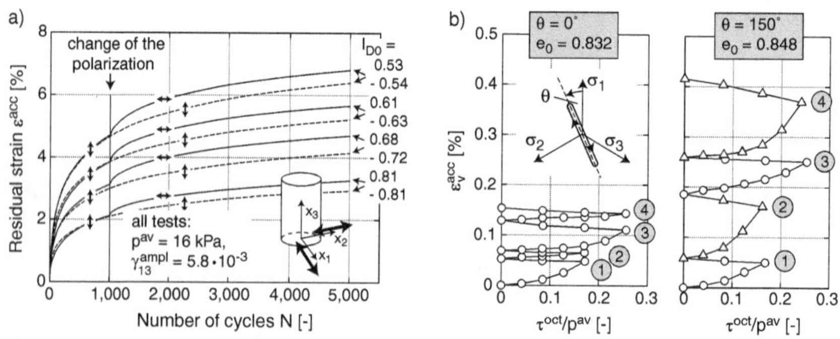


**Figure 28.** Increase of the rate of strain accumulation due to a sudden change of the polarization of the cycles: a) multidimensional simple shear tests of Wichtmann et al. (2007a), b) true triaxial tests of Yamada and Ishihara (1982)

Different shapes of the accumulation curves $\varepsilon^{\text{acc}}(N)$ have been reported in the literature. Some researchers (e.g. Lentz and Baladi, 1980; Duku et al., 2008) found curves obeying $\varepsilon^{\text{acc}}(N) \sim \ln(N)$ (Fig. 29a) whereas other studies (e.g. Marr and Christian, 1981) suggest an increase of $\varepsilon^{\text{acc}}$ faster than proportional to $\ln(N)$. Sometimes a power-law relationship $\varepsilon^{\rm acc} \sim N^a$ with an exponent $a$ is used. Fig. 29b presents the accumulation curves $\varepsilon^{\rm acc}(N)$ measured by Wichtmann et al. (2005a) in different test series. The data was divided by the functions $\bar{f}_{\rm ampl}$, $\bar{f}_e$, $f_p$, $f_Y$ and $f_\pi = 1$ of the HCA model (Tab. 1) in order to purify it from the influences of stress amplitude, initial density, average mean pressure and average stress ratio. The purified data fall together into a band which can be approximated by the function $f_N$ defined in Eq. (14). It consists of a logarithmic and a linear portion. The logarithmic portion is pre-dominant up to $N = 10^4$ while the linear portion is necessary to describe the curves $\varepsilon^{\rm acc}(N)$ for larger numbers of cycles.

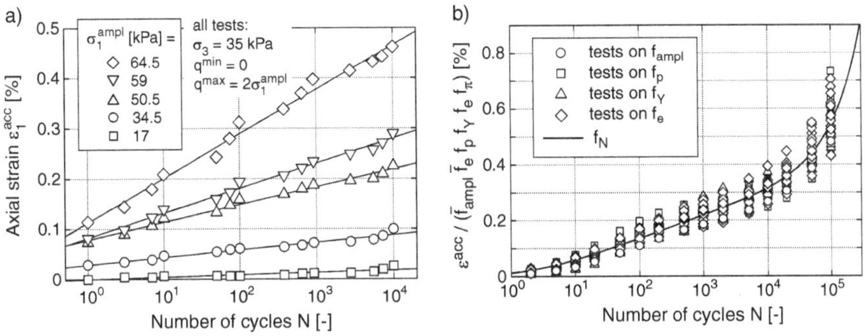


**Figure 29.** a) Curves $\varepsilon_1^{\rm acc}(N)$ in cyclic triaxial tests of Lentz and Baladi (1980), b) Accumulation curves $\varepsilon^{\rm acc}(N)$ divided by the functions $\bar{f}_{\rm ampl}$, $\bar{f}_e$, $f_p$ and $f_Y$ of the HCA model (Wichtmann et al., 2005a)

Tests on the eight sands shown in Fig. 33a (Wichtmann et al., 2009) revealed that the shape of the accumulation curves $\varepsilon^{\rm acc}(N)$ depends on the grain size distribution curve, in particular on the coefficient of uniformity $C_u = d_{60}/d_{10}$. Fig. 30 shows data for six of the sands. For the uniform sands Nos. 1 to 6, the residual strain increases almost proportionally to $\ln(N)$ up to $N \approx 10^4$ and over-proportionally for larger $N$-values. For the sands Nos. 7 and 8, having higher $C_u$-values, the residual strain grows faster than according to $\varepsilon^{\rm acc} \sim \ln N$ already from the beginning of the test. The differences between the curves $\varepsilon^{\rm acc}(N)$ reported by different researchers may be due due to the different grain size distribution curves of the tested materials.

The large influence of a cyclic preloading is illustrated in Fig. 31a. It presents the evolution of void ratio $e$ with the number of cycles $N$ in three cyclic triaxial tests with identical stresses but slightly different initial void ratios $e_0$. Considering a state with identical void ratio and identical stress

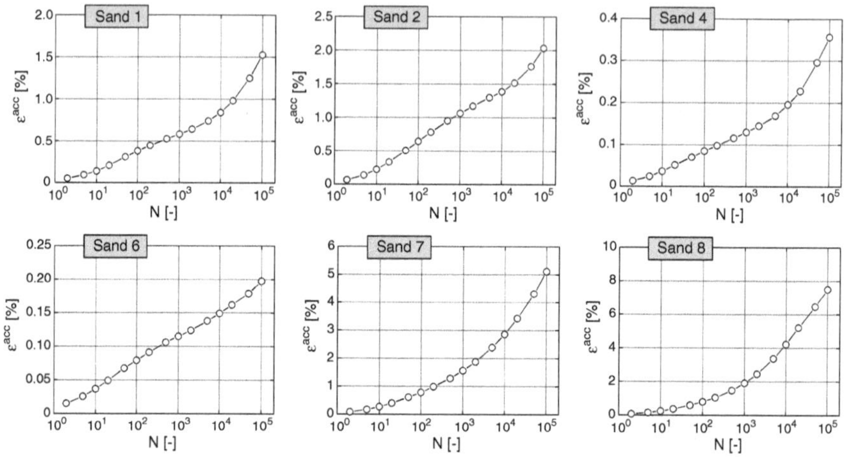


**Figure 30.** Accumulation curves $\varepsilon^{\mathrm{acc}}(N)$ for six of the sands tested by Wichtmann et al. (2009) (see the grain size distribution curves in Fig. 33a)

(as marked by the horizontal line in Fig. 31a) the compaction rate $\dot{e}_1$ of a freshly pluviated sample is significantly larger than the rate $\dot{e}_3$ of a sample which was preloaded with 40,000 load cycles. Thus, the accumulation rate is significantly reduced by a cyclic preloading. For a good prediction with a HCA model the knowledge of void ratio and stress alone is not sufficient. An information about the cyclic preloading of the soil is necessary. Unfortunately, it cannot be directly measured in situ but has to be determined from correlations. Despite considerable efforts a clear correlation of cyclic preloading with dynamic soil properties (e.g. P- and S-wave velocities) could not be established (Wichtmann and Triantafyllidis, 2004a,b, see Section 5.2). A correlation with the liquefaction resistance, however, could be formulated (Wichtmann et al., 2005c, see Fig. 72) but its practical application has still to be proven. A correlation of cyclic preloading with acoustic emissions seems to be rather insufficient (Niemunis et al., 2007b). As an alternative, the cyclic preloading could be determined by cyclic test loadings in situ (some ideas are explained by Wichtmann, 2005).

In many practical problems the amplitude of the cycles is not constant but varies from cycle to cycle. Such random cyclic loadings could be replaced by packages of cycles each with a constant amplitude (Fig. 31b) if the sequence of application would not affect the residual strain, that means if Miner's rule (Miner, 1945) were applicable to soil. The test results of Kaggwa et al. (1991) on calcareous sand (Fig. 32a) and our own test results

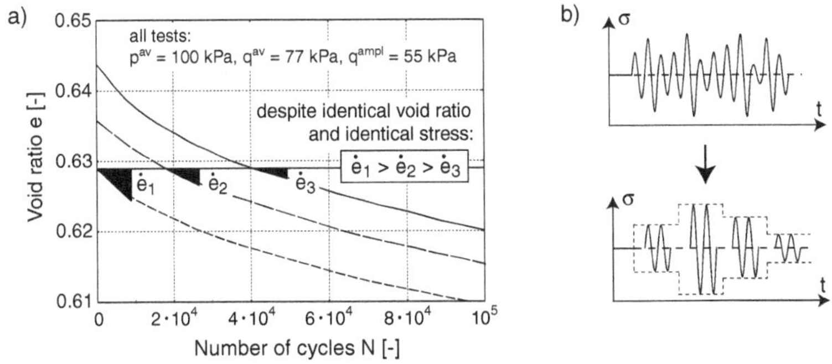


**Figure 31.** a) Curves $e(p)$ in cyclic triaxial tests with slightly different initial void ratios: different compaction rates $\dot{e}$ for $e, p =$ constant due to a different cyclic preloading (Wichtmann et al., 2005a), b) Decomposition of a cyclic loading with varying amplitudes into packages of cycles with constant amplitude

(Fig. 32b) demonstrate that Miner's rule can be assumed to be approximately valid for sand, at least for a constant polarization of the cycles. In the six tests shown in Fig. 32b four packages each with 25,000 cycles were applied in succession. The amplitudes $q^{\mathrm{ampl}} = 20$, 40, 60 and 80 kPa were applied in different sequences. Irrespectively of the chosen sequence the residual strains at the end of the tests were quite similar.

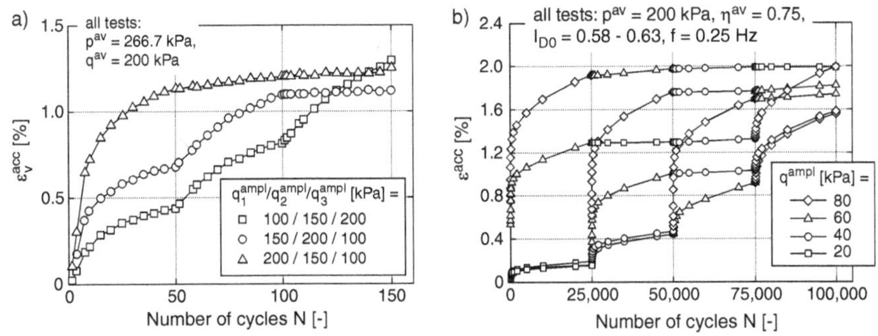


**Figure 32.** Accumulation curves $\varepsilon_v^{\mathrm{acc}}(N)$ or $\varepsilon^{\mathrm{acc}}(N)$, respectively, in cyclic triaxial tests with packages of cycles with varying amplitudes, a) Kaggwa et al. (1991), b) Wichtmann et al. (2006b)

The influence of the grain size distribution curve on the rate of strain accumulation was studied by Wichtmann et al. (2009) in approx. 200 drained cyclic triaxial tests on the eight sands shown in Fig. 33a. It was found that for a similar relative density the rate of strain accumulation decreases with increasing mean grain size $d_{50}$ and that it considerably grows with increasing coefficient of uniformity $C_u = d_{60}/d_{10}$ (Fig. 33b).

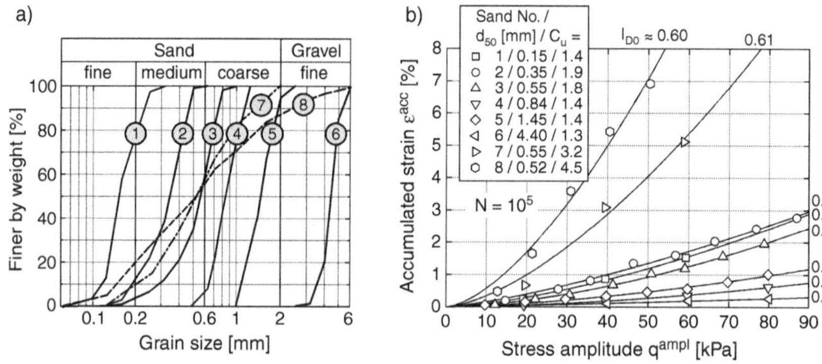


**Figure 33.** a) Eight grain size distribution curves tested by Wichtmann et al. (2009), b) Residual strain after $10^5$ cycles as a function of stress amplitude for the eight tested sands (Wichtmann et al., 2009)

The functions $f_{\text{ampl}}$, $f_e$, $f_p$, $f_Y$ and $f_N$ of the HCA model were found useful for all of the eight tested sands. Wichtmann et al. (2009) formulated correlations of the material constants of the HCA model with granulometric properties ($d_{50}$, $C_u$) or with index properties ($e_{\min}$). Some of them are shown in Fig. 34. They may be used for a simplified determination of a set of material constants.

**Stiffness $\mathsf{E}$ of the HCA Model** The elastic stiffness $\mathsf{E}$ used in the basic constitutive equation (6) of the HCA model interrelates the rates of stress and strain accumulation. The bulk modulus $K$ can be experimentally obtained from a comparison of the rate $\dot{u}$ of pore pressure accumulation in an undrained cyclic triaxial test and the rate $\dot{\varepsilon}_v$ of volumetric strain accumulation in a drained cyclic test. Both specimens should be prepared with similar initial void ratios. The initial stress and the stress amplitude should be the same in both tests. For isotropic stress conditions ($q = 0$, $\dot{q} = 0$, $m_q = 0$) Eq. (17) takes either the form of isotropic relaxation

$$\dot{p} \quad = \quad -K \; \dot{\varepsilon}^{\text{acc}} \; m_v \tag{19}$$

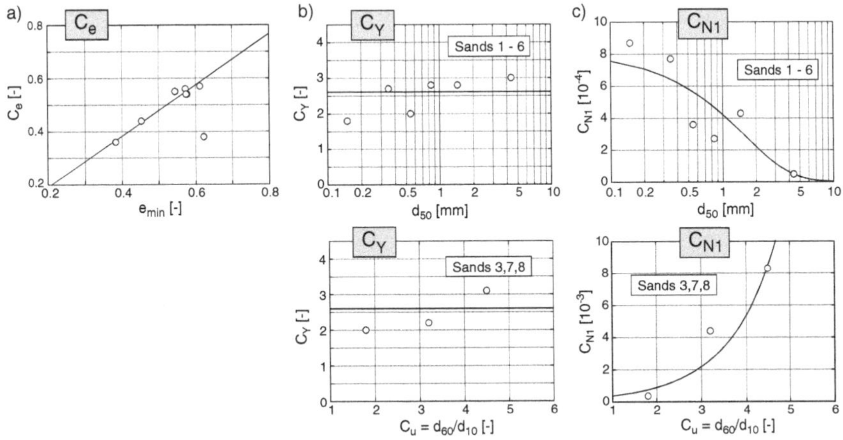


**Figure 34.** Correlations between the constants of the HCA model and index properties of the different sands tested by Wichtmann et al. (2009) (see the grain size distribution curves in Fig. 33a)

under undrained conditions ($\dot{\varepsilon}_v = 0$) or the form of volumetric creep

$$\dot{\varepsilon}_v = \dot{\varepsilon}^{\rm acc}\, m_v \tag{20}$$

under drained conditions ($\dot{p} = 0$). Comparing these equations one may eliminate $\dot{\varepsilon}^{\rm acc}\, m_v$ and obtain

$$K = -\frac{\dot{p}}{\dot{\varepsilon}_v} \quad \text{or} \quad \frac{\dot{u}}{\dot{\varepsilon}_v} \tag{21}$$

For the determination of Poisson's ratio $\nu$ the effective stress evolution ($\dot{p}$,$\dot{q}$) observed in a strain-controlled undrained cyclic triaxial test commenced at an anisotropic initial stress may be compared with the prediction of Eq. (17). For $\dot{\varepsilon}_v = 0$ and $\dot{\varepsilon}_1 = 0$ and therefore $\dot{\varepsilon}_q = 0$ one obtains:

$$\begin{bmatrix} \dot{p} \\ \dot{q} \end{bmatrix} = \begin{bmatrix} K & 0 \\ 0 & 3G \end{bmatrix} \begin{bmatrix} -\dot{\varepsilon}^{\rm acc}\, m_v \\ -\dot{\varepsilon}^{\rm acc}\, m_q \end{bmatrix} \tag{22}$$

The ratio of the relaxation rates

$$\frac{\dot{q}}{\dot{p}} = \frac{3G}{K}\frac{m_q}{m_v} = \frac{9(1-2\nu)}{2(1+\nu)}\, \frac{2\eta^{\rm av}}{M^2 - (\eta^{\rm av})^2} \tag{23}$$

depends on $\nu$. Stress paths for different $\nu$-values are plotted exemplary in Fig. 37a. The stress relaxates until $\boldsymbol{\sigma}' = \mathbf{0}$ is reached. The Poisson's ratio

$\nu$ can be determined from the measured $\dot{q}/\dot{p}$ or judged by a curve-fitting of the experimental data using Fig. 37a.

The pressure-dependent bulk modulus $K$ was quantified from fifteen pairs of drained and undrained cyclic tests on a medium coarse sand with different consolidation pressures and stress amplitudes (Wichtmann et al., 2010b). Some of the curves $u(N)$ of pore water pressure versus the number of cycles measured in the undrained cyclic tests are given in Fig. 35a. The corresponding curves $\varepsilon_v^{\text{acc}}(N)$ of volumetric strain accumulation in the drained tests are shown in Fig. 35b.

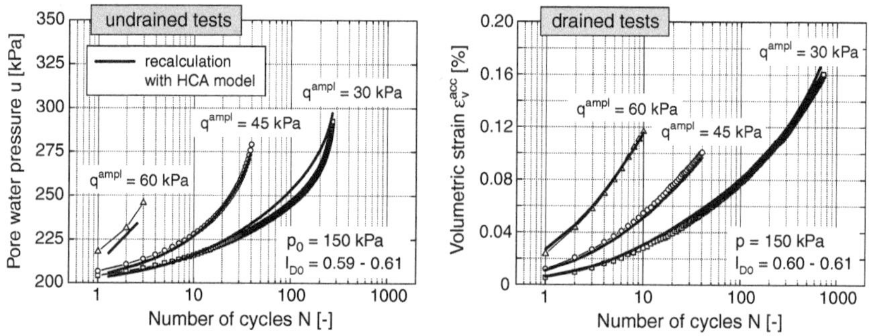


**Figure 35.** Accumulation of pore water pressure $u$ or volumetric strain $\varepsilon_v^{\text{acc}}$ in undrained or drained cyclic triaxial tests, respectively (Wichtmann et al., 2010b)

Membrane penetration effects in the undrained tests were considered by applying the correction method proposed by Tokimatsu (1990) (see Section 4). During an undrained cyclic test with $q^{\text{ampl}}$ = constant the average mean pressure decreases and the strain amplitude $\varepsilon^{\text{ampl}}$ increases considerably. In the drained test $p^{\text{av}}$ is constant and the strain amplitude does not change much, but the void ratio $e$ decreases with $N$. In order to evaluate $\dot{u}$ and $\dot{\varepsilon}_v^{\text{acc}}$ for exactly the same test conditions that means same values of $\varepsilon^{\text{ampl}}$, $e$, $p^{\text{av}}$ and $g^A$, the rate $\dot{\varepsilon}_v^{\text{acc}}$ from the drained tests was corrected as described by Wichtmann et al. (2010b). Fig. 36 presents the bulk modulus $K$ versus average mean pressure $p^{\text{av}}$. The data from all 15 pairs of tests fall into a concentrated cloud of points. The obvious pressure-dependence of $K$ can be approximated by

$$K \quad = \quad A\, p_{\text{atm}}^{1-n}\, p^n \tag{24}$$

with $A = 467$ and $n = 0.46$ (the fat solid line in Fig. 36). Wichtmann et al. (2010b) demonstrated that for a simplified determination procedure, $K$ can be estimated from the un- and reloading curve of an oedometric compression test (Fig. 36). A quite similar relationship $K(p)$ was found in a more recent test series on a uniform fine sand (Wichtmann et al., 2010c).

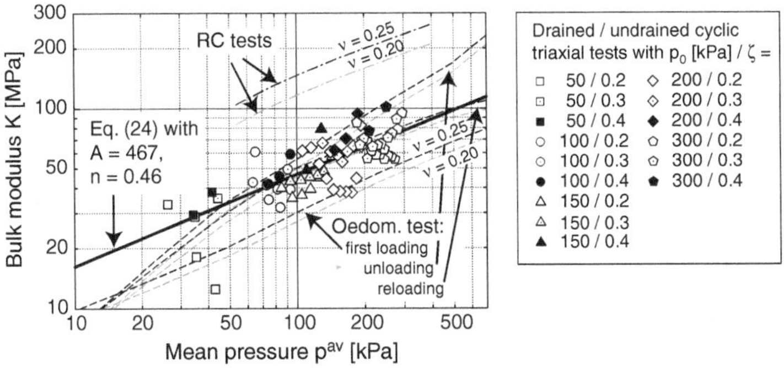


**Figure 36.** Bulk modulus $K = \dot{u}/\dot{\varepsilon}_v^{\text{acc}}$ derived from 15 pairs of drained and undrained cyclic tests on a medium coarse sand (Wichtmann et al., 2010b)

A typical effective stress path in an undrained cyclic triaxial test with a constant strain amplitude $\varepsilon_1^{\text{ampl}} = 6 \cdot 10^{-4}$ commenced at an anisotropic initial stress is shown in Fig. 37b. The effective stress relaxates until $\boldsymbol{\sigma}' = \mathbf{0}$ is reached. Plotting the course of the average effective stress (the stress after each cycle, that means at $\varepsilon_1 = 0$), the diagrams in Fig. 38 were obtained. No influence of the strain amplitude (Fig. 38a) and of the void ratio on the curves $q^{\text{av}}(p^{\text{av}})$ could be detected (Wichtmann et al., 2010c) within the studied range $\varepsilon^{\text{ampl}}$- and $e$-values. Fig. 38b shows the paths for different initial mean pressures $p_0$ while the initial stress ratio $\eta_0$ was kept constant. Paths for different initial stress ratios $\eta_0$ and for $p_0 = 200$ kPa are plotted in Fig. 38c.

For low to intermediate initial stress ratios -0.75 $\leq \eta_0 \leq$ 0.75 Poisson's ratio is approximately 0.30 while larger $\nu$-values are necessary in order to reproduce the measured stress paths for larger stress ratios in triaxial compression ($\eta_0$ = 1.0). The test results reveal that for large values of $\eta$ an isotropic elastic stiffness E may not be sufficient. However, in order to keep the HCA model as simple as possible, at present the use of an isotropic elasticity with $K(p)$ from Eq. (24) and with $\nu = 0.30$ is recommended.

### 3.4 Applications of the HCA Model

**Recalculation of Element Tests** The HCA model was used to recalculate the element tests documented by Wichtmann et al. (2005a). The quite

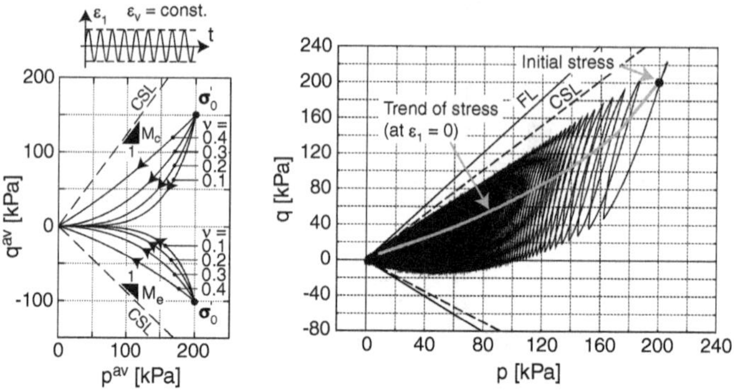


**Figure 37.** Determination of Poisson's ratio $\nu$ from the average effective stress path in cyclic triaxial tests with constant strain amplitude commenced at an anisotropic initial stress (Wichtmann et al., 2010c)

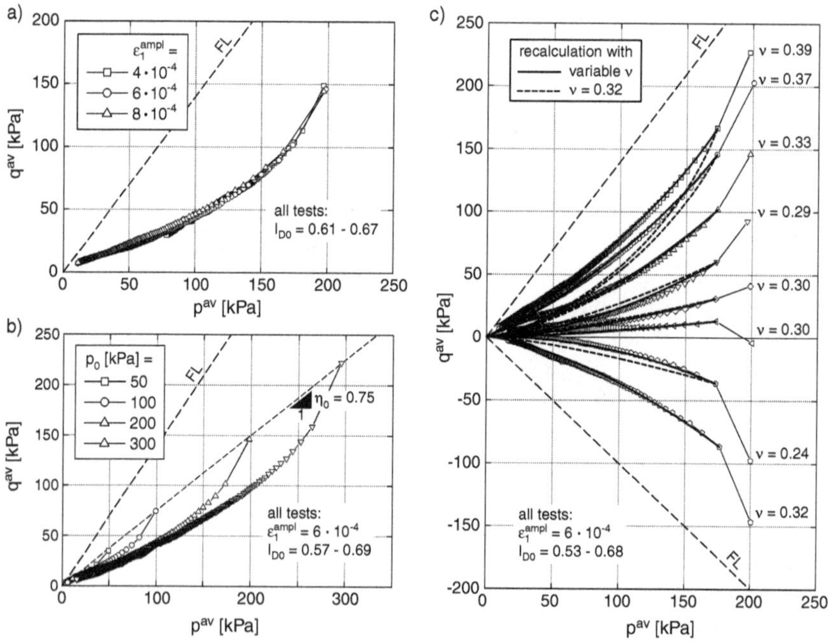


**Figure 38.** Paths of the average effective stress in cyclic triaxial tests with $\varepsilon_1^{\mathrm{ampl}}$ = constant commenced at an anisotropic initial stress for a) different strain amplitudes, b) different initial mean pressure $p_0'$ and c) different initial stress ratio $\eta_0$ (Wichtmann et al., 2010c)

good agreement between the test results and the curves predicted by the HCA model is shown in Fig. 39 for the test series with different average stress ratios. The HCA model also predicts quite well the accumulation of strain in cyclic triaxial tests with packages of cycles with different amplitudes applied in different sequences (Fig. 40). The recalculation of drained and undrained cyclic tests performed for the determination of the bulk modulus $K$ is given in Fig. 35. The HCA model with a single set of constants in combination with the bulk modulus calculated from Eq. (24) with $A = 467$ and $n = 0.46$ describes well both, the accumulation of pore water pressure in undrained cyclic tests and the accumulation of volumetric strain in drained cyclic tests. Recalculations of other test series are presented by Wichtmann (2005).

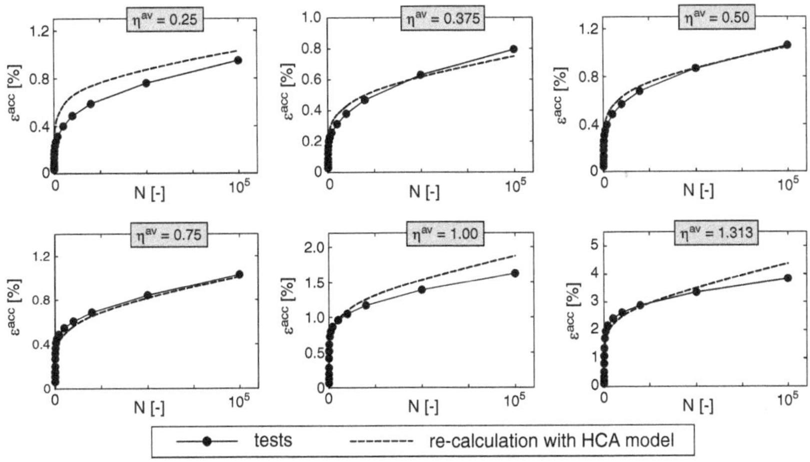


**Figure 39.** Recalculation of cyclic triaxial tests with different average stress ratios using the HCA model (Wichtmann, 2005)

**Foundations with Cyclic Loading** In order to confirm the prediction of the HCA model for boundary value problems a centrifuge model test of Helm et al. (2000) was re-calculated using the finite element method. In the model test (acceleration level 30g) a strip foundation was placed without embedding on a freshly pluviated dense fine sand and afterwards loaded with $N = 10^5$ stress cycles. The dimensions in the prototype scale and the loading are given in Figure 41a. In order to re-calculate the model test, the material constants of the fine sand used in the centrifuge test were determined (Wichtmann, 2005). Fig. 41a presents the FE discretisation.

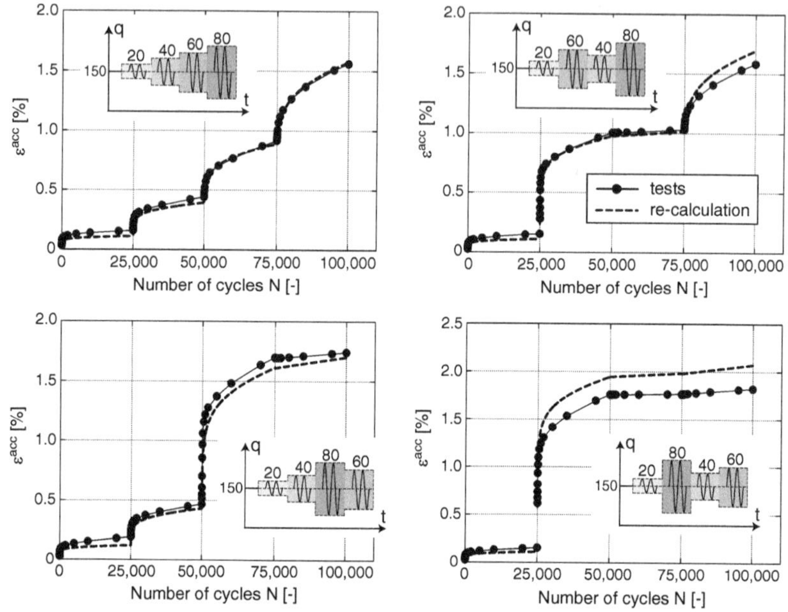


**Figure 40.** Comparison of curves $\varepsilon^{\mathrm{acc}}(N)$ measured in cyclic triaxial tests with packages of cycles with different amplitudes applied in different sequences and curves predicted by the HCA model (Wichtmann, 2005)

Fig. 41b compares the settlement curves $s(N)$ measured in the model test and obtained from the FE calculation. A good congruence of the curves can be observed. Therefore, the HCA model was confirmed to deliver realistic settlement predictions.

Parametric studies of strip foundations under cyclic loading were performed by Wichtmann et al. (2005b). The variables of the soil (initial density $I_{D0}$, coefficient of lateral earth pressure $K_0 = \sigma_h/\sigma_v$, historiotropic variable $g_0^A$), the loading of the foundation (average value $\sigma^{\mathrm{av}}$, amplitude $\sigma^{\mathrm{ampl}}$) and the geometry of the foundation (depth t of embedding, width b) were varied. Two examples showing the increase of the settlement with increasing number of cycles are presented in Fig. 42. They demonstrate that the settlement grows with increasing load amplitude and with decreasing depth of embedding.

Niemunis et al. (2005a) calculated the differential settlements of two neighboured foundations. The spatial distribution of the void ratio $e(\mathbf{x})$ was stochastically generated with three different correlation lengths. 30 dif-

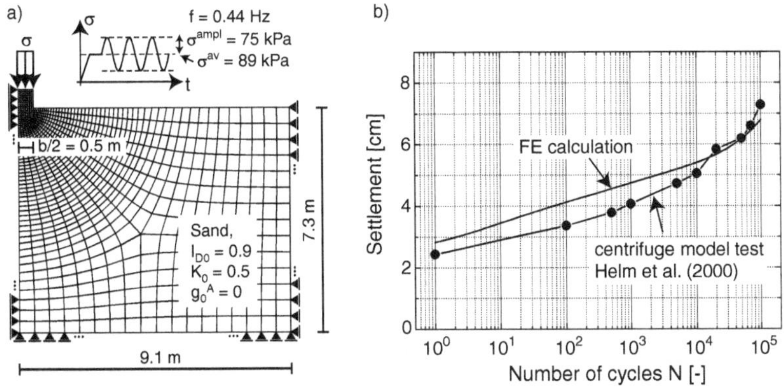


**Figure 41.** Recalculation of a centrifuge model test performed by Helm et al. (2000) using the HCA model (Wichtmann, 2005)

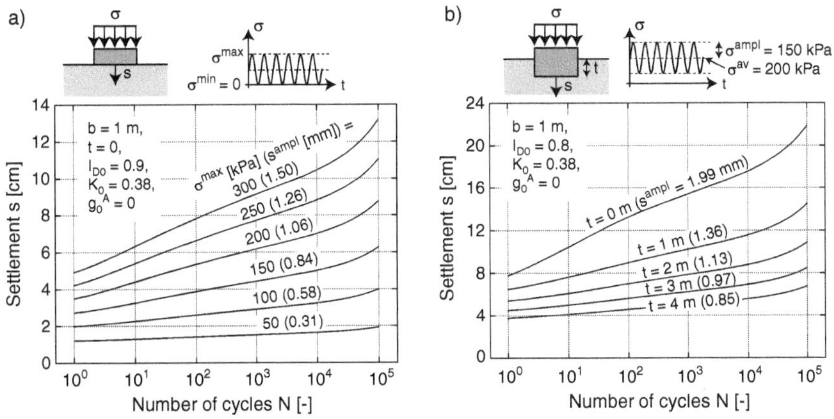


**Figure 42.** Parametric studies of shallow foundations under cyclic loading (Wichtmann, 2005; Wichtmann et al., 2005b)

ferent fields $e(\mathbf{x})$ (see an example in Fig. 43a) were tested. Let $s_l$ and $s_r$ be the settlements of the left and the right foundation, respectively (Fig. 43a). The differential settlement $\Delta s = |s_l - s_r|$ was divided by the mean value $\bar{s} = (s_l + s_r)/2$. The ratio $(\Delta s/\bar{s})_{\text{stat}}$ due to static loading up to $\sigma^{\text{av}}$ was compared to the ratio $(\Delta s/\bar{s})_{\text{cyc}}$ describing the additional differential settlement accumulated during the subsequent $10^5$ cycles. Independently of the correlation length the differential settlement $(\Delta s/\bar{s})_{\text{cyc}}$ resulting from cyclic loading was observed to be approximately three times larger than $(\Delta s/\bar{s})_{\text{stat}}$ caused by static loading (Fig. 43b).

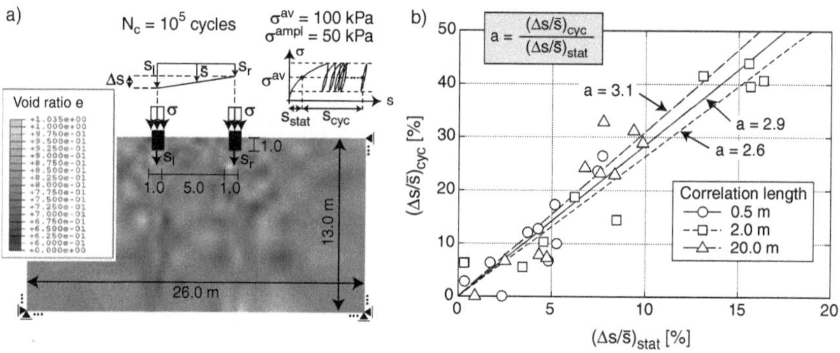


**Figure 43.** Differential settlement of two neighbored shallow foundations on a soil with a stochastically distributed void ratio (Niemunis et al., 2005a)

Calculations of a statically indeterminate frame structure on a soil with a stochastically fluctuating void ratio have been presented by Wichtmann et al. (2007b) (Fig. 44). A positive feedback effect between the settlement and the settlement rate could be demonstrated. A stress relaxation takes place under the middle foundation where the soil is relatively loose compared to the soil below the outer foundations. While the contact normal force under the middle column significantly decreases with $N$ it consequently increases under the outer columns (Fig. 44). For $N > 10^4$ even tension forces occur in the middle column which are carried by friction. The middle foundation loses its function resulting in large bending moments in the structure. The rate of stress relaxation significantly increases after each control cycle because the strain amplitudes in the soil do so. This is due to a decrease of the effective stress while the load amplitudes do not significantly change with $N$. An additional acceleration of accumulation may be due to an increase of $\dot{\varepsilon}^{\text{acc}}$ with decreasing $p^{\text{av}}$ (function $f_p$ in the HCA model).

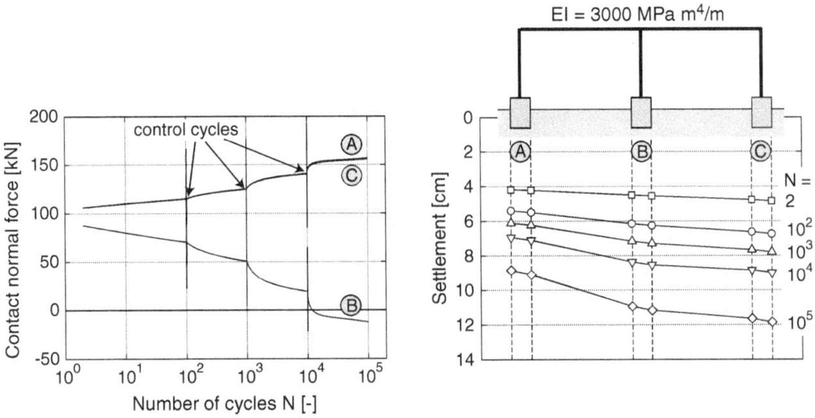


**Figure 44.** Settlement and re-distribution of forces in a statically indeterminate structure on an inhomogeneous soil (Wichtmann et al., 2007b)

Fig. 45 presents a study of the long-term deformations of a monopile foundation of an offshore wind power plant (Wichtmann et al., 2008). Since an uni-directional cyclic loading was studied the symmetry of the system could be utilized for the FE discretisation (Fig. 45a). Fig. 45b shows the lateral deflection of the pile as a function of depth after $10^6$ cycles. As expected the permanent tilting of the monopile foundation increases with increasing amplitude of the applied torsional moment.

Geogrid-reinforced soil structures under cyclic loading were studied by Arwanitaki and Triantafyllidis (2006) (Fig. 46a). In particular, a geogrid-reinforced embankment on piles in soft ground was investigated. In such systems the vertical loads are conducted into the piles via stress arches developing in the base layer. The cyclic loading was applied on the soil surface simulating a traffic loading. Arwanitaki and Triantafyllidis (2006) demonstrated that the cyclic loading leads to a weakening of the stress arches causing large settlements. A reduction of the accumulated settlements with increasing number of geogrid layers was observed (Fig. 46b).

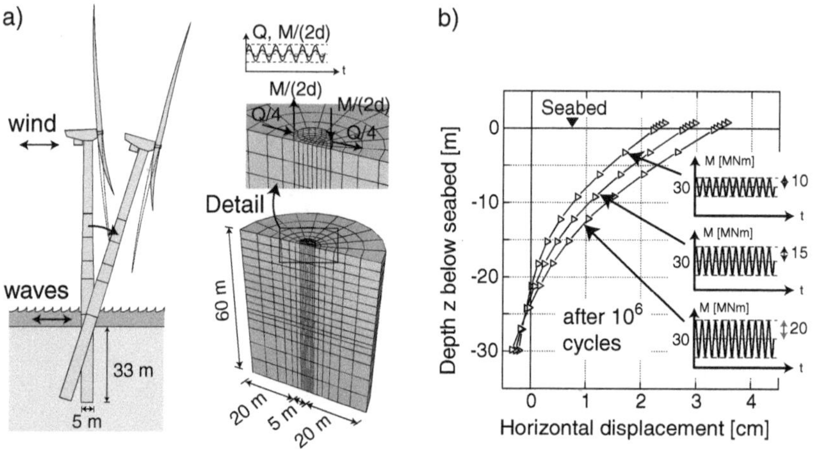


**Figure 45.** Prediction of the tilting of a monopile foundation of an offshore wind power plant (Wichtmann et al., 2008)

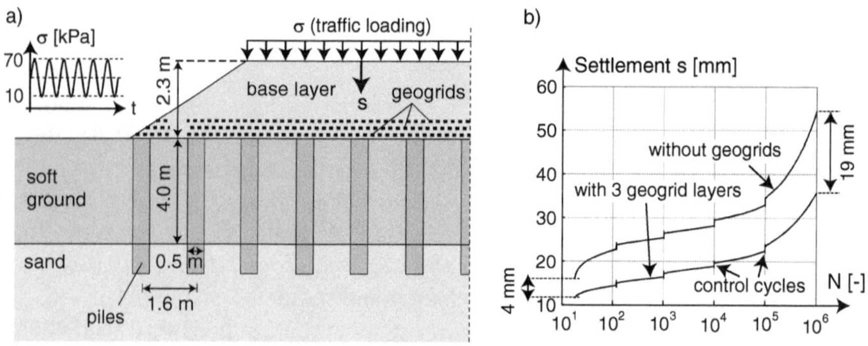


**Figure 46.** Prediction of the reduction of cyclically reduced settlements due to geogrid reinforcement (Arwanitaki and Triantafyllidis, 2006)

# 4 Pore Water Pressure Accumulation, Liquefaction and Nonlinear Phenomena Due to Undrained Cyclic Loading

## 4.1 Introduction

As is well known, an earthquake event causes the propagation of body and surface waves, but these latter are of secondary importance in relation to the problem of soil liquefaction (Ishihara, 1995). The body waves may be distinguished in compression (primary, P-) waves and shear (secondary, S-) waves. Due to the decrease in the soil stiffness with decreasing depth, the body waves propagate almost vertically near the ground surface (Towhata (2008), Fig. 47).

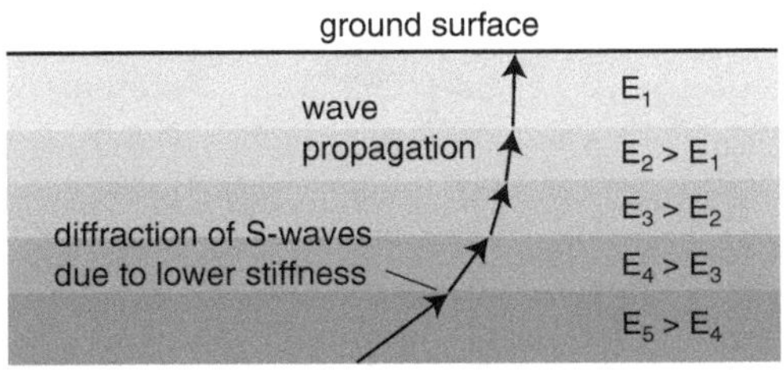


**Figure 47.** Nearly vertical propagation of body waves near the ground surface due to pressure-dependent soil stiffness

If a P-wave passes an element of saturated soil the total vertical stress $\sigma_v$ and the total horizontal stress $\sigma_h$ change by almost the same amount, that means $\sigma_v^{\text{ampl}} \approx \sigma_h^{\text{ampl}}$ and therefore $q^{\text{ampl}} = \sigma_v^{\text{ampl}} - \sigma_h^{\text{ampl}} \approx 0$ (Ishihara, 1995, Fig. 48). This almost isotropic increase in to the total stress results in an increase in the pore water pressure of the same magnitude but does not change the effective stresses $\sigma'_v = \sigma_v - u$ or $\sigma'_h = \sigma_h - u$, respectively. Therefore, the P-wave is usually not considered when estimating the liquefaction risk of an in-situ soil deposit.

On the contrary, soil liquefaction may be caused by shear waves. Fig. 2b shows the stresses acting on an element of soil before and during the transition of a shear wave. The shear wave causes cyclic shear stresses with an irregular time history $\tau(t)$. Due to the relatively fast application of the cyclic shearing partly drained or nearly undrained conditions can be assumed. Instead of the soil compaction that would have been observed during a drained cyclic shearing, the pore water pressure accumulates with each cycle.

The increase in pore water pressure $u$ is accompanied by a reduction

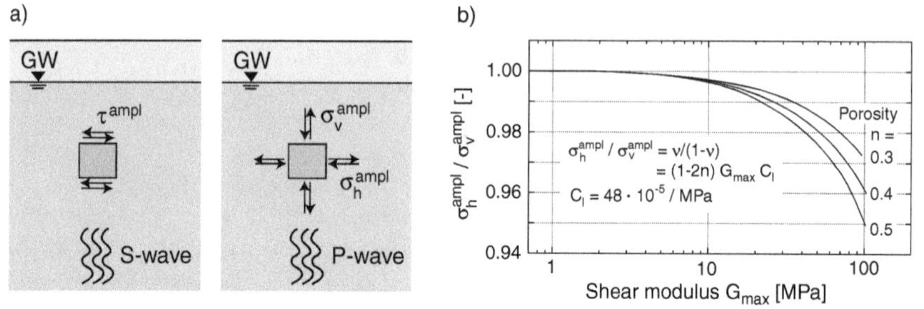


**Figure 48.** a) Stress amplitudes during the transition of S- and P-waves, b) Amplitude ratio $\sigma_v^{\rm ampl}/\sigma_h^{\rm ampl}$ as a function of small-strain shear modulus $G_{\rm max}$ (Ishihara, 1995)

in the effective stresses $\sigma_v'$ and $\sigma_h'$. Consequently, the stiffness and the shear strength of the soil are reduced. In the extreme case, the pore water pressure reaches the total vertical stress $\sigma_v$. In this case, also the effective lateral stress vanishes so that $\sigma_v' = \sigma_h' = 0$ (Fig. 49). At that time, most of the contacts among adjacent grains have got lost. The soil behaves like a suspension composed of water and grains without any shear strength. Therefore, the soil is said to have "liquified". A soil liquefaction can result into a loss of the bearing capacity of foundations, in failure of slopes or in an uplift of underground structures like tanks or sewers (Ishihara, 1995; Towhata, 2008).

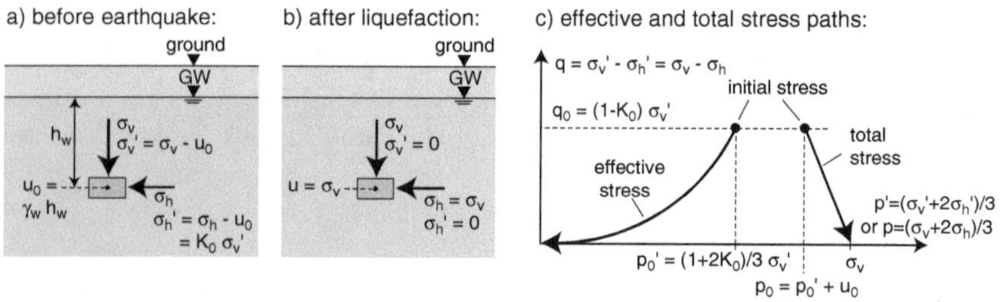


**Figure 49.** Change of total and effective stress components due to the build-up of excess pore water pressure

Different types of laboratory tests may be used to study the liquefaction susceptibility of a soil. The undrained cyclic simple shear test most closely reproduces the in-situ conditions during a cyclic shearing caused by a shear wave. However, the stress and strain fields may be inhomogeneous. The undrained cyclic triaxial test with an isotropic consolidation of the specimen and with symmetrical cycles applied in the vertical direction is frequently used. Sometimes the specimens are also consolidated anisotropically. The suitability of the undrained cyclic triaxial test has been demonstrated by Seed and Lee (1966). For level ground, Ishihara (1995) recommends the torsional shear test with an anisotropic consolidation and with a prevention of lateral strains (Fig. 50a). In that test the total lateral stress is increased to maintain $\varepsilon_3 = 0$. For a sloping ground Ishihara (1995) prefers a torsional shear test with anisotropic consolidation and with no constraints for the lateral strain ($\varepsilon_3 \neq 0$, Fig. 50b). The undrained cyclic triaxial test with an anisotropic consolidation and with an oscillation of the axial stress is considered to be equivalent (Ishihara, 1995).

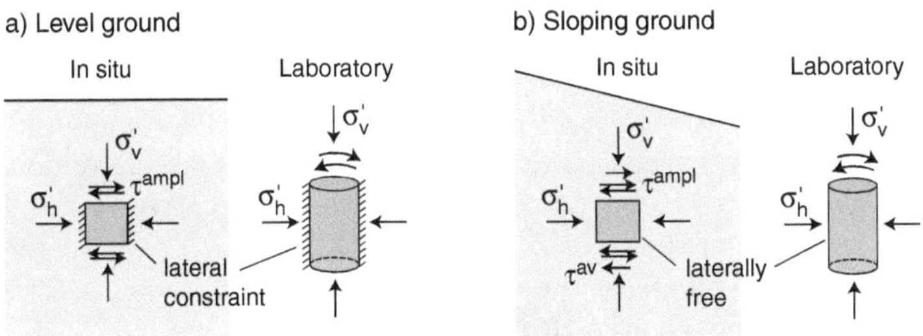


**Figure 50.** Appropriate laboratory tests to reproduce the behaviour of an element of soil below level or sloping ground (Ishihara, 1995)

Typical effective stress paths during undrained *monotonic* triaxial tests on air-pluviated specimens are shown in Fig. 51. A reduction in the effective mean stress $p'$ corresponds with a contractive soil behaviour while an increase is equivalent to dilatancy. For a certain initial pressure, the shape of the effective stress paths mainly depends on density. A loose sand (curve A) shows a contractive behaviour. When the deviatoric stress $q$ has passed a maximum it decreases until a "steady state (SS)" is reached. Then the effective stress does not change any more although the shearing is continued.

In the case of a medium dense sand (curve B) the deviatoric stress first increases until a local maximum is reached. Afterwards it decreases up to a local minimum (so-called "quasi steady state (QSS)"). Up to this point the behaviour is contractive, that means $p'$ decreases. If the shearing is continued the material behaviour changes from contractive to dilative (so-called "phase transformation", PT). The subsequent increase in $p'$ is accompanied by an increase in the deviatoric stress. During this last phase of the test the effective stress path in the $p'$-$q$-plane follows an almost linear curve.

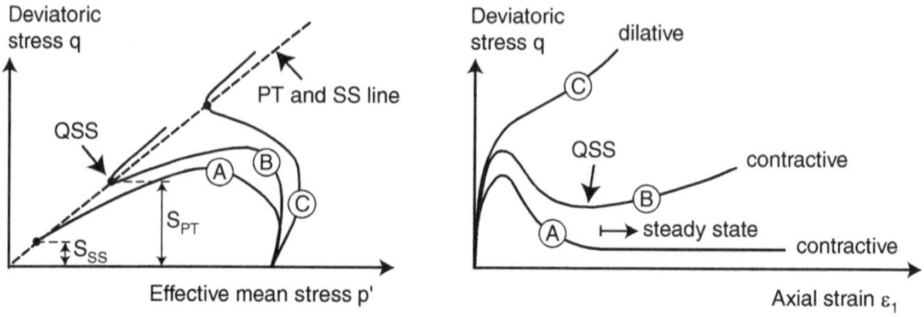


**Figure 51.** Effective stress paths in the $p'$-$q$-plane typically measured in undrained monotonic triaxial tests on loose, medium dense and dense sand samples (Vaid et al., 1981)

The curve C for dense sand is similar to curve B for medium coarse sand with the only exception that the deviatoric stress continuously increases during the test. The transition from the contractive to the dilative behaviour is similar for curves B and C. A QSS is not observed for dense sand. If the elbow points of the stress paths B and C are connected they lay on the same linear curve through the origin (PT line). The QSS is a special case of a PT.

For practical engineering problems the minimum values of the undrained shear strength are of main interest, that means the SS value for loose sand and the QSS value for medium dense sand (Vaid and Sivathayalan, 2000). The SS state which a medium coarse sand reaches after considerable shear deformations ($q_{SS} > q_{QSS}$) is of little practical interest (Vaid and Sivathayalan, 2000).

Fig. 52 presents own undrained monotonic triaxial tests performed on a fine sand (Wichtmann et al., 2010c). Triaxial compression and extension tests were performed with different initial mean pressures $p_0'$ (Fig. 52a) and with different initial densities $I_{D0}$ (Fig. 52b). The shape of the measured effective stress paths agrees well with the curves shown in Fig. 51. The behaviour of soil subjected to an undrained monotonic loading and the most important influencing parameters are discussed in more detail in Section 4.2.

The principle behaviour of a non-cohesive soil subjected to an undrained *cyclic* loading is explained in Figs. 53 and 54 by means of a cyclic triaxial test performed on a fine sand (Wichtmann et al., 2010c). The medium dense specimen was prepared by air pluviation and consolidated under an isotropic effective stress of $\sigma_1' = \sigma_3' = 300$ kPa using a back pressure of 200 kPa. The cycles were applied with an amplitude of $q^{\mathrm{ampl}} = 75$ kPa. Fig. 53a

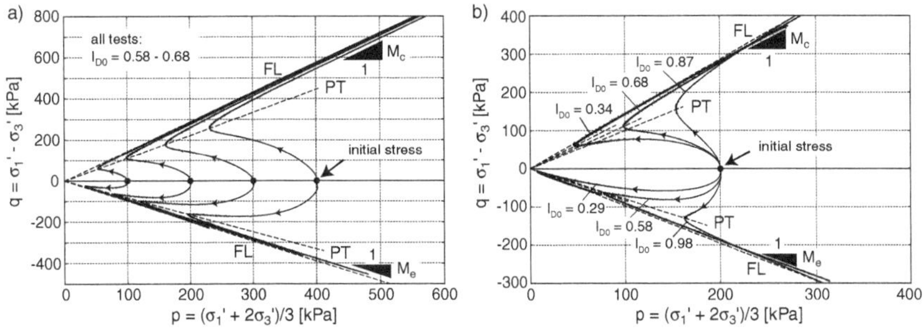


**Figure 52.** Undrained monotonic triaxial tests on a fine sand (Wichtmann et al., 2010c) with a) different initial pressures $p_0'$ and b) different initial densities $I_{D0}$

shows the increase in the pore water pressure with each cycle. After approx. 90,000 seconds the pore water pressure $u$ reached the total lateral stress $\sigma_3$ for the first time, that means $\sigma_3' = 0$ holds. This state of a test is usually referred to as "initial liquefaction". A phase of "cyclic mobility" followed: this is characterized by a pore water pressure reaching condition $u = \sigma_3$ twice within a cycle, passing two minima of different magnitude. When the effective stress components within a cycle temporarily become zero, one speaks of a "partial liquefaction".

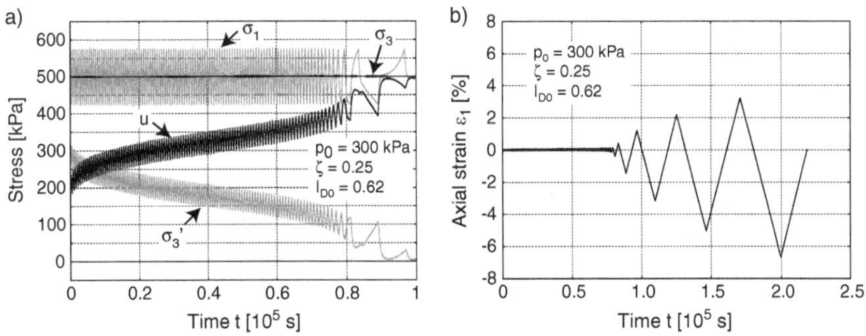


**Figure 53.** Development of a) total and effective stress components and b) of axial strain with time in an undrained cyclic triaxial test on a fine sand (Wichtmann et al., 2010c)

While the amplitude of axial strain $\varepsilon_1^{\mathrm{ampl}}$ remained small during the first cycles, it considerably increased after the initial liquefaction (Fig. 53b).

Afterwards $\varepsilon_1^{\text{ampl}}$ increased with each subsequent cycle. The specimen is usually said to be "fully liquefied" or failed when a certain value of the axial strain amplitude is reached. Different limiting value (e.g. $2\varepsilon_1^{\text{ampl}} = 2$, 5 or 10 %) are used in the literature. The choice of a failure criterion will depend on the tested material and on the type of foundation which has to be designed avoiding liquefaction. Due to the different strength in triaxial compression and extension symmetrical $q$-cycles around the $p$-axis usually cause a failure on the extension side. In the test shown in Fig. 53b the "initial" and the "full liquefaction" lay close to each other which is typical for loose or medium dense clean sand.

The deviatoric stress $q$ is plotted versus the axial strain $\varepsilon_1$ in Fig. 54a. Before initial liquefaction the stress-strain-hystereses are narrow and nearly linear. After initial liquefaction the curves run parallel to the $\varepsilon_1 = 0$-axis over a wide range of strain, implying that the soil does not mobilize any shear resistance. Once both a certain axial strain is reached and the pore water pressure drops due to dilatancy, a shear resistance is regained. The length of these phases without a considerable shear strength increases at increasing numbers of cycles.

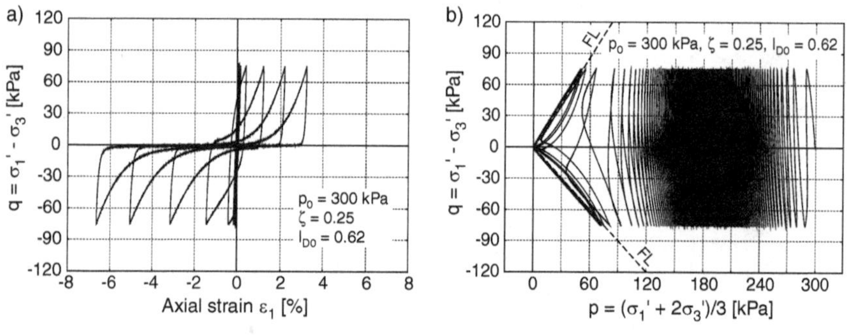


**Figure 54.** a) $q$-$\varepsilon_1$-hystereses and b) effective stress path in the $p$-$q$-plane in an undrained cyclic triaxial test on a fine sand (Wichtmann et al., 2010c)

Fig. 54b shows the effective stress path in the $p'$-$q$-plane. Due to the increase in the pore water pressure and the accompanying decrease in the effective mean pressure $p'$ associated with the increase in the number of cycles, the stress path shifts to the left. During the phase of cyclic mobility the stress path takes a butterfly-like shape.

Usually several such tests with different stress amplitudes are conducted for a given material. The number of cycles to liquefaction decreases with increasing amplitude. A common representation of the test results is given in

Fig. 55. The amplitude $q^{ampl}$ or the "cyclic stress ratio" $CSR = q^{ampl}/(2p'_0)$ with the initial effective mean pressure $p'_0$ is plotted versus the number of cycles necessary to reach initial liquefaction or a predetermined failure criterion. For torsional shear tests $CSR = \tau^{ampl}/p'_0$ and for simple shear tests $CSR = \tau^{ampl}/\sigma'_{1,0}$ are often used on the ordinate of diagrams similar to that shown in Fig. 55. The amplitude-pressure ratio fulfilling a given failure criterion (e.g. $2\varepsilon_1^{ampl} = 2, 5$ oder 10 %) in a certain number of cycles (usually $N = 10$, 15 or 20) is termed "undrained cyclic strength". It can be read off from a diagram as that shown in Fig. 55.

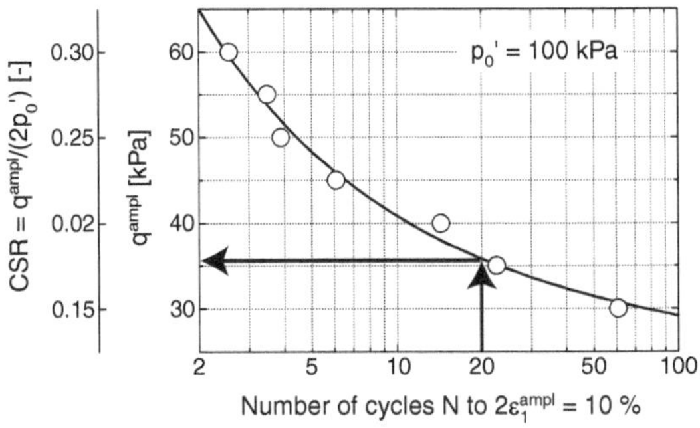


**Figure 55.** Stress amplitude $q^{ampl}$ or "cyclic stress ratio" $q^{ampl}/(2p'_0)$ versus number of cycles to failure (here defined at $2\varepsilon_1^{ampl} = 10\%$), results of undrained cyclic triaxial tests on a medium coarse sand (Wichtmann et al., 2005c)

With respect to an undrained cyclic loading, Vaid and Chern (1985) (see also Vaid and Sivathayalan, 2000 and Robertson et al., 1992) distinguished three different cases depending on soil density. A specimen looser than critical will show a behaviour as illustrated in Fig. 56a. After a few cycles a "steady state" is reached: the final stage of the test is similar to a test with an undrained monotonic loading on loose sand specimens. The type of failure is denoted as "flow liquefaction". A sand with the critical void ratio will show a "limited liquefaction" (Fig. 56b). After some cycles the stress path passes a "quasi-steady state" during which large deformations are generated. The test ends with a phase of cyclic mobility. If the soil is denser than critical (Fig. 56c) stress paths as those already shown in Fig. 54b are observed.

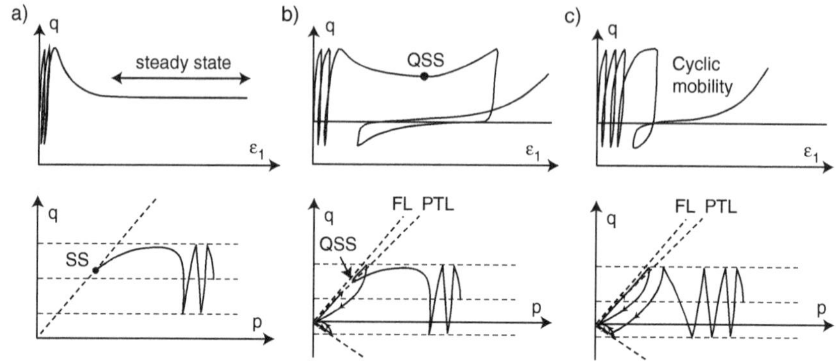


**Figure 56.** Schematic illustration of typical curves $q(\varepsilon_1)$ (Vaid and Sivathayalan, 2000) and $q(p)$ in cyclic triaxial tests on a) very loose, b) loose and c) medium dense or dense sand

### 4.2 Parameters Affecting Soil Response to Undrained Cyclic Loading

It has been already shown in Fig. 55 that an increase in the stress amplitude causes a decrease in the number of cycles necessary to reach a certain failure criterion. The faster accumulation of pore water pressure with increasing shear strain amplitudes becomes also obvious from the results of displacement-controlled undrained cyclic triaxial tests performed by Dobry et al. (1982). In Fig. 57a the ratio $\Delta u/\sigma'_{30}$ of the excess pore water pressure and the initial effective lateral stress is plotted versus the shear strain amplitude. No build-up of excess pore water pressure was observed below a certain threshold amplitude which was approximately $\gamma_{tv} = 10^{-4}$, independently of the method of sample preparation, of the effective consolidation pressure $p'_0$ and of the type of sand (Fig. 57b). However, if a larger number of cycles had been applied, a moderate accumulation of pore water pressure would probably have been observed also for amplitudes $\gamma^{\text{ampl}} < 10^{-4}$ (compare Fig. 20a).

One of the most important influencing parameters regarding the accumulation in pore water pressure under undrained conditions is the density of the soil. This has been clearly demonstrated by the pioneer work of Seed and Lee (1966) and Lee and Seed (1967). For a certain stress amplitude, a much larger number of cycles is necessary to cause a liquefaction in a dense sand compared to a loose one (Fig. 58). That means that at increasing soil densities curves $CSR(N)$ are shifted upwards and to the right. The differences between the curves $CSR(N)$ for the different densities are even

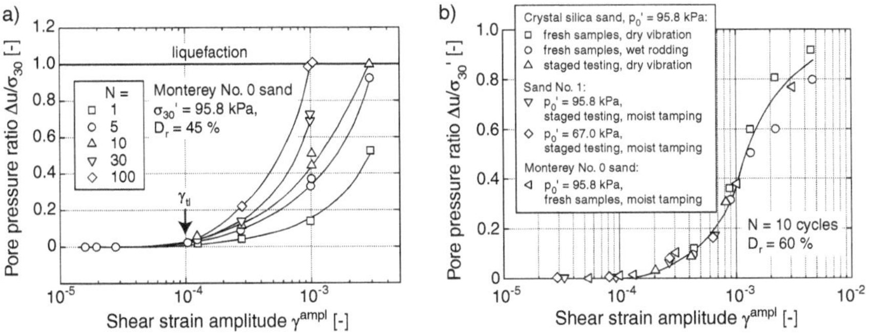


**Figure 57.** Simple shear tests with different shear strain amplitudes (Dobry et al., 1982)

more pronounced for a failure criterion $\varepsilon_1^{\text{ampl}} = 20$ % (Fig. 58b) than for the initial liquefaction (Fig. 58a). This is due to the fact, that after initial liquefaction the increase in the strain amplitude with each subsequent cycle is much smaller in the case of a dense sand than for a loose one (Fig. 59). While for a loose sand the initial and the full liquefaction lay close to each other, a considerable number of cycles may lay between them for a dense sand.

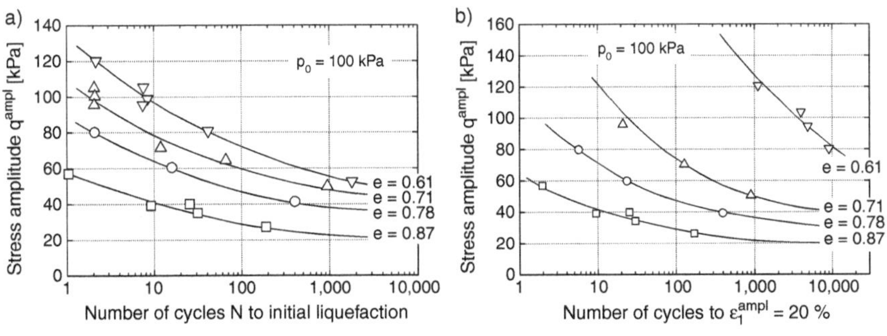


**Figure 58.** Influence of void ratio on the number of cycles to a) initial liquefaction and b) a strain amplitude $\varepsilon_1^{\text{ampl}} = 20\%$, undrained cyclic triaxial tests of Lee and Seed (1967)

Another example for the increase in the liquefaction resistance with increasing density are the test results of Mori et al. (1978) in Fig. 60a. Tatsuoka et al. (1986b) found a linear relationship between the liquefaction

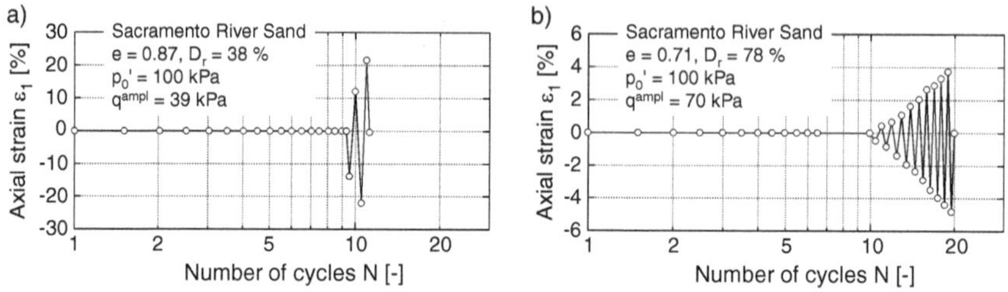


**Figure 59.** Development of axial strain with time in undrained cyclic triaxial tests on a) loose and b) dense sand (Lee and Seed, 1967)

resistance and the relative density up to $D_r \approx 75$ % (Fig. 60b). For larger relative densities the cyclic stress ratio causing liquefaction increases over-linear with $D_r$.

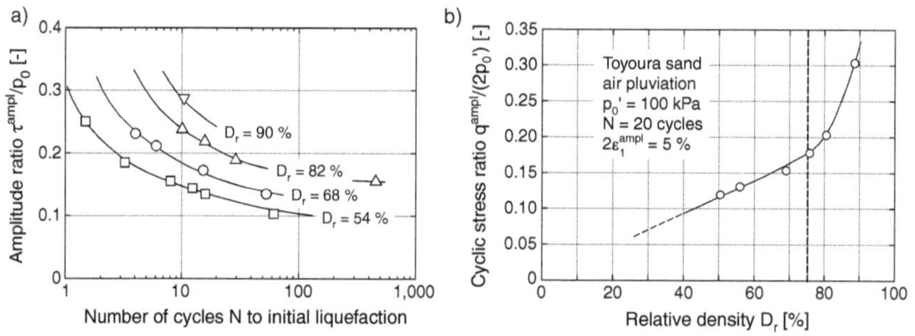


**Figure 60.** Increase of the liquefaction resistance with increasing density: tests of a) Mori et al. (1978) and b) Tatsuoka et al. (1986b)

Lee and Seed (1967) demonstrated that the influence of the initial effective mean pressure $p_0'$ is nearly eliminated if data are plotted using the amplitude-pressure ratio $q^{\mathrm{ampl}}/(2p_0')$ (lower row of diagrams in Fig. 61) on the ordinate instead of the stress amplitude $q^{\mathrm{ampl}}$ (upper row of diagrams in Fig. 61). A slight decrease in the curves $CSR(N)$ with increasing pressure may be concluded from the diagram for $D_r = 100$ % in Fig. 61. Such a decrease has been also reported by other researchers. Ishihara (1995) collected the data of Seed and Harder (1990), Kokusho et al. (1983), Frydman et al. (1980) and Vaid and Thomas (1994) and proposed a pressure-dependent correction factor $K_\sigma(p_0')$ (solid line in Fig. 62) with which the liquefaction resistance measured for a pressure $p_0' = 100$ kPa can be reduced for larger pressures. Ishihara (1995) stated that $K_\sigma$ strongly depends on the type of soil. Furthermore, Vaid and Thomas (1995) demonstrated that $K_\sigma$ also depends on density. For loose specimens $K_\sigma \approx 1$ was observed almost independent of $p_0'$.

In displacement-controlled undrained cyclic triaxial tests Dobry et al. (1982) found a decrease in the rate of pore water pressure accumulation with increasing initial effective mean pressure $p_0'$. These observations are in good accordance with our results from drained cyclic tests (Fig. 25b).

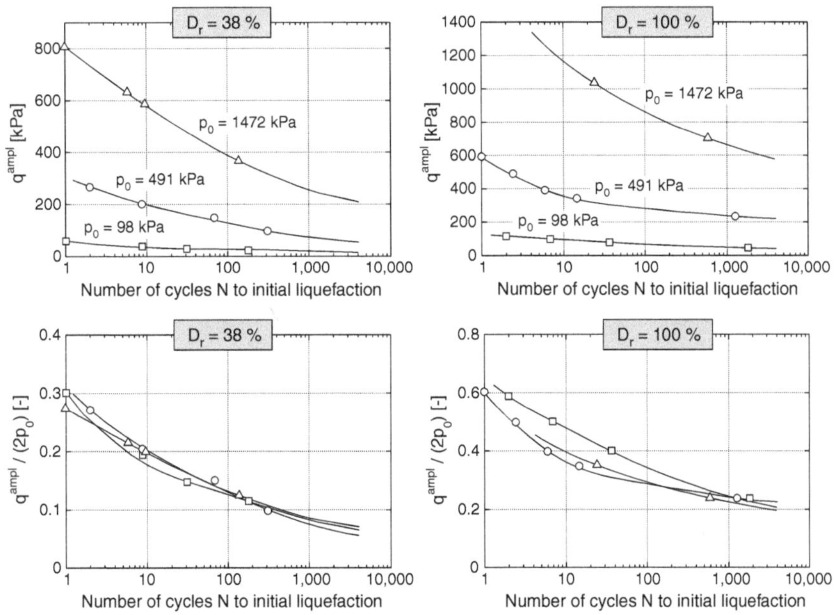


**Figure 61.** The use of an amplitude-pressure ratio $q^{\text{ampl}}/(2p_0')$ on the ordinate purifies the data from the influence of the initial effective mean pressure $p_0'$ (Lee and Seed, 1967)

Ishihara (1995) studied the effect of the lateral pressure coefficient $K_0$ in undrained torsional shear tests. Three different values of $K_0$ (0.5, 1.0 and 1.5) were tested. The total lateral stress was increased after each cycle in order to keep the lateral strain zero ($\varepsilon_3 = 0$). Therefore, with increasing number of cycles the stress approached an isotropic state. While a large influence of $K_0$ can be observed if the data are plotted with the ratio $\tau^{\text{ampl}}/\sigma_{10}'$ on the ordinate (Fig. 63a), the curves for the different $K_0$-values fall together if the ratio $\tau^{\text{ampl}}/p_0'$ with the initial effective mean effective

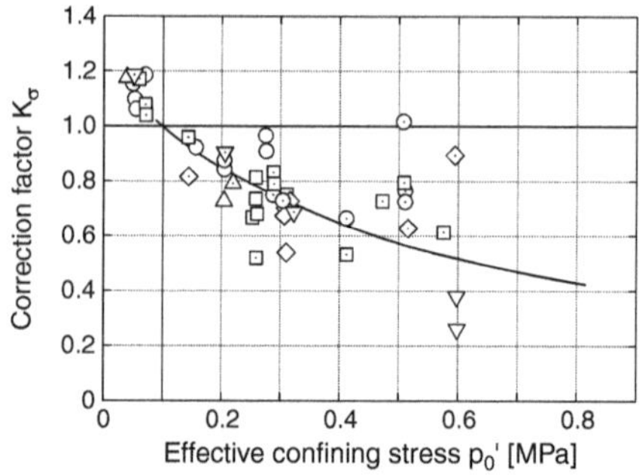


**Figure 62.** Correction factor $K_\sigma$ for a consideration of the decrease of the cyclic stress ratio $q^{\mathrm{ampl}}/(2p_0')$ causing liquefaction with increasing pressure $p_0'$ (Ishihara, 1995)

pressure $p_0' = (1 + 2K_0)\sigma_{10}'/3$ is used (Fig. 63b). Therefore, the coefficient $K_0$ does not influence the liquefaction resistance as long as the data are analyzed in terms of $p_0'$ instead of $\sigma_{1,0}'$. It seems justified to replace an anisotropic in-situ stress state by an isotropic one in the cyclic triaxial test as long as the initial effective mean pressure is the same.

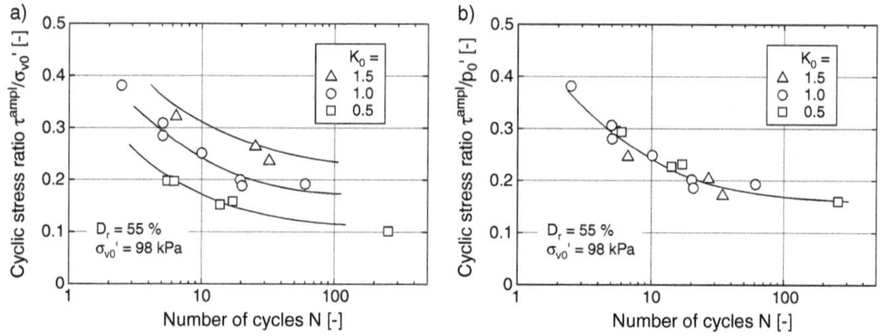


**Figure 63.** Elimination of the effect of $K_0$ by analyzing the data with an amplitude-pressure ratio $q^{\mathrm{ampl}}/(2p_0')$ with the initial effective mean pressure $p_0'$ on the ordinate instead of $q^{\mathrm{ampl}}/\sigma_{1,0}'$ (Ishihara, 1995)

A static shear stress $\tau^{\mathrm{av}}$ has to be considered for example in the case of sloping ground or below the edges of foundations. The static shear stress can be considered in simple shear or torsional shear tests in which the cycles are applied around an average (static) value. Cyclic triaxial tests with a consolidation under anisotropic stress conditions are an alternative.

Hyodo et al. (1994) performed cyclic triaxial tests with an anisotropic consolidation at different values of $q^{av}$. They found an increase in the liquefaction resistance with increasing average deviatoric stress for $q^{av} \leq 40$ kPa (Fig. 64a). When $q^{av}$ was further increased, a significant reduction in the liquefaction resistance was observed. In many of the tests of Hyodo et al. (1994) large permanent axial strains developed (called "flow" deformation by Hyodo et al., 1994) accompanied by a more or less sudden increase in the pore water pressure (Fig. 65). Hyodo et al. (1994) assumed that the flow occurs when the effective stress path in the cyclic tests reaches a line in the $p'$-$q$-plane which is often referred to as the "instability line" (IL) in the literature. This line passes through the local maxima of the effective stress paths observed in undrained monotonic tests on medium dense specimens (Fig. 64b). However, not all of the stress paths measured in the cyclic tests of Hyodo et al. (1994) can be explained by this theory.

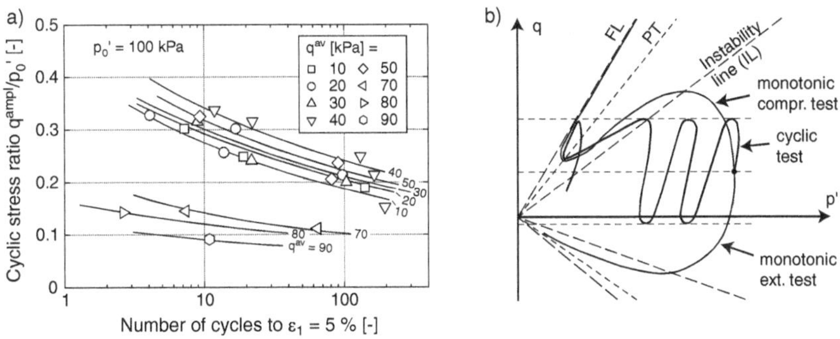


**Figure 64.** a) $CSR(N)$-diagram from cyclic triaxial tests with anisotropic consolidation stresses (Hyodo et al., 1994), b) Comparison of effective stress paths in undrained monotonic and cyclic triaxial tests (Hyodo et al., 1994)

Simple shear tests of Vaid and Finn (1979) showed that the effect of a static shear stress depends on the chosen failure criterion. Fig. 66a presents the cyclic stress ratio causing a certain shear strain amplitude $\gamma^{ampl} = 2$, 5 or 10 % in 10 cycles as a function of the static stress ratio $\alpha = \tau^{av}/\sigma'_{1,0}$. With increasing $\alpha$ a lower cyclic stress ratio is necessary to cause a shear strain amplitude $\gamma^{ampl} = 2$ %, but a larger amplitude is needed for $\gamma^{ampl} =$ 5 and 10 %. Considering a constant shear stress amplitude, with increasing static shear stress a lower number of cycles is necessary to reach $\gamma^{ampl} = 2$ % but a larger one is needed for the larger shear strain amplitudes.

The undrained cyclic triaxial tests of Rollins and Seed (1990) revealed

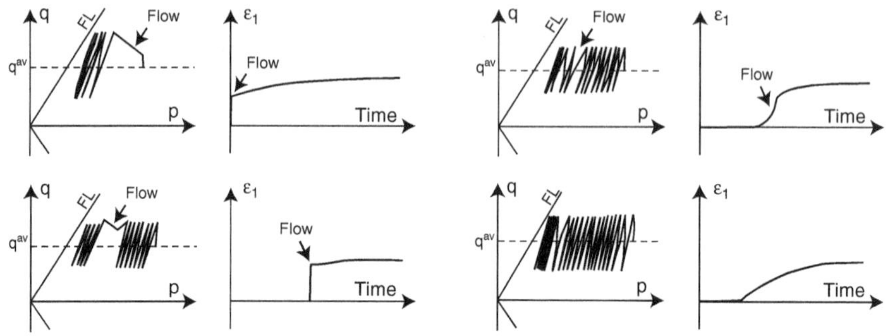


**Figure 65.** Different kinds of flow deformation observed by Hyodo et al. (1994) in undrained cyclic triaxial tests with anisotropic consolidation stresses

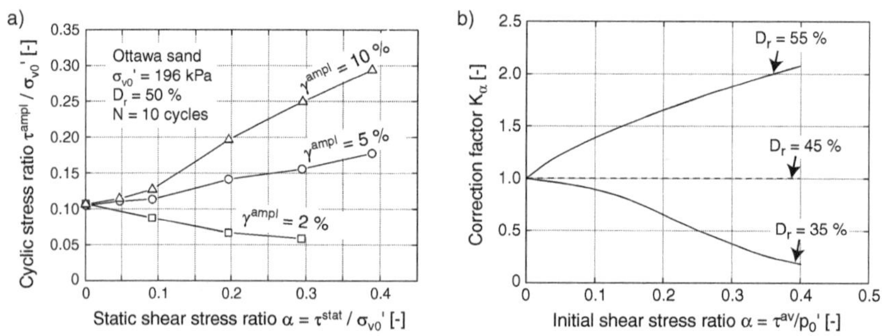


**Figure 66.** Influence of a static initial shear stress on the liquefaction resistance depends on a) the chosen failure criterion (Vaid and Finn, 1979) and b) on soil density (Rollins and Seed, 1990)

that the effect of a static shear stress depends on density. The correction factor $K_\alpha$ in Fig. 66b is defined as the cyclic stress ratio causing failure ($2\varepsilon_1^{\text{ampl}} = 5$ %) for a certain static shear stress $\tau^{\text{av}}$ divided by the cyclic stress ratio causing failure at $\tau^{\text{av}} = 0$. For loose soil ($D_r = 35$ %), the liquefaction resistance decreases with increasing static stress ratio $\alpha = \tau^{\text{av}}/p'_0$. For $D_r =$ 45 % there is almost no influence of $\tau^{\text{av}}$ and for $D_r = 55$ % the liquefaction resistance increases with $\alpha$. $K_\alpha$-values for other relative densities can be interpolated. A similar density-dependence of $K_\alpha$ was found by Stedman (1997) (see also Vaid and Sivathayalan, 2000). Ishihara (1995) and Vaid and Sivathayalan (2000) stated that the correction factors $K_\alpha$ shown in Figure

66b can be regarded as conservative, especially for loose sands.

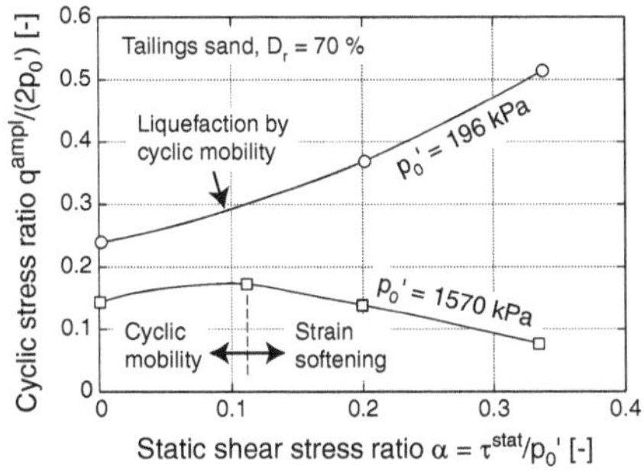


**Figure 67.** Influence of a static initial shear stress on the liquefaction resistance depends on initial mean effective stress $p_0'$ (Vaid and Chern, 1985)

The tests of Vaid and Chern (1985) on dense sand (Fig. 67) demonstrate furthermore, that the influence of a static shear stress also depends on the effective consolidation pressure $p_0'$. Based on the data in Fig. 67 Vaid and Sivathayalan (2000) concluded that the influence of $p_0'$ (correction factor $K_\sigma$, see Fig. 62) and the influence of $\tau^{av}$ (correction factor $K_\alpha$) cannot be treated independently of each other. A reduction in the liquefaction resistance by the product $K_\alpha \cdot K_\sigma$ would drastically underestimate the liquefaction resistance, in particular for loose sands. For a certain sand Vaid and Sivathayalan (2000) recommended to study the combined influence of $p_0'$ and $\tau^{av}$ in laboratory tests.

Several studies in the literature show that the loading frequency does not influence the liquefaction resistance measured in laboratory tests on clean sand. In the undrained cyclic triaxial tests of Yasuda and Soga (1984) the same liquefaction resistance was measured for different loading frequencies between 0.05 and 1 Hz (Fig. 68a). In the regime of higher loading frequencies ($1 \leq f \leq 12$ Hz) Yoshimi and Oh-Oka (1975) could also not detect any frequency-influence in their undrained cyclic torsional ring shear tests (Fig. 68b). Similar conclusions were drawn by Wong et al. (1975), Tatsuoka et al. (1986a) and Kokusho et al. (2004).

The fabric of the grain skeleton in situ (orientation of the grains and of the grain contacts, etc.) depends on the depositional and on the deformation history of the soil. In the laboratory different initial fabrics can be obtained by different specimen preparation methods. Several researchers (Ladd, 1974, 1977; Park and Silver, 1975; Tatsuoka et al., 1986b; Porcino et al., 2004) observed a large influence of the preparation method

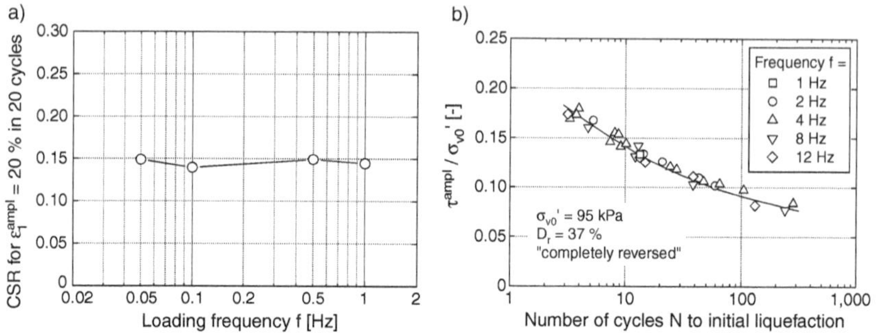


**Figure 68.** No influence of frequency on the liquefaction resistance: tests a) of Yasuda and Soga (1984) with 0.05 Hz $\leq f \leq$ 1 Hz and b) of Yoshimi and Oh-Oka (1975) with 1 Hz $\leq f \leq$ 12 Hz

on the liquefaction resistance of sand specimens. Exemplary, the results of Mulilis et al. (1975, 1977) are shown in Fig. 69. Although all specimens were prepared with the same relative density of $D_r = 50$ %, the different specimen preparation methods lead to significant differences in the liquefaction resistance. Specimens prepared in the moist condition tended to be characterized by a higher liquefaction resistance than those prepared dry. Specimens which were compacted after preparation (by vibration, tamping or rodding) showed a higher liquefaction resistance than specimens directly prepared with $D_r = 50$ %. Water pluviation tended to produce a slightly higher resistance than air pluviation. Most studies in the literature agree that air pluviation leads to the lowest liquefaction resistance. However, the literature is somewhat ambiguous concerning the effect of preparation methods working with moist material (Towhata, 2008). In principle, the method of sample preparation in the laboratory should reflect the depositional history of the soil in situ. Dobry and Ladd (1980) showed that the influence of the method of specimen preparation may be disregarded when the data are analyzed in terms of the strain amplitude instead of the stress amplitude (Fig. 57b).

Many authors (Mulilis et al., 1975, 1977; Peck, 1979; Tokimatsu and Hosaka, 1986; Hatanaka et al., 1988; Seki et al., 1992; Porcino et al., 2004) reported about a significantly larger liquefaction resistance of high-quality undisturbed specimens than of specimens of the same disturbed material but reconstituted in the laboratory. The increased liquefaction resistance is probably due to aging or cyclic preloading effects (discussed later in this section). Therefore, testing of reconstituted samples may significantly un-

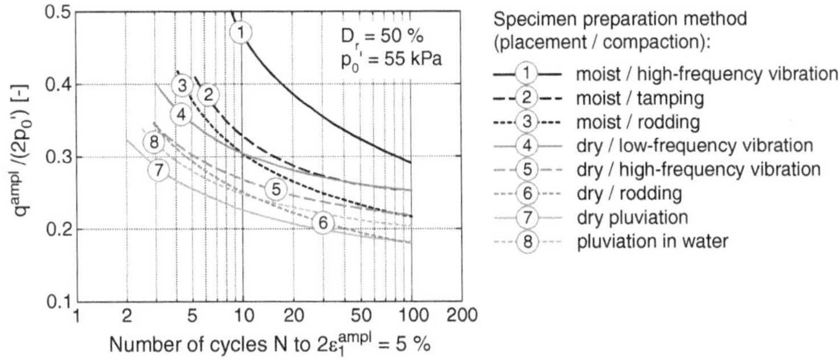


**Figure 69.** Large influence of the specimen preparation method on the liquefaction resistance, data of Mulilis et al. (1975)

derestimate the liquefaction resistance of an in-situ soil deposit as long as the specimen preparation method does not reproduce the in-situ fabric of the soil. High-quality undisturbed specimens can be obtained for example by the ground freezing technique. Such undisturbed samples have to be handled very carefully since the slight fabric effects can easily be destroyed (Seed, 1979).

As an example test results of Yoshimi et al. (1984, 1989) are shown in Fig. 70. They used the ground freezing technique to take undisturbed samples from a natural sand deposit in Niigata. For comparison purposes, an artificial sand deposit (4 m $\times$ 6 m $\times$ 5 m) was prepared by water pluviation. Specimens were taken by ground freezing also from this deposit. The undrained cyclic triaxial tests revealed a significantly higher liquefaction resistance of the specimens taken from the natural sand layer compared to those from the artificial deposit. This may be taken as a hint that samples obtained by the ground freezing technique preserve their in-situ fabric of the grain skeleton. Other test series in the literature (Vaid et al., 1999; Ghionna and Porcino, 2006), however, found a similar liquefaction resistance for undisturbed samples taken by ground freezing and reconstituted samples prepared by water pluviation. These results support the thesis that ground freezing does *not* preserve the in-situ fabric. The contradictory results may be due to a different handling of the frozen specimens in the laboratory. A thawing under reduced effective in-situ stresses (Vaid et al., 1996; Vaid and Sivathayalan, 2000; Hydro, 1993; Singh et al., 1982) seems to be advantageous compared to a thawing under full in-situ stresses (Hofmann et al., 1995; Konrad and Pouliet, 1997). The latter method may lead to a disuniform stress distribution, to changes of the void ratio and thus to fabric changes (Wu and Chang, 1982; Sivathayalan and Vaid, 1999). However, due to creep or relaxation effects in the ice phase also samples obtained by ground freezing will not be fully undisturbed (Vaid and Sivathayalan, 2000).

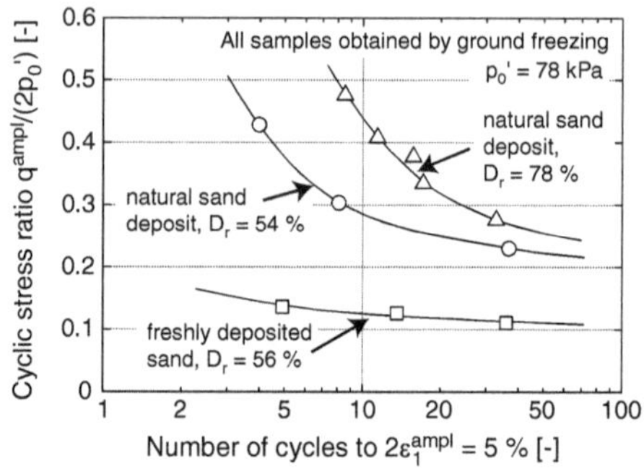


**Figure 70.** Comparison of the liquefaction resistance of undisturbed samples obtained by ground freezing from a natural sand deposit and from a freshly deposited (artificial) sand layer (Yoshimi et al., 1989)

It is well-known that a cyclic preloading with small amplitudes increases the liquefaction resistance of a granular material (Finn et al., 1970; Park and Silver, 1975; Suzuki and Toki, 1984; Seed et al., 1988; Teachavorasinskun et al., 1994; Emery et al., 1973; Oda et al., 2001). Seed et al. (1977) performed shaking table tests in which a water-saturated sand layer was subjected to an undrained simple shearing. Five packages with a small number of cycles were applied in succession. They should simulate small earthquakes. After each package the soil was reconsolidated. The accumulation of pore water pressure became lower from package to package, that means with increasing cyclic preloading (Fig. 71a). Finally, the soil was cyclically sheared until liquefaction occurred. Despite similar density, the curves $CSR(N)$ of specimens subjected to a cyclic preshearing were found to lay significantly higher than the curves of freshly pluviated samples (Fig. 71b).

Another example is shown in Fig. 72 where the results of undrained cyclic triaxial tests of Wichtmann et al. (2005c) are presented. Parts of the samples were subjected to a drained cyclic preloading with 10 or 100 cycles and with an amplitude of $q^{\mathrm{ampl}} = 30$ or 50 kPa. Fig. 72a presents the development of pore water pressure with time for both a preloaded and a freshly pluviated sample. Although both samples had a similar density and were subjected to the same average and cyclic stresses, the accumulation of pore water pressure was much smaller in the test on the preloaded sample.

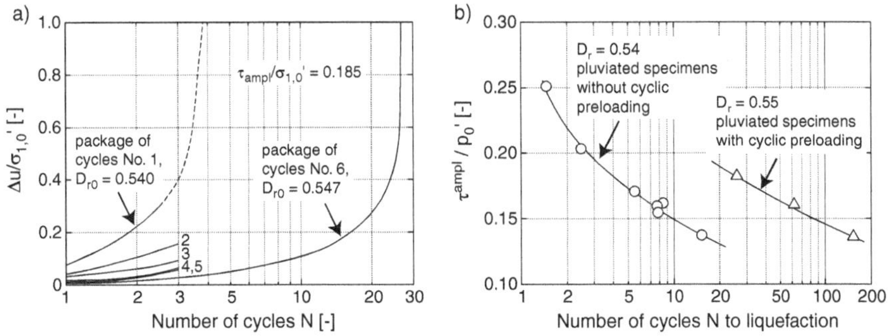


**Figure 71.** Increase in the liquefaction resistance due to a cyclic preloading with small amplitudes: shaking table tests of Seed et al. (1977)

The curves $CSR(N)$ given in Fig. 72b also demonstrate the increase in the liquefaction resistance with increasing intensity of the cyclic preloading.

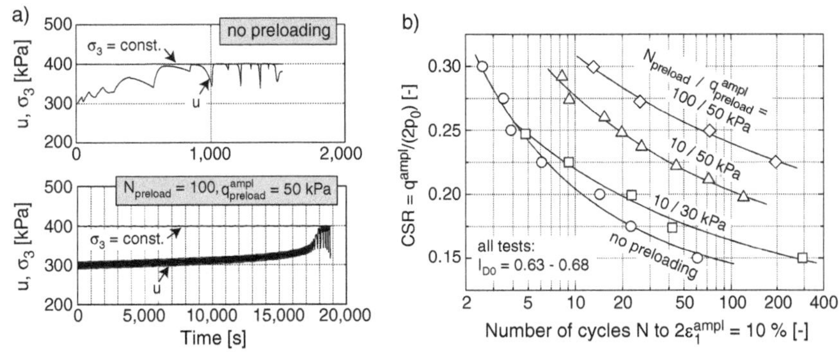


**Figure 72.** Increase in the liquefaction resistance due to a cyclic preloading with small amplitudes: undrained cyclic triaxial tests with a drained cyclic preloading (Wichtmann et al., 2005c)

Ishihara and Okada (1978, 1982) distinguished between a small and a large preloading. A small preloading is defined as a preloading where the effective stress path does not exceed the phase transformation line. The path goes beyond the PT line in the case of a large preloading. In their undrained cyclic triaxial tests Ishihara and Okada (1978, 1982) also found an increased liquefaction resistance due to a small cyclic preloading (Fig. 73a). In the case of a large preloading the behaviour depends on whether the

PT line is exceeded in the triaxial compression or extension regime. If it is exceeded during triaxial compression and the specimen is reconsolidated afterwards, than during a second cyclic loading a slower accumulation of pore water pressure is observed above the $p$-axis while it is faster below the $p$-axis (Fig. 73c). It is the other way around for a preloading applied in the triaxial extension regime (Fig. 73b).

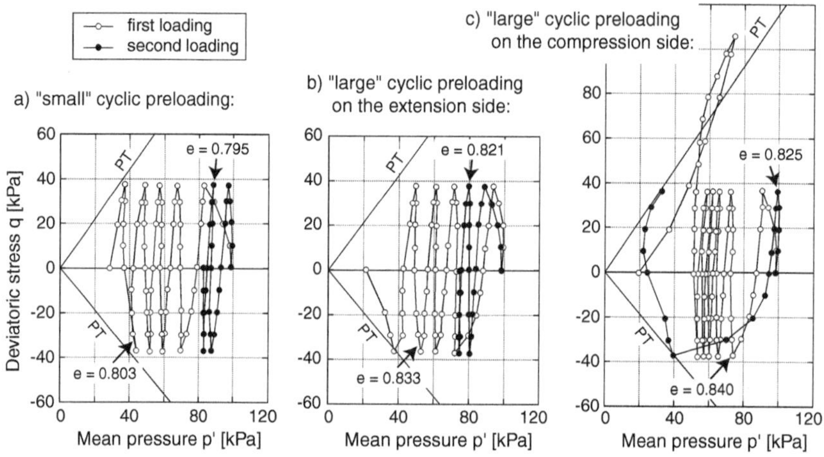


**Figure 73.** Effect of a "small" and a "large" preloading on the effective stress paths in undrained cyclic triaxial tests (Ishihara and Okada, 1978)

The liquefaction resistance of soil specimens usually increases with the time after stress application (so-called "aging"). Fig. 74 presents the results of undrained cyclic triaxial tests performed by Seed (1979). The consolidation time (1 to 100 days) was varied for the different specimens. In Fig. 74 the cyclic stress ratio of "aged" specimens causing a liquefaction in a certain number of cycles is divided by the respective $CSR$-value of a specimen which was subjected to the undrained cyclic loading immediately after preparation. Obviously, the liquefaction resistance increases with increasing consolidation time, for example by approx. 25 % after 100 days. An extrapolation of the test results to larger consolidation times could explain the larger liquefaction resistance usually observed for in-situ sand deposits (Fig. 74). A similar increase in the liquefaction resistance with time was reported by Tatsuoka et al. (1988).

Next, the influence of the grain size distribution curve on the liquefaction resistance is discussed. Fig. 75a presents data of Lee and Fitton (1969) concerning the influence of the mean grain size $d_{50}$. The cyclic stress ratio

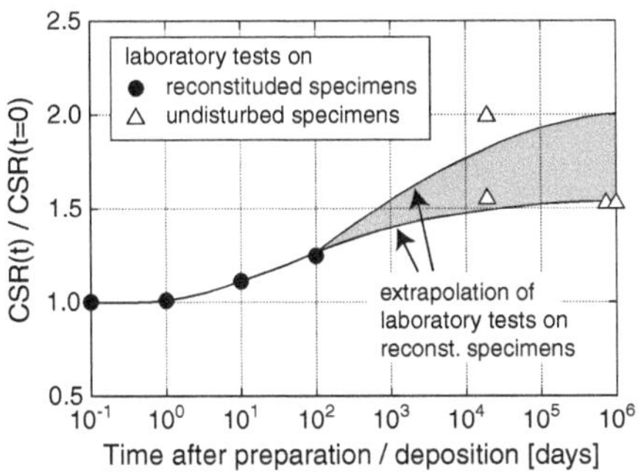


**Figure 74.** Increase in the liquefaction resistance with time: cyclic loading of laboratory specimens after different consolidation times (Seed, 1979)

causing a liquefaction in 30 cycles is presented as a function of $d_{50}$. The curve $CSR(d_{50})$ shows a minimum at a mean grain size of $d_{50} = 0.1$ mm. For smaller and larger values of $d_{50}$ the liquefaction resistance increases with decreasing or increasing $d_{50}$, respectively. Seed and Idriss (1971) collected data from tests performed by several researchers on different kinds of sands. They reported about a significant decrease in the liquefaction resistance with decreasing mean grain size. Seed and Idriss (1971) stated that the liquefaction resistance is lowest for uniform fine sands with $d_{50} \approx 0.08$ mm. Also Castro and Poulos (1977) and Dupla and Canou (2003) (Fig. 75b) found an acceleration of the accumulation of pore water pressure with decreasing mean grain size. In situ a decrease in the grain size is accompanied by a decrease in the permeability of a soil. The poor drainage during an earthquake loading increases the liquefaction susceptibility of a fine sand.

Vaid et al. (1990b) studied the influence of the coefficient of uniformity $C_u = d_{60}/d_{10}$ of the grain size distribution curve on the liquefaction resistance. The three grain size distribution curves shown in Fig. 76a were mixed from a natural river sand. They had different $C_u$-values (1.5, 3 and 6) but the same mean grain size $d_{50} = 0.42$ mm. Specimens were prepared using the "slurry deposition method" proposed by Kuerbis and Vaid (1988). In Fig. 76b the cyclic stress ratio causing a liquefaction in 10 cycles is shown as a function of relative density. Obviously, for $D_r =$ constant the liquefaction resistance is similar for the three tested sands, that means there is hardly any dependence on the coefficient of uniformity.

The liquefaction resistance is decreased in the presence of a non-plastic (non-cohesive) fines content while it is increased by a plastic (cohesive) fines content.

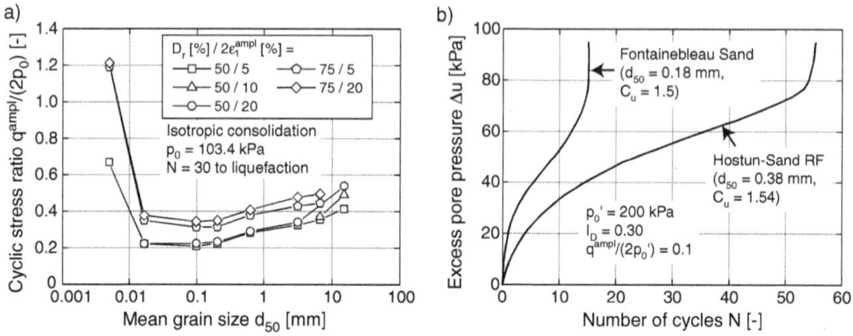


**Figure 75.** Influence of the mean grain size $d_{50}$ on the liquefaction resistance: undrained cyclic tests of a) Lee and Fitton (1969) and b) Dupla and Canou (2003)

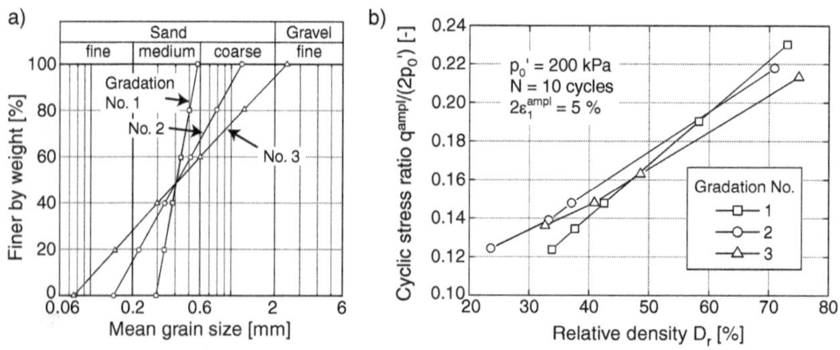


**Figure 76.** No influence of the coefficient of uniformity $C_u = d_{60}/d_{10}$ for a constant relative density, undrained cyclic tests of Vaid et al. (1990b) on sands with different grain size distribution curves

Chien et al. (2002) tested a fine sand mixed with different amounts of non-plastic fines (grains with $d < 0.075$) in the range $0 \leq FC \leq 30$ %. Specimens were prepared by moist tamping in five layers. Fig. 77a shows that for a constant relative density the liquefaction resistance decreases with increasing fines content. The effect of the fines content is even more pronounced if the data is analyzed for a constant void ratio $e$ (Fig. 77b). Similar test results were reported by Kokusho et al. (2004).

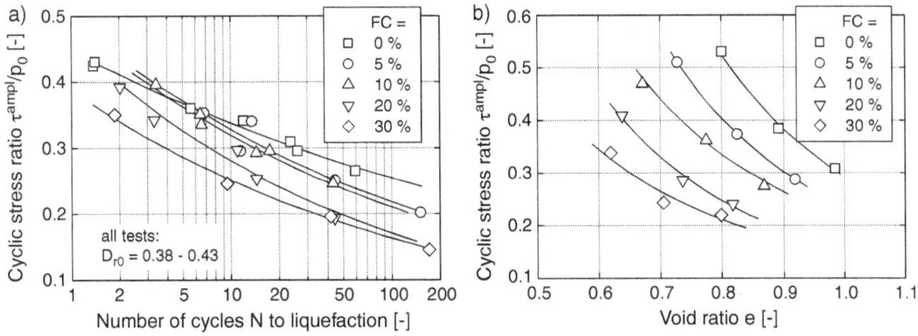


**Figure 77.** Reduction of the liquefaction resistance in the presence of non-plastic fines, tests of Chien et al. (2002)

With respect to Seed and Idriss (1982) a clayey sand is susceptible to liquefaction if the grains with $d < 0.005$ mm are less than 15 %, if the liquid limit of the fines is $w_L < 35$ % and if the natural water content is $w > 0.9\ w_L$ (so-called "chinese criteria"). Kuwano et al. (1996) performed undrained cyclic triaxial tests on mixtures of Toyoura sand with Fujinomori clay (plasticity index $I_P = 37.7$ %). The specimens were prepared either by pluviation or by consolidation out of a slurry. Fig. 78a shows the cyclic stress ratio causing a double axial strain amplitude of $2\varepsilon^{\text{ampl}} = 5$ % in 20 cycles as a function of the clay content. Obviously, the liquefaction resistance increases with increasing clay content, independently of the sample preparation method. Ishihara et al. (1978) reported an even faster increase of the liquefaction resistance with the clay content for an over-consolidated soil than for a normally consolidated one. The undrained cyclic triaxial tests of Ishihara and Koseki (1989) (Fig. 78b) demonstrated that the increase of the liquefaction resistance due to a plastic fines content is the larger, the higher the plasticity index $I_P$ of the fines is. For small plasticity indices $I_P < 10$ the liquefaction resistance is similar to that of clean sand.

During earthquakes also gravelly soil deposits have been observed to liquefy (Youd et al., 1985; Andrus and Youd, 1987; Ishihara, 1995; Towhata, 2008). Despite their large permeability such a liquefaction can occur if the

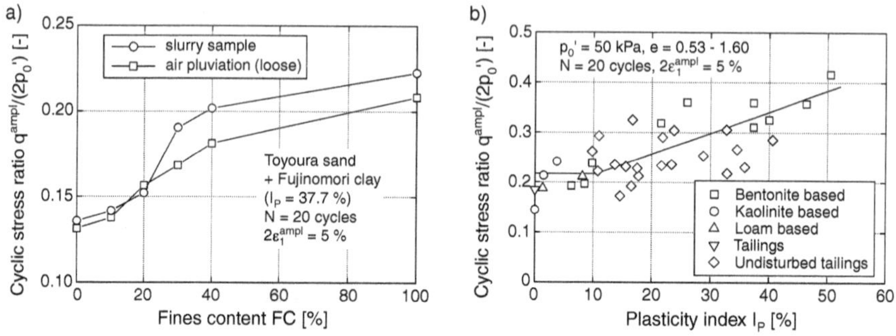


**Figure 78.** Increase of the liquefaction resistance by plastic fines: a) study of Kuwano et al. (1996), b) data collected by Ishihara and Koseki (1989)

gravelly layers are encompassed by layers with a much lower permeability. In most laboratory studies on undisturbed or reconstituted samples of gravelly materials (Wong et al., 1975; Kokusho and Tanaka, 1994; Goto et al., 1994; Hatanaka et al., 1988; Tanaka et al., 1991; Konno et al., 1994; Ishihara, 1995) a higher liquefaction resistance has been measured as usually obtained for sand. This is in good agreement with the $d_{50}$-dependence shown in Fig. 75. Ishihara (1995) stated that the liquefaction resistance of gravelly soils can strongly vary in dependence of their depositional history. Towhata (2008) warned that the higher liquefaction resistance usually measured for gravelly soils may be partly due to membrane penetration effects (see a discussion below).

The important role of the degree of saturation $S_r$ is demonstrated by the data of Xia and Hu (1991) in Fig. 79. They prepared specimens of a fine sand with four different degrees of saturation $S_r$ = 100.0, 99.5, 99.1 and 97.8, corresponding to $B$-values of 1.0, 0.928, 0.857 and 0.714. The test results in Fig. 79 reveal a significant decrease of the liquefaction resistance with increasing $S_r$.

The effect of the polarization of the cycles was tested by Choi and Arduino (2004) in true triaxial tests on a gravelly material. They applied shear cycles in different directions of the deviatoric plane. The pore water pressure accumulation was found similar for all tested directions (Fig. 80a). Therefore, the polarization of the cycles seems not to influence the liquefaction resistance as long as is not changed. This is in accordance with the results from drained cyclic tests (Fig. 22b).

Yamada and Ishihara (1982) tested the influence of polarization changes in undrained true triaxial tests. The same loading as used in the correspond-

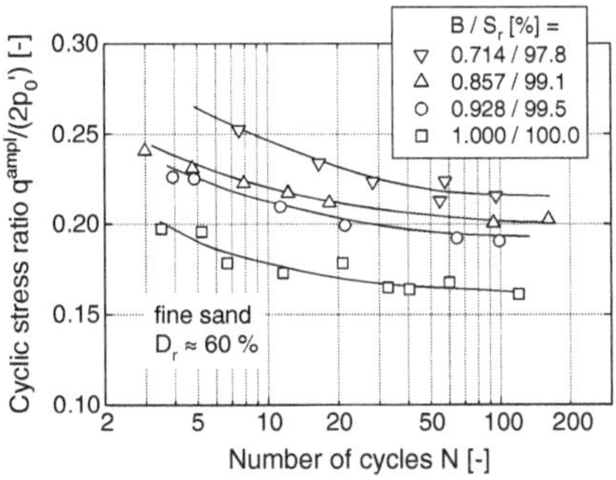


**Figure 79.** Decrease of the liquefaction resistance with increasing degree of saturation, undrained cyclic triaxial tests of Xia and Hu (1991), curves are given for initial liquefaction

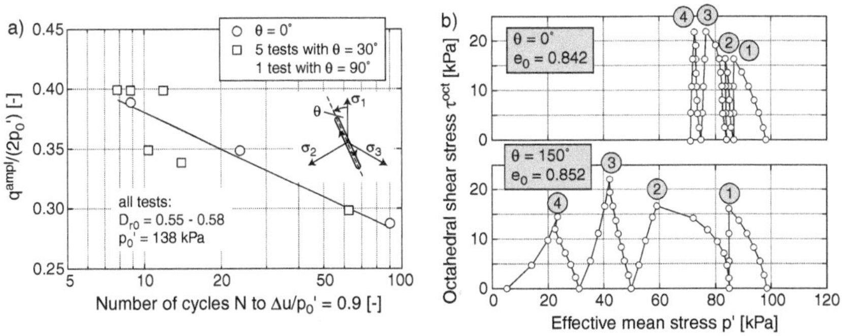


**Figure 80.** a) No influence of the polarization of the cycles in the deviatoric plane (Choi and Arduino, 2004), b) Increase of the rate of pore water pressure accumulation due to a change of the polarization of the cycles (Yamada and Ishihara, 1982)

ing drained tests (Fig. 28b) was applied. Similar as in the drained tests a significant increase of the rate of pore water pressure accumulation was observed due to a change of the polarization of the cycles (Fig. 80b). The effect was stronger for larger angles $\theta$ between the subsequent polarizations.

Ishihara and Yamazaki (1980) studied the influence of the shape of the cycles in multiaxial undrained simple shear tests. They tested elliptical and cross-shaped cycles (Fig. 81). The shear stress amplitude $\tau_1^{\text{ampl}}$ in one direction was kept constant while the amplitude $\tau_2^{\text{ampl}}$ in the orthogonal direction was varied in the range $0 \leq \tau_2^{\text{ampl}} \leq \tau_1^{\text{ampl}}$. For both, the elliptical and the cross-shaped cycles the accumulation of pore water pressure increased with increasing ratio $\tau_2^{\text{ampl}}/\tau_1^{\text{ampl}}$. A slightly faster accumulation of pore water pressure was observed for the elliptical cycles compared to the cross-shaped ones. This strong influence of the shape of the cycles is in good agreement with the results from drained cyclic tests (Fig. 23).

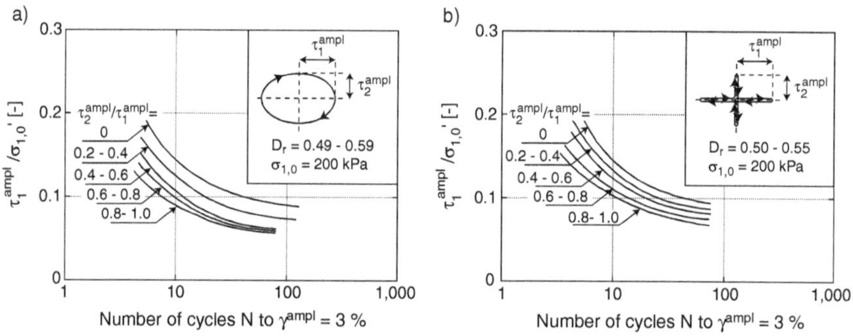


**Figure 81.** Influence of the shape of the cycles: multi-directional simple shear tests of Ishihara and Yamazaki (1980) with elliptical and cross-shaped stress loops

The time history of the shear stress $\tau(t)$ during an earthquake is irregular (Fig. 2b). For an estimation of the liquefaction risk (see Section 4.4) this irregular load pattern is usually replaced by an equivalent number of regular cycles with a constant amplitude (Seed et al., 1975b; Seed and Idriss, 1971). Equivalent means that both, the irregular and the regular load pattern cause the same final value of the excess pore water pressure. The amplitude of the regular cycles is chosen as a certain fraction of the maximum shear stress during the irregular time history. For an appropriate choice of this factor, several researchers have compared an irregular and a regular cyclic loading (Ishihara and Yasuda, 1972, 1975; Ishihara and Nagase, 1988). Exemplary, results from hollow cylinder triaxial tests of Ishihara and Yasuda

(1975) are shown in Fig. 82. Signals from real earthquakes were scaled and applied as shear stress $\tau(t)$ to the laboratory specimens. Different types of signals were tested. Ishihara and Yasuda (1975) distinguished between "shock-type" (e.g. Fig. 82a) and "vibration-type" signals (e.g. Fig. 82b). A "shock-type" signal is characterized by only a few peaks (reaching at least 60 % of $\tau^{\max}$) prior and after the maximum shear stress $\tau^{\max}$. Otherwise the signal is classified to be of the "vibration-type". Fig. 82c presents the maximum stress ratio $\tau^{\max}/p_0'$ as a function of the normalized pore water pressure $\Delta u/p_0'$ for both types of signals. From such diagrams the lowest ratio $\tau^{\max}/p_0'$ causing a liquefaction ($\Delta u/p_0' = 1$) is read off. A comparison with the amplitudes of regular cycles causing a liquefaction in 20 cycles lead to a ratio $\tau^{\max}_{\text{regular}}/\tau^{\max}_{\text{irregular}} = 0.55$ for the shock-type signals and to $\tau^{\max}_{\text{regular}}/\tau^{\max}_{\text{irregular}} = 0.70$ for the vibration-type ones. Seed and Idriss (1971) used a factor of 0.65. Whitman (1971) proposed to use a factor of 0.67 for a strong earthquake with 20 seconds duration and larger values for a longer duration.

Ishihara and Nagase (1988) performed multiaxial simple shear tests and applied irregular cyclic loading histories in two perpendicular directions. A comparison with a one-dimensional regular loading with constant amplitude lead to correction factors that consider both, the irregularity and the multidimensionality of the cyclic loading.

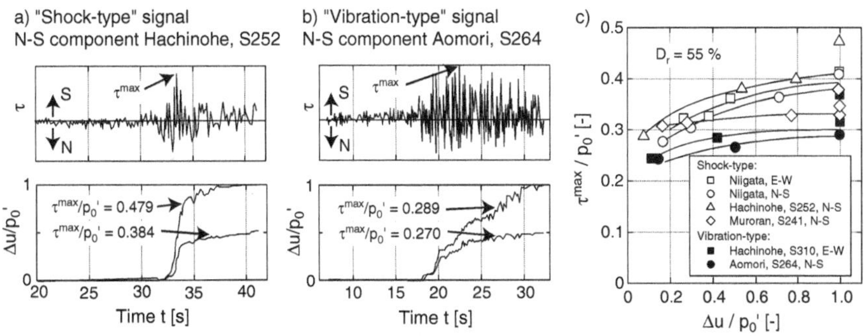


**Figure 82.** Comparison of regular and irregular cyclic loadings (Ishihara and Yasuda, 1975)

The results of undrained cyclic tests may be erroneous due to membrane penetration effects. If the pore water pressure increases due to an undrained cyclic loading the effective lateral stress $\sigma_3'$ decreases and thus the penetration of the membrane into the voids at the boundary of the specimen decreases (Fig. 83a). Therefore, the volume occupied by the pore water increases. Consequently, the pore water pressure accumulation is lower than in the case $\varepsilon_v = 0$ and a larger number of cycles is necessary to reach liquefaction. Therefore, membrane penetration effects lead to an overestimation of the liquefaction resistance which may result in an unsafe design for structures.

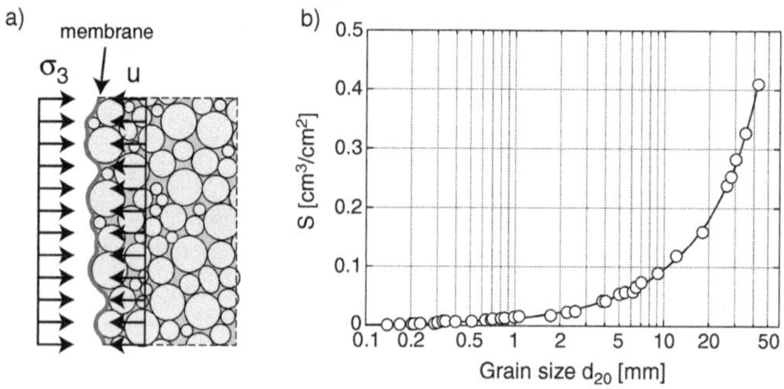


**Figure 83.** a) Schematic illustration of membrane penetration effects (Wichtmann, 2005), b) Increase of membrane penetration with increasing grain size $d_{20}$ according to Nicholson et al. (1993)

Membrane penetration effects increase with increasing grain size of the tested material. This becomes clear from the diagram of Nicholson et al. (1993) (Fig. 83b) which shows the volume change due to membrane penetration per unit area of the membrane as a function of the grain size $d_{20}$.

Some researchers developed a special control of undrained cyclic tests in order to compensate membrane penetration effects and to guarantee a constant volume. Tokimatsu and Nakamura (1986) compared the results of such tests performed with a special control (without system compliance) with results from conventional tests (with system compliance). A sand with a mean grain size of $d_{50} = 1$ mm was tested. Fig. 84 shows a much faster accumulation of pore water pressure and thus a significantly lower liquefaction resistance in the tests without system compliance. Therefore, the liquefaction resistance may be strongly over-estimated for coarse materials if membrane penetration effects are not taken into account.

Several methods have been developed to purify the results of conventional undrained cyclic tests from membrane penetration effects (Martin et al., 1978; Tokimatsu and Nakamura, 1987; Tokimatsu, 1990). A simple method was proposed by Tokimatsu (1990). The system compliance ratio

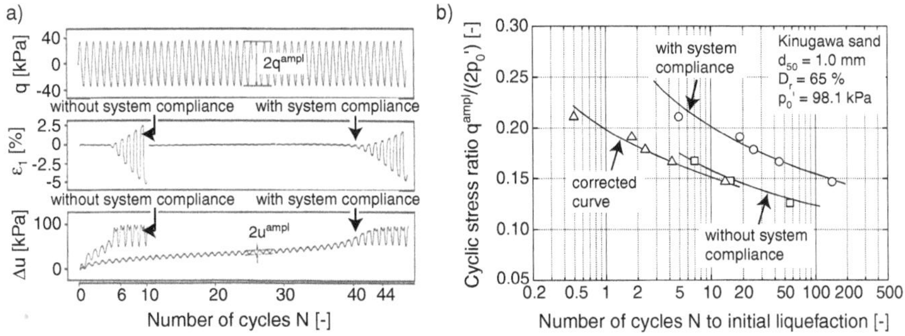


**Figure 84.** Effect of membrane penetration (system compliance) on the liquefaction resistance: tests of Tokimatsu and Nakamura (1986) on a sand with $d_{50} = 1$ mm and $C_u = 1.5$

$C_R$ can be estimated from

$$C_R = \frac{B}{3\frac{u^{\mathrm{ampl}}}{q^{\mathrm{ampl}}}} - 1 \tag{25}$$

with Skempton's $B$-value and with the amplitudes $u^{\mathrm{ampl}}$ and $q^{\mathrm{ampl}}$ of the pore water pressure and the deviatoric stress taken from the middle stage of a test as shown in Fig. 84a. With $C_R$ as an input value the parameter $C_N$ can be read off from the correlation given in Fig. 85a. $C_N$ may be used for a correction of the curves $CSR(N)$ as shown in Fig. 85b. $C_N$ is the ratio between the number of cycles $N_c$ necessary to reach liquefaction for a certain cyclic stress ratio in a test with system compliance and the corresponding number of cycles $N_0$ in a test without system compliance. Tokimatsu (1990) could demonstrate that the curves $CSR(N)$ from the conventional tests corrected by membrane penetration effects using the described method agreed well with the curves from the specially controlled tests without system compliance (Fig. 84b).

### 4.3 Parameters Affecting Soil Response to Undrained Monotonic Loading

The combined influence of void ratio and effective consolidation stress can be demonstrated based on undrained monotonic triaxial tests performed by Verdugo and Ishihara (1996) (Fig. 86). The specimens of Toyoura sand were prepared by moist tamping. Fig. 86 presents the $q(\varepsilon_1)$-curves and the effective stress paths in the $p'$-$q$-plane of four tests performed on specimens

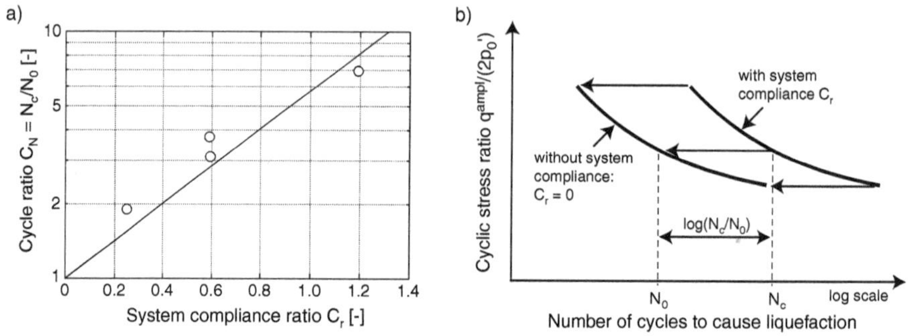


**Figure 85.** a) Correlation between $C_R$ and $C_N$, b) correction of the curves $CSR(N)$ measured in conventional undrained cyclic triaxial tests considering membrane penetration effects (Tokimatsu, 1990)

with the same relative density $D_r = 37.9$ % but with different isotropic consolidation stresses. All specimens reached the same final stress ("steady state"). The specimens consolidated at a small pressure $p_0'$ showed a dilative behaviour and a continuous increase of the deviatoric stress. A contractive behaviour and a strain softening was observed for the specimens consolidated at a large $p_0'$.

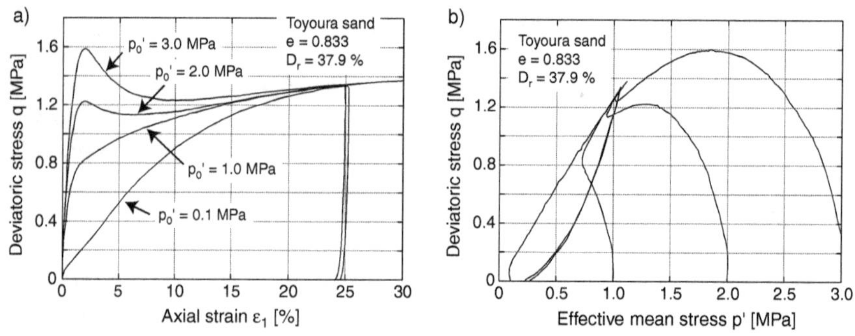


**Figure 86.** a) Curves $q(\varepsilon_1)$ and b) effective stress paths measured in undrained monotonic tests performed on specimens with similar density but consolidated at different pressures $p_0'$ (Verdugo and Ishihara, 1996)

Doanh et al. (2006) compared the effective stress paths for specimens consolidated isotropically (Fig. 87a) or anisotropically (Fig. 87b). The very loose specimens of Hostun RF sand were prepared by moist tamping. All specimens showed a contractive behaviour reaching a "steady state" at low pressures. In the case of the anisotropically consolidated specimens the maximum deviatoric stress was reached near the consolidation stress and at a small axial strain. Although the shape of the stress paths is different for an isotropic and an anisotropic consolidation, the inclination of the PT and the failure line does not depend on the type of consolidation.

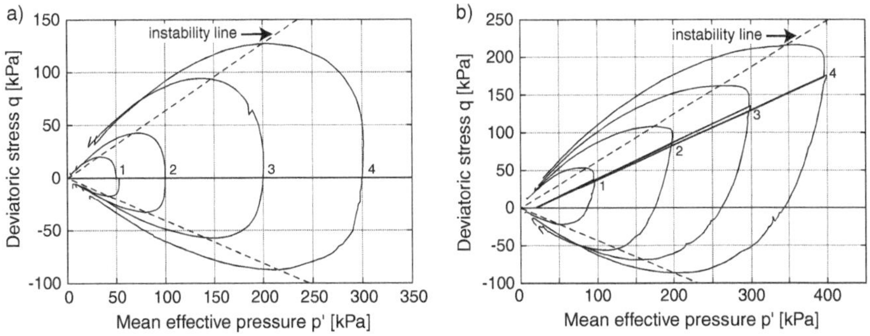


**Figure 87.** Effective stress paths in undrained monotonic tests with a) isotropic and b) anisotropic consolidation stresses (Doanh et al., 2006)

The soil response to an undrained monotonic shearing depends strongly on the initial fabric of the grain skeleton, that means on the specimen preparation method. Fig. 88a presents results from undrained simple shear tests performed by Vaid et al. (1995). The specimens were prepared by air pluviation, water pluviation or moist tamping. Despite similar initial densities and although the axial effective consolidation stress $\sigma'_{1,0}$ was equal, the curves $q(\varepsilon_1)$ look quite different. After having reached a maximum deviatoric stress, the specimen prepared by moist tamping showed a strong softening behaviour. A continuous increase of the deviatoric stress was measured for the water-pluviated sample. The curve for the specimen prepared by air pluviation lays in between. Triaxial tests of Vaid et al. (1999) showed a similar large influence of the sample preparation technique (Fig. 88b). While the specimen preparation method affects the point where the effective stress path reaches the PT- or QSS line, it does not influence the inclination of the PT- or QSS line in the $p'$-$q$-plane (Ishihara, 1993, 1995). Furthermore, the "steady state" is independent of the preparation method.

The effect of a monotonic preloading was studied by Doanh et al. (2006). They performed tests with an isotropic, $K_0$- or deviatoric preloading applied under drained or undrained conditions. Results from tests with a drained deviatoric preloading applied on the triaxial compression side are given in

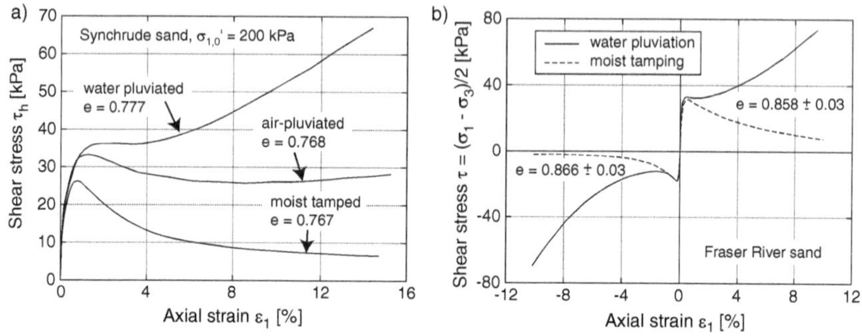


**Figure 88.** Influence of the specimen preparation method on undrained monotonic response in a) simple shear tests (Vaid et al., 1995) and b) triaxial tests (Vaid et al., 1999)

Fig. 89a. If an undrained triaxial compression test was performed after this preloading, the maximum deviatoric stress reached during the test increased with increasing preloading. In the initial phase of a test the stress path tended to follow the stress path applied during preloading. Although the specimens were still very loose after preloading, the stress path passed a pronounced QSS state which is untypical for loose sand. However, the inclination of the PT- or QSS-line seems to be unaffected by the monotonic preloading. If an undrained triaxial extension test was performed on a specimen which was preloaded on the triaxial compression side, the minimum deviatoric stress reached during the test was reduced compared to a non-preloaded specimen. For a preloading on the extension side the behaviour is the other way around (Fig. 89b). Similar conclusions concerning the effect of a monotonic preloading were drawn by Vaid et al. (1989).

The influence of the angle $\alpha$ between the direction of the largest principal stress $\sigma_1'$ and the axis of sedimentation was studied by Vaid and Sivathayalan (2000). For triaxial compression $\alpha = 0°$ and for triaxial extension $\alpha = 90°$ holds. Considering the different curves $q(\varepsilon_1)$ in triaxial compression and extension tests (Fig. 90a) it becomes clear that $\alpha$ influences the undrained monotonic response. While the triaxial compression tests showed a strain hardening, a strain softening behaviour was observed in the triaxial extension tests. Similar test results have been reported elsewhere (Bishop, 1971; Hanzawa, 1980; Miura and Toki, 1982; Riemer and Seed, 1997; Vaid et al., 1990a). The tests of Uthayakumar and Vaid (1998) showed that the transition from a strain hardening to a strain softening behaviour occurs continuously with increasing $\alpha$ (Fig. 90b). The observed dependence of the

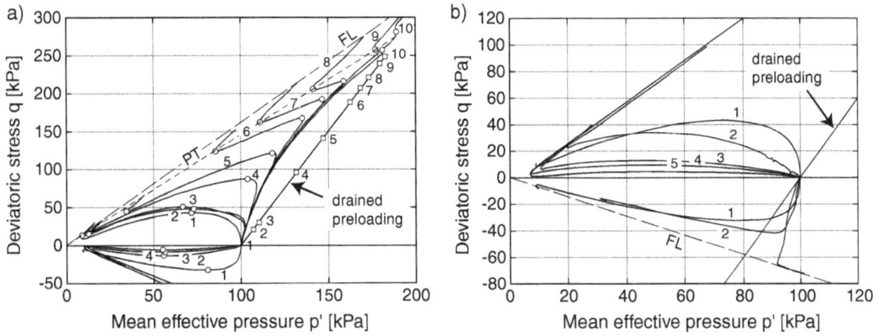


**Figure 89.** Undrained monotonic triaxial tests of Doanh et al. (2006) on very loose samples prepared by moist tamping: effect of a drained deviatoric preloading with a) triaxial compression and b) triaxial extension

undrained monotonic response on the direction of loading is due to an inherent anisotropy of the soil samples induced during specimen preparation.

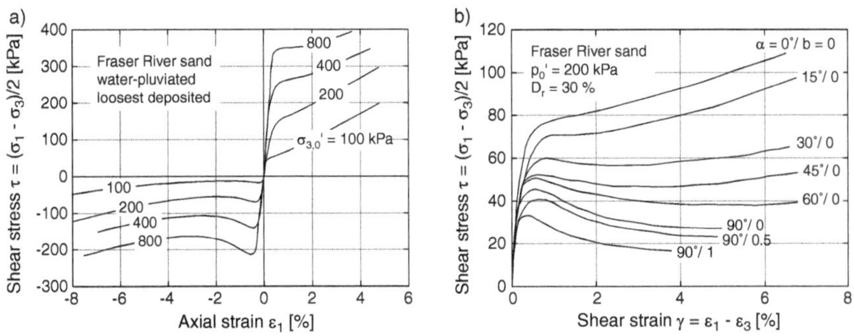


**Figure 90.** Effect of the loading direction on undrained monotonic response: a) Comparison of triaxial compression and extension tests (Vaid and Sivathayalan, 2000), b) Tests with varying $\alpha$-values (Uthayakumar and Vaid, 1998)

The influence of the intermediate principal stress $\sigma_2'$ can be quantified by the parameter

$$b = \frac{\sigma_2' - \sigma_3'}{\sigma_1' - \sigma_3'} \qquad (26)$$

with $\sigma_1'$ and $\sigma_3'$ being the largest and the smallest principal stress, respectively. For triaxial compression ($\sigma'_2 = \sigma'_3$) $b = 0$ holds while $b = 1$ applies during triaxial extension ($\sigma'_2 = \sigma'_1$). In the tests of Uthayakumar and Vaid (1998) (Fig. 90b) the loading direction was kept constant to $\alpha = 90°$ while the parameter $b$ was varied between 0 and 1. Obviously, the strain softening becomes more pronounced with increasing $b$.

If a monotonic undrained triaxial test is interrupted by several un- and reloading cycles, the curve $q(\varepsilon_1)$ and the effective stress path in the $p'$-$q$-plane are similar as in a test without such cycles (Fig. 91).

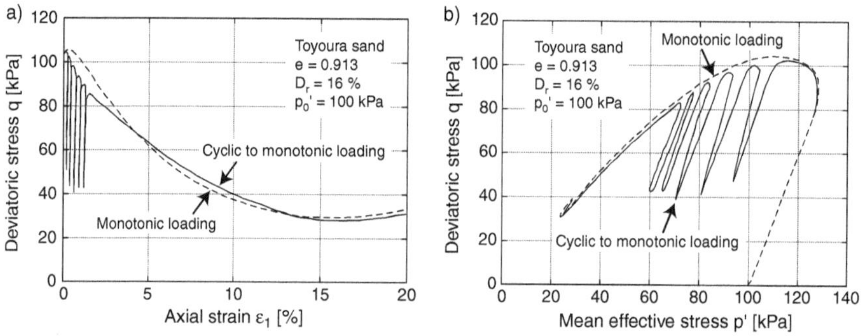


**Figure 91.** Undrained monotonic triaxial test with several un- and reloading cycles compared to a test without such cycles (Ishihara, 1995)

Vaid and Eliadorani (1998) compared the soil behaviour in monotonic triaxial tests with partly drained or fully undrained conditions, respectively (Fig. 92). In the partly drained test the volumetric strain was controlled to fulfill $\varepsilon_v = 0.6\varepsilon_1$. While in the fully undrained test a dilative and strain hardening behaviour was observed, the partly drained test showed a contractive and strain softening response. Vaid and Eliadorani (1998) concluded that even small volume changes in situ could lead to a strain softening and consequently fully undrained tests were on the unsafe side.

## 4.4 Estimation of the Liquefaction Risk in Situ

The liquefaction risk for an in-situ soil deposit may be estimated by employing the following methods (Towhata, 2008):

1. Comparison of sounding resistances (e.g. SPT blow count or CPT tip resistance) with critical values
2. Estimation via a safety factor

$$F_L = \frac{\text{Resistance}}{\text{Action}}, \tag{27}$$

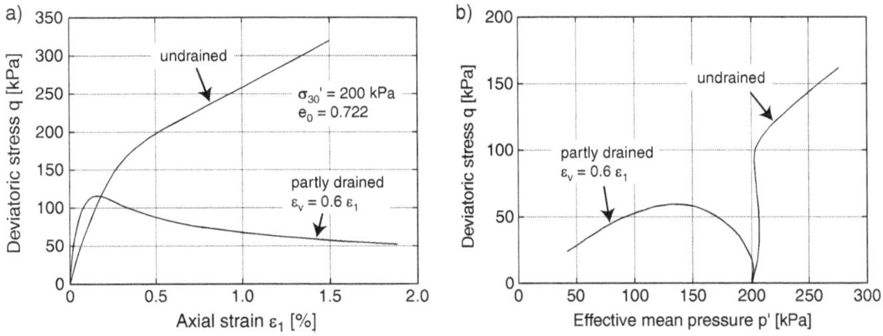


**Figure 92.** Comparison of a partly drained and a fully undrained monotonic triaxial test (Vaid and Eliadorani, 1998)

which is evaluated as a function of depth $z$ below ground surface. The action can be obtained using either

a) an estimation based on the maximum acceleration at the ground surface using a semi-empirical equation or

b) a dynamic calculation of wave propagation in which the time history $u_h(t)$ of the horizontal displacement is prescribed at a certain depth. The reduction of soil stiffness due to a pore water pressure build-up is disregarded in these calculations.

The resistance of the soil against liquefaction can be assessed from:

a) cyclic undrained laboratory tests on undisturbed or reconstituted specimens or from

b) correlations with sounding resistances (e.g. SPT blow count or CPT tip resistance) or with the shear wave velocity or from

c) large-scale in-situ tests

Both, the action and the resistance may be either analyzed in terms of stresses ("stress approach", Seed and Idriss, 1971; Seed et al., 1975a; Seed, 1979) or in terms of strains ("strain approach", Dobry et al., 1982). A liquefaction is likely when the acting shear stress amplitude $\tau_{\text{act}}^{\text{ampl}}$ is larger than the soil resistance $\tau_{\text{res}}^{\text{ampl}}$ (Fig. 93a). If the analysis is performed with strains, it has to be examined first if the acting shear strain amplitude $\gamma_{\text{act}}^{\text{ampl}}$ is larger than a threshold value $\gamma_{tv}^{\text{ampl}}$ (Fig. 93b). In case $\gamma_{\text{act}}^{\text{ampl}} > \gamma_{tv}^{\text{ampl}}$ the pore water pressure ratio $\Delta u/p_0'$ has to be estimated from a diagram as shown in Fig. 57a. A liquefaction is likely to occur if $\Delta u/p_0'$ is equal to one.

3. Dynamic calculation of wave propagation using a constitutive model which describes the stress-strain hysteresis and considers the accumu-

lation of volumetric strain in the drained case and the build-up of pore water pressure in the undrained case (e.g. Osinov, 2003a,b).

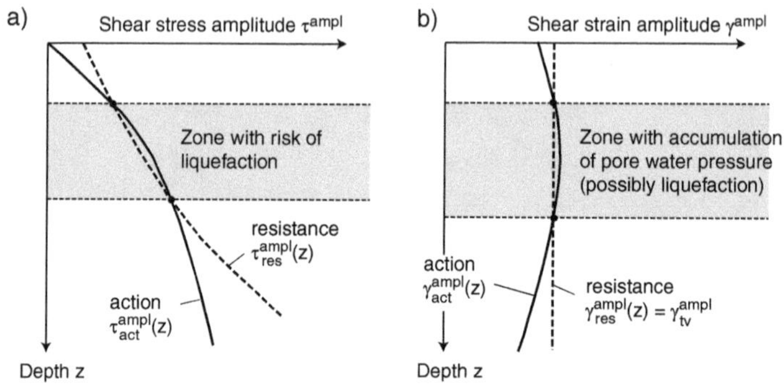


**Figure 93.** Comparison of acting a) shear stress or b) shear strain amplitudes with the corresponding resistance of the soil against liquefaction

The profile $\tau_{\text{act}}^{\text{ampl}}(z)$ of the acting shear stress amplitude with depth can be estimated from the maximum acceleration $a^{\max}$ at the ground surface using (Seed and Idriss, 1971)

$$\frac{\tau_{\text{act}}^{\text{ampl}}(z)}{\sigma_{v0}'(z)} = 0.65 \, \frac{a^{\max}}{g} \, \frac{\sigma_{v0}(z)}{\sigma_{v0}'(z)} \, r_d(z) \tag{28}$$

with $\sigma_{v0}(z)$ and $\sigma_{v0}'(z)$ being the total or effective vertical stress before the earthquake and $g$ the acceleration of gravity. Eq. (28) was derived from the assumption that the upper soil layers move like a rigid block (Fig. 94a), although the displacement of the shear waves with depth follows a trigonometric function. The error made by this assumption is assumed to be compensated by the reduction factor $r_d(z)$ in Eq. (28). It takes the value 1 at the ground surface and decreases with depth (Fig. 94b). In standard codes the reduction factor is often approximated by a linear function, e.g. $r_d = 1 - 0.015z$. Modern design codes (e.g. the Japanese "Highway Bridge Design Code") use modifications of Eq. (28) and specify the maximum acceleration $a^{\max}$ that has to be used for design (e.g. $a^{\max} = 0.15g$).

The irregular cyclic loading history has to be replaced by an equivalent number of regular cycles with constant amplitude. The amplitude of the regular cycles is usually chosen as $\tau^{\text{ampl}} = 0.65\tau^{\max}$ with $\tau^{\max}$ being the maximum shear stress in the irregular load pattern. In order to determine the equivalent number of cycles, all cycles in the irregular load pattern which have an amplitude $\tau^{\text{ampl}} \neq 0.65\tau^{\max}$ are weighted by a factor which

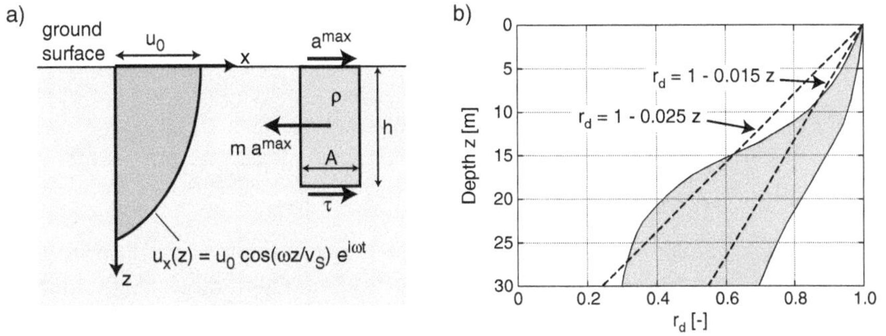


**Figure 94.** a) Analogy of a rigid block as basis for Eq. (28), b) Correction factor $r_d(z)$

can be taken from a diagram as shown in Fig. 95a (e.g. one cycle with $\tau^{\text{ampl}}/\tau^{\text{max}} = 1$ counts like three cycles with $\tau^{\text{ampl}}/\tau^{\text{max}} = 0.65$ and one cycle with $\tau^{\text{ampl}}/\tau^{\text{max}} = 0.45$ counts like 0.1 cycles with $\tau^{\text{ampl}}/\tau^{\text{max}} = 0.65$). The sum of these weighted numbers of cycles is the equivalent number of cycles $N_{\text{eq}}$. The procedure may be applied separately to the peaks above and below the time axis, taking the mean value as $N_{\text{eq}}$. A simplified determination of $N_{\text{eq}}$ based on the earthquake magnitude has also been developed (Fig. 95b). Modern design codes usually specify the number of cycles for which the design has to be made (often $N = 20$, Towhata, 2008).

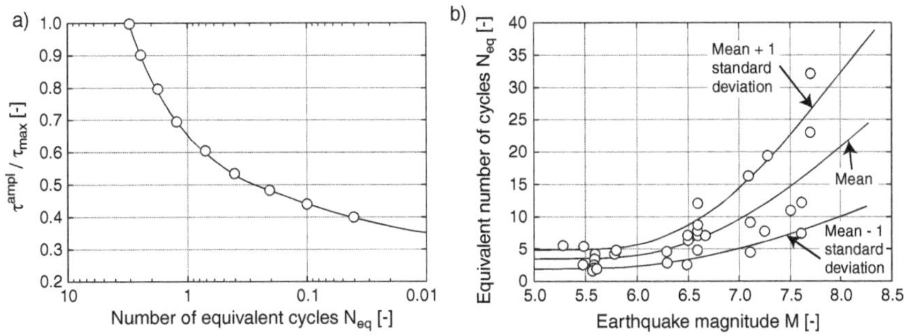


**Figure 95.** a) Weighting factor for amplitudes $\tau^{\text{ampl}} \neq 0.65\ \tau^{\text{max}}$ (DeAlba et al., 1975), b) Estimation of an equivalent number of cycles $N$ in dependence of the earthquake magnitude $M$ (Seed et al., 1975b)

Using the strain approach, a profile of the acting shear strain amplitude $\gamma_{act}^{ampl}(z)$ may be estimated from the maximum acceleration $a^{max}$ at the ground surface using (Dobry et al., 1982)

$$\gamma_{act}^{ampl}(z) = 0.65 \frac{a^{max}}{g} \frac{\sigma_0(z)\ r_d(z)}{G_{max}(z)\ \frac{G}{G_{max}}(\gamma_{act}^{ampl}(z))} \tag{29}$$

wherein $G_{max}$ is the small-strain shear modulus at shear strain amplitudes $\gamma^{ampl} \leq 10^{-6}$ and $G/G_{max}$ is the amplitude-dependent modulus reduction factor (see Section 5). $G_{max}$ can be determined in situ from measurements of the shear wave velocity. The modulus reduction factor can be measured in resonant column tests in the laboratory. Since $G/G_{max}$ depends on $\gamma_{act}^{ampl}$, Eq. (29) has to be solved by iteration.

The liquefaction resistance can be determined in laboratory tests on high-quality undisturbed specimens or on reconstituted specimens that reproduce the in-situ fabric of the grain skeleton. The profile of the shear stress amplitude $\tau_{res}^{ampl}$ causing a liquefaction in $N$ equivalent cycles can be estimated from (Seed and Idriss, 1971)

$$\frac{\tau_{res}^{ampl}}{\sigma_{v0}'(z)} = \left(\frac{q^{ampl}}{2p_0'}\right)_{50} C_r\ \frac{D_r(z)}{50} \tag{30}$$

Therein $(q^{ampl}/(2p_0'))_{50}$ is the cyclic stress ratio causing a liquefaction in $N$ cycles in undrained cyclic triaxial tests on specimens with a relative density of $D_r = 50$ %. It should be considered that the linear relationship between the liquefaction resistance and $D_r$ implemented in Eq. (30) is valid only for $D_r < 75$ % (Fig. 60b). In order to transfer the results from undrained cyclic triaxial tests to in-situ conditions the density-dependent correction factor $C_r$ (Fig. 96) has been introduced in Eq. (30). It mainly considers the differences between the isotropic consolidation in the laboratory test and the $K_0$ stress state in situ.

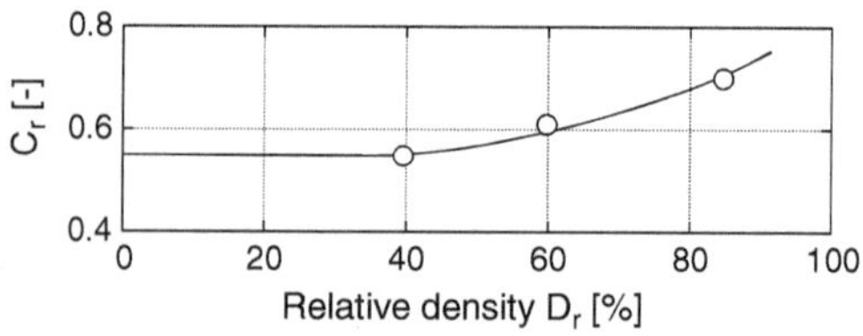


**Figure 96.** Correction factor $C_r$ for the application of laboratory test results to in-situ conditions (Seed and Idriss, 1971)

The threshold shear strain amplitude $\gamma_{tv}^{\rm ampl}$ necessary for the analysis in terms of strains can be determined from strain-controlled cyclic tests with different amplitudes (Fig. 57).

The liquefaction resistance is often estimated using correlation diagrams with the SPT blow count or the CPT tip resistance as input values. An example for an SPT-based correlation is given in Fig. 97a. For the development of such correlations informations about the sounding resistance of a sand deposit, the shear stresses developed during an earthquake and whether the soil has liquefied or not (judged by visible signs on the ground surface) are needed. Combinations of sounding resistance and cyclic shear stress ratio having caused a liquefaction are marked by a filled symbol. Those without visible signs of a liquefaction are marked by an empty symbol. The correlation curve separates the region with filled symbols from that with empty symbols (see the solid curve in Fig. 97a). Usually a pressure-normalized sounding resistance is used on the abscissa:

$$N_{SPT,1} = c_N \, N_{SPT} \tag{31}$$

Different functions are employed for the factor $c_N$. Towhata (2008) recommends to calculate it from

$$c_N = \frac{1.7}{\sigma'_v[\text{kPa}]/98 + 0.7} \tag{32}$$

Several correlations between the liquefaction resistance and the SPT blow count have been developed for various kinds of sands (Seed and Idriss, 1971; Castro, 1975; Christian and Swiger, 1975; Seed et al., 1975a, 1977; Seed, 1979, 1987; Tatsuoka et al., 1980; Seed et al., 1985; Robertson et al., 1992). Some of these correlations consider the influence of the mean grain size $d_{50}$ or the influence of the fines content. Some authors specify different curves for different amounts of shear strain. Some correlation curves have been collected by Ishihara (1993) in Fig. 97b. Since the SPT method used in Japan induces a 1.2 times larger energy compared with the method applied in the USA, the SPT blow count values of correlations developed based on US data have been multiplied by $1/1.2 = 0.833$ in Fig. 97b for comparison purposes. Some few correlations for gravelly materials have also been proposed (Tokimatsu, 1988; Andrus and Youd, 1989; Ishihara, 1995). If the $N_{\rm SPT}$-values are lower than the critical values specified by Ishihara (1995) in Fig. 98a a flow liquefaction is likely to occur.

Beside the SPT-based diagrams, also several correlations of the liquefaction resistance with the CPT tip resistance can be found in the literature (Robertson and Campanella, 1983, 1985; Seed and de Alba, 1986; Shibata and Teparaksa, 1988; Sladen and Hewitt, 1989; Mitchell and Tseng, 1990;

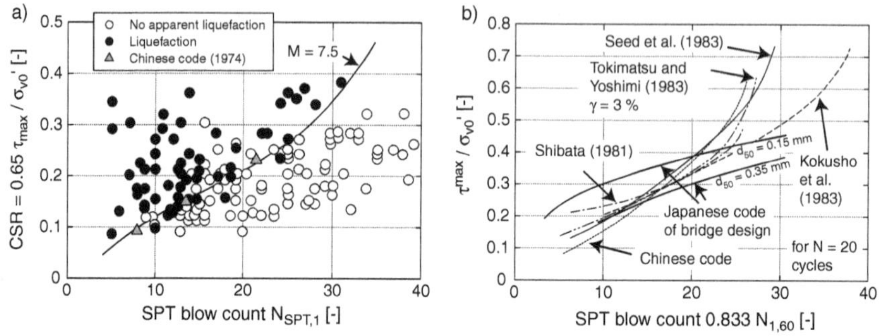


**Figure 97.** Correlations of the liquefaction resistance with pressure-normalized blow count $N_1$ from SPT soundings: a) data collected by Towhata (2008), b) correlations compared by Ishihara (1993)

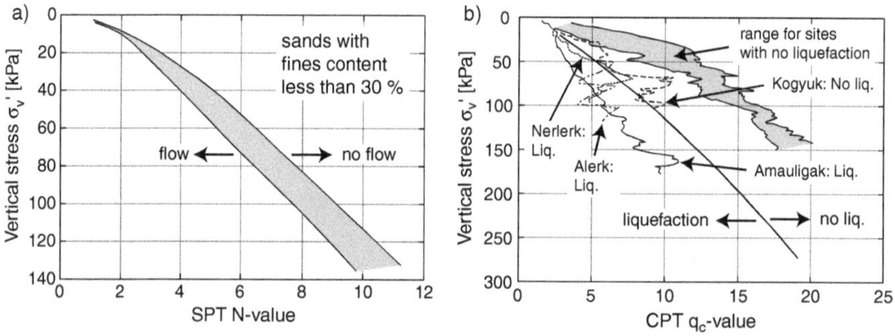


**Figure 98.** Critical values for flow liquefaction (Ishihara, 1995): a) SPT blow count $N$, b) CPT tip resistance $q_c$

Olsen and Koester, 1995; Suzuki et al., 1995; Stark and Olson, 1995; Robertson et al., 1992; Robertson and Wride, 1998; Robertson and Fear, 1995). An example is given in Fig. 99a. These correlations of Shibata and Teparaksa (1988) use the pressure-normalized tip resistance

$$q_{c1} = c_N \, q_c \qquad \text{with} \qquad c_N = \frac{0.17}{\sigma'_{v0}[\text{MPa}] + 0.07} \tag{33}$$

Furthermore, another scaling factor $C_2 = d_{50}/0.25$ has been introduced on the $x$-axis for grain sizes $d_{50} < 0.25$ mm in order to describe the data for various mean grain sizes by the same correlation curve (Fig. 99a). Several correlations between the cyclic stress ratio and $q_{c1}$ have been collected by

Ishihara (1993) in Fig. 99b. If the $q_c$-values are lower than the critical values specified by Ishihara (1995) in Fig. 98b a flow liquefaction is likely to occur.

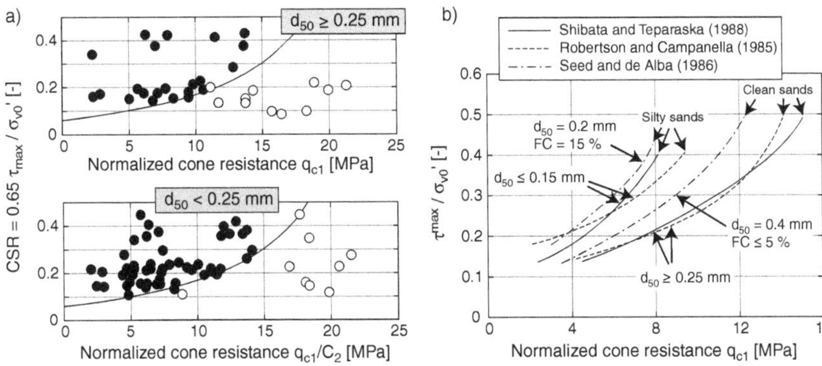


**Figure 99.** Correlations of the liquefaction resistance with pressure-normalized CPT tip resistance $q_{c1}$: a) data collected by Shibata and Teparaksa (1988), b) correlations compared by Ishihara (1993)

Correlation diagrams between the liquefaction potential of a soil and the sounding resistance can also be established based on a combination of in-situ soundings and undrained cyclic laboratory tests on high-quality undisturbed specimens (Seed, 1979; Ishihara, 1995). The procedure is illustrated schematically in Fig. 100.

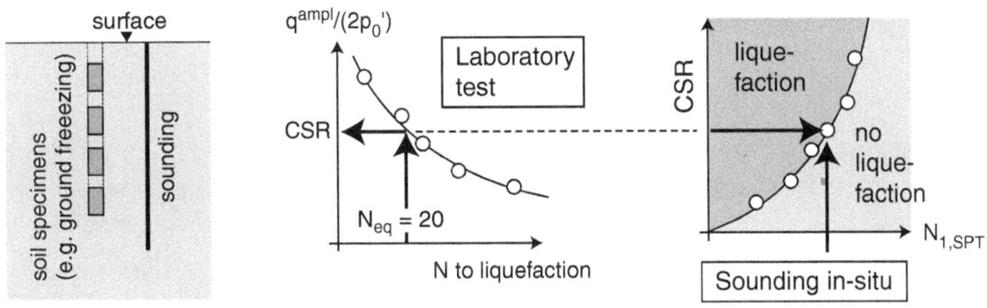


**Figure 100.** Development of correlation diagrams based on soundings and undrained cyclic laboratory tests on high-quality undisturbed specimens

Another group of correlations uses the shear wave velocity $v_S$ or the small-strain shear modulus $G_{max}$ as input values (Bierschwale and Stokoe, 1984; Robertson et al., 1992). Fig. 101a presents a correlation of Robertson et al. (1992). The pressure-normalized shear wave velocity $v_{S1}$ is calculated from

$$v_{S1} = c_N \, v_S \qquad \text{with} \qquad c_N = \left(\frac{p_{\text{atm}}}{\sigma'_{v0}}\right)^{0.25} \tag{34}$$

Another correlation formulated by Tokimatsu and Uchida (1990) in terms of $G_{\max}$ is shown in Fig. 101b. The void ratio function used in this correlation is defined as $F(e_{\min}) = (2.17 - e_{\min})^2/(1 + e_{\min})$ with the minimum void ratio $e_{\min}$. It has to be critically remarked that fabric effects (aging, cyclic preloading) significantly change the liquefaction resistance while they hardly affect $v_S$ or $G_{\max}$ (as discussed in Section 5). Therefore, correlations between $v_S$ and the liquefaction resistance are questionable.

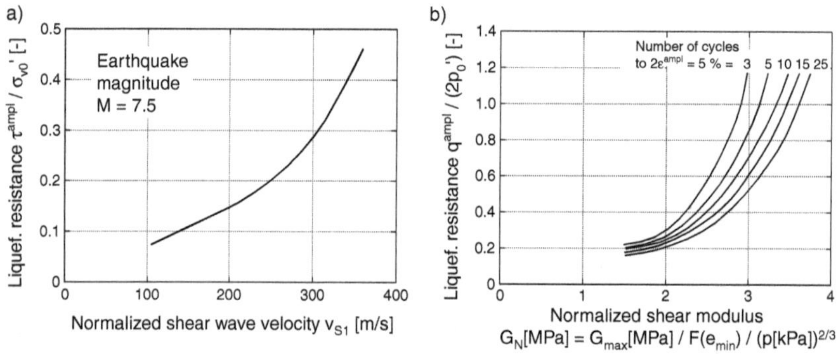


**Figure 101.** Correlations of the liquefaction resistance with a) pressure-normalized shear wave velocity $v_{S1}$ (Robertson et al., 1992) and b) pressure- and void ratio-normalized small strain shear modulus $G_N$ (Tokimatsu and Uchida, 1990)

Correlations between the volumetric strain accumulated in a drained cyclic pressuremeter test and the liquefaction resistance have been developed by Dupla and Canou (2003). They performed cavity expansion tests in a thick-walled hollow cylinder specimen and in a calibration chamber and correlated the results with those of undrained cyclic triaxial tests. Such a correlation seems promising since both, the accumulation of strain in the drained case and the accumulation of pore water pressure under undrained conditions are "two sides of the same medal". Therefore they are influenced by parameters like density, aging, cyclic preloading etc. in a similar manner. However, this method has not been confirmed by in-situ tests yet.

The liquefaction resistance can also be estimated by means of an in-situ test loading. For that purpose a water-saturated soil deposit must be instrumented with geophones and pore water pressure transducers. Then a dynamic loading is applied on the soil surface (e.g. by a truck equipped with a hydraulic shaker, Stokoe II et al., 2004) or in a certain depth. The amplitude is step-wise increased. With the shear strains estimated from the geophone signals and the measured pore water pressure (Fig. 102a) a diagram as shown in Fig. 102b can be developed. It looks similar to that derived from displacement-controlled cyclic laboratory tests (Fig. 57) and may be applied for an analysis of the liquefaction risk using the strain approach (Dobry et al., 1982).

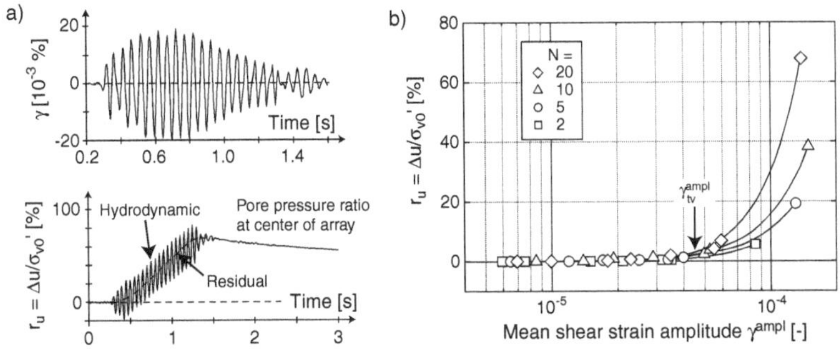


**Figure 102.** Development of correlations $\Delta u$ - $\gamma^{\mathrm{ampl}}$ from in-situ test loadings (Stokoe II et al., 2004)

# 5 Secant Stiffness During Cyclic Loading (Small-Strain Stiffness)

## 5.1 Introduction

A schematic illustration of a closed shear stress-shear strain-hysteresis is given in Fig. 103a. The secant shear modulus $G = \tau^{\mathrm{ampl}}/\gamma^{\mathrm{ampl}}$ corresponds with the inclination of a straight line passing through the two extreme points of the hysteresis. Below a certain shear strain amplitude (for sand usually below $\gamma^{\mathrm{ampl}} = 10^{-5}$) the secant shear modulus $G$ takes its maximum $G_{\max}$ and is independent of the shear strain amplitude. In the literature $G_{\max}$ is also addressed as "dynamic shear modulus" and termed $G_{\mathrm{dyn}}$ or $G_0$. For $\gamma^{\mathrm{ampl}} > 10^{-5}$, the secant shear modulus $G$ decreases with increasing shear strain amplitude (Fig. 103b).

The secant shear modulus $G$ or the shear wave velocity $v_S = \sqrt{G/\varrho}$ may be used in a dynamic analysis of foundations, in order to represent the soil response by a combination of a spring and a dashpot as "lumped" pa-

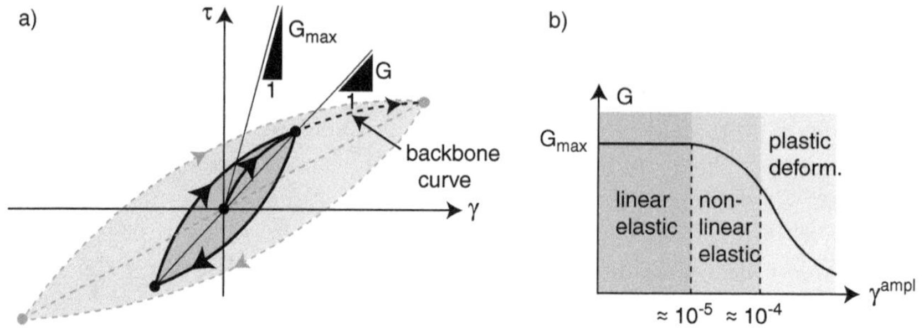


**Figure 103.** a) Definition of the secant shear modulus $G$, b) Decrease of $G$ with increasing shear strain amplitude $\gamma^{\text{ampl}}$

rameters. Common formulas for the spring and dashpot constants (Richart et al., 1970; Triantafyllidis and Prange, 1987) take $G_{\max}$ and Poisson's ratio $\nu$ as input variables. The elastic soil parameters are also necessary for wave propagation analyses. Furthermore, as previously commented, a high-cycle accumulation model needs the resilient portion of deformation to be defined, in particular the spatial field of the strain amplitude $\varepsilon^{\text{ampl}}$ as input. The "dynamic" soil properties may be also used in order to prove the stability of railways (Hu et al., 2003, 2004).

The "dynamic" soil properties can be determined from in-situ measurements (e.g. cross-hole or down-hole measurements). For feasibility studies and for final design calculations in small projects dynamic soil properties are often estimated from empirical equations (Gazetas, 1991). Correlation diagrams between "dynamic" and "static" stiffness moduli or between $G_{\max}$ and sounding resistances are also applied.

### 5.2 Parameters Affecting the Small-Strain Stiffness of Non-Cohesive Soils

The influence of pressure on the small-strain stiffness may be derived from the examination of the contact of two particles. Fig. 104a shows the contact of two perfect elastic spheres. The stiffness $E$ of this system can be calculated from (Hertz, 1881):

$$E = \frac{3}{2}\left[\frac{2\bar{G}}{3(1-\bar{\nu})}\right]^{\frac{2}{3}} \sigma^{\frac{1}{3}} \tag{35}$$

with $\bar{G}$ and $\bar{\nu}$ being the shear modulus and the Poisson's ratio of the sphere material. $\sigma = F/d^2$ is the stress in the axial direction, with $F$ being the axial contact force and $d$ the diameter of the spheres. According to Eq. (35), the P-wave velocity $v_P = \sqrt{E/\varrho}$ should increase proportionally to $\sigma^{1/6}$. Goddard (1990) derived the stiffness of a contact of a sphere and a conus (inclination angle $\alpha$, see Fig. 104b):

$$E = \left(\frac{\bar{G}}{1-\bar{\nu}}\right)^{\frac{1}{2}} \left(\frac{6}{\pi\alpha}\right)^{\frac{1}{2}} \sigma^{\frac{1}{2}} \qquad (36)$$

Eq. (36) delivers a proportionality $v_P \sim \sigma^{1/4}$. Goddard assumed Equation (36) to be valid below a transition stress $\sigma^*$, while Eq. (35) holds for $\sigma > \sigma^*$ (Fig. 104c). $\sigma^*$ can be derived from setting equal Equations (35) and (36). It strongly depends on $\alpha$:

$$\sigma^* = \frac{1}{96} \frac{\bar{G}}{1-\bar{\nu}} \pi^3 \alpha^3 \qquad (37)$$

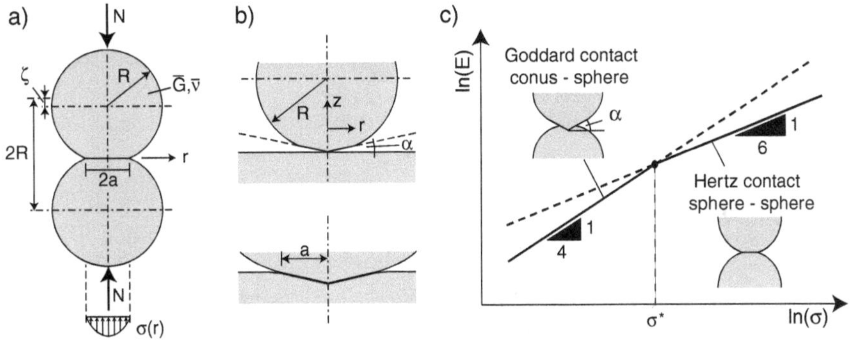


**Figure 104.** a) Contact of two elastic spheres (Hertz, 1881), b) Contact of a conus and a sphere (Goddard, 1990), c) Transition from Goddard to Hertz contact with increasing contact pressure (Goddard, 1990)

One of the first experimental studies on the stress-dependence of wave velocities was performed by Duffy and Mindlin (1957). They measured $v_P$ in the longitudinal direction of rods composed of stainless steel spheres having a single diameter $d = 3.175$ mm. Spheres with two different tolerances of the diameter ($\pm 12,7 \cdot 10^{-4}$ mm and $\pm 2,5 \cdot 10^{-4}$ mm) were tested. They were encompassed by a membrane and stabilized by vacuum. For the spheres with the larger tolerances Duffy and Mindlin (1957) observed $v_p \sim p^{0.25}$ over the whole tested range of pressures $0 \leq p \leq 100$ kPa (Fig. 105a), which is in accordance with the contact theory described by Goddard (1990).

For the more perfect spheres (i.e. with smaller tolerance) $v_p \sim p^{0.195}$ was obtained for $p < 35$ kPa while $v_p \sim p^{0.167}$ was measured for larger pressures. The latter value is in good congruence with the theory of Hertz (1881). The experiments demonstrate that small deviations from the shape of ideal spheres cause a significant increase in the exponent $n$ of the relationship $v_p \sim p^n$.

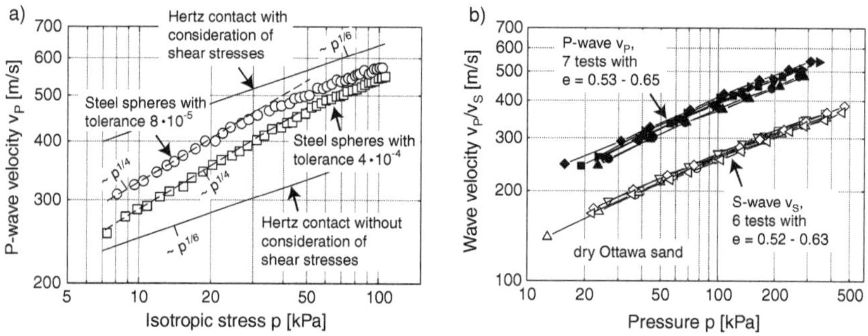


**Figure 105.** a) P-wave velocity measured in rods of steel spheres with different tolerances (Duffy and Mindlin, 1957), b) Increase of P- and S-wave velocities of dry Ottawa sand with increasing pressure (Hardin and Richart Jr., 1963)

In a pioneer work Hardin and Richart Jr. (1963) determined P- and S-wave velocities in dry and water-saturated sands by means of RC tests. Round-sphaped Ottawa sand and a crushed quartz sand were tested. Curves $v_S(p)$ and $v_P(p)$ for dry Ottawa sand are given in Fig. 105b. The pressure-dependence could be approximated by $v_P \sim p^n$ or $v_S \sim p^n$, respectively. For large pressures $p \geq 95$ kPa all exponents $n$ were around 0.25, which is in accordance with the contact theory of Goddard (1990). Larger $n$-values $(0.29 \leq n \leq 0.42)$ were measured for Ottawa-Sand at small pressures $p < 95$ kPa. For a uniform quartz sand with subangular grain shape Wichtmann (2005) obtained exponents $n$ between 0.22 and 0.24.

The decrease in $v_S$ or $G_{\max}$ with increasing void ratio $e$ is demonstrated in Figure 106 which shows test results of Hardin and Richart Jr. (1963) and Wichtmann and Triantafyllidis (2004b). Hardin and Black (1966) proposed an approximation of $G_{\max}(e,p)$ by

$$G_{\max} = A \underbrace{\frac{(a-e)^2}{1+e}}_{F(e)} p^n \qquad (38)$$

The constants $A$, $a$ and $n$ were derived from the tests of Hardin and Richart Jr. (1963). Using $G_{\max}$ in [MPa] and $p$ in [kPa], for Ottawa sand and small pressures $p \leq 95$ kPa the constants $A = 4.804$, $a = 2.12$ and $n = 0.6$ were recommended while $A = 6.911$, $a = 2.17$ and $n = 0.5$ were determined for $p \geq 95$ kPa. For crushed quartz sand the constants were $A = 3.231$, $a = 2.97$ and $n = 0.5$. Eq. (38) is also often used in a version with dimensionless constants

$$G_{\max} \quad = \quad A\ F(e)\ {p_{\mathrm{atm}}}^{1-n}\ p^n \tag{39}$$

using the atmospheric pressure $p_{\mathrm{atm}} = 100$ kPa.

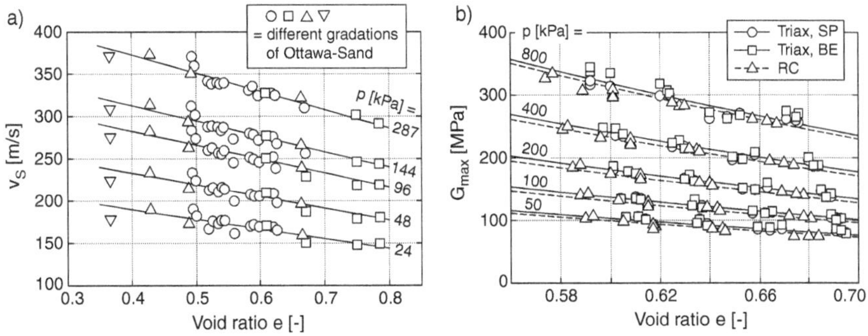


**Figure 106.** Decrease in shear wave velocity $v_S$ or small-strain shear modulus $G_{\max}$ with increasing void ratio: tests of a) Hardin and Richart Jr. (1963) and b) Wichtmann and Triantafyllidis (2004b)

An alternative empirical formula using a dimensionless modulus coefficient $K_2$ is frequently used in the USA (Seed et al., 1986) (here converted to SI units):

$$G \quad = \quad 218.8\ K_2\ p^{0.5} \tag{40}$$

with $G$ and $p$ in [kPa]. At very low strain amplitudes $\gamma^{\mathrm{ampl}} \approx 10^{-6}$ the modulus coefficient is denoted as $K_{2,\max}$. Seed et al. (1986) stated that $K_{2,\max}$-values obtained from laboratory tests range from about 30 for loose sands to about 75 for dense sands. For gravelly soils, somewhat higher values of $K_{2,\max}$ were measured.

Curves $G(\gamma^{\mathrm{ampl}})/G_{\max}$ given by Seed et al. (1986) for different pressures are presented in Fig. 107a. Obviously, the decrease in $G/G_{\max}$ with increasing shear strain amplitude $\gamma^{\mathrm{ampl}}$ becomes larger with decreasing pressures $p$. Similar test results were presented by Kokusho (1980) and Wichtmann and

Triantafyllidis (2005a). Based on several experimental studies in the literature Seed et al. (1986) specify a typical range of the curves $G(\gamma^{\mathrm{ampl}})/G_{\mathrm{max}}$ for sand (Fig. 107a).

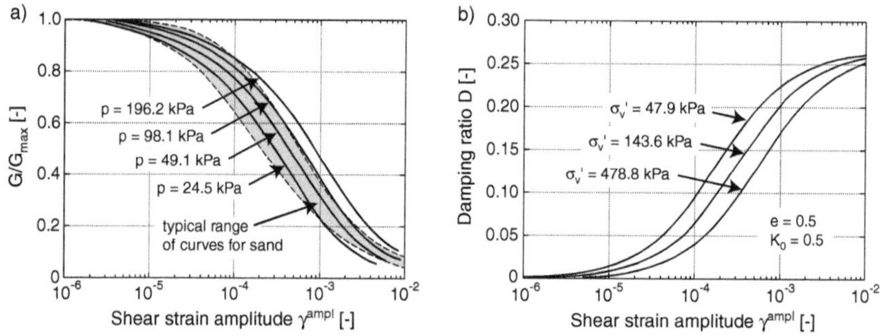


**Figure 107.** a) Decrease in secant shear modulus $G$ (Seed et al., 1986) and b) increase in damping ratio (Hardin and Drnevich, 1972a) with increasing shear strain amplitude $\gamma^{\mathrm{ampl}}$

The effect of pressure on the curves of damping ratio $D(\gamma^{\mathrm{ampl}})$ may be judged from the experimental results of Hardin and Drnevich (1972a) shown in Fig. 107b. The increase in the damping ratio with increasing shear strain amplitudes is larger for smaller pressures.

Kokusho (1980) could not find an influence of void ratio on the curves $G(\gamma^{\mathrm{ampl}})/G_{\mathrm{max}}$ and $D(\gamma^{\mathrm{ampl}})$ (Fig. 108). This is in accordance with test results obtained by the authors (Wichtmann and Triantafyllidis, 2005a) shown in Fig. 109.

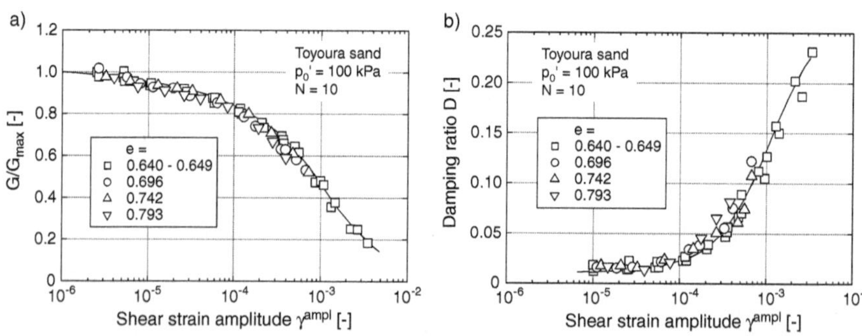


**Figure 108.** No influence of density on the curves $G(\gamma^{\mathrm{ampl}})/G_{\mathrm{max}}$ and $D(\gamma^{\mathrm{ampl}})$ (Kokusho, 1980)

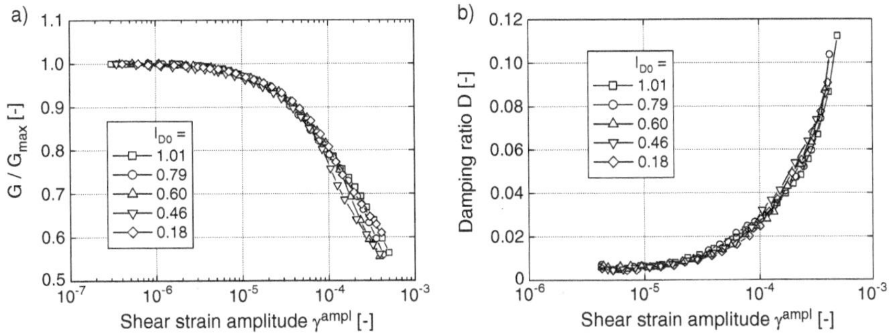


**Figure 109.** No influence of density on the curves $G(\gamma^{\text{ampl}})/G_{\max}$ and $D(\gamma^{\text{ampl}})$ (Wichtmann and Triantafyllidis, 2005a)

The onset of settlements in a resonant column test is usually observed at shear strain amplitudes between $\gamma^{\text{ampl}} = 5 \cdot 10^{-5}$ and $\gamma^{\text{ampl}} = 1 \cdot 10^{-4}$ (Fig. 110a).

If the secant shear modulus $G$ at various shear strain amplitudes $\gamma^{\text{ampl}}$ is plotted versus mean pressure $p$ (Fig. 110b, Wichtmann (2005)), the pressure-dependence of $G$ is found to depend on the shear strain amplitude. The exponent $n$ of the relationship $G \sim p^n$ increases with increasing $\gamma^{\text{ampl}}$. Ishihara (1995) collected the exponents $n$ measured in several experimental studies in the literature (Hardin and Richart Jr., 1963; Drnevich et al., 1967; Silver and Seed, 1971b; Kuribayashi et al., 1975; Kokusho, 1980). In Fig. 111 $n$ is plotted as a function of the shear strain amplitude $\gamma^{\text{ampl}}$. Obviously, $n$ increases with $\gamma^{\text{ampl}}$.

Based on the test results presented by Hardin and Drnevich (1972a), Hardin and Drnevich (1972b) proposed equations to describe the curves $G(\gamma^{\text{ampl}})/G_{\max}$ and $D(\gamma^{\text{ampl}})$:

$$\frac{G}{G_{\max}} = \frac{1}{1 + \frac{\gamma^{\text{ampl}}}{\gamma_r}} \quad \text{and} \quad \frac{D}{D_{\max}} = \frac{\frac{\gamma^{\text{ampl}}}{\gamma_r}}{1 + \frac{\gamma^{\text{ampl}}}{\gamma_r}} \tag{41}$$

Therein $D_{\max}$ is an asymptotic value of $D$ at large strain amplitudes. The reference shear strain $\gamma_r = \tau_{\max}/G_{\max}$ is defined as shown in Fig. 112a. Eq. (41) is based on a hyperbolic approximation of the "backbone curve" $\tau(\gamma)$ observed in a monotonic shear test (see also Fig. 103a). $\tau_{\max}$ is the asymptotic value of $\tau$ at large strains (shear strength). The normalization with $\gamma_r$ intends to eliminate the influences of mean stress and void ratio.

As an alternative, Hardin and Drnevich (1972b) also proposed more

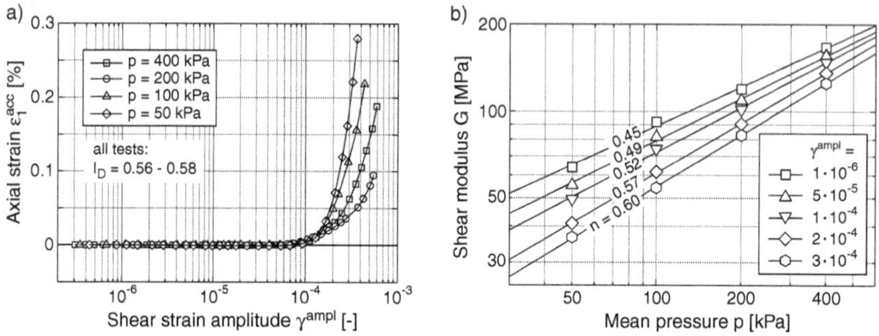


**Figure 110.** a) Onset of permanent deformations when a threshold shear strain amplitude $\gamma_{tv}^{\text{ampl}}$ is surpassed (Wichtmann and Triantafyllidis, 2005a), b) Increase of the exponent $n$ of the relationship $G \sim p^n$ with increasing shear strain amplitude (Wichtmann, 2005)

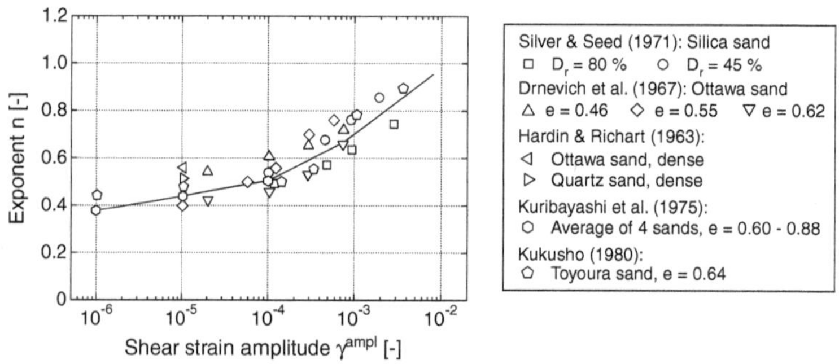


**Figure 111.** Increase of the exponent $n$ of the relationship $G \sim p^n$ with increasing shear strain amplitude $\gamma^{\text{ampl}}$, data collected by Ishihara (1995)

flexible relationships:

$$\frac{G}{G_{\max}} = \frac{1}{1+\gamma_h} \quad \text{and} \quad \frac{D}{D_{\max}} = \frac{\gamma_h}{1+\gamma_h} \tag{42}$$

with

$$\gamma_h = \frac{\gamma^{\text{ampl}}}{\gamma_r}\left[1 + a\,\exp\left(-b\,\frac{\gamma^{\text{ampl}}}{\gamma_r}\right)\right] \tag{43}$$

wherein $a$ and $b$ are fitting parameters. Another description of the curves $G(\gamma^{\text{ampl}})/G_{\max}$ is provided by the model of Ramberg and Osgood (1943).

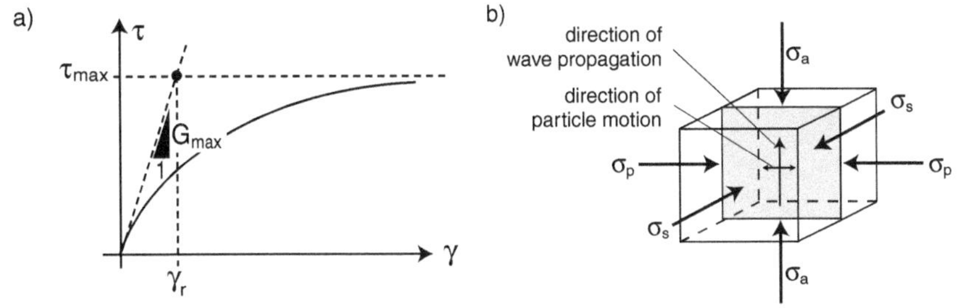


**Figure 112.** a) Definition of reference shear strain $\gamma_r$, b) Definition of stress components $\sigma_a$, $\sigma_p$ and $\sigma_s$ (Roesler, 1979)

As an alternative to Eq. (38) empirical formulas were formulated with the effective stress components $\sigma_a$ and $\sigma_p$ in the directions of the shear wave propagation and polarization, respectively, instead of the effective mean stress $p$ (Fig. 112b, e.g. Roesler, 1979; Knox et al., 1982; Bellotti et al., 1996):

$$G_{\max} \quad = \quad A\ F(e)\ {\sigma_a}^n\ {\sigma_p}^m \tag{44}$$

Roesler (1979) measured the shear wave velocity $v_S$ in cubical specimens (30 × 30 × 30 cm) of dry fine sand. Torsional exciters and receivers were placed within the specimen. First an isotropic stress was applied using vacuum. Then an external load was applied in the vertical direction. After unloading the cube samples were carefully rotated by 90° and a loading was applied in the second direction. Afterwards, the procedure was repeated for the third direction. Similar measurements at a smaller isotropic pressure followed. Fig. 113 illustrates the shear wave velocity as a function of the varied stress component. While $v_S$ increases with increasing values of $\sigma_a$ and $\sigma_p$ (Fig. 113a,b), the stress component $\sigma_s$ acting perpendicular to the plane of wave propagation obviously does not influence $v_S$ (Fig. 113c). Similar conclusions were drawn by Bellotti et al. (1996) with respect to the P-wave velocity (Fig. 114a). Therefore, $\sigma_s$ has not to be considered in Eq. (44).

Regarding Eq. (44), Gazetas (1991) critically remarked that S-waves may propagate in all directions away from a dynamically loaded foundation. It may not be readily evident which are the $a$ and $b$ directions. Thus, the use of $p$ may be as advantageous.

The influence of stress anisotropy was also tested by Yu and Richart Jr. (1984) in RC tests (Fig. 114b). The ratio $\sigma_1/\sigma_3$ of the largest and the smallest principal stresses was increased along different stress paths, some of them with triaxial compression and others with triaxial extension. The

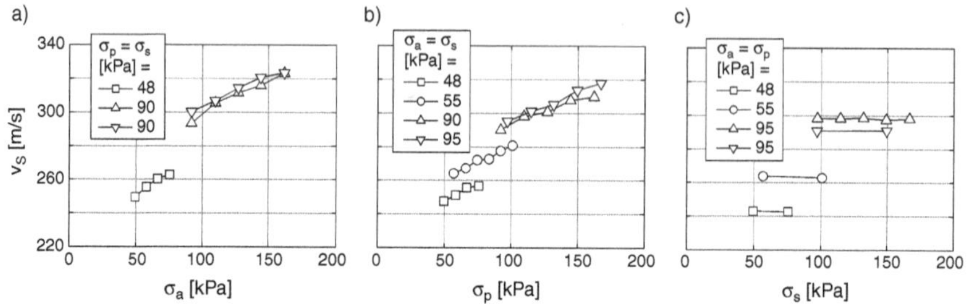


**Figure 113.** Variation of shear wave velocity $v_S$ with an increase of the stress components $\sigma_a$, $\sigma_p$ or $\sigma_s$, respectively (Roesler, 1979)

reduction in the small-strain shear modulus with increasing ratio $\sigma_1/\sigma_3$ does not significantly depend on the stress path (Fig. 114b). Only 10 % decrease was observed up to a stress ratio of $\sigma_1/\sigma_3 = 2.5$. Therefore, the effect of a stress anisotropy becomes significant only for stresses near failure. Yu and Richart Jr. (1984) proposed a reduction factor for Eqs. (39) or (44), respectively:

$$F\left(\frac{\sigma_1}{\sigma_3}\right) = 1 - a\left[\frac{\frac{\sigma_1}{\sigma_3} - 1}{\left(\frac{\sigma_1}{\sigma_3}\right)_{\max} - 1}\right]^b \qquad (45)$$

with $(\sigma_1/\sigma_3)_{\max}$ being the stress ratio at failure and with constants $a$ and $b$.

Similar to the liquefaction resistance the small-strain stiffness increases with increasing time (aging effects). Afifi and Woods (1971) and Afifi and Richart (1973) found an almost linear increase in $G_{\max}$ value with the logarithm of time (Fig. 115a), with a slight increase in the inclination at $t \approx 1000$ min. The changes of $G_{\max}$ could not be attributed to compaction. Since a similar stiffening was measured in tests with an application of a continuous vibration, the vibration applied for the $G_{\max}$-measurements is not responsible for the increase in $G_{\max}$. The curves $G_{\max}(t)$ can be described by

$$G_{\max}(t) = G_{\max}(t_0)\left[1 + \left(N_G \log\left(\frac{t}{t_0}\right)\right)\right] \qquad (46)$$

with $t_0$ being a reference time which is usually chosen as $t_0 = 1000$ min and with $G_{\max}(t_0)$ being the corresponding shear modulus. According to Afifi and Woods (1971) and Afifi and Richart (1973), for dry sand the inclination

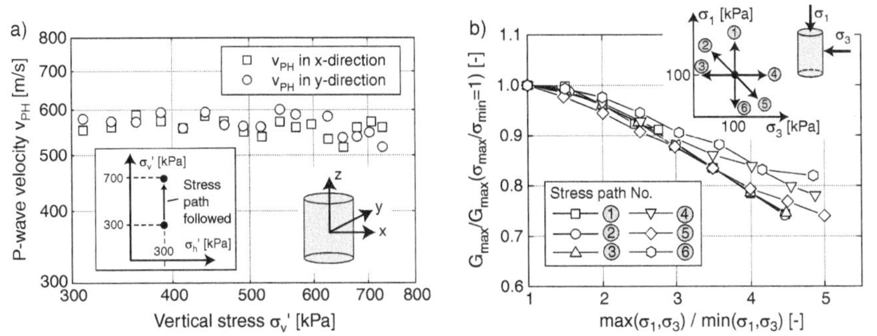


**Figure 114.** a) No influence of the stress component $\sigma_s$ acting perpendicular to the plane of wave propagation (Bellotti et al., 1996), b) Decrease in small strain shear modulus $G_{\max}$ with increasing stress anisotropy $\sigma_1/\sigma_3$ at $p$ = constant (Yu and Richart Jr., 1984)

factor $N_G$ lays within the range $2\% < N_G < 5\%$. An increase in $G_{\max}$ with time was also measured by Baxter (1999) (Fig. 115b). He found significantly larger $N_G$-values ($N_G = 5$ to 12 %) for carbonate sands than for silica sands ($N_G = 1.1$ to 3.5 %). In our own tests a noticeable influence of pressure or density on the curves $G_{\max}(t)/G_{\max}(t=0)$ can not be observed (Fig. 116a). The increase in $G_{\max}$ with time in the RC tests of Li and Yang (1998) was accompanied by a decrease in the damping ratio (Fig. 116b).

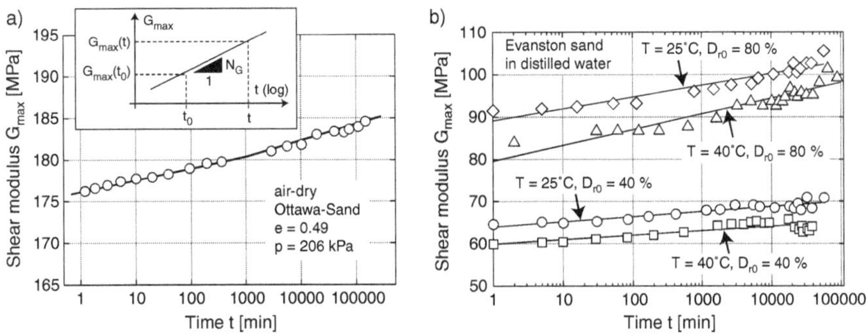


**Figure 115.** Increase in $G_{\max}$ with time under sustained pressure (aging effects): tests of a) Afifi and Woods (1971) and b) Baxter (1999)

Tatsuoka et al. (1979) performed drained RC tests and torsional shear tests on specimens of Toyoura sand prepared by different methods (Fig. 117a).

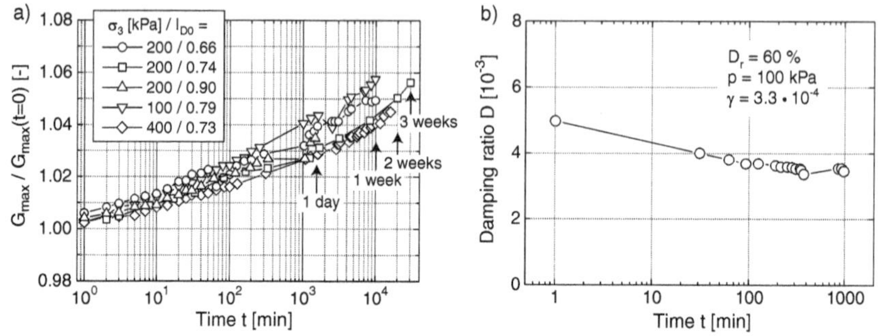


**Figure 116.** a) Increase in $G_{\max}$ (Wichtmann and Triantafyllidis, 2005a) and b) decrease of damping ratio $D$ with time (Li and Yang, 1998)

Dry specimens were prepared by air pluviation without further compaction or by spooning in five layers with a subsequent compaction of each layer using different methods (vibration, tamping, tapping, rodding). Water-saturated specimens were prepared by spooning the sand into de-aired water and subsequent compaction by different methods. Partly saturated specimens were obtained by spooning of moist material followed by compaction. Some special preparation methods including freezing and thawing were also tested. Fig. 117b shows the dependence of $G_{\max}$ divided by the void ratio function $F(e) = (2.17-e)^2/(1+e)$ on the isotropic stress for dry specimens. The $G_{\max}$-values are only moderately affected by the specimen preparation method. The curves $G(\gamma^{\text{ampl}})$ and $D(\gamma^{\text{ampl}})$ measured for different specimen preparation techniques are given in Figs. 118 and 119. The $G$-values are divided by a reference value $G^*_{\max}$ which is the small-strain shear modulus calculated from Eq. (38). Obviously, the curves $G(\gamma)$ and $D(\gamma)$ are not significantly influenced by the method of specimen preparation, that means they are almost independent of the initial fabric of the specimen. This may also imply the conclusion that fabric changes of the specimens cannot be detected by $G_{\max}$- or $v_S$-measurements (compare the discussion of the determination of the historiotropic variable $g^A$ of the HCA model for in-situ soils, Section 3).

Laboratory tests often show smaller $G_{\max}$-values than in-situ measurements (Yokota and Konno, 1985; Kokusho, 1987; Ishihara, 1995; Toki et al., 1995; Stokoe and Santamarina, 2000). Fig. 120a presents data collected by Yasuda and Yamaguchi (1985). The ratio of the shear modulus determined in the laboratory and the shear modulus measured in-situ is plotted versus the in-situ value. Most of the undisturbed specimens were retrieved by tube

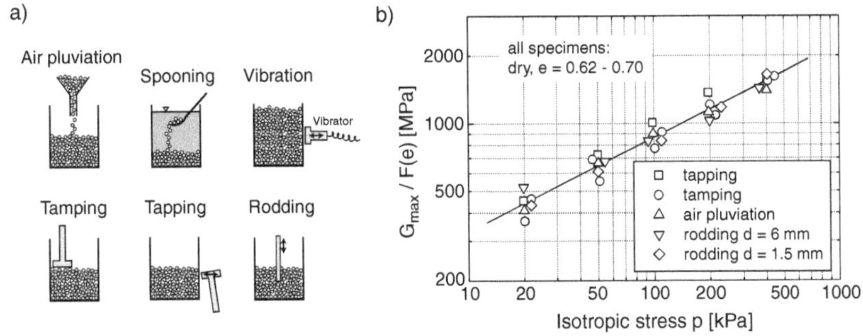


**Figure 117.** Only moderate influence of the specimen preparation method on small strain shear modulus $G_{\max}$ (Tatsuoka et al., 1979)

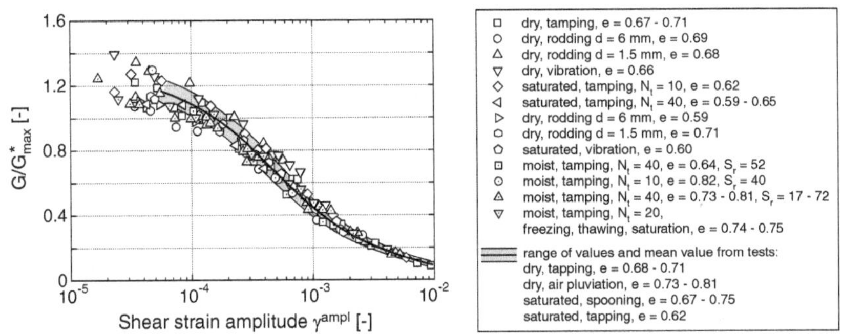


**Figure 118.** Hardly no influence of the specimen preparation method on the modulus degradation curves $G(\gamma^{\mathrm{ampl}})/G_{\max}$ (Tatsuoka et al., 1979)

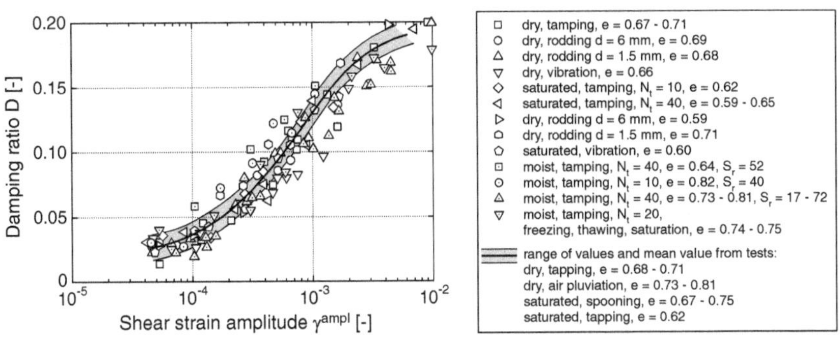


**Figure 119.** Hardly no influence of the specimen preparation method on damping ratio $D(\gamma^{\mathrm{ampl}})$ (Tatsuoka et al., 1979)

sampling. The ratio $G_{\max}^{\text{lab}}/G_{\max}^{\text{in situ}}$ decreases with increasing stiffness of the soil.

Kokusho and Tanaka (1994) tested undisturbed specimens of diluvial gravelly sands obtained by different sampling methods. The $G_{\max}$-values of specimens obtained by tube sampling were generally smaller than those of specimens recovered by ground freezing. In Figure 120b the laboratory $G_{\max}$-data with $G_{\max}$-values measured in-situ using the downhole method are compared. Even for ground freezing the laboratory $G_{\max}$-values were about 50 % smaller than the respective values measured in the field. Therefore, both sampling methods seem to delete subtle fabric effects (e.g. aging) that lead to the higher in-situ stiffness. The disturbance may be larger for the tube sampling method. However, other researchers (Toki et al., 1995; Stokoe and Santamarina, 2000) concluded that undisturbed specimens obtained by ground freezing reproduce the in-situ $G_{\max}$-values. As outlined in Section 4.2 the different results may be due to a different treatment of the frozen samples during thawing.

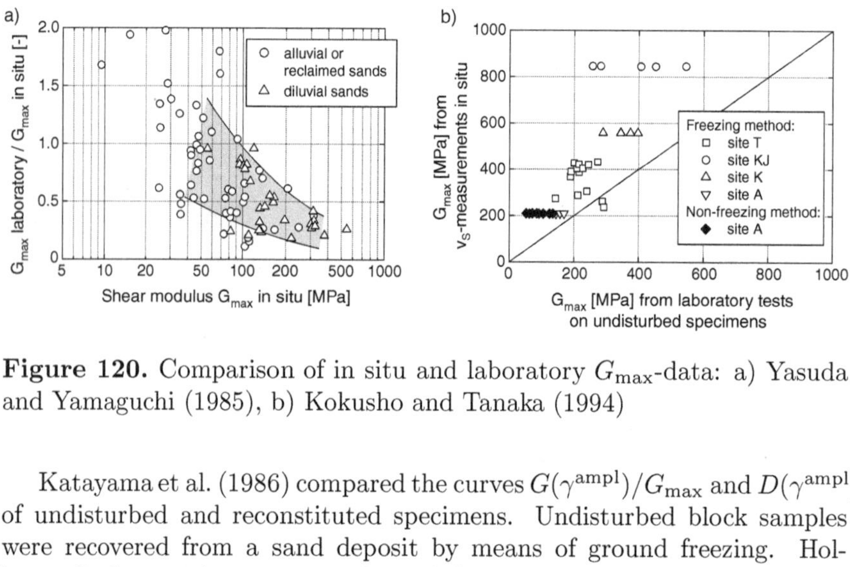


**Figure 120.** Comparison of in situ and laboratory $G_{\max}$-data: a) Yasuda and Yamaguchi (1985), b) Kokusho and Tanaka (1994)

Katayama et al. (1986) compared the curves $G(\gamma^{\text{ampl}})/G_{\max}$ and $D(\gamma^{\text{ampl}})$ of undisturbed and reconstituted specimens. Undisturbed block samples were recovered from a sand deposit by means of ground freezing. Hollow cylinder specimens were trimmed from the blocks and subjected to undrained torsional shear. Reconstituted specimens with similar void ratios were prepared from the same material. While the $G_{\max}$-values of the undisturbed specimens were twice as large as those of the reconstituted samples (Fig. 121a), the curves $G(\gamma^{\text{ampl}})$ nearly coincided at large strain amplitudes $\gamma^{\text{ampl}} > 10^{-3}$. The differences in the curves $G(\gamma^{\text{ampl}})/G_{\max}$ were rather small (Fig. 121b). Furthermore, hardly any differences in the

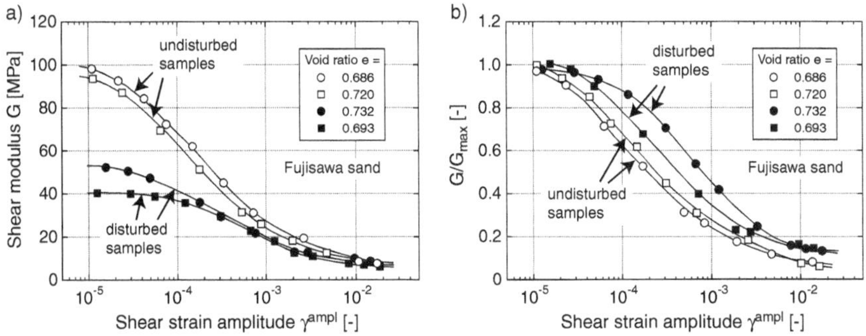


**Figure 121.** Comparison of curves $G(\gamma^{\text{ampl}})/G_{\text{max}}$ of undisturbed and reconstituted specimens (Katayama et al., 1986)

curves $D(\gamma^{\text{ampl}})$ between the undisturbed and the reconstituted samples could be found. Similar conclusions were drawn by Hatanaka et al. (1988) from tests on gravelly soils.

The literature is ambiguous concerning the influence of a cyclic or dynamic preloading on shear modulus and damping ratio. While some researchers reported on a significant change in $G$ and $D$, others could hardly find any influence.

A significant increase in $G_{\text{max}}$ was measured by Drnevich and Richart (1970) in hollow cylinder RC tests (Fig. 122a). Specimens were subjected to a torsional prestraining with a certain amplitude $\gamma^{\text{ampl}}_{\text{prestrain}}$. From time to time the shear strain amplitude was reduced to measure the curves $G(\gamma^{\text{ampl}})$ and $D(\gamma^{\text{ampl}})$ in the range $\gamma^{\text{ampl}} \leq \gamma^{\text{ampl}}_{\text{prestrain}}$. For small prestraining amplitudes $\gamma^{\text{ampl}}_{\text{prestrain}} \approx 10^{-4}$ no change in the curves $G(\gamma^{\text{ampl}})$ and $D(\gamma^{\text{ampl}})$ occurred. Despite negligible changes in the void ratio, a significant increase in $G_{\text{max}}$ was observed for larger prestraining amplitudes. In the test shown in Fig. 122a this increase was close to a factor 3 during $N = 2.2 \cdot 10^7$ cycles. This stiffening was more pronounced for larger pressures $p$. In contrast, the shear stiffness $G(\gamma^{\text{ampl}}_{\text{prestrain}})$ did hardly change. The experimental observations were explained with abrasion effects. Similar results had been previously obtained by Drnevich et al. (1967).

In contrast, Teachavorasinskun et al. (1994) could not find an influence of a cyclic preloading on the small-strain stiffness. They studied the change in Young's modulus $E_{\text{max}}$ during drained cyclic triaxial tests on air-pluviated samples of Toyoura and Ticino sand. The cyclic loading was applied with large strain amplitudes $\varepsilon_1^{\text{ampl}} = 10^{-3}$. From time to time this loading was

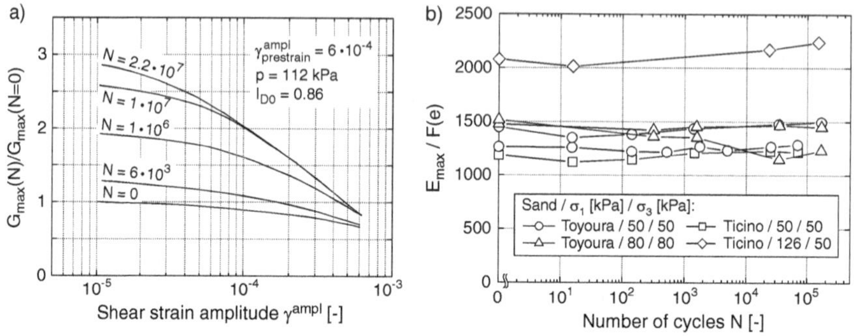


**Figure 122.** a) Significant increase in $G_{\max}$ due to a torsional prestraining (Drnevich and Richart, 1970), b) Hardly any change of the secant Young's modulus during cyclic loading (Teachavorasinskun et al., 1994)

interrupted and $E_{\max}$ was measured at small strain amplitudes $\varepsilon_1^{\text{ampl}} < 10^{-5}$. $E_{\max}$ did hardly change during $10^5$ cycles (Fig. 122b). In Fig. 122b the influence of compaction has been considered by dividing $E_{\max}$ by a void ratio function $F(e)$.

Li et al. (1998), Li and Yang (1998) and Li and Cai (1999) studied the effect of a dynamic torsional vibration in RC tests on dry, air-pluviated fine sand. The specimens were prestrained with $N_{\max}$ cycles of a certain amplitude $\gamma^{\text{ampl}}_{\text{prestrain}}$. Finally, the curves $G(\gamma^{\text{ampl}})$ and $D(\gamma^{\text{ampl}})$ were recorded over the whole range of applicable shear strain amplitudes. While the small strain shear-modulus $G_{\max}$ was only moderately increased due to prestraining, Li and Yang (1998) observed the development of plateaus in the curves $G(\gamma^{\text{ampl}})$ and $D(\gamma^{\text{ampl}})$ located around the prestraining amplitude $\gamma^{\text{ampl}}_{\text{prestrain}}$ (Fig. 123a). The plateaus became more pronounced with increasing intensity of the prestraining, that is with increasing number of cycles and with increasing amplitudes. Similar plateaus were also observed in the RC tests of Wichtmann and Triantafyllidis (2004a) as a consequence of a torsional prestraining.

Wichtmann and Triantafyllidis (2004a) applied a torsional prestraining with large shear strain amplitudes ($10^{-3} \leq \gamma^{\text{ampl}}_{\text{prestrain}} \leq 10^{-2}$) to hollow cylinder specimens. Afterwards the specimens were tested in the RC device. The small-strain shear modulus $G_{\max}$ was only moderately affected by the cyclic prestraining (Fig. 123b).

Wichtmann and Triantafyllidis (2004b) performed cyclic triaxial tests in which the oscillation of the axial stress was interrupted after definite numbers of cycles in order to measure $v_P$ and $v_S$. The average stress $\boldsymbol{\sigma}^{\text{av}}$,

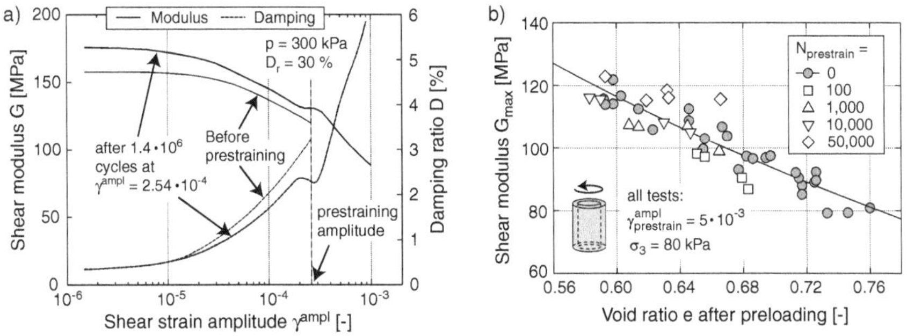


**Figure 123.** a) Development of plateaus in the curves $G(\gamma^{\text{ampl}})$ and $D(\gamma^{\text{ampl}})$ due to torsional prestraining (Li and Yang, 1998), b) Hardly any changes of $G_{\max}$ due to a large torsional prestraining of hollow cylinder specimens (Wichtmann and Triantafyllidis, 2004a)

the stress amplitude $q^{\text{ampl}}$ and the initial density $I_{D0}$ were varied from test to test. Although a significant accumulation of residual strain occurred in some tests, hardly any changes of the small-strain stiffness could be observed (Fig. 124). The curves in Fig. 124 were purified from the effect of compaction by a normalization with the void ratio function $F(e)$. These test results fit well those reported by Teachavorasinskun et al. (1994) (Fig. 122b).

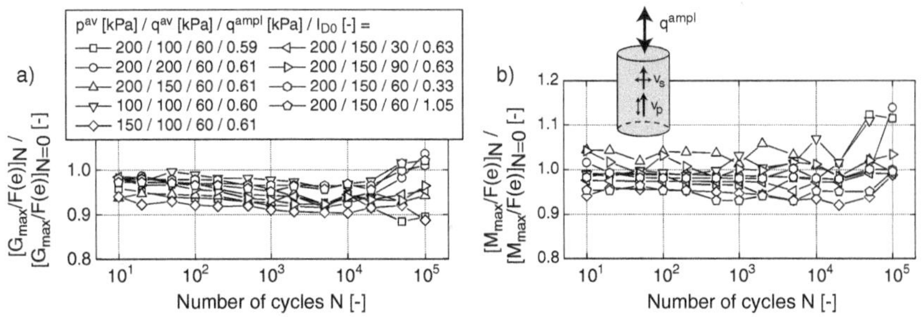


**Figure 124.** Hardly any changes of a) S- and b) P-wave velocities during cyclic triaxial tests with various initial conditions and cyclic loading (Wichtmann and Triantafyllidis, 2004b)

Based on RC tests, Iwasaki and Tatsuoka (1977) reported that $G_{\max}$ is strongly affected by the grain size distribution curve. For poorly graded sands ($C_u < 1.8$, $0.16 \text{ mm} \leq d_{50} \leq 3.2 \text{ mm}$) without a fines content (i.e. no grains smaller than $d = 0.074$ mm) the values of $G_{\max}(e)$ did not depend on $d_{50}$ (Fig. 125a). Furthermore, Iwasaki and Tatsuoka (1977) could not observe a significant influence of the grain shape. Similar $G_{max}$-values were measured for sands with round, subangular and angular grains (Fig. 125a). Fig. 125b presents tests of Iwasaki and Tatsuoka (1977) on grain size distribution curves with different coefficients of uniformity and different fines contents. The measured shear moduli $G_{max}$ were normalized by the values of the poorly-graded sands without a fines content at same values of $e$ and $p$. Obviously, $G_{max}$ decreases significantly with increasing $C_u$ and with increasing $FC$. The decrease in $G_{max}$ with increasing $FC$ becomes even more clear from Fig. 126a. However, only one test was performed for most of the sands and Iwasaki and Tatsuoka (1977) did not extend Eq. (38) by the influence of $C_u$ or $FC$.

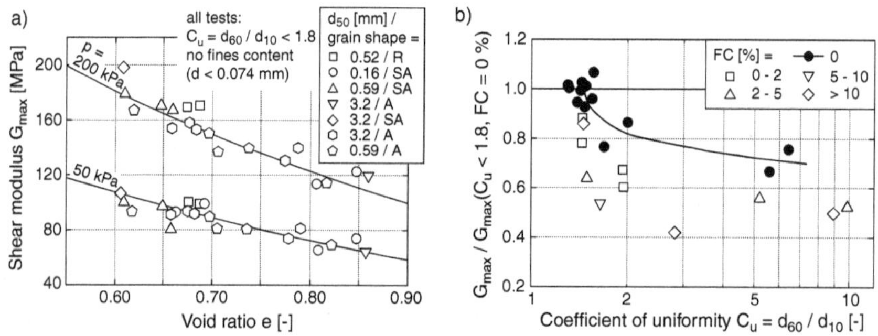


**Figure 125.** a) $G_{max}$ does not depend on mean grain size $d_{50}$ but b) strongly decreases with increasing coefficient of uniformity $C_u$ and with increasing fines content $FC$, RC tests of Iwasaki and Tatsuoka (1977)

RC tests on four different quartz sands have been presented by Wichtmann and Triantafyllidis (2005b). For a constant void ratio, significantly lower $G_{max}$-values were measured for a well-graded material than for three uniform grain size distribution curves (Fig. 126b). Hardin's equation, (38) with its commonly used constants, was found to strongly overestimate the small-strain shear modulus of well-graded soils (Fig. 126b).

The test results shown in Fig. 126b served as a motivation for a more detailed experimental study (Wichtmann and Triantafyllidis, 2009a). The grain size distribution curves shown in Fig. 127 were tested for that purpose. They were linear in the semi-logarithmic scale and had different mean grain sizes in the range 0.1 mm $\leq d_{50} \leq$ 6 mm and different coefficients of uniformity in the range $1.5 \leq C_u \leq 8$.

In Fig. 128a the $G_{max}$-values for the eight materials L1 to L8 (see the

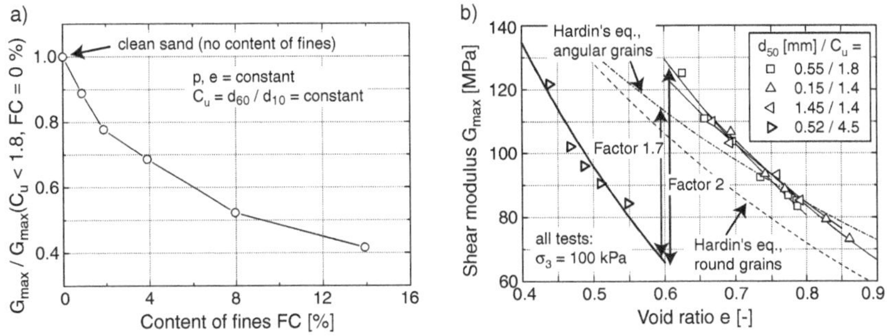


**Figure 126.** Strong decrease in $G_{\max}$ with a) increasing fines content $FC$ (Iwasaki and Tatsuoka, 1977), and b) increasing coefficient of uniformity $C_u$ (Wichtmann and Triantafyllidis, 2005b)

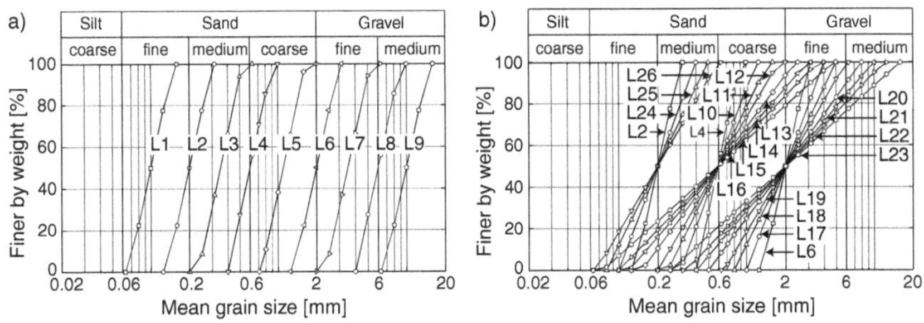


**Figure 127.** Grain size distribution curves tested by Wichtmann and Triantafyllidis (2009a)

grain size distribution curves in Fig. 127a) with the same $C_u = 1.5$ but with different mean grain sizes 0.1 mm $\leq d_{50} \leq$ 6 mm are compared. For a given pressure $p$ the data points for the seven sands L1 to L7 lay on a unique curve, demonstrating that for a constant void ratio the variation in $d_{50}$ does not lead to changes in $G_{\max}$. The lower values for the gravel L8 are due to an insufficient interlocking between the tested material and the end plates which were glued with coarse sand (Martinez, 2007). Therefore, in accordance with the test results of Iwasaki and Tatsuoka (1977), the mean grain size needs not to be considered in an empirical equation for $G_{\max}$, at least with respect to the tested range of $d_{50}$-values.

In Fig. 128b the $G_{\max}$-data for the sands with different $C_u$-values (see the grain size distribution curves in Fig. 127b) are given for a constant

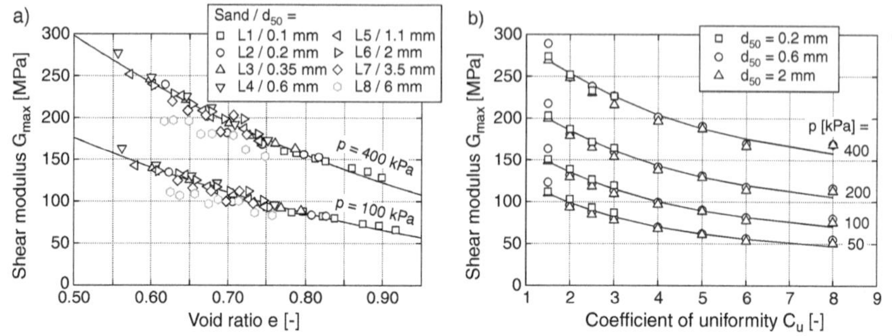


**Figure 128.** Results of RC tests performed by Wichtmann and Triantafyllidis (2009a): a) No influence of mean grain size $d_{50}$, b) Strong decrease of $G_{\max}$ with increasing $C_u$ for a constant void ratio $e = 0.55$

void ratio $e = 0.55$. Obviously, $G_{\max}$ significantly decreases with increasing coefficient of uniformity. This decrease is similar for different mean grain sizes. The strong decrease in $G_{\max}$ with increasing $C_u$ agrees well with the observations reported by Iwasaki and Tatsuoka (1977).

In order to describe the influence of $C_u$ on $G_{\max}$ in an empirical equation, the parameters $A$, $a$ and $n$ in Eq. (39) have been correlated with $C_u$ (Fig. 129):

$$a = 1.94 \exp(-0.066\ C_u) \tag{47}$$
$$n = 0.40\ {C_u}^{0.18} \tag{48}$$
$$A = 1563 + 3.13\ {C_u}^{2.98} \tag{49}$$

Fig. 130 shows the good agreement of the $G_{\max}$-values predicted by Eq. (39) with (47) to (49) with the measured $G_{\max}$-data. A correlation of the small-strain shear modulus with relative density $D_r$ appeared to be less accurate. In future, it is planned to extend the correlations (47) to (49) by the influence of the fines content. Furthermore, the applicability of the correlations to other shapes of the grain size distribution curve will be inspected.

Usually, larger shear moduli are reported for gravelly soils in comparison to sand. It may be demonstrated by the curves $K_2(\gamma^{\text{ampl}})$ measured by Seed et al. (1986) for various sands and gravels (Fig. 131a). Another example are the curves $G_{\max}(e)$ collected by Ishihara (1995) for different types of materials (Fig. 131b). Also Hardin and Kalinski (2005) found larger small-strain shear moduli for different gradations of river gravel compared to Ottawa sand although the coefficients of uniformity of the gravels were larger.

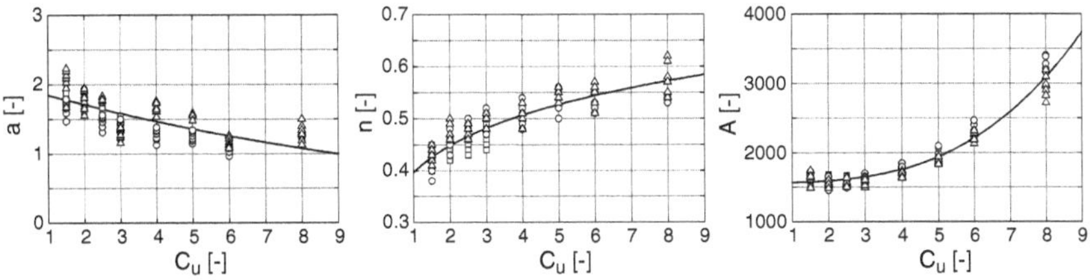


**Figure 129.** Correlations between the parameters $A$, $a$ and $n$ of Hardin's equation (39) with the coefficient of uniformity $C_u$ (Wichtmann and Triantafyllidis, 2009a)

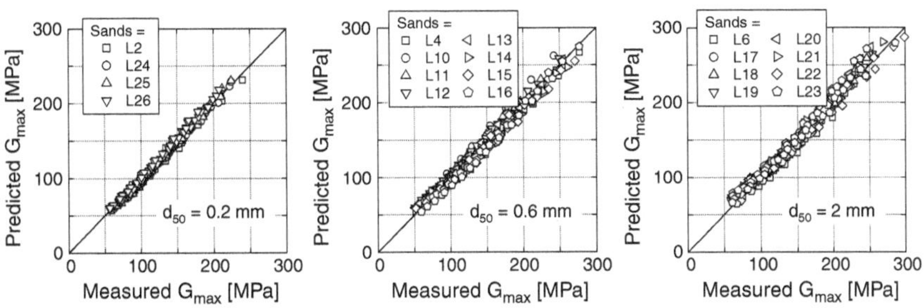


**Figure 130.** Small strain shear moduli $G_{\max}$ predicted by Hardin's equation (39) in combination with the new correlations (47) to (49) plotted versus the measured data (Wichtmann and Triantafyllidis, 2009a)

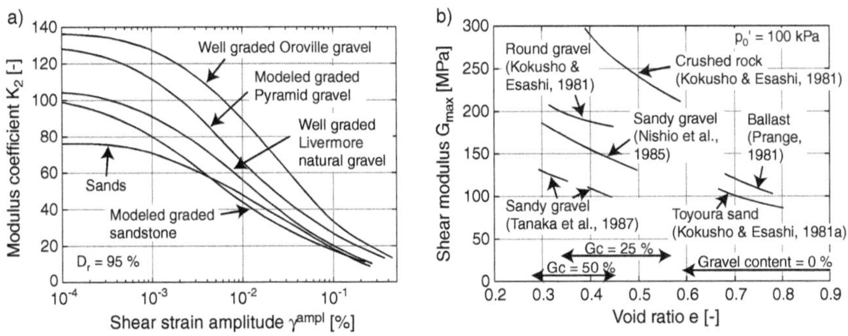


**Figure 131.** Comparison of a) the curves $K_2(\gamma^{\mathrm{ampl}})$ (Seed et al., 1986) and b) the small-strain shear modulus $G_{\max}$ (Ishihara, 1995) for different kinds of sands and gravels

In partly saturated soils, the capillary pressure $p_c$ causes an increase in the effective stress and thus an increase in the small-strain shear stiffness compared with dry or fully saturated soils (Wu et al., 1984; Qian et al., 1993). The effect can be considered by increasing the effective pressure $p$ in an empirical equations for $G_{\max}$ by the capillary pressure $p_c$. Qian et al. (1993) presented curves of $G_{\max}/G_{\max}^{\text{dry}}$ versus the degree of saturation $S_r$ (Fig. 132a) with $G_{\max}^{\text{dry}}$ being the small strain shear modulus of dry specimens. These curves were given for different types of sands (variation of grain shape, grain size distribution and fines content), void ratios and confining stresses and may also be a useful tool for correcting the laboratory data to in-situ conditions.

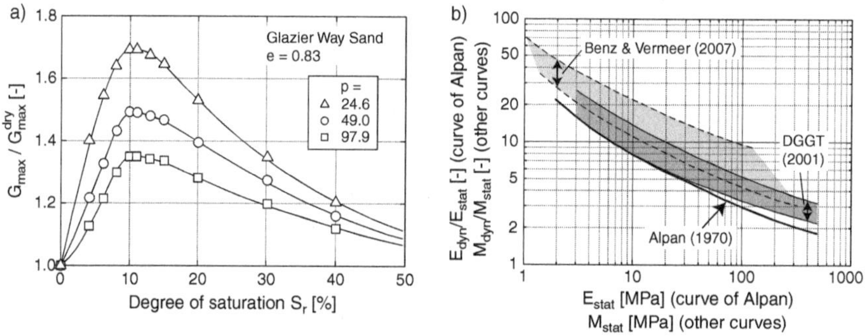


**Figure 132.** a) $G_{\max}$ in dependence of the degree of saturation $S_r$ (Qian et al., 1993), b) Comparison of different correlations between "dynamic" and "static" moduli

## 5.3 Estimation of the Small-Strain Stiffness From Correlations

**Correlations of "Static" With "Dynamic" Moduli** In engineering practice, for smaller projects or feasibility studies, the small-strain stiffness ("dynamic" stiffness) is often estimated by means of a diagram correlating "dynamic" with "static" stiffness moduli. The "static" values are determined from either conventional oedometric or triaxial tests. The curve originally proposed by Alpan (1970) (Fig. 132b) describes a relationship between "static" and "dynamic" Young's moduli ($E_{\text{stat}}$, $E_{\text{dyn}} = E_{\max}$). The static values are obtained from triaxial test data. Alpan (1970) did not clearly state if the "static" Young's modulus corresponds with either the stiffness at first loading (e.g. $E_{50}$, defined as the inclination of the initial phase of the curve $q(\varepsilon_1)$ up to $0.5q_{\max}$) or the secant stiffness $E_{ur}$ during a large un- and re-loading cycle (see the discussion by Benz and Vermeer (2007) and Wichtmann and Triantafyllidis (2007)).

The correlation provided by DGGT (2001) (Fig. 132b) has been derived from the curve of Alpan (1970), probably with the assumption $E_{\text{stat}} = E_{50}$. The correlation is given in terms of modulus $M$ for oedometric compression. Upper and lower boundaries are specified. Since no un- and reloading cycles are mentioned by DGGT (2001), the input parameter $M_{\text{stat}}$ probably corresponds with the stiffness during first loading. An alternative correlation (Fig. 132b) which was derived from the curve of Alpan (1970) using the assumption $E_{\text{stat}} = E_{ur} \approx 3E_{50}$ is provided by Benz and Vermeer (2007).

The authors have inspected the different correlations for four sands with different grain size distribution curves (Wichtmann and Triantafyllidis, 2009b). Oedometric compression and drained monotonic triaxial tests were performed in order to determine the "static" moduli $M_{\text{stat}}$ and $E_{\text{stat}}$. RC tests and measurements of the P-wave velocity were performed for $G_{\text{dyn}} = G_{\max}$ and $M_{\text{dyn}}$.

Fig. 133 presents the correlations between $M_{\text{stat}}$ and $M_{\text{dyn}} = M_{\max}$. For the fine to medium coarse sand No. 1, most of the values $M_{\text{dyn}}/M_{\text{stat}}$ derived from the test data lay within the range specified by DGGT (2001). For the medium to coarse sand No. 2, the values lay at the upper boundary of that range. The experimentally obtained data for the coarse sand No. 3 exceeds the range recommended by DGGT (2001). For the well-graded sand No. 4, the bandwidth specified by DGGT (2001) strongly underestimates the experimentally obtained ratios $M_{\text{dyn}}/M_{\text{stat}}$ (factor 1.5 to 3). Thus, based on Fig. 133, the correlation given by DGGT (2001) seems to be applicable only for poorly-graded fine to medium coarse sands. The small-strain stiffness of poorly-graded coarse sands and of well-graded sands may be strongly underestimated by this correlation. The correlation of Benz and Vermeer (2007) fits somewhat better the experimental data.

The determination of $G_{\text{dyn}}$ from $M_{\text{stat}}$ may be abbreviated using direct correlations of both quantities (Fig. 134). It is recommended to use the correlations shown in Fig. 134 since they are based on experimental results and take into account the grain size distribution curve.

A correlation between $E_{\text{stat}}$ and $E_{\text{dyn}}$ was also inspected. The assumptions $E_{\text{stat}} = E_{50}$ (Fig. 135a) and $E_{\text{stat}} = E_{ur} \approx 3E_{50}$ (Fig. 135b) have been compared. In contrast with the data in Fig. 133 the $E_{\text{dyn}}/E_{\text{stat}}$-values of the four different sands do not differ much. While the curve of Alpan (1970) significantly overestimates the "dynamic" stiffness if $E_{\text{stat}} = E_{50}$ is assumed (factor 1.5 to 2.5, Fig. 135a), the experimental data fit quite well the curve of Alpan (1970) in the analysis with $E_{\text{stat}} = 3E_{50}$ (Fig. 135b). However, the assumption $E_{ur} = 3E_{50}$ seems quite rough since the factor

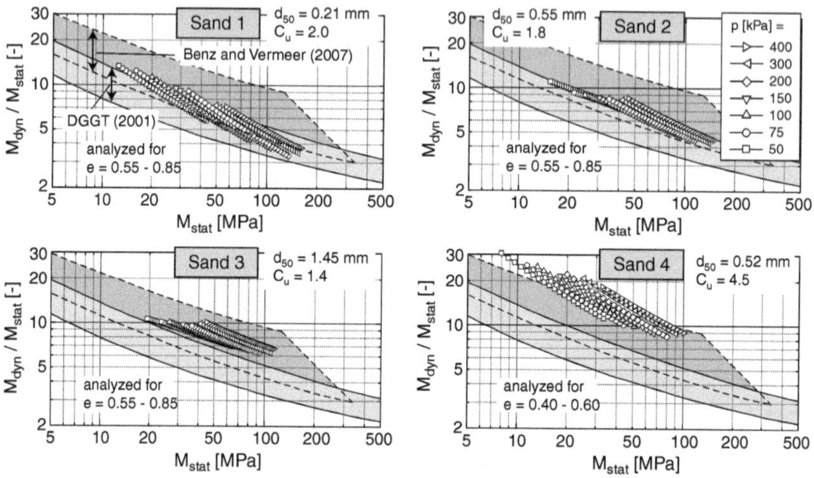


**Figure 133.** Inspection of correlations between the small-strain ("dynamic") constrained elastic modulus $M_{\mathrm{dyn}} = M_{\max}$ and the oedometric ("static") modulus $M_{\mathrm{stat}}$ for four different sands (Wichtmann and Triantafyllidis, 2009b)

$E_{ur}/E_{50}$ depends on the axial strain $\varepsilon_1$, at which the un- and reloading cycle is performed, on the distance $\Delta\varepsilon_1$ between two subsequent cycles, on the minimum deviatoric stress $q_{\min}$ during the cycle, on the void ratio $e$ and on the effective lateral stress $\sigma_3$ (Wu, 1992) (Fig. 136). If a more accurate correlation between $E_{\mathrm{dyn}}$ and $E_{ur}$ is intended, the procedure for the performance of the un- and reloading cycles has to be clearly defined.

Similar to Fig. 134 for $M_{\mathrm{stat}}$, a direct correlation of the small-strain shear modulus $G_{\max} = G_{\mathrm{dyn}}$ with Young's modulus $E_{\mathrm{stat}} = E_{50}$ from monotonic triaxial compression tests has been established (Fig. 137). This may be used for a more simple and accurate estimation of the small-strain stiffness, compared with the correlation given by DGGT (2001).

**Correlations With Sounding Resistances** In addition to the empirical formulas discussed in Section 5.2 and the correlations between "static" and "dynamic" moduli, the small-strain stiffness can also be estimated from correlations with sounding resistances. Figure 138a exemplary presents a correlation between the shear wave velocity and the SPT $N$-value developed by Imai and Yoshimura (1970). A collection of different correlations between $G_{\max}$ and $N_{\mathrm{SPT}}$ has been provided by Ishihara (1995) (Figure 138b).

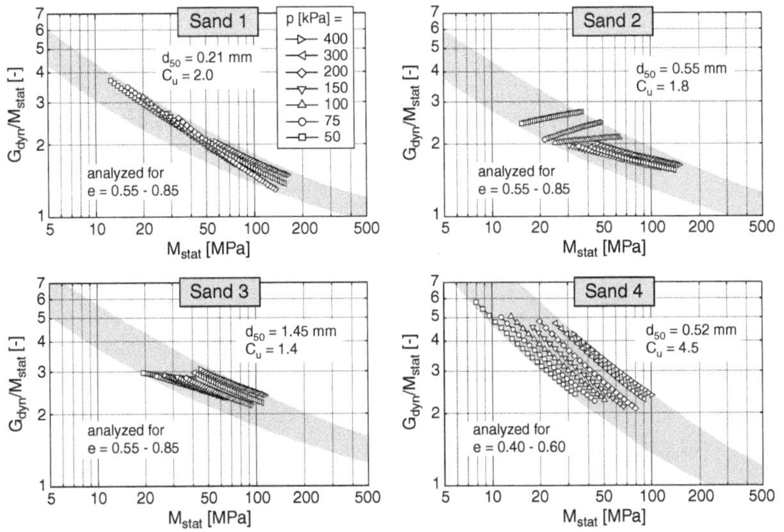


**Figure 134.** New correlations between the small-strain ("dynamic") shear modulus $G_{\text{dyn}} = G_{\max}$ and the oedometric ("static") modulus $M_{\text{stat}}$ for four different sands (Wichtmann and Triantafyllidis, 2009b)

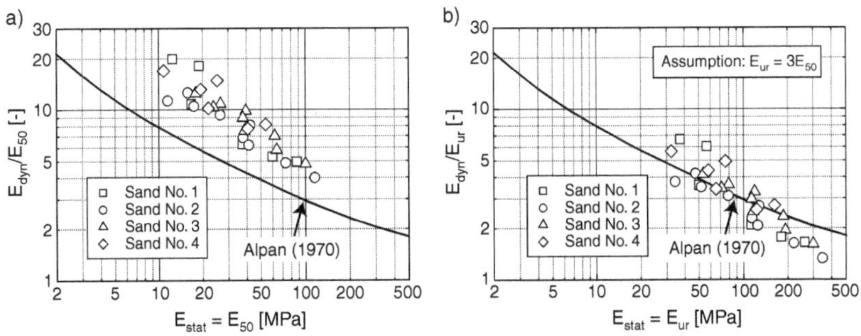


**Figure 135.** New correlations between the small-strain ("dynamic") Young's modulus $E_{\text{dyn}} = E_{\max}$ and the "static" Young's modulus $E_{\text{stat}}$ in the triaxial test, assuming a) $E_{\text{stat}} = E_{50}$ and b) $E_{\text{stat}} = E_{ur} \approx 3E_{50}$ (Wichtmann and Triantafyllidis, 2009b)

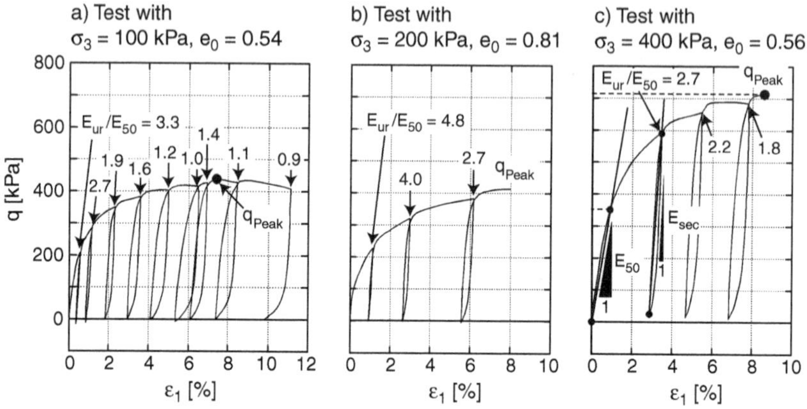


**Figure 136.** Strong dependence of $E_{ur}$ on the axial strain $\varepsilon_1$ where the un- and reloading cycle is performed (Wu, 1992)

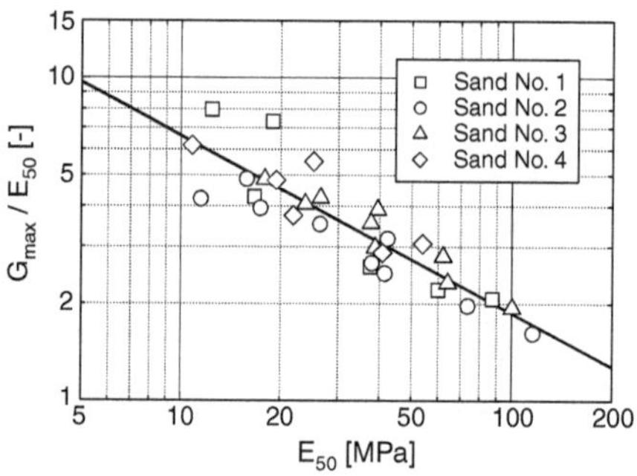


**Figure 137.** Direct correlation between $G_{\text{dyn}} = G_{\max}$ and $E_{\text{stat}} = E_{50}$ (Wichtmann and Triantafyllidis, 2009b)

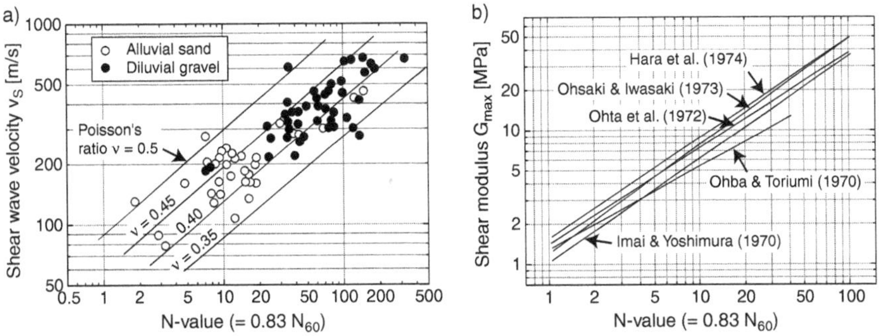


**Figure 138.** Correlations of $v_S$ or $G_{\max}$ with sounding resistances: a) data of Imai and Yoshimura (1970), b) curves collected by Ishihara (1995)

## List of Symbols

| | | |
|---|---|---|
| acc | accumulation component | §3 |
| act | acting | §4.4 |
| ampl | amplitude | §1, §3, §4, §5 |
| av | average | §3, §4, §5.2 |
| cyc | cyclic loading conditions | §3.4 |
| dyn | dynamic loading conditions | §5 |
| in situ | measured in-situ | §5.2 |
| lab | determined in the laboratory | §5.2 |
| min | minimum value | §3, §4, §5 |
| max | maximum value | §3, §4, §5 |
| pl | plastic component | §3.2 |
| prestrain | prestrained | §5.2 |
| res | resistance | §4.4 |
| stat | static loading conditions | §3.4, §5.3 |
| $QSS$ | quasi steady state | §4 |
| $SS$ | steady state | §4 |
| 0 | initial value | §3 |
| $\dot{\sqcup}$ | rate | §3, §4 |
| $\sqcup'$ | effective | §4 |
| $\sqcup^*$ | deviatoric part | §3.2 |
| $\bar{\sqcup}$ | mean value over $N$ | §3 |
| $a$ | parameter | §2, §5.2 |
| $a^{\max}$ | maximum acceleration at the ground surface | §4.4 |
| $A$ | constant | §5.2 |
| b | foundation width | §3.4 |
| $b$ | $= \dfrac{\sigma_2' - \sigma_3'}{\sigma_1' - \sigma_3'}$ | §4.3 |
| $b$ | parameter | §5.2 |
| $B$ | Skempton's $B$-value | §4.2 |
| $c_N$ | factor | §4.4 |

| | | |
|---|---|---|
| $C_i$ | $i = N1, N2, N3, p, Y, e$ material constant | §3.2 |
| $C_N$ | $= N_c/N_0$ | §4.2 |
| $C_r$ | density-dependent correction factor | §4.4 |
| $C_R$ | system compliance ratio | §4.2 |
| $C_u$ | coefficient of uniformity | §3.3, §4.2, §5.2 |
| $C_2$ | scaling factor | §4.4 |
| $d$ | grain diameter | §4.2 |
| $d$ | diameter of the spheres | §5.2 |
| $d_{50}$ | mean grain size | §3.3, §4.2, §5.2 |
| $D$ | damping ratio | §2, §5 |
| $D_r$ | relative density | §4, §5.2 |
| $e$ | void ratio | §3, §5 |
| $\dot{e}$ | compaction rate | §3.3 |
| $\mathsf{E}$ | elastic stiffness | §3 |
| $E$ | stiffness of a two particles system | §5.2 |
| $E_{\text{dyn}}$ | dynamic Young's modulus | §5.3 |
| $E_{\max}$ | small-strain Young's modulus | §5 |
| $E_{\text{stat}}$ | static Young's modulus | §5.3 |
| $E_{ur}$ | secant stiffness during a large un- and re-loading cycle | §5.3 |
| $E_{50}$ | stiffness at first loading | §5.3 |
| $f$ | frequency | §1, §4 |
| $f_i$ | $i = 1, 2$ frequencies at $f = 1/\sqrt{2} f_R$ | §2 |
| $f_j$ | $j = \text{ampl}, N, e, p, Y, \pi$ auxiliary function | §3 |
| $\dot{f}_N^i$ | $i = A, B$ auxiliary function | §3.2 |
| $f_R$ | resonant frequency | §2 |
| $F$ | parameter | §3.2 |
| $F$ | axial contact force | §5.2 |
| $F_L$ | safety factor | §4.4 |
| $F(e)$ | void ratio function | §5.2 |
| $F\left(\frac{\sigma_1}{\sigma_1}\right)$ | reduction factor | §5.2 |
| $FC$ | fines content | §4.2, §5.2 |

| | | |
|---|---|---|
| $g$ | acceleration of gravity | §1, §4.4 |
| $g^A$ | preloading (historiotropic) variable | §3 |
| $G$ | secant shear modulus | §2, §3, §5 |
| $\bar{G}$ | shear modulus of the sphere material | §5.2 |
| $G_{\max}$ | small-strain shear modulus | §2, §4.4, §5 |
| $G_{\max}^{\text{dry}}$ | small-strain shear modulus of dry specimens | §5.2 |
| $G_N$ | void ratio-normalized small-strain shear modulus | §4.4 |
| $h$ | height of the specimen | §2 |
| $I_{D0}$ | initial density | §3.4, §4, §5.2 |
| $I_i$ | $i = 1, 2, 3$ basic invariant of the stress | §3.2 |
| $I_P$ | plasticity index | §4.2 |
| $J$ | mass moment of inertia of specimen | §2 |
| $J_0$ | mass moment of inertia of the base mass | §2 |
| $J_L$ | mass moment of inertia of the top mass | §2 |
| $K$ | bulk modulus | §3 |
| $K_\alpha$ | correction factor | §4.2 |
| $K_\sigma$ | pressure-dependent correction factor | §4.2 |
| $K_0$ | coefficient of lateral earth pressure | §3.4, §4 |
| $K_2$ | dimensionless modulus coefficient | §5.2 |
| $\mathbf{m}$ | direction of strain accumulation | §3 |
| $m_v$ | volumetric portion of the flow rule | §3 |
| $m_q$ | deviatoric portion of the flow rule | §3 |
| $M$ | torsional moment | §2 |
| $M$ | modulus for oedometric compression | §5.3 |
| $M_c$ | $= \dfrac{6 \sin \varphi_c}{3 - \sin \varphi_c}$ | §3 |
| $M_e$ | $= -\dfrac{6 \sin \varphi_c}{3 + \sin \varphi_c}$ | §3 |
| $M_{\max}$ | small-strain constrained elastic modulus | §2 |
| $n$ | constant | §5.2 |
| $N$ | number of cycles | §1, §3, §4 |
| $N_c$ | number of cycles prior to liquefaction in a test with system compliance | §4.2 |

| | | |
|---|---|---|
| $N_{\mathrm{eq}}$ | equivalent number of cycles | §4.4 |
| $N_G$ | inclination factor | §5.2 |
| $N_{SPT}$ | sounding resistance | §4.4 |
| $N_{SPT,1}$ | pressure-normalized sounding resistance | §4.4 |
| $N_0$ | number of cycles prior to liquefaction in a test without system compliance | §4.2 |
| $p$ | $=(\sigma_1+2\sigma_3)/3$ mean stress | §2, §3, §5 |
| $p_{\mathrm{atm}}$ | atmospheric pressure | §3.3, §4.4, §5.2 |
| $p_c$ | capillary pressure | §5.2 |
| $P$ | $=\sqrt{3}p$ isomorphic stress invariant | §3.3 |
| $q$ | $=\sigma_1-\sigma_3$ distortional stress | §2, §3, §4, §5 |
| $q_c$ | tip resistance | §4.4 |
| $q_{c1}$ | pressure-normalized tip resistance | §4.4 |
| $Q$ | $=\sqrt{2/3}q$ isomorphic stress invariant | §3.3 |
| $r_d$ | reduction factor | §4.4 |
| $s_l$ | settlements of the left foundation | §3.4 |
| $s_r$ | settlements of the right foundation | §3.4 |
| $\bar{s}$ | $=(s_l+s_r)/2$ | §3.4 |
| $\Delta s$ | $=\lvert s_l-s_r\rvert$ differential settlement | §3.4 |
| $S_r$ | degree of saturation | §4, §5.2 |
| $t$ | time | §1, §3, §4, §5 |
| $\mathrm{t}$ | foundation embedding depth | §3.4 |
| $\mathrm{u}$ | displacement | §1 |
| $u$ | pore water pressure | §3, §4 |
| $u_h$ | horizontal displacement | §4.4 |
| $v_P$ | compression wave velocity | §2, §5 |
| $v_S$ | shear wave velocity | §2, §4.4, §5 |
| $v_{S1}$ | pressure-normalized shear wave velocity | §4.4 |
| $w$ | water content | §4.2 |
| $w_L$ | liquid limit | §4.2 |
| $W_e$ | specimen elastic energy under maximum shear strain | §2 |
| $\Delta W$ | area of the Lissajous figure | §2 |

| | | |
|---|---|---|
| $Y$ | $= -\frac{I_1 I_2}{I_3} = \frac{27(3+\eta)}{(3+2\eta)(3-\eta)}$ | §3 |
| $\bar{Y}$ | $= \frac{Y-9}{Y_c-9}$ | §3 |
| $Y_c$ | $= \frac{9-\sin^2\varphi_c}{1-\sin^2\varphi_c}$ | §3.3 |
| $z$ | depth | §4.4 |
| $\alpha$ | static stress ratio | §4.2 |
| $\alpha$ | angle between the direction of the largest principal stress and the axis of sedimentation | §4.3 |
| $\alpha$ | inclination angle | §5.2 |
| $\alpha_{PQ}$ | $= \arctan(Q/P)$ polarization angle | §3.3 |
| $\gamma$ | shear strain | §2, §3, §4, §5 |
| $\gamma_h$ | $= \frac{\gamma^{\text{ampl}}}{\gamma_r}\left[1 + a\ \exp\left(-b\ \frac{\gamma^{\text{ampl}}}{\gamma_r}\right)\right]$ | §5.2 |
| $\gamma_r$ | $= \tau_{\max}/G_{\max}$ reference shear strain | §5.2 |
| $\gamma_{tv}$ | threshold shear strain amplitude | §4 |
| $\Delta_1$ | inclination of a log-linear plot of the strain amplitude versus the number of cycles | §2 |
| $\varepsilon$ | $= \sqrt{(\varepsilon_1)^2 + 2(\varepsilon_3)^2}$ | §3.3 |
| $\boldsymbol{\varepsilon}$ | strain | §3 |
| $\varepsilon^{\text{acc}}$ | accumulated strain | §3 |
| $\varepsilon^{\text{ampl}}$ | strain amplitude | §1, §3, §5 |
| $\varepsilon_q$ | $= 2/3\,(\varepsilon_1 - \varepsilon_3)$ deviatoric strain | §3 |
| $\varepsilon_v$ | $= \varepsilon_1 + 2\varepsilon_3$ volumetric strain | §3, §4.2 |
| $\varepsilon_i$ | $i = 1, 2, 3$ principal strain | §3, §4, §5 |
| $\zeta$ | amplitude ratio | §3.3 |
| $\eta$ | $= q/p$ stress ratio | §3.2 |
| $\theta$ | load angle | §3.3 |
| $\nu$ | Poisson's ratio | §3, §5 |
| $\bar{\nu}$ | Poisson's ratio of the sphere material | §5.2 |
| $\varrho$ | soil density | §2, §5 |
| $\sigma$ | stress in the axial direction | §5.2 |
| $\boldsymbol{\sigma}$ | stress | §3 |

| | | |
|---|---|---|
| $\sigma^*$ | transition stress | §5.2 |
| $\sigma_a$ | stress component in the direction of the shear wave propagation | §5.2 |
| $\sigma_h$ | total horizontal stress | §4 |
| $\sigma_p$ | stress component in the direction of the shear wave polarization | §5.2 |
| $\sigma_s$ | stress component acting perpendicular to the plane of wave propagation | §5.2 |
| $\sigma_v$ | total vertical stress | §4 |
| $\sigma_i$ | $i = 1, 2, 3$ principal stress | §2, §3, §4, §5 |
| $\sigma_{3a}$ | outer pressure | §2 |
| $\sigma_{3i}$ | inner pressure | §2 |
| $\tau$ | shear stress | §2, §3, §4, §5 |
| $\tau_{\mathrm{oct}}$ | octahedral shear stress | §3.3 |
| $\varphi_c$ | critical friction angle | §3 |
| $\omega$ | angular frequency | §1 |

## Bibliography

M. Abdelkrim, P. De Buhan, and G. Bonnet. A general method for calculating the traffic load-induced residual settlement of a platform, based on a structural analysis approach. *Soils and Foundations*, 46(4):401–414, 2006.

S.S. Afifi and Jr. Richart, F.E. Stress-history effects on shear modulus of soils. *Soils and Foundations*, 13(1):77–95, 1973.

S.S. Afifi and R.D. Woods. Long-term pressure effects on shear modulus of soils. *Journal of the Soil Mechanics and Foundations Division, ASCE*, 97 (SM10):1445–1460, 1971.

I. Alpan. The Geotechnical Properties of Soils. *Earth Science Reviews, Elsevier*, 6(6):5–49, 1970.

R.D. Andrus and T.L. Youd. Subsurface investigation of a liquefaction-induced lateral spread Thousand Springs Valley, Idaho. In *Miscellaneous Paper (GL-87-8), US Army Corps of Engineers*, 1987.

R.D. Andrus and T.L. Youd. Penetration tests on liquefiable gravels. In *Proc. of the 12th International Conference on Soil Mechanics and Foundation Engineering*, volume 1, pages 679–682, 1989.

A. Arwanitaki and Th. Triantafyllidis. Numerische Simulation mehrlagig mit Geogittern bewehrter Erdkörper über pfahlartigen Gründungselementen unter statischer und zyklischer Einwirkung. *Bautechnik*, 83(10): 695–707, 2006.

C.D.P. Baxter. *An experimental study on the aging of sands.* PhD thesis, Faculty of the Virginia Polytechnic Institute and State University, July 1999.

R. Bellotti, M. Jamiolkowski, D.C.F. Lo Presti, and D.A. O'Neill. Anisotropy of small strain stiffness in Ticino sand. *Géotechnique*, 46 (1):115–131, 1996.

T. Benz and P.A. Vermeer. Discussion of "On the correlation of oedometric and "dynamic" stiffness of non-cohesive soils" by T. Wichtmann and Th. Triantafyllidis (Bautechnik 83, No. 7, 2006) (in German). *Bautechnik*, 84 (5):361–364, 2007.

J.G. Bierschwale and K.H. Stokoe. Analytical evaluation of liquefaction potential of sands subjected to the 1981 Westmorland earthquake. Technical Report Geotechnical Engineering Report GR-84-15, Civil Engineering Department, University of Texas, Austin, 1984.

A.W. Bishop. Shear strength parameters for undisturbed and remoulded soil specimens. In *Proceedings of the Roscoe Memorial Symposium, Cambridge University, Cambridge, Mass.*, pages 3–58, 1971.

G. Bouckovalas, R.V. Whitman, and W.A. Marr. Permanent displacement of sand with cyclic loading. *Journal of Geotechnical Engineering, ASCE*, 110(11):1606–1623, 1984.

M. Budhu. Nonuniformities imposed by simple shear apparatus. *Canadian Geotechnical Journal*, 20:125–137, 1984.

M. Budhu and A. Britto. Numerical analysis of soils in simple shear devices. *Soils and Foundations*, 27(2):31–41, 1987.

G. Castro. Liquefaction and cyclic mobility of saturated sands. *Journal of the Geotechnical Engineering Division, ASCE*, 101(GT6):551–569, 1975.

G. Castro and S.J. Poulos. Factors affecting liquefaction and cyclic mobility. *Journal of the Geotechnical Engineering Division, ASCE*, 103(GT6):501–516, 1977.

J.L. Chaboche. Modelling of ratchetting: evaluation of various approaches. *European Journal of Mechanics*, 13(4):501–781, 1994.

C.S. Chang and R.V. Whitman. Drained permanent deformation of sand due to cyclic loading. *Journal of Geotechnical Engineering, ASCE*, 114 (10):1164–1180, 1988.

L.-K. Chien, Y.-N. Oh, and C.-H. Chang. Effects of fines content on liquefaction strength and dynamic settlement of reclaimed soil. *Canadian Geotechnical Journal*, 39:254–265, 2002.

C. Choi and P. Arduino. Behavioral characteristics of gravelly soils under general cyclic loading conditions. In T. Triantafyllidis, editor, *Cyclic behaviour of soils and liquefaction phenomena, Proc. of CBS04*, pages 115–122. CRC Press, Taylor & Francis Group, London, 2004.

J.T. Christian and W.F. Swiger. Statistics of Liquefaction and SPT results. *Journal of the Geotechnical Engineering Division, ASCE*, 101(GT11): 1135–1150, 1975.

P. DeAlba, C.K. Chan, and H.B. Seed. Determination of soil liquefaction characteristics by large scale laboratory tests. Technical Report 75-14, Earthquake Engineering Research Center, University of California, 1975.

DGGT. Empfehlungen des Arbeitskreises 1.4 "Baugrunddynamik" der Deutschen Gesellschaft für Geotechnik e.V. , 2001.

V.A. Diyaljee and G.P. Raymond. Repetitive load deformation of cohesionless soil. *Journal of the Geotechnical Engineering Division, ASCE*, 108 (GT10):1215–1229, 1982.

T. Doanh, Z. Finge, S. Boucq, and Ph. Dubujet. Histotropy of Hostun RF loose sand. In W. Wu and H.-S. Yu, editors, *Modern Trends in Geomechanics*, volume 106, pages 399–411. Springer, 2006.

R. Dobry and R. Ladd. Discussion to "Soil liquefaction and cyclic mobility evaluation for level ground during earthquakes" by H.B. Seed and "Liquefaction potential: Science versus Practice" by R.B. Peck. *Journal of the Geotechnical Engineering Division, ASCE*, 106(GT6):720–724, 1980.

R. Dobry, R.S. Ladd, F.Y. Yokel, R.M. Chung, and D. Powell. Prediction of pore pressure buildup and liquefaction of sands during earthquakes by the cyclic strain method. Technical Report 138, U.S. Department of Commerce, National bureau of standards, 1982. NBS Building science series.

V.P. Drnevich and F.E. Richart. Dynamic prestraining of dry sand. *Journal of the Soil Mechanics and Foundations Division, ASCE*, 96(SM2):453–467, 1970.

V.P. Drnevich, J.R.Jr. Hall, and F.E.Jr. Richart. Effects of amplitude of vibration on the shear modulus of sand. In *Proc. Int. Symp. on Wave Propagation and Dynamic Properties of Earth Mat.*, pages 189–199, 1967.

J. Duffy and R.D. Mindlin. Stress-strain relations and vibrations of a granular medium. *Journal of Applied Mechanics*, pages 585–593, 1957.

P.M. Duku, J.P. Stewart, D.H. Whang, and E. Yee. Volumetric strains of clean sands subject to cyclic loads. *Journal of Geotechnical and Geoenvironmental Engineering, ASCE*, 134(8):1073–1085, 2008.

J.-C. Dupla and J. Canou. Cyclic pressuremeter loading and liquefaction properties of sands. *Soils and Foundations*, 43(2):17–31, 2003.

J.J. Emery, W.D.L. Finn, and K.W. Lee. Uniformity of saturated sand specimen. Technical Report STP523, pp. 182-194, ASTM, 1973.

W.D.L. Finn, P.L. Bransby, and D.J. Pickering. Effect of strain history on liquefaction of sand. *Journal of the Soil Mechanics and Foundations Division, ASCE*, 96(SM6):1917–1934, 1970.

S. Frydman, D. Hendron, H. Horn, J. Steinbach, R. Baker, and B. Shaal. Liquefaction study of cemented sand. *Journal of Geotechnical Engineering, ASCE*, 106(GT3):275–297, 1980.

G. Gazetas. *Foundation Engineering Handbook, 2nd Edition*, chapter 15: Foundation vibrations, pages 553–593. 1991.

V.N. Ghionna and D. Porcino. Liquefaction resistance of undisturbed and reconstituted samples of a natural coarse sand from undrained triaxial tests. *Journal of Geotechnical and Geoenvironmental Engineering, ASCE*, 132(2):194–201, 2006.

J.D. Goddard. Nonlinear elasticity and pressure-dependent wave speeds in granular media. *Proceedings of the Royal Society London*, 430:105–131, 1990.

S. Goto, S. Nishio, and Y. Yoshimi. Dynamic properties of gravels sampled by ground freezing. In *Proceedings of Session on Ground Failures under Seismic Conditions, ASCE Convention, Atlanta*, pages 141–157, 1994.

A. Gotschol. Veränderlich elastisches und plastisches Verhalten nichtbindiger Böden und Schotter unter zyklisch-dynamischer Beanspruchung. Dissertation, Universität Gh Kassel, 2002.

G. Gudehus. Nichtlineare Bodendynamik in der Geotechnik. In *Vorträge der Baugrundtagung 2000 in Hannover*, pages 263–270. DGGT, 2000.

H. Hanzawa. Undrained strength and stability analysis for a quick sand. *Soils and Foundations*, 20(2):17–29, 1980.

B.O. Hardin and W.L. Black. Sand stiffness under various triaxial stresses. *Journal of the Soil Mechanics and Foundations Division, ASCE*, 92 (SM2):27–42, 1966.

B.O. Hardin and V.P. Drnevich. Shear modulus and damping in soils: measurement and parameter effects. *Journal of the Soil Mechanics and Foundations Division, ASCE*, 98(SM6):603–624, 1972a.

B.O. Hardin and V.P. Drnevich. Shear modulus and damping in soils: design equations and curves. *Journal of the Soil Mechanics and Foundations Division, ASCE*, 98(SM7):667–692, 1972b.

B.O. Hardin and M.E. Kalinski. Estimating the shear modulus of gravelly soils. *Journal of Geotechnical and Geoenvironmental Engineering, ASCE*, 131(7):867–875, 2005.

B.O. Hardin and F.E. Richart Jr. Elastic wave velocities in granular soils. *Journal of the Soil Mechanics and Foundations Division, ASCE*, 89 (SM1):33–65, 1963.

M. Hatanaka, Y. Suzuki, T. Kawasaki, and M. Endo. Cyclic undrained shear properties of high quality undisturbed Tokyo gravel. *Soils and Foundations*, 28(4):57–68, 1988.

J. Helm, J. Laue, and Th. Triantafyllidis. Untersuchungen an der RUB zur Verformungsentwicklung von Böden unter zyklischen Belastungen. In *Beiträge zum Workshop: Boden unter fast zyklischer Belastung: Erfahrungen und Forschungsergebnisse, Veröffentlichungen des Institutes für Grundbau und Bodenmechanik, Ruhr-Universität Bochum, Heft Nr. 32*, pages 201–222, 2000.

H. Hertz. Über die Berührung fester elastischer Körper. *Journal reine und angewandte Mathematik*, 92:156–171, 1881.

B.A. Hofmann, D.C. Sego, and P.K. Robertson. Undisturbed sampling of loose sand using ground freezing. In *Proceedings of the 48th Canadian Geotechnical Conference, Vancouver*, pages 197–204, 1995.

Y. Hu, E. Gartung, H. Prühs, and B. Müllner. Bewertung der dynamischen Stabilität von Erdbauwerken unter Eisenbahnverkehr. *Geotechnik*, 26(1), 2003.

Y. Hu, W. Haupt, and B. Müllner. ResCol-Versuche zur Prüfung der dynamischen Langzeitstabilität von TA/TM-Böden unter Eisenbahnverkehr. *Bautechnik*, 81(4):295–306, 2004.

B.C. Hydro. Liquefaction assessment and seismic stability of Duncan Dam. Technical Report H2599, B.C. Hydro Hydroelectric Engineering Division, Geotechnical Department, 1993.

M. Hyodo, H. Tanimizu, N. Yasufuku, and H. Murata. Undrained cyclic and monotonic triaxial behaviour of saturated loose sand. *Soils and Foundations*, 34(1):19–32, 1994.

T. Imai and M. Yoshimura. Elastic shear wave velocity and mechanical characteristics of soft soil deposits (in Japanese). *Tsuchi to Kiso*, 18(1): 17–22, 1970.

K. Ishihara. Liquefaction and flow failure during earthquakes. The 33rd Rankine Lecture. *Géotechnique*, 43(3):351–415, 1993.

K. Ishihara. *Soil Behaviour in Earthquake Geotechnics.* Oxford Science Publications, 1995.

K. Ishihara and J. Koseki. Cyclic shear strength of fines-containing sands, earthquake geotechnical engineering. In *Proceedings of the Discussion Session on Influence of Local Conditions on Seismic Response, 12th IC-SMFE Rio de Janeiro*, pages 101–106, 1989.

K. Ishihara and H. Nagase. Multi-directional irregular loading tests on sand. *Soil Dynamics and Earthquake Engineering*, 7:201–212, 1988.

K. Ishihara and S. Okada. Effects of stress history on cyclic behavior of sand. *Soils and Foundations*, 18(4):31–45, 1978.

K. Ishihara and S. Okada. Effects of large preshearing on cyclic behavior of sand. *Soils and Foundations*, 22(3):109–125, 1982.

K. Ishihara and F Yamazaki. Cyclic simple shear tests on saturated sand in multi-directional loading. *Soils and Foundations*, 20(1):45–59, 1980.

K. Ishihara and S. Yasuda. Sand liquefaction due to irregular excitation. *Soils and Foundations*, 12(4):65–77, 1972.

K. Ishihara and S. Yasuda. Sand liquefaction in hollow cylinder torsion under irregular excitation. *Soils and Foundations*, 15(1):29–45, 1975.

K. Ishihara, M. Sodekawa, and Y. Tanaka. Effects of overconsolidation on liquefaction characteristics of sands containing fines. *Dynamic Geotechnical Testing, ASTM*, STP 654:246–264, 1978.

T. Iwasaki and F. Tatsuoka. Effects of grain size and grading on dynamic shear moduli of sands. *Soils and Foundations*, 17(3):19–35, 1977.

W.S. Kaggwa, J.R. Booker, and J.P. Carter. Residual strains in calcareous sand due to irregular cyclic loading. *Journal of Geotechnical Engineering, ASCE*, 117(2):201–218, 1991.

I. Katayama, F. Fukui, M. Goto, Y. Makihara, and K. Tokimatsu. Comparison of dynamic deformation characteristics of dense sand between undisturbed and disturbed samples (in Japanese). In *Proceedings of the 21st Annual Conference of JSSMFE*, pages 583–584, 1986.

D.P. Knox, K.H.II Stokoe, and S.E. Kopperman. Effects of state of stress on velocity of low-amplitude shear wave propagating along principal stress directions in dry sand. Technical Report GR 82-23, University of Texas, Austin, 1982.

T. Kokusho. Cyclic triaxial test of dynamic soil properties for wide strain range. *Soils and Foundations*, 20(2):45–59, 1980.

T. Kokusho. In situ dynamic soil properties and their evaluation. In *Proceedings of the 8th Asian Regional Conference on Soil Mechanics and Foundation Engineering, Kyoto, Japan*, volume 2, pages 215–435, 1987.

T. Kokusho and Y. Tanaka. Dynamic properties of gravel layers investigated by in situ freezing sampling. In *Proc. of Session on Ground Failures under Seismic Conditions, ASCE Convention, Atlanta*, pages 121–140, 1994.

T. Kokusho, Y. Yoshida, K. Nishi, and Y. Esashi. Evaluation of seismic stability of dense sand layer (part 1) - dynamic strength characteristics of dense sand (in Japanese). Technical Report 383025, Electric Power Central Research Institute, Japan, 1983.

T. Kokusho, T. Hara, and R. Hiraoka. Undrained shear strength of granular soils with different particle gradations. *Journal of Geotechnical and Geoenvironmental Engineering, ASCE*, 130(6):621–629, 2004.

T. Konno, M. Hatanaka, K. Ishihara, Y. Ibe, and S. Iizuka. Gravelly soil properties evaluation by large scale in situ cyclic shear tests. In *Proceedings of Session on Ground Failures under Seismic Conditions, ASCE Convention, Atlanta*, pages 177–200, 1994.

J.-M. Konrad and M. Pouliet. Ultimate state of reconstituted and intact samples of deltaic sand. *Canadian Geotechnical Journal*, 34:737–748, 1997.

R. Kuerbis and Y.P. Vaid. Sand sample preparation - the slurry deposition method. *Soils and Foundations*, 28(4):107–118, 1988.

E. Kuribayashi, T. Iwasaki, F. Tatsuoka, and S. Horiuchi. Effects of particle characteristics on dynamic deformational properties of soils. In *Proceedings of the 5th Asian Regional Conference on Soil Mechanics and Foundation Engineering, Bangalore, India*, pages 361–367, 1975.

K. Kuwano, H. Nakazawa, K. Sugihara, and H. Yabe. Undrained cyclic shear strength of sand containing fines (in Japanese). In *Proc. 11th Japan Nat. Conf. Geotech. Eng.*, volume 1, pages 993–994, 1996.

R.S. Ladd. Specimen preparation and liquefaction of sands. *Journal of the Geotechnical Engineering Division, ASCE*, 100(GT10):1180–1184, 1974.

R.S. Ladd. Specimen preparation and cyclic stability of sands. *Journal of the Geotechnical Engineering Division, ASCE*, 103(GT6):535–547, 1977.

K.L. Lee and J.A. Fitton. Factors affecting the cyclic loading strength of soil. In *Vibration Effects of Earthqakes on Soils and Foundations, ASTM Special Technical Publication 450*, pages 71–95, 1969.

K.L. Lee and H.B. Seed. Cyclic stress conditions causing liquefaction of sand. *Journal of the Soil Mechanics and Foundations Division, ASCE*, 93(SM1):47–70, 1967.

R.W. Lentz and G.Y. Baladi. Simplified procedure to characterize permanent strain in sand subjected to cyclic loading. In *International Symposium on soils under cyclic and transient loading*, pages 89–95, Januar 1980.

X.S. Li and Z.Y. Cai. Effects of low-number previbration cycles on dynamic properties of dry sand. *Journal of Geotechnical and Geoenvironmental Engineering, ASCE*, 125(11):979–987, 1999.

X.S. Li and W.L. Yang. Effects of vibration history on modulus and damping of dry sand. *Journal of Geotechnical and Geoenvironmental Engineering, ASCE*, 124(11):1071–1081, 1998.

X.S. Li, W.L. Yang, C.K. Chen, and W.C. Wang. Energy-injecting virtual mass resonant column system. *Journal of Geotechnical and Geoenvironmental Engineering, ASCE*, 124(5):428–438, 1998.

M.P. Luong. Mechanical aspects and thermal effects of cohesionless soils under cyclic and transient loading. In *Proc. IUTAM Conf. on Deformation and Failure of Granular materials, Delft*, pages 239–246, 1982.

J. Malkus. Untersuchung des Bodenverhaltens im Kontaktbereich zyklisch axial belasteter zylindrischer Gründungselemente. In *Beiträge zum Workshop: Boden unter fast zyklischer Belastung: Erfahrungen und Forschungsergebnisse*, pages 93–108, 2000. Veröffentlichungen des Institutes für Grundbau und Bodenmechanik, Ruhr-Universität Bochum, Heft Nr. 32.

W.A. Marr and J.T. Christian. Permanent displacements due to cyclic wave loading. *Journal of the Geotechnical Engineering Division, ASCE*, 107(GT8):1129–1149, 1981.

G.R. Martin, W.D.L. Finn, and H.B. Seed. Effects of system compliance on liquefaction tests. *Journal of the Geotechnical Engineering Division, ASCE*, 104(GT4):463–479, 1978.

R. Martinez. Influence of the grain size distribution curve on the stiffness and the damping ratio of non-cohesive soils at small strains (in German). Diploma thesis, Institute of Soil Mechanics and Foundation Engineering, Ruhr-University Bochum, 2007.

H. Matsuoka and T. Nakai. A new failure criterion for soils in three-dimensional stresses. In *Deformation and Failure of Granular Materials*, pages 253–263, 1982. Proc. IUTAM Symp. in Delft.

M. Miner. Cumulative damage in fatigue. *Transactions of the American Society of Mechanical Engineering*, 67:A159–A164, 1945.

J.K. Mitchell and D.-J. Tseng. Assessment of liquefaction potential by cone penetration resistance. In J.M. Duncan, editor, *Proceedings of the H. Bolton Seed Memorial Symposium, Berkeley, Calif.*, volume 2, pages 335–350. Bitech Publishers, 1990.

S. Miura and S. Toki. A sample preparation method and its effect on static and cyclic deformation-strength properties of sand. *Soils and Foundations*, 22(1):61–77, 1982.

K. Mori, H.B. Seed, and C.K. Chan. Influence of sample disturbance on sand response to cyclic loading. *Journal of the Geotechnical Engineering Division, ASCE*, 104(GT3):323–339, 1978.

Z. Mróz, V.A. Norris, and O.C. Zienkiewicz. An anisotropic hardening model for soils and its application to cyclic loading. *Int. J. Numer. Anal. Meth. Geomech.*, 2:203–221, 1978.

J.P. Mulilis, C.K. Chan, and H.B. Seed. The effects of method of sample preparation on the cyclic stress-strain behavior of sands. Technical Report EERC 75-18, Earthquake Engineering Research Center, University of California, Berkeley, 1975.

J.P. Mulilis, H.B. Seed, C.K. Chan, J.K. Mitchell, and K. Arulanandan. Effects of sample preparation on sand liquefaction. *Journal of the Geotechnical Engineering Division, ASCE*, 103(GT2):91–108, 1977.

P.G. Nicholson, R.B. Seed, and H.A. Anwar. Elimination of membrane compliance in undrained triaxial testing. I. Measurement and evaluation. *Canadian Geotechnical Journal*, 30:727–738, 1993.

A. Niemunis. Extended hypoplastic models for soils. Habilitation, Veröffentlichungen des Institutes für Grundbau und Bodenmechanik, Ruhr-Universität Bochum, Heft Nr. 34, 2003. available from www.pg.gda.pl/~aniem/an-liter.html.

A. Niemunis and I. Herle. Hypoplastic model for cohesionless soils with elastic strain range. *Mechanics of Cohesive-Frictional Materials*, 2:279–299, 1997.

A. Niemunis, T. Wichtmann, Y. Petryna, and Th. Triantafyllidis. Stochastic modelling of settlements due to cyclic loading for soil-structure interaction. In G. et al. Augusti, editor, *Proc. of 9th International Conference on Structural Safety and Reliability, ICOSSAR 2005, Rom*, page 263, 2005a.

A. Niemunis, T. Wichtmann, and T. Triantafyllidis. A high-cycle accumulation model for sand. *Computers and Geotechnics*, 32(4):245–263, 2005b.

A. Niemunis, T. Wichtmann, and Th. Triantafyllidis. On the definition of the fatigue loading for sand. In *International Workshop on Constitutive Modelling - Development, Implementation, Evaluation, and Application, 12-13 January 2007, Hong Kong*, 2007a.

A. Niemunis, T. Wichtmann, and Th. Triantafyllidis. Statial stress fluctuations: acoustic evidence and numerical simulations. In G. Pande and S. Pietruszczak, editors, *Numerical Models in Geomechanics: Proceedings of the Tenth International Symposium on Numerical Models in Geomechanics (NUMOG X), Rhodes, Greece, 25-27 April 2007*, pages 159–166. Taylor & Francis, 2007b.

M. Oda, K. Kawamoto, K. Suzuki, H. Fujimori, and M. Sato. Microstructural interpretation on reliquefaction of saturated granular soils under cyclic loading. *Journal of Geotechnical and Geoenvironmental Engineering, ASCE*, 127(5):416–423, 2001.

R.S. Olsen and J.P. Koester. Prediction of liquefaction resistance using the CPT. In *Proceedings of the International Symposium on Cone Penetration Testing, CPT'95, Linkoping, Sweden*, volume 2, pages 251–256, 1995.

V.A. Osinov. A numerical model for the site response analysis and liquefaction of soil during earthquakes. In O. Natau, E. Fecker, and E. Pimentel, editors, *Geotechnical Measurements and Modelling*, pages 475–481. Swets & Zeitlinger, Lisse, 2003a.

V.A. Osinov. Cyclic shearing and liquefaction of soil under irregular loading: an incremental model for the dynamic earthquake-induced deformation. *Soil Dynamics and Earthquake Engineering*, 23:533–548, 2003b.

T. Park and M.L. Silver. Dynamic soil properties required to predict the dynamic behavior of elevated transportation structures. Technical Report DOT-TST-75-44, U.S. Dept. of Transportation, 1975.

R.B. Peck. Liquefaction potential: science versus practice. *Journal of the Geotechnical Engineering Division, ASCE*, 105(GT3):393–398, 1979.

D. Porcino, G. Cicciù, and V.N. Ghionna. Laboratory investigation of the undrained cyclic behaviour of a natural coarse sand from undisturbed and reconstituted samples. In T. Triantafyllidis, editor, *Cyclic Behaviour of Soils and Liquefaction Phenomena, Proc. of CBS04*, pages 187–192. CRC Press, Taylor & Francis Group, London, 2004.

T.B.S. Pradhan, F. Tatsuoka, and Y. Sato. Experimental stress-dilatancy relations of sand subjected to cyclic loading. *Soils and Foundations*, 29 (1):45–64, 1989.

R. Pyke, H.B. Seed, and C.K. Chan. Settlement of sands under multidirectional shaking. *Journal of the Geotechnical Engineering Division, ASCE*, 101(GT4):379–398, 1975.

X. Qian, D.H. Gray, and R.D. Woods. Voids and granulometry: effects on shear modulus of unsaturated sands. *Journal of Geotechnical Engineering, ASCE*, 119(2):295–314, 1993.

W. Ramberg and W.R. Osgood. Description of stress-strain curves by three parameters. Technical Note 902, National Advisory Committee for Aeronautics, Washington, D.C., 1943.

F.E.Jr. Richart, J.R.Jr. Hall, and R.D. Woods. *Vibrations of Soils and Foundations*. Prentice-Hall, Englewood Cliffs, New Jersey, 1970.

M.F. Riemer and R.B. Seed. Factors affecting apparent position of the steady state line. *Journal of the Geotechnical Engineering Division, ASCE*, 105(2):281–288, 1997.

P.K. Robertson and R.G. Campanella. Interpretation of cone penetration tests - Part I (sand). *Canadian Geotechnical Journal*, 20:718–733, 1983.

P.K. Robertson and R.G. Campanella. Liquefaction potential of sands using the cone penetration test. *Journal of Geotechnical Engineering, ASCE*, 22(3):384–403, 1985.

P.K. Robertson and C.E. Fear. Liquefaction of sands and its evaluation. keynote lecture. In K. Ishihara, editor, *IS Tokyo'95, Proceedings of the 1st International Conference on Earthquake Geotechnical Engineering.* Balkema, 1995.

P.K. Robertson and C.E. Wride. Evaluating cyclic liquefaction potential using the cone penetration test. *Canadian Geotechnical Journal*, 35:442–459, 1998.

P.K. Robertson, D.J. Woeller, and W.D.L. Finn. Seismic cone penetration test for evaluating liquefaction potential under cyclic loading. *Canadian Geotechnical Journal*, 29:686–695, 1992.

S.K. Roesler. Anisotropic shear modulus due to stress anisotropy. *Journal of the Geotechnical Engineering Division, ASCE*, 105(GT7):871–880, 1979.

K.M. Rollins and H.B. Seed. Influence of buildings on potential liquefaction damage. *Journal of Geotechnical Engineering, ASCE*, 116(GT2):165–185, 1990.

A. Sawicki and W. Świdziński. Compaction curve as one of basic characteristics of granular soils. In E. Flavigny and D. Cordary, editors, *4th Colloque Franco-Polonais de Mechanique des Sols Appliquee*, volume 1, pages 103–115, 1987. Grenoble.

A. Sawicki and W. Świdziński. Mechanics of a sandy subsoil subjected to cyclic loadings. *Int. J. Numer. Anal. Meth. Geomech.*, 13:511–529, 1989.

H.B. Seed. Soil liquefaction and cyclic mobility evaluation for level ground during earthquakes. *Journal of the Geotechnical Engineering Division, ASCE*, 105(GT2):201–255, 1979.

H.B. Seed. Design problems in soil liquefaction. *Journal of Geotechnical Engineering, ASCE*, 113(8):827–845, 1987.

H.B. Seed and P. de Alba. Use of SPT and CPT tests for evaluating the liquefaction resistance of sands. In S.P. Clemence, editor, *Use of in situ tests in geotechnical engineering*, volume Geotechnical Special Publication 6, pages 281–302, 1986.

H.B. Seed and I.M. Idriss. Simplified procedure for evaluating soil liquefaction potential. *Journal of the Soil Mechanics and Foundations Division, ASCE*, 97(SM9):1249–1273, 1971.

H.B. Seed and I.M. Idriss. Ground motions and soil liquefaction during earthquakes. *Earthquake Engineering Research Institute, University of California, Berkeley, Monograph series, ISBN 0-943198-24-0*, page 110ff, 1982.

H.B. Seed and K.L. Lee. Liquefaction of saturated sands during cyclic loading. *Journal of the Soil Mechanics and Foundations Division, ASCE*, 92(SM6):105–134, 1966.

H.B. Seed, I. Arango, and C.K. Chan. Evaluation of soil liquefaction potential during earthquakes. Technical Report EERC 75-28, Earthquake Engineering Research Center, University of California, 1975a.

H.B. Seed, I.M. Idriss, F. Makdisi, and N. Banerjee. Representation of irregular stress time histories by equivalent uniform stress series in liquefaction analyses. Technical Report EERC 75-29, Univ. of California, Berkeley, Calif, 1975b.

H.B. Seed, K. Mori, and C.K. Chan. Influence of seismic history on liquefaction of sands. *Journal of the Geotechnical Engineering Division, ASCE*, 103(GT4):257–270, 1977.

H.B. Seed, K. Tokimatsu, L.F. Harder, and R.M. Chung. Influence of SPT procedures in soil liquefaction resistance evaluations. *Journal of Geotechnical Engineering, ASCE*, 111(12):1425–1445, 1985.

H.B. Seed, R.T. Wong, I.M. Idriss, and K. Tokimatsu. Moduli and damping factors for dynamic analyses of cohesionless soil. *Journal of Geotechnical Engineering, ASCE*, 112(11):1016–1032, 1986.

R.B. Seed and L.F. Harder. SPT-based analysis of cyclic pore pressure generation and undrained residual strength. In *Proceedings of the B. Seed Memorial Symposium*, volume 2, pages 351–376, 1990.

R.B. Seed, S.R. Lee, and H.-L. Jong. Penetration and liquefaction resistances: prior seismic history effects. *Journal of Geotechnical Engineering, ASCE*, 114(6):691–697, 1988.

S. Seki, S. Mori, and H. Tachibana. Study on liquefaction and ageing effects in archaeological sites at Yuhigahama of Kamakura (in Japanese). In *Proc. 47th Ann. Conv. JSCE*, volume 3, 1992.

M.J. Shenton. Deformation of Railway Ballast under repeated loading conditions. Railroad track mechanics and technology. *Pergamon Press*, pages 405–425, 1978.

T. Shibata and W. Teparaksa. Evaluation of liquefaction potential of soils using cone penetration tests. *Soils and Foundations*, 28(2):49–60, 1988.

M.L. Silver and H.B. Seed. Volume changes in sands during cyclic loading. *Journal of the Soil Mechanics and Foundations Division, ASCE*, 97 (SM9):1171–1182, 1971a.

M.L. Silver and H.B. Seed. Deformation characteristics of sands under cyclic loading. *Journal of the Soil Mechanics and Foundations Division, ASCE*, 97(SM8):1081–1098, 1971b.

S. Singh, H.B. Seed, and C.K. Chan. Undisturbed sampling of saturated sands by freezing. *Journal of the Geotechnical Engineering Division, ASCE*, 108(2):247–264, 1982.

S. Sivathayalan and Y.P. Vaid. Ultimate state of reconstituted and intact samples of deltaic sand: Discussion (of the paper by J.M. Konrad and N. Pouliot, CGJ 34:737-748). *Canadian Geotechnical Journal*, 36:173–175, 1999.

J.A. Sladen and K.J. Hewitt. Influence of placement method of the in situ density of hydraulic and fills. *Canadian Geotechnical Journal*, 26:453–466, 1989.

T.D. Stark and S.M. Olson. Liquefaction resistance using CPT and field case histories. *Journal of Geotechnical Engineering, ASCE*, 121(12):856–869, 1995.

J.D. Stedman. Effects of confining pressure and static shear on liquefaction resistance of fraser rivers sand. Master's thesis, University of British Columbia, Vancouver, 1997.

K.H. Stokoe and J.C. Santamarina. Seismic-wave-based testing in geotechnical engineering. In *Proc. GeoEng 2000: An International Conference on Goetechnical and Geological Engineering, Melbourne*, volume 1, pages 1490–1536, 2000.

K.H. Stokoe II, E.M. Rathje, B.R. Cox, and W.J. Chan. Using large hydraulic shakers to induce liquefaction in the field. In Th. Triantafyllidis, editor, *Cyclic Behaviour of Soils and Liquefaction Phenomena, Proc. of CBS04, Bochum*, pages 313–320. CRC Press, Taylor & Francis Group, London, 2004.

T. Suzuki and S. Toki. Effects of preshearing on liquefaction characteristics of saturated sand subjected to cyclic loading. *Soils and Foundations*, 24 (2):16–28, 1984.

Y. Suzuki, K. Tokimatsu, K. Koyamada, Y. Taya, and Y. Kubota. Field correlation of soil liquefaction based on CPT data. In *Proceedings of the International Symposium on Cone Penetration Testing, CPT'95, Linkoping, Sweden*, volume 2, pages 583–588, 1995.

Y. Tanaka, K. Kokusho, K. Kudo, and Y. Yoshida. Dynamic strength of gravelly soils and its relation to the penetration resistance. In *Proceedings of the 2nd International Conference on Recent Advances in Geotechnical Earthquake Engineering and Soil Dynamics, St. Louis*, volume 1, pages 399–406, 1991.

F. Tatsuoka, T. Iwasaki, S. Yoshida, S. Fukushima, and H. Sudo. Shear modulus and damping by drained tests on clean sand specimen reconstituted by various methods. *Soils and Foundations*, 19(1):39–54, 1979.

F. Tatsuoka, T. Iwasaki, K.-I. Tokida, S. Yasuda, M. Hirose, T. Imai, and M. Kon-No. Standard penetration tests and soil liquefaction potential evaluation. *Soils and Foundations*, 20(4):95–111, 1980.

F. Tatsuoka, S. Maeda, K. Ochi, and S. Fujii. Prediction of cyclic undrained strength of sand subjected to irregular loadings. *Soils and Foundations*, 26(2):73–89, 1986a.

F. Tatsuoka, K. Ochi, S. Fujii, and M. Okamoto. Cyclic undrained triaxial and torsional shear strength of sands for different sample preparation methods. *Soils and Foundations*, 26(3):23–41, 1986b.

F. Tatsuoka, H. Kimura, and T.B.S. Pradhan. Liquefaction strength of sands subjected to sustained pressure. *Soils and Foundations*, 28(1):119–131, 1988.

S. Teachavorasinskun, F. Tatsuoka, and D.C.F. Lo Presti. Effects of cyclic prestraining on dilatancy characteristics and liquefaction of sand. In Shibuya, Mitachi, and Miura, editors, *Pre-failure deformation of geomaterials*, pages 75–80, 1994.

S. Toki, S. Shibuya, and S. Yamashita. Standardization of laboratory test methods to determine the cyclic deformation properties of geomaterials in Japan. In S. Shibuya, T. Mitachi, and S. Miura, editors, *Pre-failure Deformations of Geomaterials*, volume 2, pages 741–784, 1995.

K. Tokimatsu. Penetration tests for dynamic problems. In *Proc. of the First International Symposium of Penetration Testing, ISOPT-1*, pages 117–136, 1988.

K. Tokimatsu. System compliance correction from pore pressure response in undrained triaxial tests. *Soils and Foundations*, 30(2):14–22, 1990.

K. Tokimatsu and Y. Hosaka. Effects of sample disturbance on dynamic properties of sand. *Soils and Foundations*, 26(1):53–64, 1986.

K. Tokimatsu and K. Nakamura. A liquefaction test without membrane penetration effects. *Soils and Foundations*, 26(4):127–138, 1986.

K. Tokimatsu and K. Nakamura. A simplified correction for membrane compliance in liquefaction tests. *Soils and Foundations*, 27(4):111–122, 1987.

K. Tokimatsu and A. Uchida. Correlation between liquefaction resistance and shear wave velocity. *Soils and Foundations*, 30(2):33–42, 1990.

I. Towhata. *Geotechnical Earthquake Engineering*. Springer, 2008.

Th. Triantafyllidis and B. Prange. Dynamic subsoil-coupling between rigid rectangular foundations. *Soil Dyn. Earthquake Eng..*, 1987.

M. Uthayakumar and Y.P. Vaid. Liquefaction of sands under multiaxial stresses. *Canadian Geotechnical Journal*, 35:273–283, 1998.

Y. Vaid and W.D.L. Finn. Static shear and liquefaction potential. *Journal of the Geotechnical Engineering Division, ASCE*, 105(GT10):1233–1246, 1979.

Y.P. Vaid and J.C. Chern. Cyclic and monotonic undrained response of sands. In *Proceedings of Advances in the Art of Testing Soils under Cyclic Loading Conditions, Detroit*, pages 120–147, 1985.

Y.P. Vaid and A. Eliadorani. Instability and liquefaction of granular soils under undrained and partially drained states. *Canadian Geotechnical Journal*, 35:1053–1062, 1998.

Y.P. Vaid and S. Sivathayalan. Fundamental factors affecting liquefaction susceptibility of sands. *Canadian Geotechnical Journal*, 37:592–606, 2000.

Y.P. Vaid and J. Thomas. Post liquefaction behaviour of sand. In *Proceedings of the 13th International Conference on Soil Mechanics and Foundation Engineering, New Delhi*, volume 1, pages 1303–1310, 1994.

Y.P. Vaid and J. Thomas. Liquefaction and postliquefaction behavior of sand. *Journal of Geotechnical Engineering, ASCE*, 121(2):163–173, 1995.

Y.P. Vaid, P.M. Byrne, and J.M.O. Hughes. Dilation angle and liquefaction potential. *Journal of the Geotechnical Engineering Division, ASCE*, 107 (GT7):1003–1008, 1981.

Y.P. Vaid, E.K.F. Chung, and R.H. Kuerbis. Preshearing and undrained response of sands. *Soils and Foundations*, 29(4):49–61, 1989.

Y.P. Vaid, E.K.F. Chung, and R.H. Kuerbis. Stress path and steady state. *Canadian Geotechnical Journal*, 27:1–7, 1990a.

Y.P. Vaid, J.M. Fisher, R.H. Kuerbis, and D. Negussey. Particle gradation and liquefaction. *Journal of Geotechnical Engineering, ASCE*, 116(4): 698–703, 1990b.

Y.P. Vaid, M. Uthayakumar, S. Sivathayalan, P.K. Robertson, and B. Hofmann. Laboratory testing of Syncrude sand. In *Proceedings of the 48th Canadian Geotechnical Conference, Vancouver*, volume 1, pages 223–232, 1995.

Y.P. Vaid, S. Sivathayalan, A. Eliadorani, and M. Uthayakumar. Characterisation of static and dynamic liquefaction of sands. Laboratory testing at U.B.C. Technical Report 04/96, CANLEX, 1996.

Y.P. Vaid, S. Sivathayalan, and D. Stedman. Influence of specimen-reconstituting method on the undrained response of sand. *Geotechnical Testing Journal, ASTM*, 22(3):187–195, 1999.

R. Verdugo and K. Ishihara. The steady state of sandy soils. *Soils and Foundations*, 36(2):81–91, 1996.

P.-A. von Wolffersdorff. A hypoplastic relation for granular materials with a predefined limit state surface. *Mechanics of Cohesive-Frictional Materials*, 1:251–271, 1996.

R.V. Whitman. Resistance of soil to liquefaction and settlement. *Soils and Foundations*, 11(4):59–68, 1971.

T. Wichtmann. Explicit accumulation model for non-cohesive soils under cyclic loading. PhD thesis, Publications of the Institute for Soil Mechanics and Foundation Engineering, Ruhr-University Bochum, Issue No. 38, available from www.rz.uni-karlsruhe.de/~gn97/, 2005.

T. Wichtmann and T. Triantafyllidis. Dynamic stiffness and damping of sand at small strains (in German). *Bautechnik*, 82(4):236–246, 2005a.

T. Wichtmann and T. Triantafyllidis. On the influence of the grain size distribution curve on dynamic and cumulative behaviour of non-cohesive soils (in German). *Bautechnik*, 82(6):378–386, 2005b.

T. Wichtmann and T. Triantafyllidis. On the influence of the grain size distribution curve of quartz sand on the small strain shear modulus $G_{\max}$. *Journal of Geotechnical and Geoenvironmental Engineering, ASCE*, 10: 1404–1418, 2009a.

T. Wichtmann and Th. Triantafyllidis. Reply to the discussion of T. Benz and P.A. Vermeer on "On the correlation of oedometric and "dynamic" stiffness of non-cohesive soils" (Bautechnik 83, No. 7, 2006) (in German). *Bautechnik*, 84(5):364–366, 2007.

T. Wichtmann and Th. Triantafyllidis. On the correlation of "static" and "dynamic" stiffness moduli of non-cohesive soils. *Bautechnik, Special Issue "Geotechnical Engineering", July 2009*, pages 28–39, 2009b.

T. Wichtmann and Th. Triantafyllidis. Influence of a cyclic and dynamic loading history on dynamic properties of dry sand, part I: cyclic and dynamic torsional prestraining. *Soil Dynamics and Earthquake Engineering*, 24(2):127–147, 2004a.

T. Wichtmann and Th. Triantafyllidis. Influence of a cyclic and dynamic loading history on dynamic properties of dry sand, part II: cyclic axial preloading. *Soil Dynamics and Earthquake Engineering*, 24(11):789–803, 2004b.

T. Wichtmann, A. Niemunis, and T. Triantafyllidis. Strain accumulation in sand due to cyclic loading: drained triaxial tests. *Soil Dynamics and Earthquake Engineering*, 25(12):967–979, 2005a.

T. Wichtmann, A. Niemunis, and T. Triantafyllidis. FE prediction of settlements of shallow foundations on sand subjected to cyclic loading (in German). *Bautechnik*, 82(12):902–911, 2005b.

T. Wichtmann, A. Niemunis, T. Triantafyllidis, and M. Poblete. Correlation of cyclic preloading with the liquefaction resistance. *Soil Dynamics and Earthquake Engineering*, 25(12):923–932, 2005c.

T. Wichtmann, A. Niemunis, and T. Triantafyllidis. Experimental evidence of a unique flow rule of non-cohesive soils under high-cyclic loading. *Acta Geotechnica*, 1(1):59–73, 2006a.

T. Wichtmann, A. Niemunis, and Th. Triantafyllidis. Is Miner's rule applicable to sand? (in German). *Bautechnik*, 83(5):341–350, 2006b.

T. Wichtmann, A. Niemunis, and T. Triantafyllidis. On the influence of the polarization and the shape of the strain loop on strain accumulation in sand under high-cyclic loading. *Soil Dynamics and Earthquake Engineering*, 27(1):14–28, 2007a.

T. Wichtmann, A. Niemunis, and Th. Triantafyllidis. Differential settlements due to cyclic loading and their effect on the lifetime of structures. In F. Stangenberg, O.T. Bruhns, D. Hartmann, and G. Meschke, editors, *Proc. of 3rd Intern. Conf. on Lifetime-oriented Design Concepts, Bochum, 12 - 14 November*, pages 357–364. Aedificatio, 2007b.

T. Wichtmann, A. Niemunis, and Th. Triantafyllidis. Prediction of long-term deformations for monopile foundations of offshore wind power plants. In *11th Baltic Sea Geotechnical Conference: "Geotechnics in Maritime Engineering", Gdańsk, Poland, 15-18 September 2008*, pages 785–792, 2008.

T. Wichtmann, A. Niemunis, and T. Triantafyllidis. Validation and calibration of a high-cycle accumulation model based on cyclic triaxial tests on eight sands. *Soils and Foundations*, 49(10):711–728, 2009.

T. Wichtmann, A. Niemunis, and T. Triantafyllidis. On the determination of a set of material constants for a high-cycle accumulation model for non-cohesive soils. *Int. J. Numer. Anal. Meth. Geomech.*, 34(4):409–440, 2010a.

T. Wichtmann, A. Niemunis, and T. Triantafyllidis. On the "elastic" stiffness in a high-cycle accumulation model for sand: a comparison of drained and undrained cyclic triaxial tests. *Canadian Geotechnical Journal*, 47 (7):791–805, 2010b.

T. Wichtmann, B. Rojas, A. Niemunis, and Th. Triantafyllidis. Stress- and strain-controlled undrained cyclic triaxial tests on a fine sand for a high-cycle accumulation model. In *Proc. of the Fifth International Conference on Recent Advances in Geotechnical Earthquake Engineering and Soil Dynamics, San Diego, USA*, 2010c.

R.T. Wong, H.B. Seed, and C.K. Chan. Cyclic loading liquefaction of gravelly soils. *Journal of the Geotechnical Engineering Division, ASCE*, 101 (6):571–583, 1975.

H.-C. Wu and G.S. Chang. Stress analysis for dummy rod method for sand specimens. *Journal of Geotechnical Engineering, ASCE*, 108(9):1192–1197, 1982.

S. Wu, D.H. Gray, and F.E. Richart Jr. Capillary effects on dynamic modulus of sands and silts. *Journal of Geotechnical Engineering, ASCE*, 110 (9):1188–1203, 1984.

W. Wu. Hypoplastizität als mathematisches Modell zum mechanischen Verhalten granularer Stoffe. Veröffentlichungen des Institutes für Boden- und Felsmechanik der Universität Fridericiana in Karlsruhe, Heft Nr. 129, 1992.

H. Xia and T. Hu. Effects of saturation and back pressure on sand liquefaction. *Journal of Geotechnical Engineering, ASCE*, 117(9):1347–1362, 1991.

Y. Yamada and K. Ishihara. Yielding of loose sand in three-dimensional stress conditions. *Soils and Foundations*, 22(3):15–31, 1982.

S. Yasuda and M. Soga. Effects of frequency on undrained strength of sands (in Japanese). In *Proc. 19th Nat. Conf. Soil Mech. Found. Eng.*, pages 549–550, 1984.

S. Yasuda and I. Yamaguchi. Dynamic shear modulus obtained in the laboratory and in situ (in Japanese). In *Proceedings of the Symposium on Evaluation of Deformation and Strength of Sandy Ground*, pages 115–118, 1985.

K. Yokota and M. Konno. Comparison of soil constants obtained from laboratory tests and in situ tests (in Japanese). In *Proceedings of the Symposium on Evaluation of Deformation and Strength of Sandy Ground*, pages 111–114, 1985.

Y. Yoshimi and H. Oh-Oka. Influence of degree of shear stress reversal on the liquefaction potential of saturated sand. *Soils and Foundations*, 15 (3):27–40, 1975.

Y. Yoshimi, K. Tokimatsu, O. Kaneko, and Y. Makihara. Undrained cyclic shear strength of a dense Niigata sand. *Soils and Foundations*, 24(4): 131–145, 1984.

Y. Yoshimi, K. Tokimatsu, and Y. Hosaka. Evaluation of liquefaction resistance of clean sands based on high-quality undisturbed samples. *Soils and Foundations*, 29:93–104, 1989.

T.L. Youd. Compaction of sands by repeated shear straining. *Journal of the Soil Mechanics and Foundations Division, ASCE*, 98(SM7):709–725, 1972.

T.L. Youd, E.L. Harp, D.K. Keefer, and R.C. Wilson. The Borah Peak, Idaho Earthquake of October 28, 1983 - Liquefaction. *Earthquake Spectra*, 2(1):71–89, 1985.

P. Yu and F.E. Richart Jr. Stress ratio effects on shear modulus of dry sands. *Journal of Geotechnical Engineering, ASCE*, 110(3):331–345, 1984.

# Constitutive Modelling

David Muir Wood *
* University of Bristol, United Kingdom

## 1 Introduction to Constitutive Modelling

Numerical modelling of any problem requires noncontroversial statements of equilibrium and of kinematics or compatibility (the definition of strain); the constitutive response which links stress change and strain change is much more uncertain.

Models that are available in most numerical analysis programs that are intended for application to geotechnical problems include elastic-perfectly plastic Mohr-Coulomb, and Cam clay. We will present these models together with modest developments to improve the range of their applicability. Engineers are more likely to make use of models which can be clearly seen as incrementally different from models with which they have some familiarity than to make use of models which adopt a completely different language.

### 1.1 Stress and Strain Variables

Some of the presentation of constitutive models will concentrate on conditions that are accessible in the conventional triaxial apparatus. Most data for calibration come from this apparatus. We choose compressive stresses and compressive strains to be positive.

The triaxial apparatus provides two degrees of freedom in control of externally applied stress states. Soils consist of more or less rigid particles separated by voids. Volume changes are an important feature of the mechanical response of soils. Undrained deformation implies constant volume deformation (the compressibility of pore fluid is usually negligible) and hence pure distortion, change in shape at constant size. We divide soil deformations into compression (change of volume) and distortion (change of shape).

The principle of effective stress proposes that it is the effective stresses that control the deformation behaviour of the soil. Given a total stress tensor $\sigma_{ij}$ (compression positive) and pore pressure $u$, the effective stress tensor $\sigma'_{ij}$ is given by

$$\sigma'_{ij} = \sigma_{ij} - u\delta_{ij} \tag{1}$$

where $\delta_{ij}$ is the Kronecker delta

$$\delta_{ij} = 1, \quad i = j; \quad \delta_{ij} = 0, \quad i \neq j \tag{2}$$

In any triaxial test we can identify the axial and radial strain increments $\delta\varepsilon_a$ and $\delta\varepsilon_r$ and corresponding axial and radial effective stresses $\sigma'_a$ and $\sigma'_r$. We choose as a strain increment variable the volumetric strain increment $\delta\varepsilon_p = \delta\varepsilon_a + 2\delta\varepsilon_r$ and corresponding work conjugate volumetric stress $p' = (\sigma'_a + 2\sigma'_r)/3$; and the distortional stress $q = \sigma_a - \sigma_r = F/A$ and corresponding work conjugate distortional strain increment $\delta\varepsilon_q = 2(\delta\varepsilon_a - \delta\varepsilon_r)/3$. The work done in a small increment of strain is then:

$$\delta W = \delta W_v + \delta W_d = p'\delta\varepsilon_p + q\delta\varepsilon_q = \sigma'_a\delta\varepsilon_a + 2\sigma'_r\delta\varepsilon_r \tag{3}$$

and the volumetric (change in size) and distortional (change in shape) effects are clearly separated.

It is also convenient to define a stress ratio $\eta = q/p'$ which is equivalent to a mobilised friction $\phi'_m$. Under conditions of triaxial compression, in which $q > 0$ and the axial stress is greater than the radial stress

$$\begin{aligned} \sigma'_a/\sigma'_r &= (1 + \sin\phi'_m)/(1 - \sin\phi'_m) = (3 + 2\eta)/(3 - \eta) \\ \sin\phi'_m &= (\sigma'_a - \sigma'_r)/(\sigma'_a + \sigma'_r) = (3\eta)/(6 + \eta) \\ \eta &= 6\sin\phi'_m/(3 - \sin\phi'_m) \end{aligned} \tag{4}$$

Under conditions of triaxial extension, in which $q < 0$ and the axial stress is less than the radial stress

$$\begin{aligned} \sigma'_a/\sigma'_r &= (1 - \sin\phi'_m)/(1 + \sin\phi'_m) = (3 + 2\eta)/(3 - \eta) \\ \sin\phi'_m &= (\sigma'_r - \sigma'_a)/(\sigma'_a + \sigma'_r) = -3\eta/(6 + \eta) \\ \eta &= -6\sin\phi'_m/(3 + \sin\phi'_m) \end{aligned} \tag{5}$$

The relationships between stress variables and between strain increment variables can be summarised in matrix form

$$\begin{pmatrix} p' \\ q \end{pmatrix} = \begin{pmatrix} \frac{1}{3} & \frac{2}{3} \\ 1 & -1 \end{pmatrix} \begin{pmatrix} \sigma'_a \\ \sigma'_r \end{pmatrix} \tag{6}$$

$$\begin{pmatrix} \sigma'_a \\ \sigma'_r \end{pmatrix} = \begin{pmatrix} 1 & \frac{2}{3} \\ 1 & -\frac{1}{3} \end{pmatrix} \begin{pmatrix} p' \\ q \end{pmatrix} \tag{7}$$

$$\begin{pmatrix} \delta\varepsilon_p \\ \delta\varepsilon_q \end{pmatrix} = \begin{pmatrix} 1 & 2 \\ \frac{2}{3} & -\frac{2}{3} \end{pmatrix} \begin{pmatrix} \delta\varepsilon_a \\ \delta\varepsilon_r \end{pmatrix} \tag{8}$$

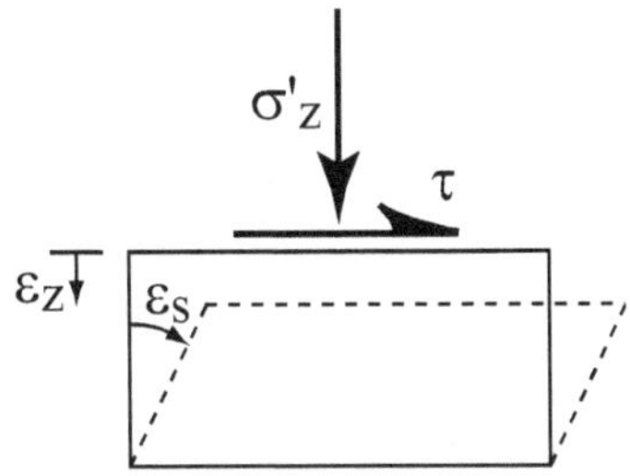


**Figure 1.** Element of soil for development of one dimensional model.

$$\begin{pmatrix} \delta\varepsilon_a \\ \delta\varepsilon_r \end{pmatrix} = \begin{pmatrix} \frac{1}{3} & 1 \\ \frac{1}{3} & -\frac{1}{2} \end{pmatrix} \begin{pmatrix} \delta\varepsilon_p \\ \delta\varepsilon_q \end{pmatrix} \tag{9}$$

$$\frac{q}{p'} = \frac{3\left[(\sigma'_a/\sigma'_r) - 1\right]}{(\sigma'_a/\sigma'_r) + 2} \tag{10}$$

$$\frac{\sigma'_a}{\sigma'_r} = \frac{1 + (2/3)\,(q/p')}{1 - (1/3)\,(q/p')} \tag{11}$$

$$\frac{\delta\varepsilon_p}{\delta\varepsilon_q} = \frac{3 + (3/2)\,(\delta\varepsilon_a/\delta\varepsilon_r)}{(\delta\varepsilon_a/\delta\varepsilon_r) - 1} \tag{12}$$

$$\frac{\delta\varepsilon_a}{\delta\varepsilon_r} = \frac{3 + (\delta\varepsilon_p/\delta\varepsilon_q)}{(\delta\varepsilon_p/\delta\varepsilon_q) - (3/2)} \tag{13}$$

### 1.2 Simple Model of Shearing

However, in parallel, we will construct models for the shearing of a sand (for example) in a shear box. We will idealise a little, but these simple models can be developed into much more sophisticated models capable of being used in numerical analysis of complete and realistic prototype geotechnical problems.

We imagine a 'simple shear' element of soil as shown in Fig. 1 which might be extracted from the central shearing region of a shear box. It is subjected to a vertical, normal effective stress $\sigma'_z$ and a shear stress $\tau$. We expect that there will be vertical strains $\varepsilon_z$ and shear strains $\varepsilon_s$. The shear strain produces a change in shape from rectangle to parallelogram. Our task will be to find general link between changes in the stresses and changes in the strains.

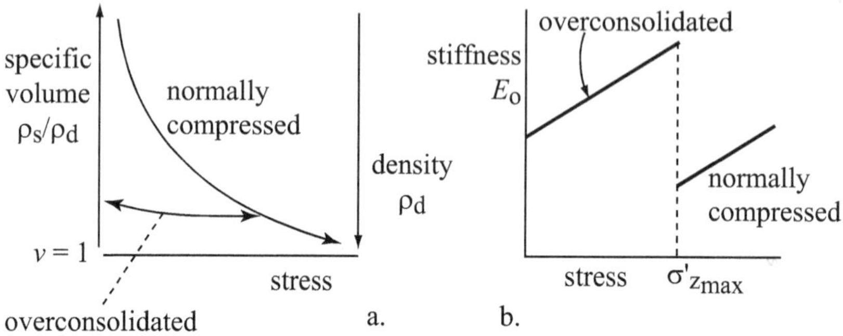


**Figure 2.** (a) One-dimensional compression and overconsolidation; (b) stiffness and overconsolidation (drawn for $\alpha = 1$).

## 2 Stiffness

If the stress $\sigma'_z$ on the element in Fig. 1 changes, there will be corresponding vertical deformation described by one-dimensional stiffness properties that might have the form:

$$\frac{E_o}{\sigma_{ref}} = \chi \left( \frac{\sigma'_z}{\sigma_{ref}} \right)^\alpha \tag{14}$$

where $\sigma_{ref}$ is a reference stress introduced to leave dimensional consistency between the two sides of (14) and $\chi$ and $\alpha$ are soil parameters describing, respectively, the magnitude of the stiffness and the way in which it depends on stress level. For any change $\delta\sigma'_z$ in vertical effective stress, the corresponding vertical strain is $\delta\varepsilon_z^\sigma$:

$$\delta\varepsilon_z^\sigma = \frac{\delta\sigma'_z}{E_o} \tag{15}$$

We will see shortly why we need to distinguish this strain increment with the superscript $^\sigma$.

In describing stiffness, we distinguish between normally compressed soils which are currently experiencing the maximum vertical stress they have ever experienced, and overconsolidated soils which have been more heavily loaded in the past (Fig. 2a). The overconsolidation ratio, $n = \sigma'_{z_{\max}}/\sigma'_z$, describes the extent of this prior loading. The rules governing one-dimensional stiffness for normally compressed and overconsolidated soils are (Fig. 2b):

1. If $\sigma'_z = \sigma'_{z_{\max}}$ and $\delta\sigma'_z > 0$, then $n = 1$, $\alpha = \alpha_{nc}$ and $\chi = \chi_{nc}$ (normally compressed);

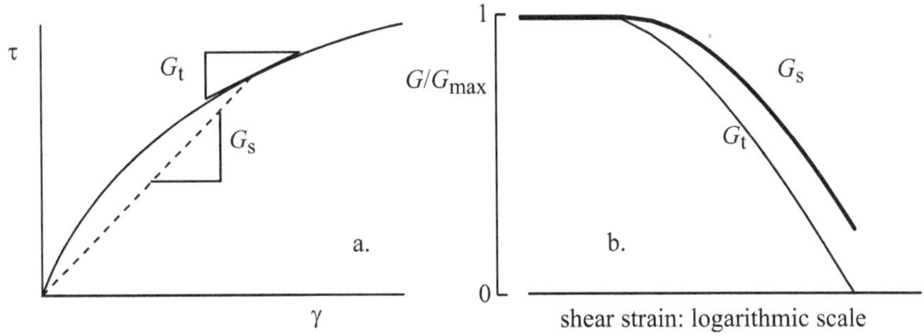


**Figure 3.** (a) Tangent and secant stiffness; (b) variation of tangent and secant shear stiffness in monotonic shearing.

2. If $\delta\sigma'_z < 0$, then $\delta n > 0$, $n \geq 1$, and $\alpha = \alpha_{oc}$ and $\chi = \chi_{oc}$ (overconsolidated);
3. If $\sigma'_z < \sigma'_{z_{\max}}$ and $\delta\sigma'_z > 0$, then $n > 1$, $\delta n < 0$, and $\alpha = \alpha_{oc}$ and $\chi = \chi_{oc}$ (overconsolidated);

where $\alpha_{nc}$ and $\alpha_{oc}$, and $\chi_{nc}$ and $\chi_{oc}$ are the values of stiffness exponent and modulus number in (14) appropriate to the normally compressed and overconsolidated states, respectively. Typically $\alpha_{nc} \approx \alpha_{oc}$. The distinction between normally consolidated and overconsolidated can be seen as a distinction between elastic+plastic (partially irrecoverable) and purely elastic (recoverable) response.

Stiffness variation in a monotonic *shearing* test, as opposed to this *confined compression*, is typically presented (Fig. 3) in a plot of shear modulus (usually, regrettably, secant modulus) against strain, with the strain plotted on a logarithmic scale because much of the initial variation of stiffness occurs at very small strains. Tangent stiffness $G_t = \mathrm{d}\tau/\mathrm{d}\gamma$ varies with strain much more rapidly than secant stiffness $G_s = \tau/\gamma$ (Fig. 3)—and if the stress:strain response reveals strain softening after some peak then the tangent modulus will actually become negative. Secant stiffness tells us about the past—tangent stiffness tells us about the present and the very short-term future. Our concern in modelling is to generate a tangent stiffness relationship for our soil.

Soil elements in geotechnical systems will be subjected to nonmonotonic paths following long term geological and shorter term construction histories and we will expect to use the laboratory testing possibilities that are available to us to explore the incremental effects of stress changes in a very general way. We need to have a coherent strategy for the conduct of this

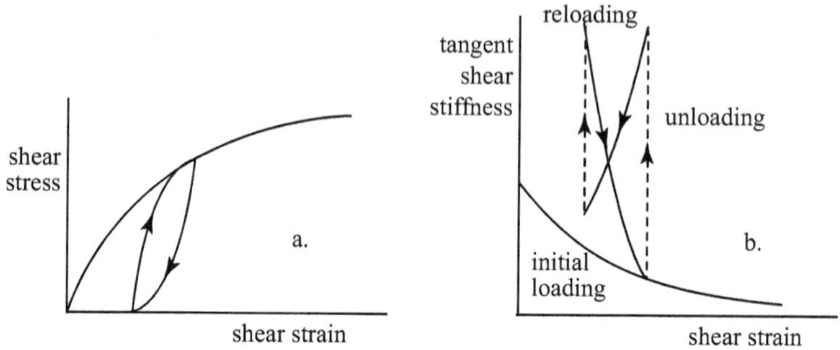


**Figure 4.** (a) Typical irreversible stress:strain response and (b) typical modulus variation for soil.

testing.

A comparison of the stress:strain response implied by a linear elastic description of soil behaviour with the actual stress:strain response of a typical soil shows that there are many features of soil response that the simple model is unable to capture (Fig. 4). In particular, it is clear that most soils show nonlinear stress:strain relationships with the stiffness falling from a high initial value. If a soil is unloaded from some intermediate, prefailure condition then it will not recover its initial state but will be left with permanent, irrecoverable deformation—which we will call plastic deformation to distinguish it from the recoverable, elastic elements of deformation. During this unloading process the tangent stiffness increases initially, typically to a value higher than the initial stiffness and then falls—a similar pattern is seen on reloading (Fig. 4).

One way of illustrating the link between strain increments and stress increments which can be useful both for planning and interpreting programmes of testing is through the generation of stress response envelopes, introduced by Gudehus (1979) as a way of illustrating the nature of the characters of response predicted by different classes of constitutive model. Thus, for axially symmetric states of stress attainable in the conventional triaxial apparatus, envelopes can be shown in terms of volumetric (mean effective) stress and distortional stress changes resulting from the application of increments of volumetric and distortional strain. If, from a given initial stress state, a series of *strain* probes of identical normalised magnitude is imposed, then the resulting envelope of *stress* responses provides a visual indication of the generalised stiffness of the soil. An example is shown in Fig. 5. The strain increments are defined in terms of volumetric strain $\delta\varepsilon_p$

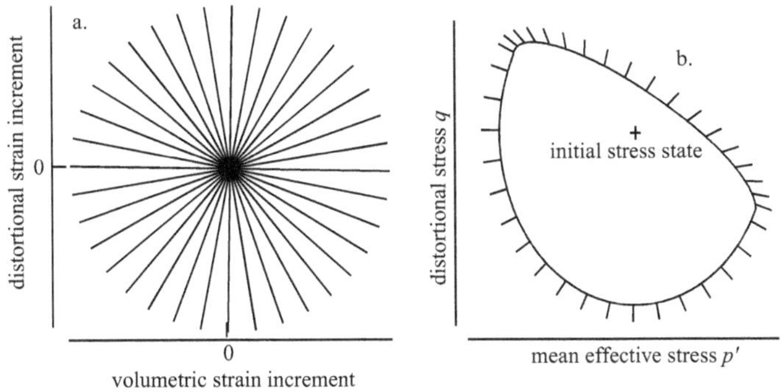


**Figure 5.** (a) Rosette of strain probes and (b) resulting stress response envelope for axisymmetric state of stress.

and distortional strain $\delta\varepsilon_q$. The rosette of strain increments of standard length is shown in Fig. 5a. A solid curve joins the resulting stress increments from the common initial stress, presented in terms of mean effective stress $p'$ and distortional stress $q$, in Fig. 5b). In this figure, at each point on the stress response envelope a little line indicates the direction of the corresponding strain probe from Fig. 5a.

Near failure we expect the stiffness for continued loading—increased distortional strain—to be considerably lower than for unloading—reversal of distortional strain—and the stress response envelope will be flattened towards the loading direction (A in Fig. 6). At lower stress ratios the envelope is likely to be more rounded (B in Fig. 6) with the generalised stiffness less dramatically influenced by the direction of the probe though still indicating lower stiffness for continued loading, higher stiffness for reversal of loading. There may be some partially unloaded states for which the response is much more independent of loading direction (C in Fig. 6). For soil in a state which tends to lead to strain softening (probably at high stress ratios) the initial stress state may lie outwith the response envelope (D in Fig. 6)—this is an indication that *all* strain increments imposed on our soil element will lead to reduction of stress ratio.

The shape of the response envelope and the location of the initial stress state relative to the response envelope will vary with history. Soil stiffness falls with monotonically increasing strain. Instrumentation does not permit us to determine response envelopes for 'zero' amplitude of strain increments. However, it is instructive, in gathering data to inspire our constitutive modelling, to look at response envelopes determined for different,

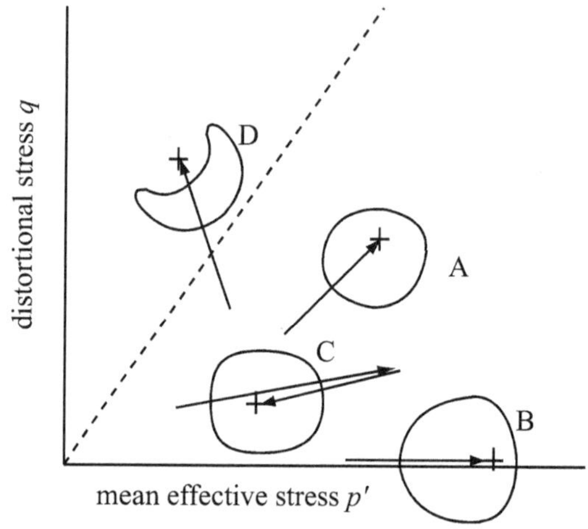


**Figure 6.** Schematic expected history dependence of stress response envelopes (points + indicate initial stress states).

finite, magnitudes of strain increment from a given initial stress state.

Stress response envelopes from true triaxial tests on kaolin are shown in a $\pi$-plane deviatoric view of stress space in Fig. 7 for two different initial histories—one (Fig. 7a) has isotropic compression to O followed by shearing (at constant mean effective stress) to A; the other (Fig. 7b) has history OA followed by unloading back to O. The response envelopes are drawn for different values of a distortional strain

$$\varepsilon = \frac{1}{\sqrt{3}}\sqrt{\left[(\varepsilon_y - \varepsilon_z)^2 + (\varepsilon_z - \varepsilon_x)^2 + (\varepsilon_x - \varepsilon_y)^2\right]} \tag{16}$$

which is proportional to the second invariant of the strain deviator tensor. The small strain envelopes in each case are closely linked to the recent stress history, OA or OAO respectively. However, as the strain magnitudes increase, the detail of the starting stress state seems to become largely irrelevant—by the 10% envelope, memory of what has gone before has been somewhat swept out.

Stress response envelopes from true triaxial tests on Leighton Buzzard sand are shown in Fig. 8 (data from Sture et al., 1988. These tests were performed in a cubical cell true triaxial apparatus in which the stresses were imposed through flexible boundary cushions: the magnitudes of strains that can be imposed while still retaining deformational uniformity are limited. Comparing deviatoric histories (imposed at constant mean stress) ABC,

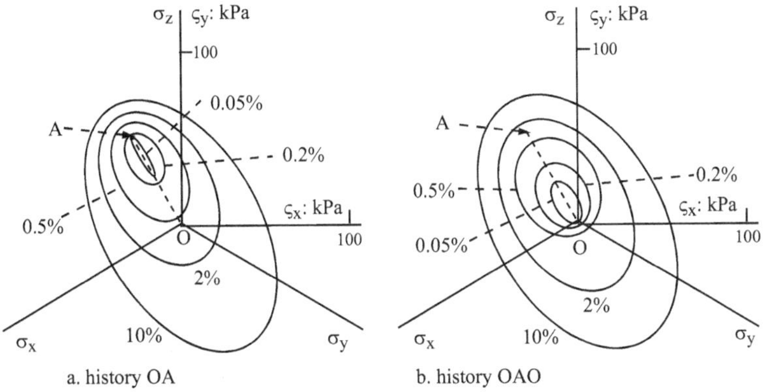


**Figure 7.** Schematic deviatoric stress response envelopes from true triaxial probing of kaolin with deviatoric history (a) OA and (b) OAO (inspired by Muir Wood, 2004).

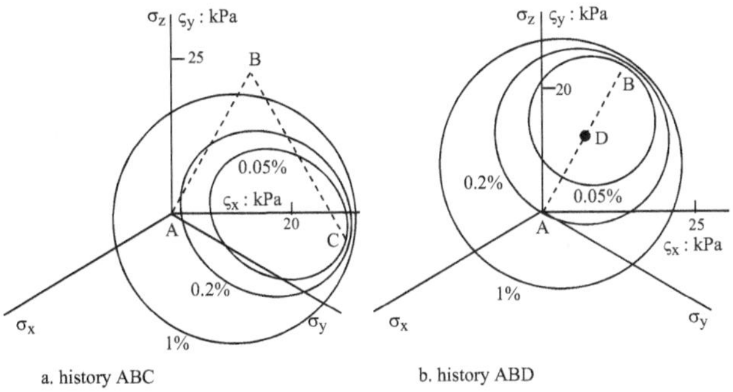


**Figure 8.** Schematic stress response envelopes from true triaxial probing of Leighton Buzzard sand with deviatoric history (a) ABC and (b) ABD (inspired by data from Sture et al., 1988).

ABD the comments made previously are reinforced. Failure is lurking in the $\pi$-plane at some finite distance from the isotropic stress axis, A, so it is to be expected that the several stress response envelopes will be closely packed together there.

The picture is always the same. There is a strongly kinematic element to the small strain response envelopes—they are carried around intimately with the most recent stress history. The larger the strain magnitude the less

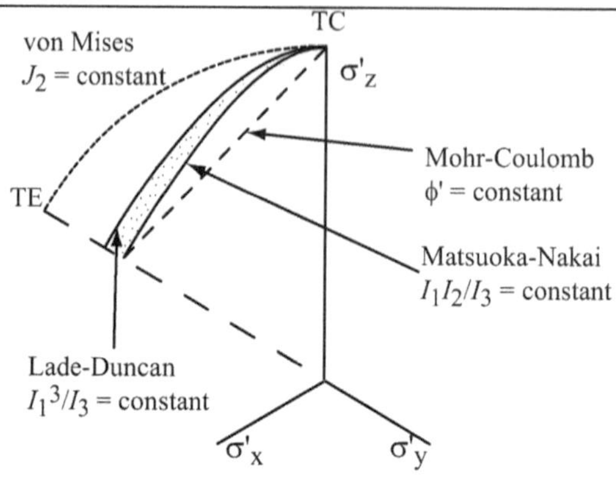


**Figure 9.** Failure criteria in $\pi$-plane deviatoric view of principal stress space (TC: triaxial compression; TE: triaxial extension).

the location of the envelope seems to care about the recent stress history and the more it is aware of other constraints on soil response—such as the limitations that failure criteria might impose.

# 3 Strength

As the shear stress $\tau$ on the element in Fig. 1 is increased, we expect eventually to reach the limit of the shear stress that can be supported—the strength of the soil. Because most of the soils with which we are dealing are, at least eventually, unbonded, we present strength as a purely *frictional* phenomenon: the strength is proportional to the stress level ($p'$, triaxial) or normal stress ($\sigma'_z$, shear box, Fig 1). This is described by a Mohr-Coulomb failure criterion with angle of shearing resistance $\phi'$:

$$\tau_{max} = \sigma'_z \tan\phi' \tag{17}$$

In the $\pi$-plane view of principal stress space, the Mohr-Coulomb failure criterion plots as an irregular hexagon for a constant mean stress section (one 60° segment is shown in Fig. 9). This suggests immediately that a model generalisation based on the second stress invariant $J_2$ will not be particularly satisfactory for soils. In fact, failure data for sand tend to lie somewhere between the Mohr-Coulomb hexagon and the $J_2$ circle. Two alternative failure criteria have been quite widely used to better describe the deviatoric failure conditions.

The Matsuoka-Nakai criterion (Matsuoka and Nakai, 2001) states that

$$\frac{I_1 I_2}{I_3} = \text{constant} \tag{18}$$

where $I_1$, $I_2$ and $I_3$ are the three invariants of the stress tensor. This gives a curved failure locus which circumscribes the Mohr-Coulomb hexagon, passing through the vertices of that hexagon.

On the other hand, many experimental data suggest that the frictional strength of sands in triaxial extension is in fact slightly higher than that in triaxial compression. Lade's (Lade and Duncan, 1975) failure criterion is expressed as

$$\frac{I_1^3}{I_3} = \text{constant} \tag{19}$$

This is also plotted in Fig. 9. Evidently the common characteristic of both these failure criteria is that, unlike the Mohr-Coulomb criterion, they include the intermediate principal stress. Different sands, tested in different laboratory devices, show strength data which favour one or other of these criteria. Both are in agreement in proposing that, for states of stress lying between triaxial compression and triaxial extension, the available friction is somewhat higher—perhaps at its maximum as much as 10% higher—than that in triaxial compression. Plane strain conditions tend to fall in this intermediate region: plane strain frictional strengths will usually be underestimated if the angle of shearing resistance is determined using triaxial compression tests.

## 4 Elastic-Perfectly Plastic Models

Most soils, if sheared to sufficiently large strains, reach a state of continuing shearing with no further change in stresses, zero incremental or tangent stiffness, at large strains. This type of behaviour, in which the tangential stiffness has fallen to zero, is described as perfect plasticity and we can describe general and particular (Mohr-Coulomb) elastic-perfectly plastic descriptions of soil response (Fig. 10).

It is assumed in these soil models that the strain increments that accompany any change in stress can be divided into elastic ($^e$) (recoverable) and plastic ($^p$) (irrecoverable) parts

$$\delta\varepsilon = \delta\varepsilon^e + \delta\varepsilon^p \tag{20}$$

This division of strain clearly reflects the observation that removal of loads from a sheared soil sample in general leaves the sample with some permanent changes in shape and size (Fig 4).

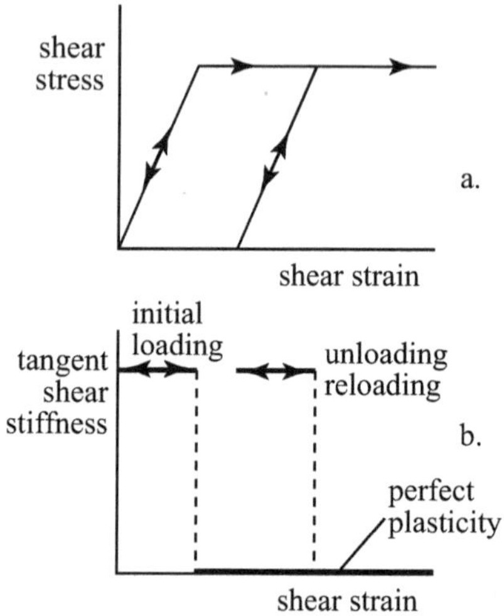


**Figure 10.** Elastic-perfectly plastic model: (a) stress:strain response and (b) modulus variation.

The strain increment tensor is thought of here as a six element *vector* of cartesian strain increment components since in this form the presentation and programming of stiffness relationships involves nothing more than straightforward matrix multiplication and manipulation. In many applications it will be a subset of this vector that will be of interest.

The elastic strain increment $\delta\boldsymbol{\varepsilon}^e$ occurs whenever there is any change in stress $\delta\boldsymbol{\sigma}$ (where the stress is also thought of as a six dimensional vector of cartesian components).

$$\delta\boldsymbol{\sigma} = \boldsymbol{D}\delta\boldsymbol{\varepsilon}^e \tag{21}$$

where $\boldsymbol{D}$ is the elastic stiffness matrix. The first ingredient of the model is therefore a description of the elastic behaviour which may be isotropic or anisotropic as appropriate.

In the elastic-perfectly plastic model there is a region of stress space which can be reached elastically, without incurring any irrecoverable deformations (Fig. 11). However, as soon as the boundary of this elastic region is reached then the material yields (or fails) at constant stress. The boundary of the elastic region is called a yield surface (Fig. 11) and is mathematically described by a yield function: this is the second ingredient of the model.

$$f(\boldsymbol{\sigma}) = 0 \tag{22}$$

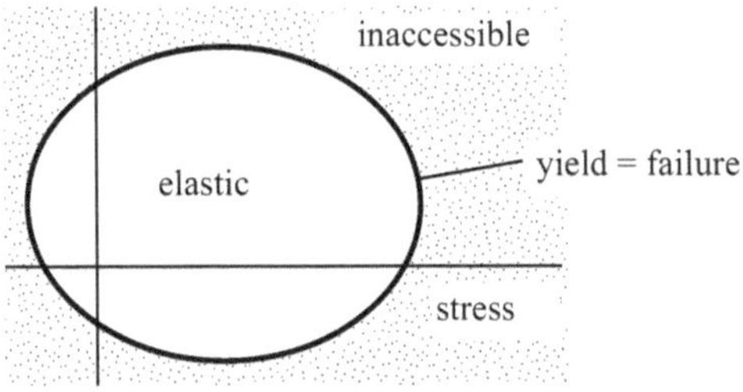


**Figure 11.** Elastic-perfectly plastic model: yield surface separating elastic and inaccessible regions of stress space.

The plastic strain increment $\delta\varepsilon^p$ (in (20)) occurs only when the stress state lies on—and remains on—the yield surface during the load increment so that

$$f(\boldsymbol{\sigma}) = 0; \quad \delta f = \frac{\partial f}{\partial \boldsymbol{\sigma}}^T \delta\boldsymbol{\sigma} = 0 \tag{23}$$

where $^T$ indicates the transpose of the vector. This relation is known as the consistency condition.

In order to be able to calculate the plastic deformations we make the assumption that there exists a plastic potential function $g(\boldsymbol{\sigma})$ which can be evaluated at the current stress state such that the plastic strain increment is given by

$$\delta\varepsilon^p = \dot{\Lambda}\frac{\partial g}{\partial \boldsymbol{\sigma}} \tag{24}$$

where $\dot{\Lambda}$ is a scalar multiplier whose magnitude is essentially arbitrary since this expression merely defines the *mechanism* of plastic deformation—the ratio of the several components of plastic deformation. It is thus only the gradient of the plastic potential function $g(\boldsymbol{\sigma})$ that is required, the actual value of the function is not relevant.

Combination of (20), (21) and (24) gives

$$\delta\boldsymbol{\sigma} = \boldsymbol{D}\delta\varepsilon - \dot{\Lambda}\boldsymbol{D}\frac{\partial g}{\partial \boldsymbol{\sigma}} \tag{25}$$

and combination of (25) with (23) allows us to determine $\dot{\Lambda}$

$$\dot{\Lambda} = \frac{\frac{\partial f}{\partial \boldsymbol{\sigma}}^T \boldsymbol{D}\delta\varepsilon}{\frac{\partial f}{\partial \boldsymbol{\sigma}}^T \boldsymbol{D}\frac{\partial g}{\partial \boldsymbol{\sigma}}} \tag{26}$$

and hence generate an expression for the elastic-plastic stiffness matrix $\boldsymbol{D}^{ep}$ giving $\delta\boldsymbol{\sigma}$ as a function of $\delta\boldsymbol{\varepsilon}$:

$$\delta\boldsymbol{\sigma} = \left[\boldsymbol{D} - \frac{\boldsymbol{D}\frac{\partial g}{\partial \boldsymbol{\sigma}}\frac{\partial f}{\partial \boldsymbol{\sigma}}^T\boldsymbol{D}}{\frac{\partial f}{\partial \boldsymbol{\sigma}}^T\boldsymbol{D}\frac{\partial g}{\partial \boldsymbol{\sigma}}}\right]\delta\boldsymbol{\varepsilon} = \boldsymbol{D}^{ep}\delta\boldsymbol{\varepsilon} \tag{27}$$

from which the stress increment can be calculated from any total strain increment that is causing yield. This stiffness relationship is written with two terms. We do not know *a priori* whether the response of the soil will be elastic (lying entirely within the yield locus) or elastic-plastic (lying on the yield locus). We can start by assuming purely elastic response and predicting the stress increment using only the elastic stiffness, $\delta\boldsymbol{\sigma} = \boldsymbol{D}\delta\boldsymbol{\varepsilon}$. If the resulting stress state lies outside the yield locus then the elastic assumption was incorrect and the elastic predicted stress increment needs to be 'corrected' back to the yield locus using the second part of the right hand side of (27). Such a two-stage process of prediction and correction is an inevitable part of the numerical implementation of any elastic-plastic model.

Note that we will expect that we can always deduce increments in stress from imposed increments in strain (the operation indicated in (27)) but that the reverse operation will not always be possible if our current state of stress is already on the yield/failure boundary of the elastic region (Fig. 11).

### 4.1 Elastic-Perfectly Plastic Mohr-Coulomb Model

To demonstrate how this final expression can be used we can look at the special case of the elastic-perfectly plastic Mohr-Coulomb soil model. We will apply the model to axisymmetric conditions.

First we define the elastic properties using an isotropic elastic model:

$$\boldsymbol{D} = \begin{pmatrix} K & 0 \\ 0 & 3G \end{pmatrix} \tag{28}$$

Next we define the yield function as (Fig. 12a)

$$f(\boldsymbol{\sigma}) = f(p', q) = q - Mp' \tag{29}$$

If $f(p', q) < 0$ the soil is behaving elastically; if $f(p', q) = 0$ the soil is yielding (failing) and generating plastic deformations. To have $f(p', q) > 0$ is impossible: this defines an inaccessible region of the $(p', q)$ stress plane (Fig. 12a). The value of the soil property $M$ can be related to the angle of shearing resistance $\phi'$ of the soil in triaxial compression:

$$M = \frac{6 \sin\phi'}{3 - \sin\phi'} \tag{30}$$

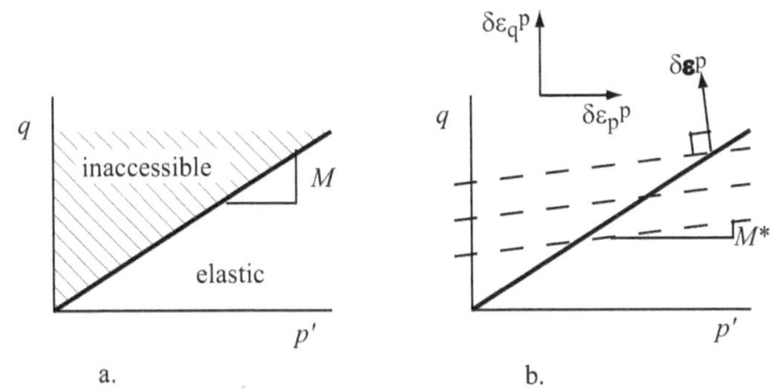


**Figure 12.** Elastic-perfectly plastic Mohr-Coulomb model: (a) yield/failure locus; (b) plastic potentials.

Most soils develop significant irrecoverable volume changes even when they are subjected only to changes in shear stress. We require some constraint on the plastic deformations in the form of a flow rule which defines the plastic deformation mechanism at the current stress state. We define a plastic potential function (Figure12b)

$$g(\boldsymbol{\sigma}) = g(p', q) = q - M^* p' + k = 0 \tag{31}$$

where $k$ is an arbitrary variable to allow the plastic potential function to be defined at the current state of stress and $M^*$ is another soil property. The plastic strain increments are given by normality to the plastic potential function at the current state of stress (Fig. 12b)

$$\begin{pmatrix} \delta\varepsilon_p^p \\ \delta\varepsilon_q^p \end{pmatrix} = \dot{\Lambda} \begin{pmatrix} \partial g/\partial p' \\ \partial g/\partial q \end{pmatrix} = \dot{\Lambda} \begin{pmatrix} -M^* \\ 1 \end{pmatrix} \tag{32}$$

where the scalar multiplier $\dot{\Lambda}$ indicates the magnitude of the plastic strain increments. The ratio of the two components of plastic strain is:

$$\frac{\delta\varepsilon_p^p}{\delta\varepsilon_q^p} = -M^* \tag{33}$$

For $M^* = 0$ plastic deformation occurs at constant volume. Soils that contract when they are sheared plastically have negative angles of dilation: $M^* < 0$ (Fig. 13c); soils that expand have positive angles of dilation: $M^* > 0$ (Fig. 13c). For all real soils $M^* < M$. A special (though physically unrealistic) case is obtained when $M^* = M$.

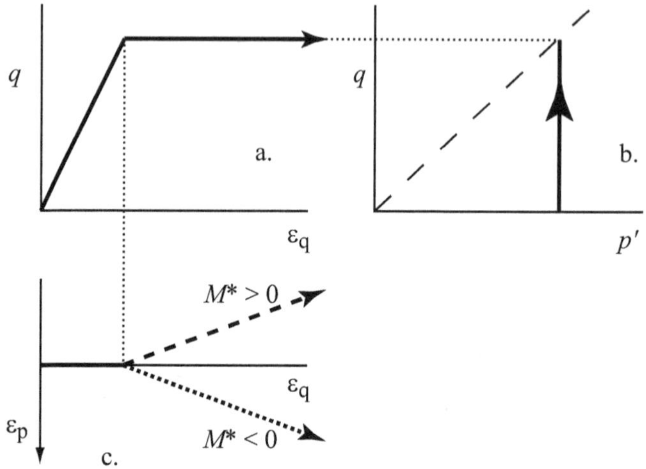


**Figure 13.** Elastic-perfectly plastic Mohr-Coulomb model: (a) stress:strain response; (b) constant $p'$ effective stress path; (c) volumetric strain, dependence on $M^*$.

The energy that is dissipated during an increment of plastic deformation is

$$\delta W^p = \boldsymbol{\sigma}^T \delta \boldsymbol{\varepsilon}^p = p' \delta \varepsilon_p^p + q \delta \varepsilon_q^p \tag{34}$$

Since the soil is yielding the stresses are related by

$$q = Mp' \tag{35}$$

and the plastic strain increments are related by (33). The plastic energy thus becomes

$$\delta W^p = (M - M^*) p' \delta \varepsilon_q^p \tag{36}$$

It is evident that if $M^* = M$ there is no plastic energy dissipation which seems likely to provide an unsatisfactory description of soil behaviour.

The complete elastic-plastic stiffness matrix (27) for this perfectly plastic model can now be generated:

$$\boldsymbol{D}^{ep} = \left[ \begin{pmatrix} K & 0 \\ 0 & 3G \end{pmatrix} - \frac{1}{KMM^* + 3G} \begin{pmatrix} MM^*K^2 & -3M^*GK \\ -3MGK & 9G^2 \end{pmatrix} \right] \tag{37}$$

The second term in (37) is only included if the soil is yielding. The link between stress increments and strain increments is then

$$\begin{pmatrix} \delta p' \\ \delta q \end{pmatrix} = \frac{3GK}{KMM^* + 3G} \begin{pmatrix} 1 & M^* \\ M & MM^* \end{pmatrix} \begin{pmatrix} \delta \varepsilon_p \\ \delta \varepsilon_q \end{pmatrix} \tag{38}$$

The elastic-plastic stiffness matrix is in general asymmetric unless $M^* = M$ which, as has been shown, is physically unreasonable. However, certain numerical analysis programs require the stiffness matrix to be symmetric for solution purposes and it is for these programs that the assumption $M^* = M$ is often forced upon the user—or else some numerical subterfuge is needed to overcome the limitation of the program.

Although it is often easier to think of stress changes producing changes in strain—and physical considerations of the behaviour of soils often encourage us to move in this direction—if we look at the diagram of the $(p', q)$ stress plane (Fig. 12a) we can see that this will not provide a secure route for analysis because a large part of the stress plane is in fact forbidden territory. On the other hand, working from strain increments to stress increments carries no such problem: all strain increments are permitted even when the current stress state sits on the yield (failure) locus. Some of these strain increments will produce purely elastic changes in stress which take the stress state away from yield; others will force the stress state to move up or down along the yield (failure) locus in such a way that the elastic component of the strain caused by the change in stress uses up that part of the total strain increment which cannot be ascribed to the plastic strain mechanism given by (33) or (24).

For each of the strain increments of a rosette of increments of similar magnitude but different direction we can use the elastic-plastic stiffness form of the model ((28) or (37) depending on whether the soil is responding elastically or elastoplastically) to calculate the stress increment response (Fig. 14). The stress response envelope thus consists of two parts.

If the strain increment can be supported by elastic unloading then the stress increment is directed away from the yield locus. For these increments the response envelope takes the form of half of an ellipse (Fig. 15). If the strain increment requires the soil to yield then the stress state has to lie on the yield locus—all plastic stress states are, in this perfectly plastic model, confined to this one line. For these increments the stress response envelope consists of a straight line tangent to (in this case coincident with) the yield locus at the given initial stress (Fig. 15).

Two limits may be noted. If the ratio of strain components is given by (33) then the stresses remain unchanged as the soil yields:

$$\begin{pmatrix} \delta\varepsilon_p \\ \delta\varepsilon_q \end{pmatrix} \propto \begin{pmatrix} -M^* \\ 1 \end{pmatrix} \quad \rightarrow \quad \begin{pmatrix} \delta p' \\ \delta q \end{pmatrix} = \begin{pmatrix} 0 \\ 0 \end{pmatrix} \tag{39}$$

It is of course possible for the stress state to move along the yield locus

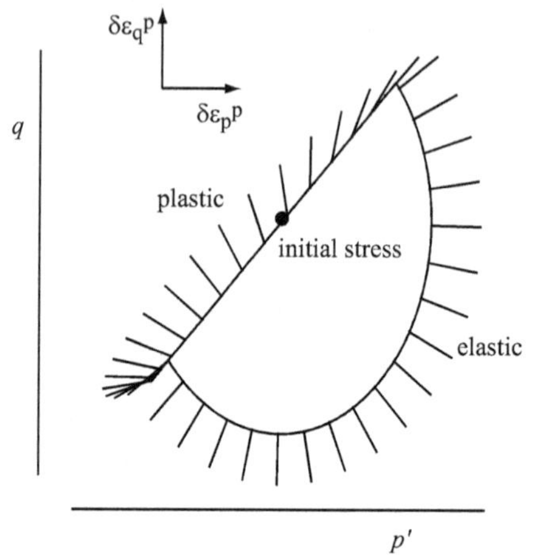


**Figure 14.** Elastic-perfectly plastic Mohr-Coulomb model: stress response envelope (calculated with $M = 1.2$, $M^* = 0.2$ and initial stresses $p' = 100$, $q = Mp' = 120$).

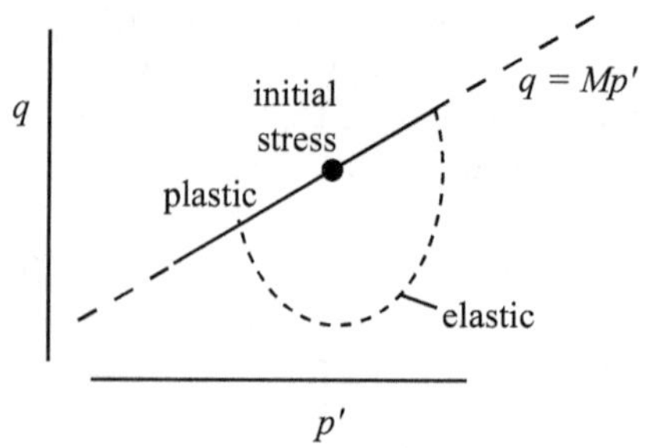


**Figure 15.** Elastic-perfectly plastic Mohr-Coulomb model: elastic and plastic sections of stress response envelope.

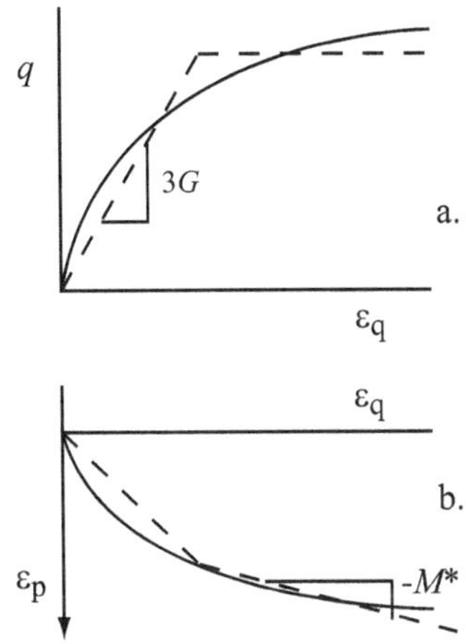


**Figure 16.** Elastic-perfectly plastic Mohr-Coulomb model compared with typical soil response in conventional drained triaxial compression test.

purely elastically without incurring plastic deformation. In this case

$$\begin{pmatrix} \delta\varepsilon_p \\ \delta\varepsilon_q \end{pmatrix} \propto \begin{pmatrix} 1/K \\ M/3G \end{pmatrix} \tag{40}$$

and this ratio defines the boundary of elastically attainable strain states in the corresponding strain increment plane. It should be clear from the stress response envelope in Fig. 14 that, not only is there a part of the stress plane that is inaccessible (anywhere implying a value of $q/p' > M$), even for stress changes which lie along the boundary of the elastic region ($\delta q/\delta p' = M$) there is an infinite number of possible causative strain increments and we cannot even tell whether the soil is behaving elastoplastically or purely elastically. The ambiguity in trying to work from stress increments to strain increments is emphasised.

A particular case is given by undrained constant volume shearing for which

$$\begin{pmatrix} \delta\varepsilon_p \\ \delta\varepsilon_q \end{pmatrix} \propto \begin{pmatrix} 0 \\ 1 \end{pmatrix} \quad \rightarrow \quad \begin{pmatrix} \delta p' \\ \delta q \end{pmatrix} \propto \frac{3GKM^*}{KMM^* + 3G} \begin{pmatrix} 1 \\ M \end{pmatrix} \tag{41}$$

and the stress path ascends or descends the line $q = Mp'$ for $M^* > 0$ (Fig. 14) or $M^* < 0$ respectively (assuming that $3G + KMM^* > 0$ always).

The elastic-perfectly plastic model forces a bilinear description of response which leads to inevitable subjectivity in choosing values of stiffness,

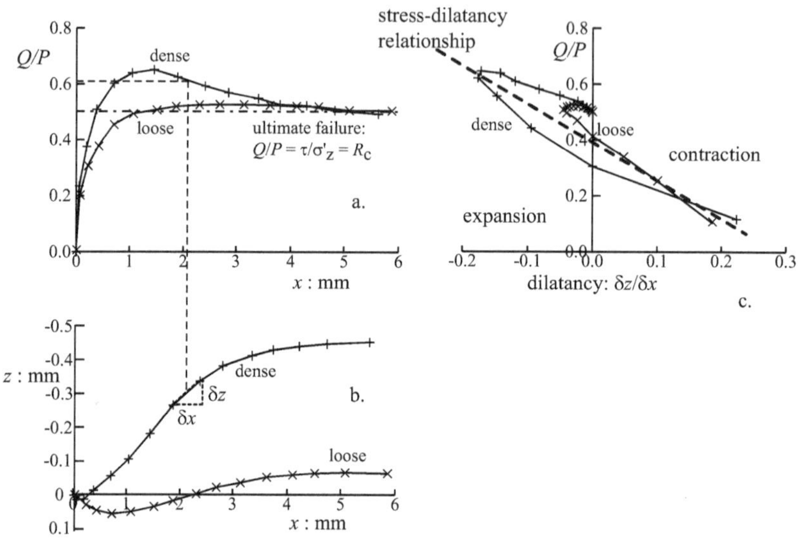


**Figure 17.** (a) Shear load:displacement response and (b) volume changes in direct shear test on Ottawa sand; (c) stress-dilatancy correlation (data from Taylor (1948).)

strength and dilatancy properties to match real soil test data (Fig. 16). Elastic-perfectly plastic models are widely used because of their simplicity. The Mohr-Coulomb model is available in every computer program that is seriously intended for numerical analysis of geotechnical problems. It requires definition of elastic properties—of which there will be at least two; and of some failure property—for example, a limiting angle of shearing resistance for a frictional model to be used for description of drained soil conditions or a limiting shear stress for a cohesive model to be used for description of undrained soil conditions; together with some statement about the volume changes that accompany failure—for example, an angle of dilation. There is obviously need for care in the selection of the elastic properties. Not all programs give the freedom to select angles of dilation which are different from the angle of shearing resistance.

## 5 Dilatancy

In presenting results of shear box tests, we can observe the volume changes that occur in sands as they are sheared (Fig. 17) and relate this dilatancy to

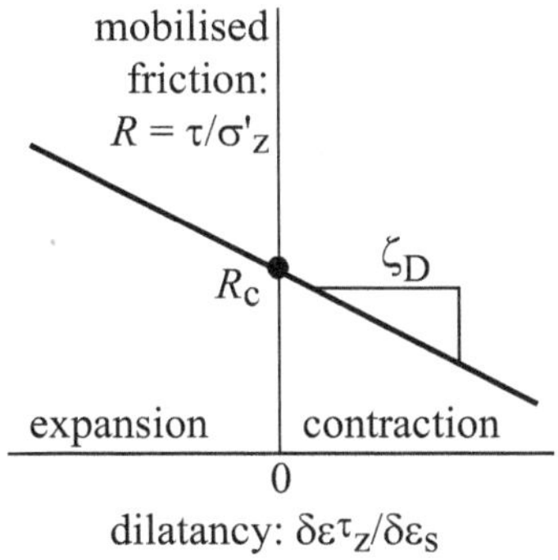


**Figure 18.** Shear box tests on sand: stress-dilatancy relationship.

the way in which granular materials are composed of rather rigid individual particles. Fig. 17c shows an interpretation of the results in Fig. 17a, b: there is a general correlation between the current slope of the volume change or vertical strain plots (Fig. 17b) and the current mobilised friction (Fig. 17a). The higher the mobilised friction, the more dramatic the rate of volume increase with continued shearing. In fact, a broad first order *stress-dilatancy* relationship could be proposed:

$$\frac{\delta z}{\delta x} = \zeta_D(R_c - R) \tag{42}$$

which tells us that the ratio of normal movement to shear movement is proportional to the difference between the current mobilised strength, $R = \tau/\sigma'_z$, and the large deformation strength $R_c$, with $\zeta_D$ being introduced as a soil parameter. For low mobilised strength, $R < R_c$, the soil compresses as it is sheared, $\delta z > 0$; for high mobilised strength, $R > R_c$, the soil expands or dilates as it is sheared, $\delta z < 0$.

We need to convert this relationship between displacements into a link between strain components for our soil element (Fig. 1), and we will write an exactly equivalent relationship (Fig. 18):

$$\frac{\delta\varepsilon_z^\tau}{\delta\varepsilon_s} = \zeta_D(R_c - R) \tag{43}$$

where the superscript $^\tau$ in $\delta\varepsilon_z^\tau$ reminds us that this is a second, distinct, route to the generation of vertical strains. The vertical strain can change either through change in vertical effective stress, $\delta\varepsilon_z^\sigma$ (15), or through shear-induced dilatancy, $\delta\varepsilon_z^\tau$, (43), or both. The total vertical strain increment is the sum of these two components:

$$\delta\varepsilon_z = \delta\varepsilon_z^{\sigma} + \delta\varepsilon_z^{\tau} \tag{44}$$

## 6 Critical States

Casagrande (1936) introduced the notion of a *critical void ratio* to which dense samples would expand and loose samples contract as they were sheared. In general it is found that the critical void ratio varies with stress level leading to a critical void ratio line is more appropriate (Wroth, 1958; Roscoe et al., 1958). The term *critical states* has been used to encapsulate all aspects of this limiting condition. *State* has the logical potential to expand to include other changing aspects of the soil that can be objectively measured: in particular, void ratio provides only a very basic indication of the current fabric of the soil.

The concept of limiting critical states has led to the development of *critical state soil mechanics* (Schofield and Wroth, 1968; Muir Wood, 1990). In its weak form (Fig. 19) it proposes a discipline for the presentation, interpretation and understanding of many aspects of the mechanical behaviour of soils—in particular bringing together treatment of compression and strength with density as a linking quantity. But density is a very incomplete indicator of the elements of state that are encapsulated within the fabric of the soil so it is not really surprising that this simple treatment should break down. In fact it is a testament to its power that this simple three part definition of state in terms of stresses and density should have demonstrated a first order usefulness.

The direct shear box being obviously inadequate for serious constitutive data gathering, Roscoe and his colleagues worked on the development of a series of simple shear apparatus—striving to capture in a single device the behaviour of the soil from the central shearing region of the shear box. These designs reached their apogee with Stroud's Simple Shear Apparatus Mark VII (Stroud, 1971). With the parallel use of radiography at Cambridge it was possible to see inside the samples and discover the extent to which the goal of internal homogeneity had indeed been attained. The soil being tested was 14/25 Leighton Buzzard sand with a rather narrow grading and a particle size $d_{50} \approx 0.8$mm. Radiographs produced by Bassett (1967) (Fig. 20) show the problem.

The radiograph detects variations in density through the absorption of X-rays so that, in these *negative* pictures, paler areas represent regions of lower density. In these areas the originally dense sample has dilated as it is sheared. With an apparatus which is 40mm high, the pattern of

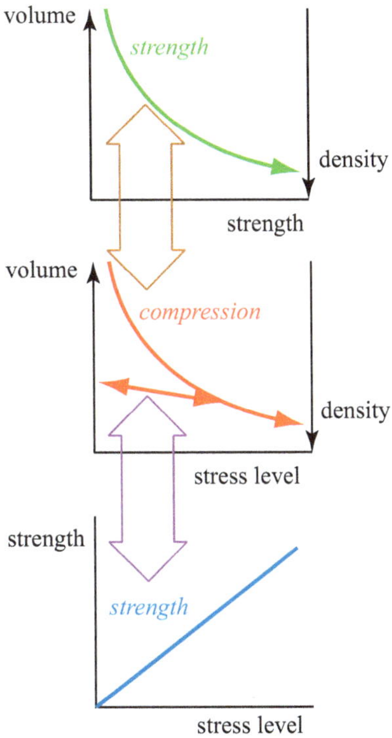


**Figure 19.** Critical state soil mechanics, *weak form*: (a) strength and density; (b) stress level and density; (c) strength.

variation of zones of dilation is evidently quite complex—and related to the deficiencies of the boundary conditions of the simple shear apparatus. As the height is reduced the pattern becomes simpler and with the lowest apparatus—20mm high—one can more or less say that there is just one dilation band, which occupies much (but not all) of the height of the sample. That this conclusion is particle size dependent is clear in Fig. 21 which shows sketches from radiographs for 14/25 Leighton Buzzard sand (Fig. 21a) and for various stages of the repeated reversal of shearing of a fine sand with particle size $d_{50} \approx 0.3$mm (Fig. 21b, c). Stroud shows that the void ratio in the dilating region appears to be tending towards a constant value as shearing proceeds—but the apparatus is only capable of imposing a shear strain of some 50%.

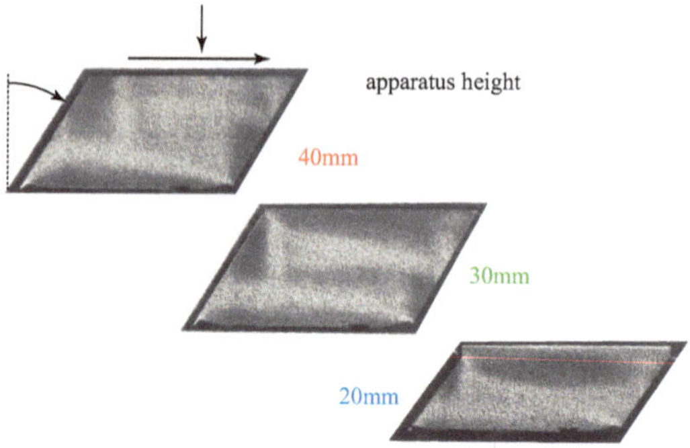


**Figure 20.** Density variations within simple shear tests on 14/25 Leighton Buzzard sand: apparatus height (a) 40mm; (b) 30mm; (c) 20mm (pale regions indicate density *decrease*)(after Bassett, 1967).

Releasing the strain constraint requires some modification such as that used by Vardoulakis (1978) (Fig. 22) using a plane strain biaxial configuration, but with the possibility of relative horizontal movement between the end platens provided by means of a lateral bearing system. There comes a point when the sample wants to behave more like a thin shear zone and an inclined band forms as a localisation within the sample. Shearing continues on this band leaving the zones that it separates more or less undeforming. Deduction of the shearing conditions in the localised region shows that, possibly, a critical state is being neared: the mobilised friction on the shear band becomes somewhat constant and the dilation, while still fluctuating, is on average around zero. The particle size of this sand is around $d_{50} \approx 0.3$mm and such localised shear bands typically have a thickness of around $10d_{50}$ so that the relative displacement indicated on the shear band corresponds to a shear strain of the order of 100% at the point when stress and density have more or less stopped changing.

The chains or columns of particles which are taking most of the load tend to be aligned with the direction of the major principal stress (Fig. 23). Such a strongly loaded pile of particles looks unstable and is indeed susceptible to buckling with the forces provided by the adjacent particles unable to do much to prevent the collapse. Buckling leads to rearrangement and irrecoverable deformation. Buckling of such a column will be exacerbated by the application of lateral forces at the ends—any attempt by the major principal stress to rotate will be very damaging. So on the one hand we can expect that testing with fixed principal axes in triaxial, biaxial, true triaxial apparatus will only explore a rather stable region of constitutive response,

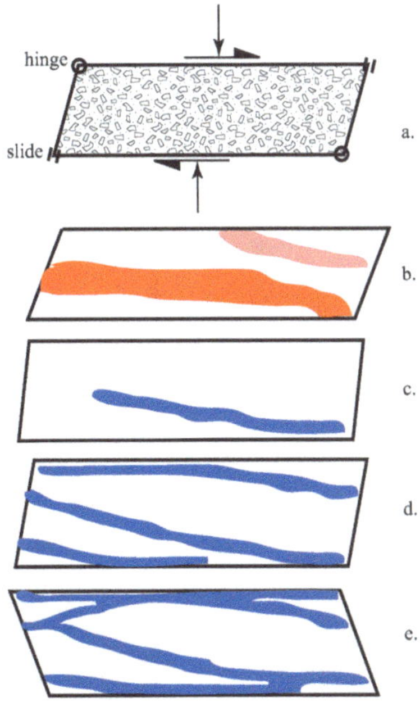


**Figure 21.** (a) Schematic diagram of simple shear apparatus (sample height 20mm); (b) regions of density decrease in simple shear test on sand with $d_{50} = 0.85$mm (original radiograph from Stroud (1971)); (c)-(e) regions of density decrease in simple shear test on sand with $d_{50} = 0.2$mm (original radiographs from Budhu (1979)) (Muir Wood, 2002).

and on the other hand if we see critical state conditions as being concerned with shearing of a band of particles then we can expect this to be a process which is particularly disruptive for the local fabric of the soil.

If stress becomes a difficult concept at the scale of individual particles, aspects of fabric of the soil are also very variable. The repeated buckling and reforming of columns of particles in the shear band which is trying to reach an asymptotic critical state leaves very large variations in void ratio (Fig. 24). There are certainly local fluctuations even if globally the density can be regarded as stationary (Kuhn, 2005) One has to conclude that heterogeneity is an inevitable property of granular materials (Fig. 25) and that tests and reporting of results only have meaning when the scale of observation is large enough relative not just to individual particles but to the patterns that form (Muir Wood, 2002; Kuhn, 2005; Muir Wood, 2006).

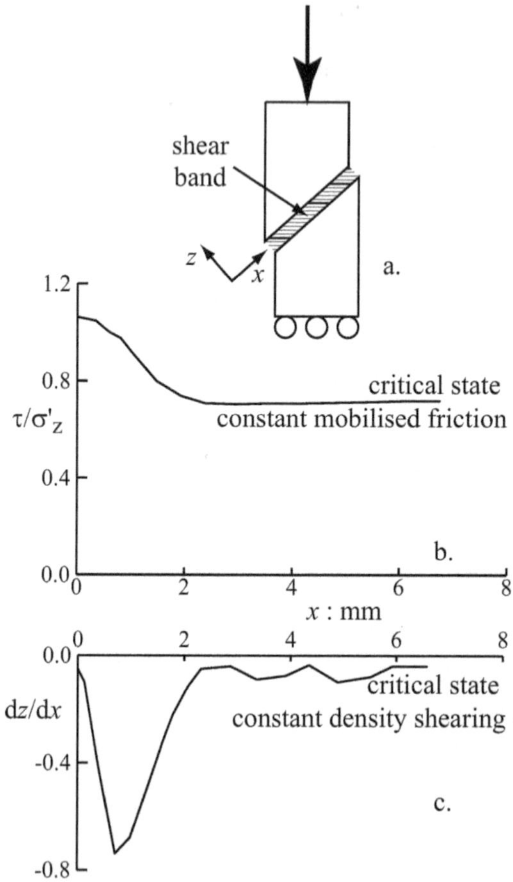


**Figure 22.** Biaxial plane strain test on Karlsruhe sand: (a) schematic illustration of mechanism to allow free formation of shear band; (b) mobilised friction and displacement on shear band; (c) dilatancy of shear band (after Vardoulakis, 1978).

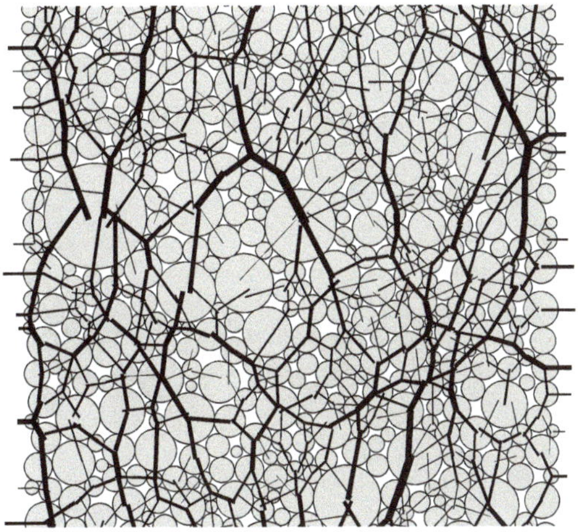

**Figure 23.** Force chains in model of particle assembly (Maeda, personal communication).

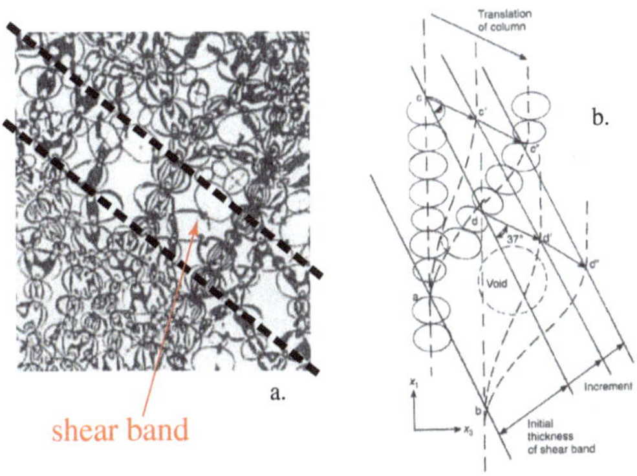


**Figure 24.** Shear band in assembly of elliptical photoelastic discs: (a) force chains and voids; (b) buckling and rotation of columns in shear band (from Oda and Kazama, 1998).

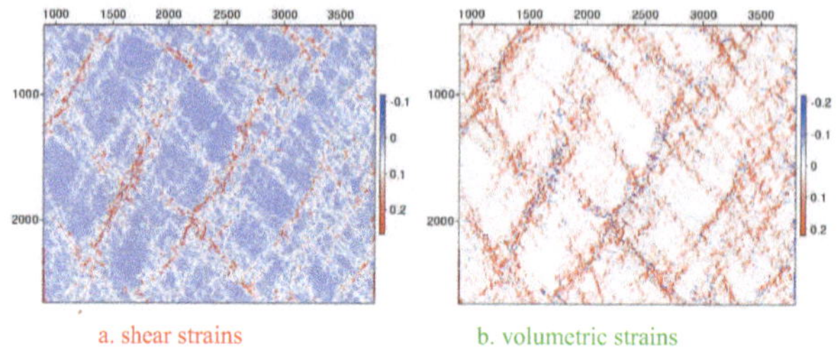


**Figure 25.** Variations of (a) shear strain and (b) volumetric strain in shearing of two dimensional rod model in $1\gamma2\varepsilon$ apparatus (Hall et al., 2010).

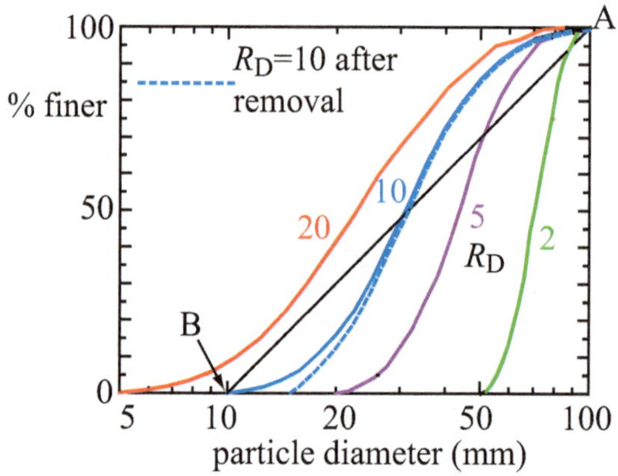


**Figure 26.** Gradings of circular discs used in DEM analyses (Muir Wood and Maeda, 2008).

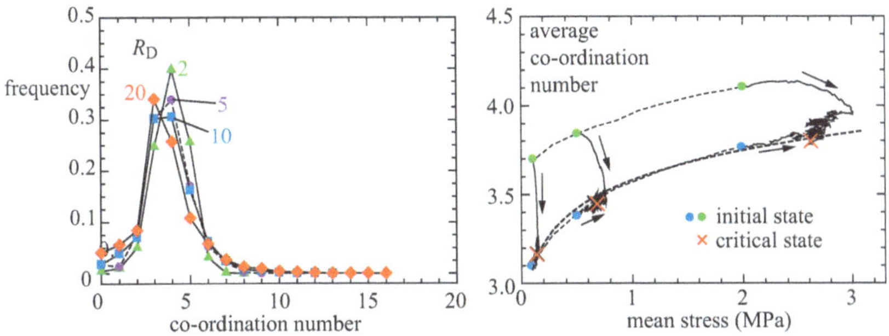


**Figure 27.** Shearing of two-dimensional assembly of discs with different gradings: (a) frequency of occurrence of coordination number; (b) variation of average coordination number with mean stress for sample with $R_D = d_{\max}/d_{\min} = 2$ (Muir Wood and Maeda, 2008).

The coordination number (the number of contacts per particle) provides some indication of the current fabric of the material. Some DEM simulations have been reported by Muir Wood and Maeda (2008) for the monotonic biaxial shearing of assemblies of discs of different sizes. The particle size distributions are shown in Fig. 26 and are characterised by $R_D = d_{\max}/d_{\min}$ where $d_{\max}$ and $d_{\min}$ are the maximum and minimum particle sizes. The spread of coordination number in samples with different gradings at the start of shearing is shown in Fig. 27a. While the infrequent occurrence of coordination numbers as high as 16 is observed for the broadest grading studied ($R_D = 20$) the dominant elements of the distribution of number of contacts are (surprisingly?) insensitive to the grading. The most frequent number of contacts is generally 4. There is a hint of a reduction towards 3 for the broadest grading and a slight increase in the number of particles with 1 contact or fewer. A partial view of the evolution of fabric during the tests is then shown in a plot of average coordination number with mean stress (Fig. 27b).

A further experimental observation relating to conditions at the critical state comes from tests performed by Gajo (quoted in Muir Wood et al., 2001) looking at the slope in effective stress terms $\delta p'/\delta q$ of small undrained unload-reload cycles performed at various stages during drained tests on Hostun sand. For an isotropic elastic soil this ratio should be zero—there is no change in mean effective stress. Departure from zero indicates some sort of elastic anisotropy. One interpretation, following Graham and Houlsby (1983), simplifies the description of transverse isotropic elasticity in a spe-

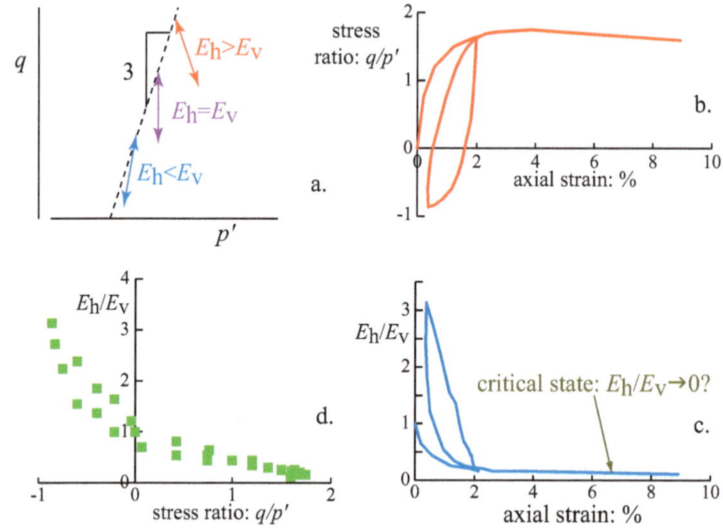


**Figure 28.** Triaxial test on Hostun sand: (a) stiffness anisotropy deduced from slope of undrained unload-reload cycle in effective stress plane; (b) stress ratio and strain; (b) deduced modulus ratio and strain; (c) modulus ratio and stress ratio (data from Gajo reported by Muir Wood et al., 2001).

cial way which implies a relationship between the slope $\delta p'/\delta q$ and the ratio of horizontal to vertical Young's moduli $E_h/E_v$ and results have been plotted using this implicit description in Fig. 28. The link between deduced modulus ratio and stress ratio in a test where the drained stress path had excursions into extension as well as compression is moderately well defined. At large strains, where one would expect critical state conditions to have been neared, the modulus ratio approaches zero, and the slope of the undrained effective stress path approaches $\delta p'/\delta q = 1/3$ implying no change in radial effective stress. Using a more complete description of transverse isotropic elasticity one would interpret this limiting stress ratio as an indication that the Poisson's ratio linking the horizontal and vertical strain increments $\nu_{vh} = 1/2$. It seems that this limiting elastic anisotropy is also attained at a modest strain level corresponding to the attainment of a steady mobilised friction. Whatever the detailed nature of this evolution of elastic anisotropy, the implied elastic-plastic coupling has an important influence on the occurrence of localisation in sands (Gajo et al., 2007) and it is probably within the localised shearing zone that critical state conditions

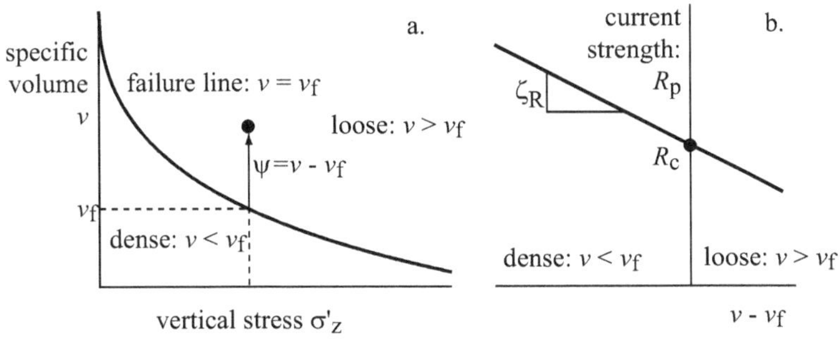


**Figure 29.** Failure line: (a) combinations of specific volume and normal stress; (b) current strength dependent on current specific volume.

will actually be reached.

## 7 State Parameter

We define the eventual relationship between specific volume and normal effective stress (Fig. 29a) by a simple relationship:

$$v_f = \breve{v} + \Delta v \exp\left[-(\sigma'_z/\sigma_{rc})^{\beta}\right] \tag{45}$$

No matter what the detailed shape of the critical state line, we can define a state parameter $\psi$ (Fig. 29)

$$\psi = v - v_f \tag{46}$$

as an indication of the distance of the current specific volume $v$ of our soil away from the critical state specific volume $v_f$ at the current stress level.

The phenomenon of dilatancy is a necessary part of the process by which a soil manages to move its density from its initial value to the value appropriate to the development of failure conditions under the current normal effective stress (Fig 30).

## 8 Strength and State Parameter

Peak strength is not a constant but is dependent on the current state of the soil. We expect that, at a given stress level, the denser the soil the higher will be the strength. Data from sands suggest that it is useful to link peak strength with state parameter $\psi$ (46) (Fig. 29) in order to introduce

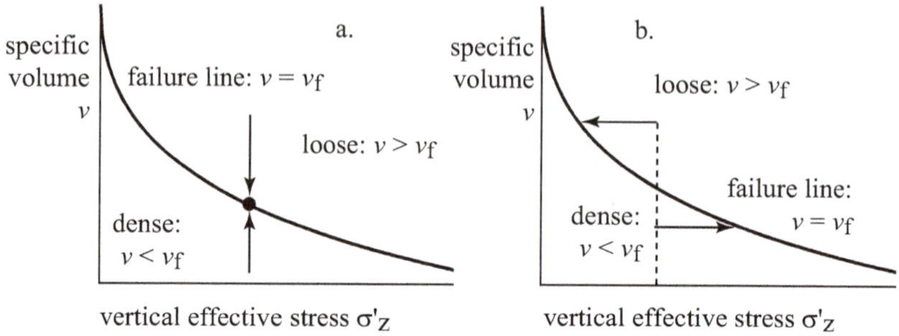


**Figure 30.** (a) Drained shearing with change of volume to failure; (b) undrained shearing with change in vertical effective stress to failure.

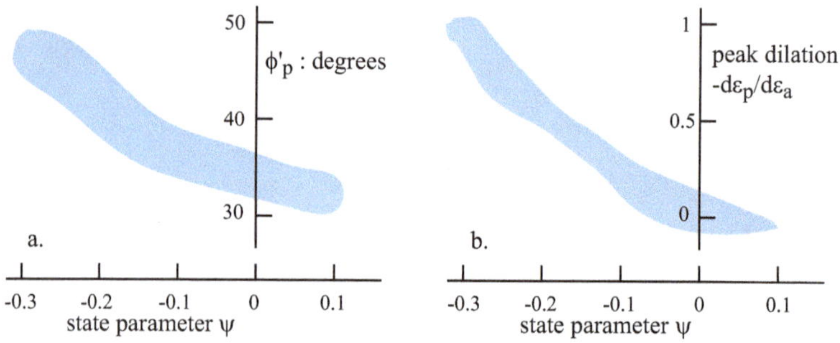


**Figure 31.** Dependence of (a) peak strength and (b) peak dilatancy of sands on state parameter (inspired by Been and Jefferies, 1986).

a link with stress level as well as with density. As the state parameter $\psi$ becomes more negative the peak strength increases (Fig. 31a)—but equally, if we shear the soil to large strains, then its strength tends to the ultimate critical state strength as the state parameter rises to zero. The peak angle of dilation also correlates well with state parameter (Fig. 31b). We will move to the implied constitutive model in two stages.

A link between peak frictional strength and state parameter implies that for soils of a given density or specific volume the strength will increase as the stress level falls and the state parameter becomes more negative (Fig. 29). If we assume a form for the critical state line (for example, (45) (though the principle of the argument is not dependent on the specific form of this relationship), then a linear relationship between peak friction $\phi'_p$ and state

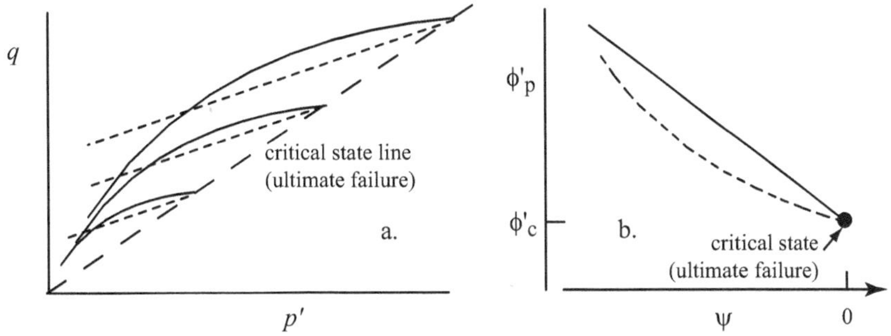


**Figure 32.** Linear dependence of strength on state parameter (solid lines) and Hvorslev strength relationship (dotted lines): (a) failure relationships in $p' : q$ plane (each line or curve describes strengths of soils with the same density or specific volume); (b) implied or assumed link between peak strength and state parameter.

parameter $\psi$ (Fig. 31, Fig. 32b: solid line) can be converted into a failure relationship in the $p' : q$ effective stress plane (Fig. 32a: solid lines). If interpreted without thought about its origin—particularly if the data do not cover a sufficiently wide range to reveal significant curvature—we might be tempted to assume that the soil is telling us that it possesses some cohesion.

We distinguish between the ultimate strength that is mobilised at large deformation and any temporary peak that may be seen on the way because the density is temporarily higher than it will be eventually. At large deformation the strength, expressed as the ratio of shear stress to normal stress $R = \tau/\sigma'_z$, has the value $R_c$. If the density of the soil is presently different from the density appropriate to ultimate failure with the present normal stress, the available strength of the soil $R_p$ is different from $R_c$. If the current specific volume is different from $v_f$, the current strength is $R_p$ (Fig. 29b):

$$R_p = R_c + \zeta_R(v_f - v) \tag{47}$$

where $v - v_f$ is the state parameter $\psi$. Thus, dense soils, with $v_f > v$, have current strength greater than the large deformation strength. Loose soils, with $v_f < v$, have current strength lower than the large deformation strength.

## 9 Elastic-Hardening Plastic Models

Perfect plasticity is rather limited in its capability for matching certain aspects of observed mechanical response of soils. Hardening plasticity opens up further modelling possibilities. Perfect plasticity enables us to reproduce the inelasticity of soil behaviour—the accumulation of irrecoverable strains. Hardening plasticity enables us in addition to describe prefailure nonlinearity. The additional feature required is that the yield function is no longer merely a function of the stresses but also introduces a hardening parameter which characterises the current size of the yield surface. An extra hardening equation is then required to define the way in which this hardening parameter changes as plastic strains occur—or in other words to define the penalty in permanent deformation of the material which is necessary in order to increase the size of the elastic region and *harden* the material.

There are four ingredients of the hardening plastic models—three of these are common to the perfectly plastic models.

1. *Elastic properties*: Whenever the stresses change elastic strains will occur. We may assume isotropic elastic behaviour for convenience but this is not essential.

$$\delta\boldsymbol{\sigma} = \boldsymbol{D}\delta\boldsymbol{\varepsilon}^e \tag{48}$$

2. *Yield criterion*: We need to define the current boundary in stress space to the region of elastic behaviour (Fig. 33). Within this region all stress changes can be applied without incurring irrecoverable deformations. The definition of the yield function allows us to answer the question: are yield and plastic deformation occurring? For a hardening model the boundary is not fixed but will depend on the history of loading of the soil.

   The yield criterion is a function of a hardening parameter $\chi$:

$$f(\boldsymbol{\sigma}, \chi) = 0 \tag{49}$$

   The current stress state cannot lie outside the current yield surface but the yield surface is able to expand in order to accommodate the imposed stress changes. The consistency condition (23), which states that the stress state must remain on the yield surface when plastic strains are being generated, now becomes:

$$f(\boldsymbol{\sigma}, \chi) = 0; \quad \delta f = \frac{\partial f}{\partial \boldsymbol{\sigma}}^T \delta\boldsymbol{\sigma} + \frac{\partial f}{\partial \chi}\delta\chi = 0 \tag{50}$$

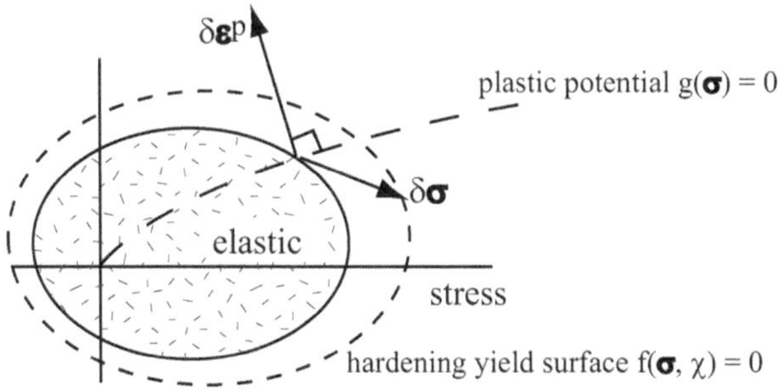


**Figure 33.** Elastic-hardening plastic model: yield surface separating elastic from plastic regions of stress space, and plastic potential for definition of plastic strain increments.

3. *Flow rule*: We require some way of describing the mechanism of plastic deformation. We can conveniently do this in just the same way as for the perfectly plastic model using a plastic potential $g(\boldsymbol{\sigma})$ to indicate the ratio of the several strain components (Fig. 33) and to show that the plastic strains are controlled by the current stresses at yield and not by the stress increment which brought the soil to yield:

$$\delta\varepsilon^p = \dot{\Lambda}\frac{\partial g}{\partial \sigma} \tag{51}$$

where $\dot{\Lambda}$ is again a scalar multiplier which we have to find. It may sometimes be convenient to assume that the functions $f$ and $g$ are the same: the material then obeys the hypothesis of associated flow (the flow is associated with the yield criterion) or normality (the strain increment vectors are normal to the yield surface at the current stress state) but this is certainly not a necessary assumption and certainly not an assumption of which soils are aware.

4. *Hardening rule*: The hardening rule links the change in size of the yield surface with the magnitude of the plastic strain and hence provides a link between $\chi$ and $\dot{\Lambda}$.

For our general hardening plasticity model we must suppose that the hardening parameter is some general function $\chi(\varepsilon^p)$ of the plastic strains. The combination of the consistency condition (50) and the flow rule (51) then gives:

$$\frac{\partial f}{\partial \sigma}^T \delta\sigma + \dot{\Lambda}\frac{\partial f}{\partial \chi}\frac{\partial \chi}{\partial \varepsilon^p}^T \frac{\partial g}{\partial \sigma} = 0 \tag{52}$$

and if we write

$$H = -\frac{\partial f}{\partial \chi}\frac{\partial \chi}{\partial \varepsilon^p}^T\frac{\partial g}{\partial \sigma} \tag{53}$$

a procedure exactly similar to that used for the perfectly plastic model can be used to generate the stiffness relationship between stress increments and total strain increments:

$$\delta\boldsymbol{\sigma} = \left[\boldsymbol{D} - \frac{\boldsymbol{D}\frac{\partial g}{\partial \boldsymbol{\sigma}}\frac{\partial f}{\partial \boldsymbol{\sigma}}^T\boldsymbol{D}}{\frac{\partial f}{\partial \boldsymbol{\sigma}}^T\boldsymbol{D}\frac{\partial g}{\partial \boldsymbol{\sigma}} + H}\right]\delta\boldsymbol{\varepsilon} = \boldsymbol{D}^{ep}\delta\boldsymbol{\varepsilon} \tag{54}$$

### 9.1 Extended Mohr-Coulomb Model

A hardening version of the Mohr-Coulomb model in which the size of the yield surface varies in some nonlinear way with the development of plastic strain is a natural extension. The hardening will be linked only with distortional strain: such a distortional hardening model is found to be quite useful for the modelling of sands where it is rearrangement of the rather hard particles that dominates the response at typical engineering stress levels and irrecoverable volumetric changes are essentially linked with this rearrangement.

1. *Elastic properties*: The elastic properties are assumed to be described by a linear isotropic elastic model which requires two stiffness properties such as shear modulus $G$ and bulk modulus $K$

$$\boldsymbol{D} = \begin{pmatrix} K & 0 \\ 0 & 3G \end{pmatrix} \tag{55}$$

In fact for many granular materials it might be reasonable to assume that the shear stiffness is not in fact constant but varies in some way with stress level—for example:

$$G \propto p'^{\frac{1}{2}} \tag{56}$$

However, cavalier introduction of nonlinear elasticity risks thermodynamic unacceptability and that if we want the shear stiffness to vary with stress level then we should really invoke an elastic strain energy function to achieve this.

2. *Yield criterion*: The yield criterion is a generalisation of the yield criterion assumed for the perfectly plastic model

$$f(\boldsymbol{\sigma}, \chi) = f(p', q, \eta_y) = q - \eta_y p' \tag{57}$$

where $\eta_y$ is a hardening parameter which indicates the current size of the yield locus (Fig. 34a). The yield locus is allowed progressively to expand until it reaches some limiting failure size.

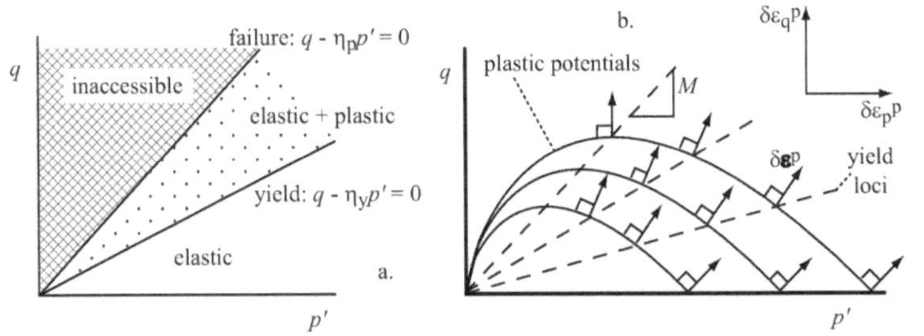


**Figure 34.** Elastic-hardening plastic Mohr-Coulomb model: (a) yield locus and failure locus separating elastic, plastic and inaccessible regions of stress plane; (b) plastic potential curves (solid lines) and yield loci (dashed lines).

3. *Flow rule*: It is not satisfactory to assume normality of plastic strain increment vectors to the current yield locus. Normality would imply

$$\frac{\delta\varepsilon_p^p}{\delta\varepsilon_q^p} = -\eta_y \tag{58}$$

and that volumetric expansion accompanies shearing at all non-zero stress ratios and that the rate of volumetric expansion—possibly characterised by an angle of dilation—increases steadily as the yield stress ratio increases. A more suitable description of the plastic volume changes can be developed from the interpretation of the results of conventional direct shear tests on sand (Fig. 17).

A stress-dilatancy equation (43) was deduced from the proposal of Taylor (1948) of a link between dilatancy and mobilised friction in a shear box test. We will now interpret this as a flow rule which controls the ratio of *plastic* strain increments:

$$\frac{\delta\varepsilon_p^p}{\delta\varepsilon_q^p} = M - \frac{q}{p'} = M - \eta_y \tag{59}$$

where $M$ is the critical state stress ratio at which constant volume shearing can occur. Evidently this flow rule is only invoked when the soil is yielding so that the stress ratio $q/p'$ is then of necessity equal to $\eta_y$.

It can be deduced that the flow rule (59) corresponds to the plastic potential function

$$g(\boldsymbol{\sigma}) = q - Mp' \ln \frac{p'_r}{p'} = 0 \tag{60}$$

where $p'_r$ is an arbitrary variable introduced in order to allow us to create a member of this general class of plastic potential curves passing through the current stress state. Then

$$\begin{pmatrix} \delta\varepsilon_p^p \\ \delta\varepsilon_q^p \end{pmatrix} = \dot{\Lambda} \begin{pmatrix} \frac{\partial g}{\partial p'} \\ \frac{\partial g}{\partial q} \end{pmatrix} = \dot{\Lambda} \begin{pmatrix} M - \eta \\ 1 \end{pmatrix} \tag{61}$$

which is consistent with (59).

These plastic potential curves are plotted in Fig. 34b together with a set of yield loci. The directions of the plastic strain increment vectors are also shown: the difference from the directions implied from normality is dramatic. Yielding at low stress ratio implies volumetric compression but the rate of volumetric compression steadily decreases as the stress ratio increases. For stress ratio $q/p' = M$ plastic deformation occurs at constant volume; for stress ratio $q/p' > M$ plastic deformation is accompanied by volumetric expansion.

4. *Hardening rule*: We will assume that the soil is a distortional hardening material so that the current size of the yield locus $\eta_y$ depends only on the plastic distortional strain $\varepsilon_q^p$. We are trying to describe using our model a mechanical behaviour in which the stiffness falls steadily as the soil is sheared towards failure. One of the simplest ways in which such a stiffness degradation can be described is using a hyperbolic relationship between stress ratio and distortional strain

$$\frac{\eta_y}{\eta_p} = \frac{\varepsilon_q^p}{\zeta_S + \varepsilon_q^p} \tag{62}$$

or incrementally

$$\delta\eta_y = \frac{(\eta_p - \eta_y)^2}{\zeta_S \eta_p} \delta\varepsilon_q^p \tag{63}$$

or

$$\begin{pmatrix} \partial\eta_y/\partial\varepsilon_p^p \\ \partial\eta_y/\partial\varepsilon_q^p \end{pmatrix} = \begin{pmatrix} 0 \\ (\eta_p - \eta_y)^2 / \zeta_S \eta_p \end{pmatrix} \tag{64}$$

where $\eta_p$ is a limiting value of stress ratio and $\zeta_S$ is a soil constant—which essentially just scales the plastic strain since (62) and (63) are actually functions of $\varepsilon_q^p/\zeta_S$.

We now have all the information that we need to produce the complete elastic-plastic stiffness relationship (54):

$$\begin{pmatrix} \delta p' \\ \delta q \end{pmatrix} =$$

$$\left[ \begin{pmatrix} K & 0 \\ 0 & 3G \end{pmatrix} - \frac{\begin{pmatrix} -K^2\eta_y(M-\eta_y) & 3GK\,(M-\eta_y) \\ -3GK\eta_y & 9G^2 \end{pmatrix}}{3G - K\eta_y\,(M-\eta_y) + p'\,(\eta_p-\eta_y)^2\,/\,(\zeta_S\eta_p)} \right] \begin{pmatrix} \delta\varepsilon_p \\ \delta\varepsilon_q \end{pmatrix} \tag{65}$$

The stiffness relation divides the stiffness into an elastic part and a plastic part. In application of the model, the elastic stiffness can be used to predict the stress change resulting from a given strain change. If this computed stress change lies outwith the current yield surface then the plastic stiffness can be applied as a corrector to bring the calculated stress state back onto the (possibly hardened) yield surface.

When the yield stress ratio reaches the asymptotic value $\eta_y = \eta_p$, this stiffness relationship becomes identical with that generated for the perfectly plastic Mohr-Coulomb model (37) if we write $\eta_y = \eta_p = M$ and $(M-\eta_y) = (M-\eta_p) = -M^*$. As then, the stiffness matrix is not symmetric because we have assumed a nonassociated flow rule: the plastic potential function (60) is quite different from the yield function (57).

The information about yielding and hardening for this model can also be used to generate the plastic compliance relationship linking plastic strain increments with stress increments.

$$\begin{pmatrix} \delta\varepsilon_p^p \\ \delta\varepsilon_q^p \end{pmatrix} = \frac{\zeta_S\eta_p}{p'\,(\eta_p-\eta_y)^2} \begin{pmatrix} -(M-\eta_y)\,\eta_y & (M-\eta_y) \\ -\eta_y & 1 \end{pmatrix} \begin{pmatrix} \delta p' \\ \delta q \end{pmatrix} \tag{66}$$

However, this relationship is not always useful because, as the stress state nears the asymptotic stress ratio $q/p' = \eta_p$, there is a region of the $(p', q)$ effective stress plane into which it is impossible for the stress increments to stray. However, as expected, we can see that as the yielding stress ratio tends to $\eta_p$ so the plastic stiffness tends to zero and the compliance tends to infinity.

The volumetric response depends on the relative values of $\eta_p$ and $M$. If $\eta_p > M$ then the model predicts compression followed by expansion (Fig. 35b). If $\eta_p = M$ then the model predicts compression reducing until a critical state of constant volume shearing is reached (Fig. 35b). (If $\eta_p < M$ the model predicts continuing volumetric compression as failure is approached: Fig. 35b.) In this simple form we cannot describe strain softening with this model. In practice, pre-peak response may be adequate since

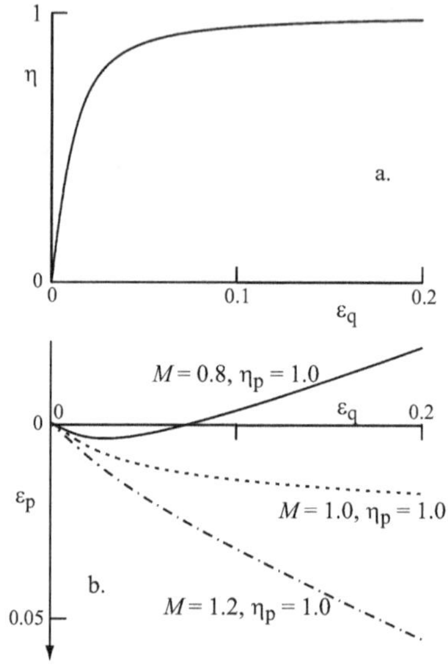


**Figure 35.** Elastic-hardening plastic Mohr-Coulomb model: triaxial compression tests with constant mean effective stress: (a) stress:strain response and (b) volumetric response for different values of $M$ and $\eta_p$ ($K = 5000$ kPa, $\nu = 0.25$, $\zeta_S = 0.005$, $p' = 100$ kPa).

working loads are not intended to produce significant amounts of failure and we are interested in pre-failure response of our geotechnical structures. If much of the soil around a structure has been brought to failure then the overall deformations of the structure are likely to be unacceptably large.

The elastic-plastic stiffness relationship (65) can be used to generate the envelope of stress responses to a rosette of applied total strain increments. These are shown in Fig. 36 for two different values of stress ratio. As the stress state approaches the peak stress ratio so the stress response envelope (which is composed of two separate elliptical sections for the elastic and elastic-plastic strain increments) becomes more and more distorted. It is evident that the stress response envelope for the elastic-perfectly plastic model (Figs. 14, 15) is a degenerate version of the response envelope for the hardening model: the elastic-plastic ellipse has collapsed to a line segment.

An example of the application of this extended Mohr-Coulomb model is provided by the calculation of the effective stress path that will be followed in an undrained test. An undrained test provides a direct deformation

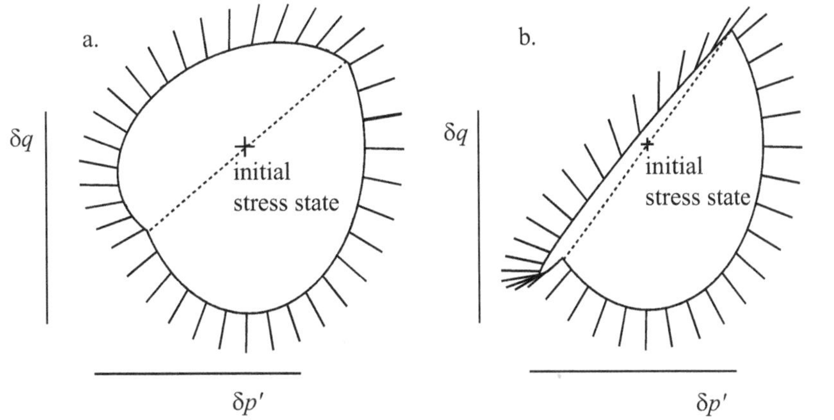


**Figure 36.** Elastic-hardening plastic Mohr-Coulomb model: stress response envelopes for (a) $\eta_y = \eta = 0.8$ and (b) $\eta_y = \eta = 1.3$ ($K = 1500$ kPa, $\nu = 0.3$, $\zeta_S = 0.01$, $M = 1.2$, $\eta_p = 1.5$, $p'_i = 100$ kPa).

constraint:

$$\delta\varepsilon_p = \delta\varepsilon_p^e + \delta\varepsilon_p^p = 0 \tag{67}$$

The sum of the elastic and plastic volumetric strain increments is zero: any tendency of the particle structure to undergo permanent rearrangement and change in volume—for example, collapse—has to be countered by a change in mean effective stress which leads to a balancing elastic volumetric expansion. (A tendency of the volume to undergo irrecoverable expansion will correspondingly be accompanied by an elastic compression.) The shape of the effective stress path can be most easily found by requiring the elastic and plastic volumetric strain increments to be equal and opposite at all stages which from (55) and (66) implies

$$\frac{\delta p'}{K} + \frac{\zeta_S \eta_p}{p'\left(\eta_p - \eta\right)^2}\left[-\left(M - \eta\right)\eta\delta p' + \left(M - \eta\right)\delta q\right] = 0 \tag{68}$$

(writing $\eta$ for $\eta_y$ since we are assuming that the soil is yielding throughout). Noting that from the definition of stress ratio $\eta$

$$\delta q = p'\delta\eta + \eta\delta p' \tag{69}$$

equation (68) can be rewritten

$$\frac{\delta p'}{K} + \frac{\zeta_S \eta_p (M - \eta) p' \delta\eta}{p' (\eta_p - \eta)^2} = 0 \tag{70}$$

and integrated to give

$$\frac{p'_i - p'}{K} = \zeta_S \eta_p \left[ \frac{(M - \eta_p)(\eta - \eta_i)}{(\eta_p - \eta)(\eta_p - \eta_i)} - \ln\left(\frac{\eta_p - \eta}{\eta_p - \eta_i}\right) \right] \tag{71}$$

where $p'_i$ and $\eta_i$ are the initial values of mean effective stress and stress ratio. For a soil which is initially isotropically compressed with $\eta_i = 0$ this can be written

$$\frac{p' - p'_i}{\zeta_S K \eta_p} = -\frac{(M/\eta_p) - 1}{(\eta_p/\eta) - 1} + \ln\left(1 - \frac{\eta}{\eta_p}\right) \tag{72}$$

and this is plotted in Fig. 37 for different values of $M$ and $\eta_p$. As $\eta \to \eta_p$ the change in mean effective stress tends to infinity but the sign of the change in mean effective stress (and hence broadly the sign of the pore pressure that develops) depends on the sign of the difference between $M$ and $\eta_p$. If $M \geq \eta_p$ then the mean stress falls steadily (and pore pressure is expected to build up). If $M < \eta_p$ then the mean stress first decreases (pore pressure build up) and then increases (pore pressure decrease). The model suggests that this increase in mean stress continues indefinitely but on the one hand the model is defective in suggesting that shearing accompanied by dilation can continue to large strains and on the other hand an undrained test would reach a physical conclusion when the pore pressure reaches a negative value of about -100 kPa and cavitation of the pore water occurs. (There is similarly a physical limit for low values of $\eta_p$ in that an effective stress of zero is reached as the pore pressure continues to increase.) These limitations simply indicate some of the deficiencies of this simple model which would need to be rectified if it were to be used for analyses in which accurate representation of such response were reckoned to be essential.

## 9.2 Mohr Coulomb Model With Strength Dependent on State Parameter

A way in which both hardening and softening can be rather simply—and elegantly—combined in a single model, which is again clearly a development from the Mohr-Coulomb family, has been described by Muir Wood et al. (1994) and by Gajo and Muir Wood (1999). This is here developed for the simple shear element of Fig 1.

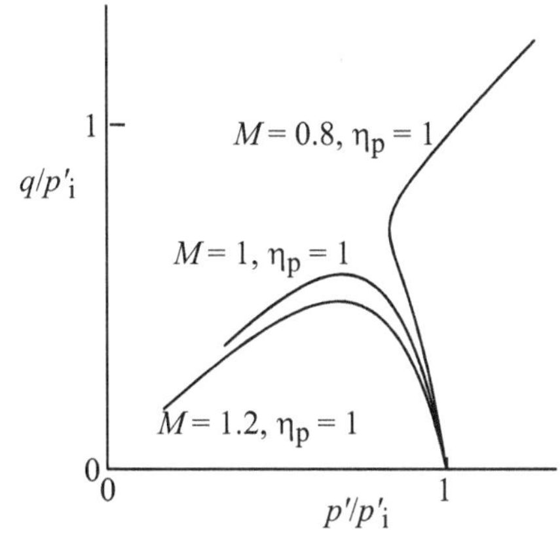


**Figure 37.** Elastic-hardening plastic Mohr-Coulomb model: undrained effective stress paths ($K = 2000$ kPa, $a = 0.02$, $p'_i = 200$ kPa).

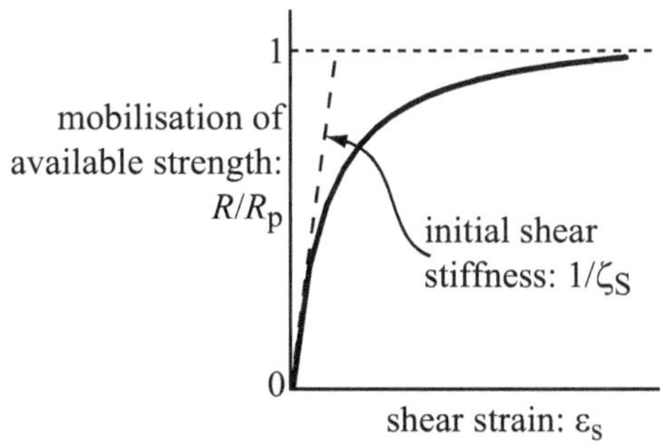


**Figure 38.** Hyperbolic mobilisation of strength.

The ratio of shear stress to vertical effective stress, $R = \tau/\sigma'_z$, is a measure of the currently mobilised friction in our soil element: it is directly equivalent to the ratio $Q/P$ for the shear box. The way in which strength is gradually mobilised as shear deformation increases is nonlinear (Fig. 17). Let us assume a simple hyperbolic relationship between mobilised friction $R$ and shear strain $\varepsilon_s$ which heads asymptotically towards the failure condition at large strains (Fig. 38):

$$\frac{R}{R_p} = \frac{\varepsilon_s}{\zeta_S + \varepsilon_s} \tag{73}$$

where $\zeta_S$ is a soil parameter which controls the initial shear stiffness of the element. Incrementally, this can be written:

$$\delta R = \frac{1}{R_p}\left[(R_p - R)^2 \frac{\delta\varepsilon_s}{\zeta_S} + R\delta R_p\right] \tag{74}$$

to remind us that the current strength $R_p$ is not constant but depends on current density from (47).

With a little manipulation, the various relationships can be written in incremental form and combined to deduce the stress increments $(\delta\sigma_z', \delta\tau)$ that result from the application of any strain increments $(\delta\varepsilon_z, \delta\varepsilon_s)$:

$$\begin{aligned} \delta\sigma_z' &= E_o[\delta\varepsilon_z - \Psi_3\delta\varepsilon_s] \\ \delta\tau &= E_o[-\Psi_1\delta\varepsilon_z + (\Psi_2 + \Psi_1\Psi_3)\delta\varepsilon_s] \end{aligned} \tag{75}$$

where:

$$\begin{aligned} \Psi_1 &= \frac{\zeta_R R}{R_p}\left[\beta(v_f - \breve{v})\left(\frac{\sigma_z'}{\sigma_{ref}}\right)^{\beta} - \frac{v\sigma_z'}{E_o}\right] - R \\ \Psi_2 &= \left[\frac{(R_p - R)^2}{\zeta_S R_p} + \frac{\zeta_R v R \Psi_3}{R_p}\right]\frac{\sigma_z'}{E_o} \\ \Psi_3 &= \zeta_D(R_c - R) \end{aligned} \tag{76}$$

The governing incremental equations (75) can be integrated numerically to generate the response to any particular loading or deformation history. Two obvious extremes to explore are the behaviour at constant vertical stress—the drained response, $\delta\sigma_z' = 0$—and the behaviour when vertical deformation is prevented—the undrained response, $\delta\varepsilon_z = 0$.

With constant vertical stress, $\delta\varepsilon_z = \Psi_3\delta\varepsilon_s$, and the vertical, volumetric strain is solely the result of dilatancy. The soil contracts if it is initially looser than the failure line $v > v_f$ or dilates if it is initially denser than the failure line $v < v_f$ (Fig. 30a). The shear stress:strain response is given by (Fig. 39):

$$\delta\tau = \Psi_2 E_o \delta\varepsilon_s \tag{77}$$

Since this will in general imply volume change, we can think of this as a drained response of the soil. We observe that, no matter what the initial density or specific volume, the stress:strain response seeks out the large deformation strength $R_c$, and the density changes, up or down, as required in order that the ultimate state of the soil should lie on the line of ultimate failure states (45).

Shearing at constant height, on the other hand, imposes a sort of conjugate mode of deformation on the soil. Whereas with constant vertical effective stress the height of the soil element will in general change (Fig. 39),

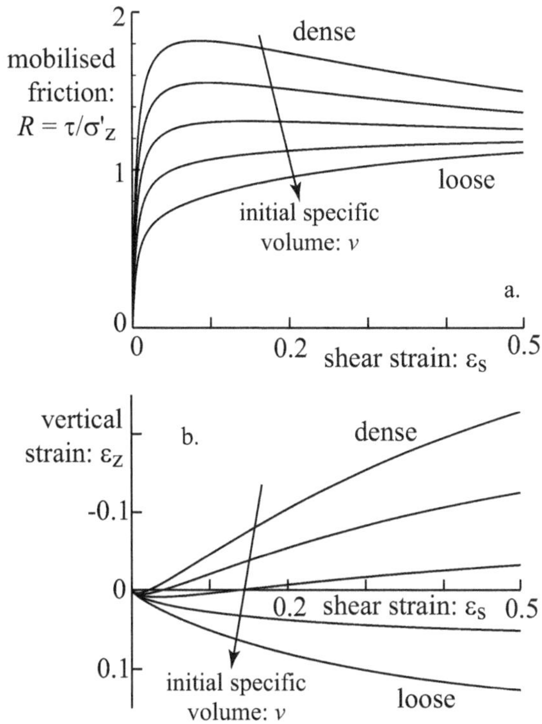


**Figure 39.** Drained shearing with different initial densities: (a) stress:strain response; (b) vertical strain.

with constant height the vertical effective stress will in general change. We can understand this by studying the component parts of (44). Our imposed constraint controls the sum of the two components not the individual components of vertical strain. The shearing produces dilatant vertical strains $\delta\varepsilon_z^\tau$ and the vertical effective stress has to change to provide vertical strains $\delta\varepsilon_z^\sigma = \delta\sigma_z'/E_o = -\delta\varepsilon_z^\tau$ to balance these, which would otherwise cause a change in height (volume) of the soil. Thus, if the soil is initially dense and is trying to expand, $\delta\varepsilon_z^\tau < 0$, the vertical effective stress has to increase to provide compensating compression, $\delta\varepsilon_z^\sigma > 0$ (Fig. 30b). Similarly, if the soil is initially loose and is trying to contract, $\delta\varepsilon_z^\tau > 0$, the vertical effective stress has to decrease to provide compensating expansion, $\delta\varepsilon_z^\sigma < 0$ (Fig. 30b). In either case, the total vertical strain increment is zero. The

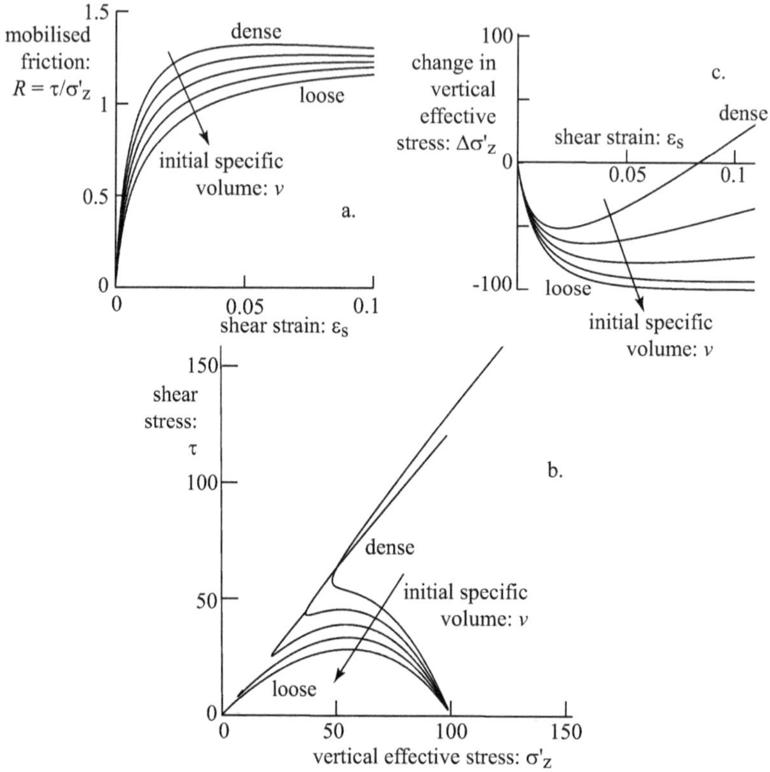


**Figure 40.** Undrained shearing with different initial densities: (a) stress:strain response; (b) effective stress path; (c) change in vertical effective stress.

shear stress:strain response is given by (Fig. 40a):

$$\delta\tau = (\Psi_2 + \Psi_1\Psi_3)E_o\delta\varepsilon_s \tag{78}$$

and the effective stress path is given by (Fig. 40b):

$$\frac{\delta\sigma_z'}{\delta\tau} = -\frac{\Psi_3}{\Psi_2 + \Psi_1\Psi_3} \tag{79}$$

This constant volume shearing is equivalent to the undrained response of the soil. The change in vertical effective stress could occur by applying a constant external vertical total stress $\delta\sigma_z = 0$, and then preventing any water from escaping from the saturated soil sample and allowing pore pressures to develop to produce the required changes in vertical effective stress: $\delta\sigma_z' = \delta\sigma_z - \delta u = -\delta u$. Alternatively, we could permit full drainage from the soil so that no pore pressures could develop, and then change the applied vertical stress to maintain the height of the sample constant. The changes in pore pressure that we discover with the first technique will exactly mirror the changes in external vertical stress that we see with the second (Fig. 40c). We can observe again that, no matter what the initial density or specific volume, the stress:strain response seeks out the large deformation strength $R_c$, and the vertical effective stress changes, up or down as required, to force the ultimate state of the soil to lie on the line of ultimate failure states (45).

This model now homes in on a critical state condition, heading always towards the current peak strength following the hyperbolic hardening law but this peak strength is itself changing as the soil compresses or dilates with shearing. Thus, even though the hardening law appears to be a simple hyperbolic monotonically increasing function of strain, nevertheless the stress:strain response is able to introduce strain softening and the accompanying smooth transition between compression and dilation. The peak strength is thus a moving target which can only be attained at infinite distortional strain (it remains the asymptote of the hardening law) by which time it is identical with the critical state strength.

The behaviour depends strongly on the initial value of state parameter: a positive initial state parameter indicates an initially loose material which tends to compress as it is sheared and shows little in the way of a peak strength; a negative initial state parameter indicates an initially dense material which dilates as soon as the critical state stress ratio is exceeded on the initial loading and then shows a peak with subsequent strain softening.

However, the model is rather powerful. It shows how the shearing response depends on the initial density of the soil. It shows how the volume changes accompanying shearing inexorably move the soil from its initial

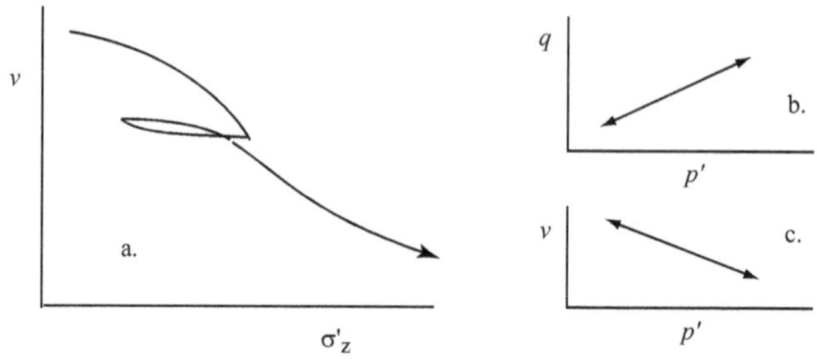


**Figure 41.** (a) Large irrecoverable volume changes in oedometer test on clay subjected to (b) typical compression stress path giving rise to (c) elastic volume changes in Mohr-Coulomb model.

condition to the failure line in drained tests, and the vertical effective stress has to change to allow the soil to move to the failure line in undrained tests.

For any complex model, there is a simpler model lurking inside which we can find by eliminating some of the special effects. Thus, we could eliminate the nonlinearity of normal stiffness by setting $\zeta_S = 0$; we could eliminate the dilatancy (the change in volume induced by shearing) by setting $\zeta_D = 0$; we could eliminate the variation of strength with density by setting $\zeta_R = 0$.

## 10 Cam Clay

Cam clay was the first *hardening* plastic model to have become generally adopted for soils. It has formed a basis for much subsequent development of soil models. Originally developed in the early 1960s, models of the Cam clay form have been widely and successfully used for analysis of problems involving the loading of soft clays. It has been less successful in describing the behaviour of sands for which models which make use of distortional hardening and nonassociated flow (§9.1) have generally been reckoned to be more satisfactory. A detailed description of the Cam clay model and of the behaviour of soils—especially clays—seen against the patterns of behaviour that the Cam clay model reveals is given by Muir Wood (1990); here we will present the model within the general framework of elastic-hardening plastic models that has been developed in section §9.

For clays, an important aspect of the observed mechanical behaviour is the large change in volume that occurs during compression (Fig. 41) when the stresses acting on a sample of soil are all increased in proportion—isotropic compression and one-dimensional compression are obvious

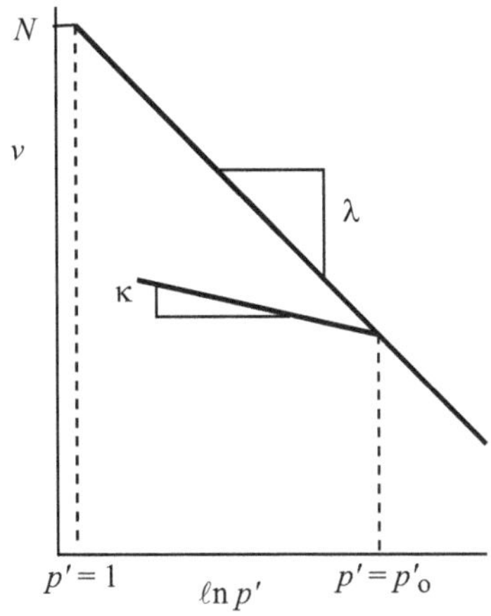


**Figure 42.** Cam clay: linear normal compression and unloading-reloading lines in semilogarithmic compression plane.

examples. Clearly if the model is to be used to reproduce the loading of soft clays then this volumetric response must be included. However, applying such proportional stress paths to any of the Mohr-Coulomb models, as presently described, will produce solely elastic response as shown in Fig. 41c (except insofar as the vertical stiffness produces vertical strains $\delta\varepsilon_z^\sigma$ whose magnitude is influenced by the overconsolidation ratio of the soil (§2)). The irrecoverability of the volumetric response suggests that a different mechanism of plastic deformation will be required. This could be achieved by adding extra yield mechanisms to the Mohr-Coulomb models; the Cam clay model that will be described here provides an elegant alternative route using a single yield surface.

As before, for simplicity we will develop the model in terms of the triaxial strain increment and stress variables and work through the several ingredients of the model in turn.

1. *Elastic properties*: We will assume that the elastic behaviour of the soil is isotropic and defined by two elastic parameters, bulk modulus $K$ and shear modulus $G$.

   Results of oedometer tests are typically presented in semilogarithmic plots because it is found that the relationships between stress and volume change then become somewhat more linear—both during loading and during unloading. Looking at the typical loading and unloading response in an oedometer (Fig. 41a) we can easily see the division of the volume changes into elastic and plastic parts. It is logical then to use the average slope $\kappa$ of an unload-reload line to characterise the elastic volumetric response (Fig. 42) and to assume that $\kappa$ is a soil constant:

$$v = v_\kappa - \kappa \ln p' \tag{80}$$

where $v_\kappa$ is a reference value of specific volume on a particular unloading-reloading relationship. We can convert this to an incremental relationship

$$\delta\varepsilon_p^e = -\frac{\delta v}{v} = \frac{\kappa}{v}\frac{\delta p'}{p'} \tag{81}$$

which implies that the bulk modulus $K$ is not constant but is dependent on stress level (and on current packing)

$$K = \frac{\delta p'}{\delta\varepsilon_p^e} = \frac{vp'}{\kappa} \tag{82}$$

Having chosen one elastic property we require one more (the elastic properties of our material may be nonlinear but the material is still assumed to be isotropic). We may often find it convenient to choose a constant shear modulus $G$ because we will see this directly from the initial behaviour in any compression test.

$$\delta\varepsilon_q^e = \frac{\delta q}{3G} \tag{83}$$

An alternative will be to choose a constant value of Poisson's ratio $\nu$, thus forcing a constant ratio of shear modulus and bulk modulus.

$$G = K\frac{3(1-2\nu)}{2(1+\nu)} \tag{84}$$

Clearly if $G$ is constant then the variation of bulk modulus $K$ with stress will lead to a varying $\nu$ (and as the effective mean stress and hence the bulk modulus fall towards zero the value of Poisson's ratio will tend towards $-1$). However, if Poisson's ratio $\nu$ is assumed to be constant then $G$ changes together with bulk modulus and there are thermodynamic problems if we make both $G$ and $K$ functions of $p'$—it becomes possible to generate or dissipate energy on supposedly elastic cycles of stress change (Zytynski *et al.*, 1978). (Zytynski et al., 1978). It is not possible to define an elastic potential which implies

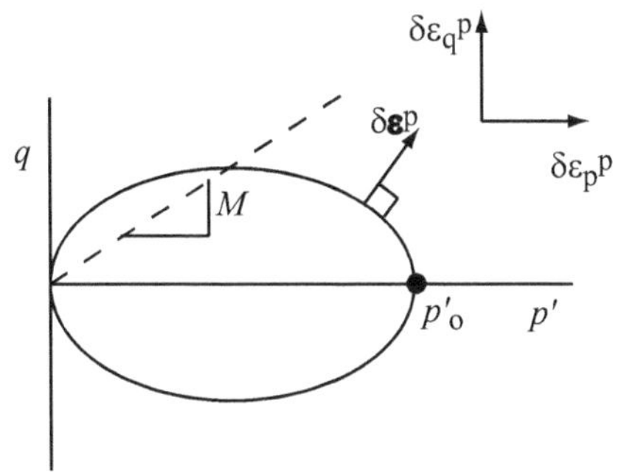


**Figure 43.** Elliptical yield locus for Cam clay model.

a constant Poisson's ratio if the bulk modulus is a function of mean stress alone.

With certain reservations then, we have the elastic stiffness and compliance relationships:

$$\begin{pmatrix} \delta p' \\ \delta q \end{pmatrix} = \begin{pmatrix} vp'/\kappa & 0 \\ 0 & 3G \end{pmatrix} \begin{pmatrix} \delta\varepsilon_p^e \\ \delta\varepsilon_q^e \end{pmatrix} \tag{85}$$

$$\begin{pmatrix} \delta\varepsilon_p^e \\ \delta\varepsilon_q^e \end{pmatrix} = \begin{pmatrix} \kappa/vp' & 0 \\ 0 & 1/3G \end{pmatrix} \begin{pmatrix} \delta p' \\ \delta q \end{pmatrix} \tag{86}$$

2. *Yield criterion*: In the triaxial stress plane $(p', q)$ it is assumed that the yield locus has an elliptical shape passing through the origin of the stress plane (Fig. 43). This introduces two variables: the aspect ratio of the ellipse $M$ which controls the shape of the ellipse, the ratio of the vertical $(q)$ axis to the horizontal $(p')$ axis; and the size of the ellipse $p'_o$ which is the hardening parameter $\chi$ for the Cam clay model. The equation of the ellipse can be presented in various different ways. To fit in with the general presentation of hardening plastic models we can write:

$$f(\boldsymbol{\sigma}, p'_o) = \frac{q^2}{M^2} - p'(p'_o - p') \tag{87}$$

so that, as usual, $f < 0$ indicates elastic behaviour, $f = 0$ indicates that yielding is occurring and $f > 0$ is not permitted.

It is often convenient to work in terms of mean stress $p'$ and stress ratio $\eta$. As the size of the yield locus changes the shape remains the same and the locus grows from the origin. Along any line at constant stress ratio $\eta = q/p'$ the angle of intersection with any yield locus is always the same.

3. *Flow rule*: It is assumed that Cam clay obeys the hypothesis of associated flow (normality) so that the plastic strain increment vector is assumed to be normal to the yield surface at the current stress state (Fig. 43). The plastic potential function then has the same form as the yield criterion:

$$g(\boldsymbol{\sigma}) = f(\boldsymbol{\sigma}, p'_o) = \frac{q^2}{M^2} - p'(p'_o - p') = 0 \tag{88}$$

The plastic strain increments are given by

$$\begin{pmatrix} \delta\varepsilon_p^p \\ \delta\varepsilon_q^p \end{pmatrix} = \dot{\Lambda} \begin{pmatrix} \frac{\partial g}{\partial p'} \\ \frac{\partial g}{\partial q} \end{pmatrix} = \dot{\Lambda} \begin{pmatrix} 2p' - p'_o \\ \frac{2q}{M^2} \end{pmatrix} \tag{89}$$

The plastic deformation mechanism can also be written:

$$\frac{\delta\varepsilon_p^p}{\delta\varepsilon_q^p} = \frac{M^2 - \eta^2}{2\eta} \tag{90}$$

showing that the mechanism of plastic deformation depends only on the stress ratio at which yielding is occurring and changes continuously as the stress ratio changes.

Several particular cases are of interest:

- for $\eta = 0$, $\delta\varepsilon_p^p/\delta\varepsilon_q^p = \infty$ which implies compression without distortion and this is appropriate for isotropic consolidation without application of distortional stresses;
- for $\eta = M$, $\delta\varepsilon_p^p/\delta\varepsilon_q^p = 0$ which implies distortion without compression—this is the critical state condition;
- yielding with low values of stress ratio $\eta < M$ gives $\delta\varepsilon_p^p/\delta\varepsilon_q^p > 0$ which implies compression plus distortion; and
- yielding with high values of stress ratio $\eta > M$ gives $\delta\varepsilon_p^p/\delta\varepsilon_q^p < 0$ which implies expansion plus distortion.

4. *Hardening rule*: The hardening rule describes the dependence of the size of the yield locus $p'_o$ on the plastic strain. Cam clay is a volumetric hardening model in which it is assumed that the size of the yield locus depends only on the plastic volumetric strain through an expression

$$\begin{pmatrix} \partial p'_o/\partial\varepsilon_p^p \\ \partial p'_o/\partial\varepsilon_q^p \end{pmatrix} = \begin{pmatrix} vp'_o/(\lambda - \kappa) \\ 0 \end{pmatrix} \tag{91}$$

This hardening rule introduces one additional soil parameter $\lambda$.

Now that all the ingredients of the model are in place the overall plastic compliance relationship can be deduced:

$$\begin{pmatrix} \delta\varepsilon_p^p \\ \delta\varepsilon_q^p \end{pmatrix} = \frac{\lambda-\kappa}{vp'\left(M^2+\eta^2\right)} \begin{pmatrix} M^2-\eta^2 & 2\eta \\ 2\eta & \frac{4\eta}{M^2-\eta^2} \end{pmatrix} \begin{pmatrix} \delta p' \\ \delta q \end{pmatrix} \tag{92}$$

and the full stiffness matrix linking the stress increments with the total strain increments can be obtained by substitution in (53) and (54). The hardening quantity $H$ is given by

$$H = -\frac{\partial f}{\partial p'_o}\frac{\partial p'_o}{\partial \varepsilon_p^p}\frac{\partial g}{\partial p'} = -\left(-p'\right)\left(\frac{vp'_o}{\lambda-\kappa}\right)\left(2p'-p'_o\right) \tag{93}$$

and the full elastic-plastic stiffness relationship is given by

$$\begin{pmatrix} \delta p' \\ \delta q \end{pmatrix} = \left[ \begin{pmatrix} K & 0 \\ 0 & 3G \end{pmatrix} - \frac{\begin{pmatrix} K^2\left(2p'-p'_o\right)^2 & \frac{6GKq\left(2p'-p'_o\right)}{M^2} \\ \frac{6GKq\left(2p'-p'_o\right)}{M^2} & \frac{36G^2q^2}{M^4} \end{pmatrix}}{\left[K\left(2p'-p'_o\right)^2 + \frac{12Gq^2}{M^4} + \frac{vp'p'_o\left(2p'-p'_o\right)}{\lambda-\kappa}\right]} \right] \begin{pmatrix} \delta\varepsilon_p \\ \delta\varepsilon_q \end{pmatrix} \tag{94}$$

where $K = vp'/\kappa$.

Whether the compliance form (92) or the stiffness form (94) is used it is evident that the controlling matrix is symmetric: this results from the assumption of associated flow in which the vectors of plastic strain increment are assumed to be normal to the yield locus at the current effective stress causing yield.

Study of (92) shows that the magnitude of the plastic strains is controlled largely by $\lambda-\kappa$. It will be the difference between these two soil parameters (rather than the absolute value of either of them) that will have to be varied in order to match available experimental data. Some qualitative statements about the nature of the stress:strain response can be made.

*What happens as $\eta \rightarrow M$?* The top line of the compliance matrix (92) shows that as the stress ratio approaches the value $M$ so the plastic volumetric strains become smaller and smaller. Since the plastic hardening depends only on the plastic volumetric strain it can be deduced correspondingly that the change in $p'_o$ in any stress increment has to tend to zero as the stress ratio approaches $M$. The bottom line of the compliance matrix shows that the shear compliance tends to infinity, or in other words the shear stiffness tends to zero. In fact, an asymptotic perfectly plastic condition is predicted in which distortional strains continue but with no further changes in size of yield locus, stresses or volumetric strains: a critical state.

$$\eta \to M: \quad \delta\varepsilon_p^p \to 0; \quad \delta p'_o \to 0; \quad \frac{\delta\varepsilon_q^p}{\delta q} \to \infty \tag{95}$$

The value of the soil parameter $M$ can therefore be related to the ultimate value of the angle of shearing resistance for the soil $\phi_c$ in triaxial compression:

$$M = \frac{6\sin\phi_c}{3 - \sin\phi_c} \tag{96}$$

The Cam clay model responds stably to yielding with stress ratio $\eta < M$ and under such conditions it does not matter whether the problem is driven by stress changes or by strain changes: it is often conceptually easier to think of the response to stress changes because the model has been described in terms of a yield locus in a stress plane. In a typical compression test the deviator stress rises steadily towards the ultimate value (low overconsolidation ratio in Fig. 44). A typical stress response envelope for this regime is shown in Fig. 45a. The two elliptical segments are now tangential to each other for stress increments which imply neutral loading with the stress increment tangential to the yield locus. This is a consequence of the assumption of associated flow.

However, if the soil is yielding with stress ratio $\eta > M$ then study of (92) shows that the distortional compliance is negative and continuing shearing with $\delta\varepsilon_q^p > 0$ implies $\delta\varepsilon_p^p < 0$, $\delta p'_o < 0$ and $\delta q < 0$ which implies strain softening (high overconsolidation ratio in Fig. 44). The stress ratio $\eta = M$ is still an ultimate asymptote but the soil now approaches this stress ratio from above rather than from below. Consideration of the equation for the yield locus (87) shows that yielding with stress ratios greater than $M$ is only possible for values of $p'/p'_o$ less than 1/2—overconsolidation ratios greater than 2. Such behaviour is characteristic of dense or heavily preloaded materials which are so tightly packed that they have to expand in order that the particles should be able to move relative to each other and allow the material to distort.

This is a real phenomenon, but as noted in §9.1, it can lead to numerical problems because of the uncertainty: does a reduction in shear stress imply an elastic unloading or a continuing plastic strain softening? The stress response envelope (Fig. 45b) illustrates this ambiguity: it is folded over on itself. All strain increments are possible and each strain increment implies an unambiguous stress increment. However, certain stress increments—

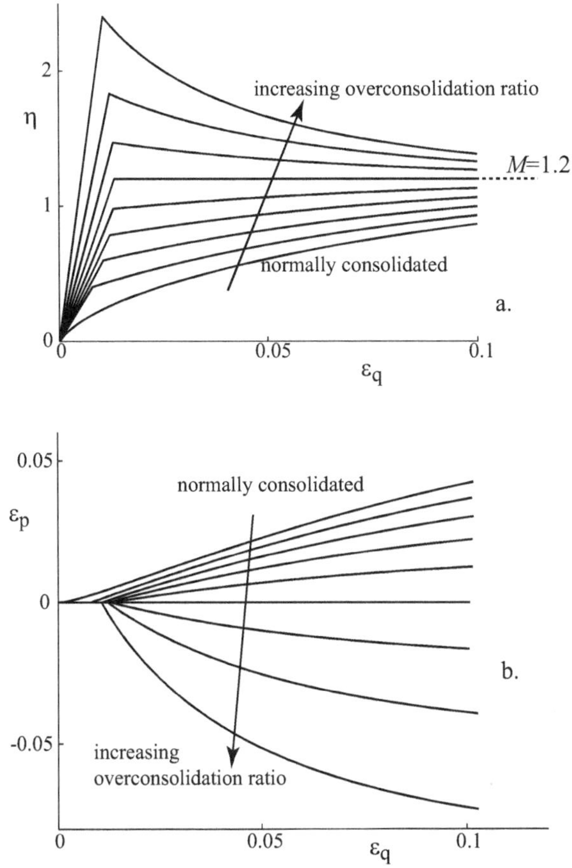


**Figure 44.** Cam clay: (a) stress:strain and (b) volumetric strain response in drained triaxial compression tests with constant mean stress ($\delta p' = 0$) ($\kappa = 0.05$, $G = 1500$ kPa, $\lambda = 0.25$, $M = 1.2$) (overconsolidation ratios $p'_o/p'_i$ in range 1-5, $p'_o = 100$ kPa).

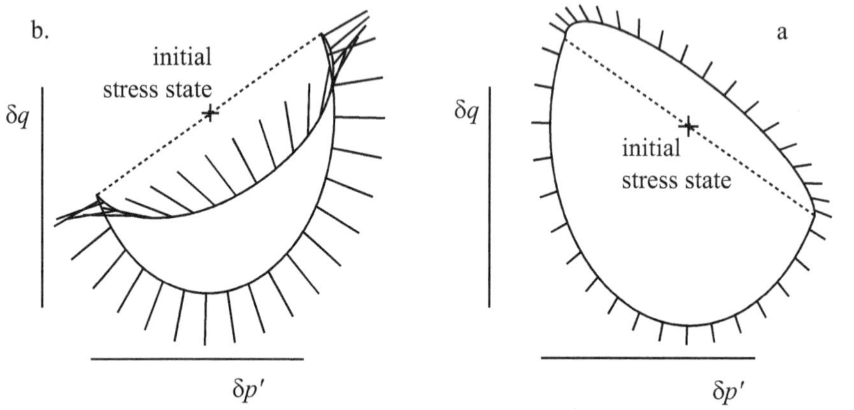


**Figure 45.** Cam clay: stress response envelopes (a) $\eta < M$, $p'_i/p'_o = 0.75$; (b) $\eta > M$, $p'_i/p'_o = 0.25$ ($\kappa = 0.1$, $\nu = 0.3$, $\lambda = 0.25$, $M = 1.2$) ($v_i = 2.5$, $p'_o = 200$ kPa).

those which attempt to escape from the *current* yield locus—are not possible and the section of the stress plane lying outside the yield locus in the region for which $\eta > M$ is inaccessible (Fig. 46). Stress changes which move inside the current yield locus can be associated with either purely elastic or with elastic plus plastic strains. In analysis of such situations the soil response has to be driven by strain increments—which will make it quite clear whether unloading or plastic softening is implied—rather than stress increments precisely because many stress increments will in fact be physically either impossible or ambiguous. Numerically and physically such behaviour leads to the occurrence of localisation: as the material softens it becomes weaker and natural inhomogeneities lead to strain concentrations and formation of ruptures or shear bands through the material.

The Cam clay model has five material properties. There are two elastic properties $\kappa$ (linked with swelling index $C_s$) and $G$ or $\nu$. There are two plastic properties $M$ and $\lambda$ which can be linked with angle of shearing resistance in triaxial compression $\phi_c$ (96) and compression index $C_c$.

The final soil parameter is a reference for volume in order that volumetric strains can be calculated. We have defined the equation of the *isotropic* normal compression line using a reference parameter $N$ to indicate the specific volume for unit mean stress (Fig. 42). However, results of predictions made using Cam clay are not usually very sensitive to plausible variations of $N$—so the reference volume can just as well be taken from one-dimensional

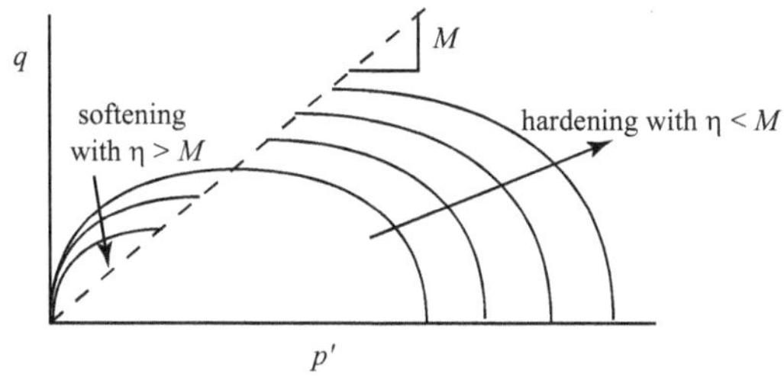


**Figure 46.** Cam clay: hardening of yield locus with $\eta < M$, softening of yield locus with $\eta > M$.

compression data.

The isotropic normal compression line defines the values of specific volume when the stress state is always at the tip of the yield locus, $p' = p'_o$ and the soil is always yielding. More generally, for stress states inside the yield locus there is some implied elastic expansion from the normal compression line (Fig. 42) and the specific volume is given by

$$v = N - \lambda \ln p'_o + \kappa \ln \left( \frac{p'_o}{p'} \right) \tag{97}$$

or, if the soil is yielding with stress ratio $\eta$

$$v = N - \lambda \ln p'_o + \kappa \ln \left( \frac{M^2 + \eta^2}{M^2} \right) \tag{98}$$

and for isotropic compression $\eta = 0$, $p' = p'_o$ and $v = N - \lambda \ln p'$. The value of $N$ depends on the units used for measurement of stress. Users of Cam clay need to be vigilant. Here we will always take the value of $N$ to correspond to a mean stress $p' = 1$ kPa.

The effective stress path followed in an undrained test can be calculated in exactly the same way as for the Mohr-Coulomb models by requiring the elastic and plastic volumetric strain increments to be always equal and opposite. From the expressions for elastic and plastic volumetric strain increments and the definitions of $p'_o$ ((87)) and of stress ratio $\eta = q/p'$:

$$\kappa \delta p' + \frac{\lambda - \kappa}{M^2 + \eta^2} \left[ \left( M^2 + \eta^2 \right) \delta p' + 2\eta p' \delta \eta \right] = 0 \tag{99}$$

Integrating this expression, from an initial yielding stress state $p'_i$ and $\eta_i$, and substituting

$$\Lambda = \frac{\lambda - \kappa}{\lambda} \tag{100}$$

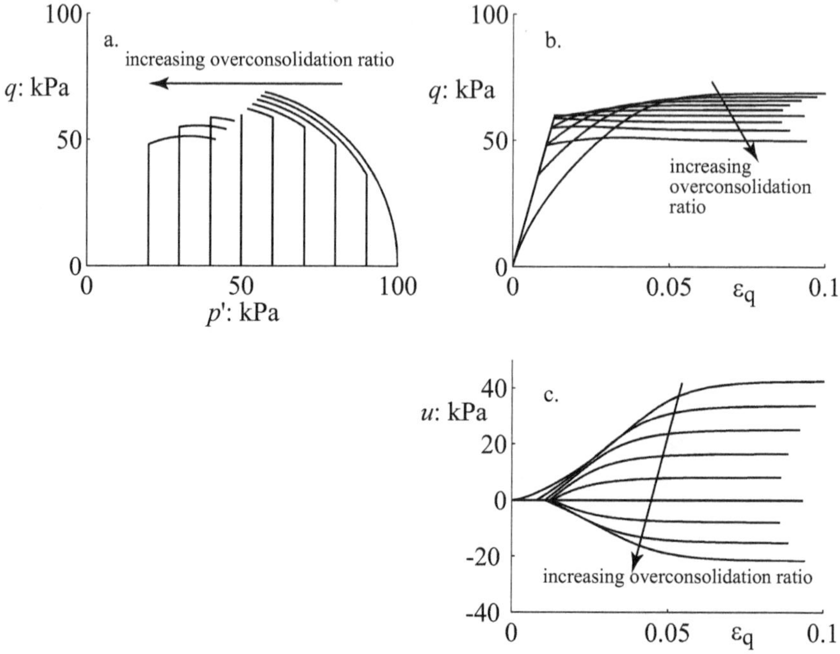


**Figure 47.** Cam clay: undrained triaxial compression tests conducted with constant total mean stress ($\delta p = 0$): (a) effective stress paths; (b) stress:strain response; (c) development of pore pressure ($\kappa = 0.05$, $G = 1500$ kPa, $\lambda = 0.25$, $M = 1.2$) (overconsolidation ratios $p'_o/p'_i$ in range 1-5, $p'_o = 100$ kPa).

the effective stress path is found to be

$$\frac{p'_i}{p'} = \left(\frac{M^2 + \eta^2}{M^2 + \eta_i^2}\right)^\Lambda \tag{101}$$

ending at a failure state with mean effective stress $p'_f$ and stress ratio equal to $M$:

$$\frac{p'_i}{p'_f} = \left(\frac{2M^2}{M^2 + \eta_i^2}\right)^\Lambda \tag{102}$$

For elastic stress changes the constant volume condition requires, for an isotropic elastic material, that there should be no change in mean effective stress. Effective stress paths for initially normally consolidated, lightly overconsolidated and heavily overconsolidated Cam clay are shown in Fig. 47. For the isotropically normally consolidated soil ($\eta_i = 0$) the ratio of mean effective stresses at the start of the test ($p'_i$) and at failure ($p'_f$) is $2^\Lambda$. The undrained strength of the soil $c_u$ is given by:

$$c_u = \frac{q_f}{2} = M\frac{p'_f}{2} = \frac{M}{2}\frac{p'_i}{2^\Lambda} \tag{103}$$

For isotropically normally consolidated soil the ratio of undrained strength to initial effective stress is a function only of soil constants $M$ and $\Lambda$:

$$\frac{c_u}{p'_i} = \frac{M}{2^{1+\Lambda}} \tag{104}$$

# 11 Response of Models to Cyclic Loading

## 11.1 Elastic-Perfectly Plastic Model

The incremental stiffness is either zero or has the full elastic value. This characteristic of the elastic-perfectly plastic model is slightly obscured when soil behaviour is described in terms of secant stiffness. It is standard practice to show variation of shear stiffness with shear strain both for monotonic testing (Fig. 48a)—where the stiffness falls as failure is approached—and for cyclic, or more generally non-monotonic testing—where the average stiffness in any cycle reflects the strain level at which the direction of loading was reversed (Fig. 48a). Typical experimental data show a reduction in average cyclic (secant) stiffness with increasing strain amplitude. An elastic-perfectly plastic model will also show a reduction in average secant stiffness with increasing strain amplitude (Fig. 49) but when the response is considered in terms of tangent stiffness it is very clear that the secant response is the combination of a constant prefailure stiffness with varying amounts of strain imposed while the soil is at failure and hence has zero tangent stiffness (Fig. 50).

If the shear strain to failure of an elastic-perfectly plastic soil with shear modulus $G$ is $\gamma_f$ then at a strain of $\gamma_m$ the shear stress is $G\gamma_f$ and the secant stiffness $G_s$ is

$$G_s = G\frac{\gamma_f}{\gamma_m} \tag{105}$$

This is plotted in Fig. 49.

The damping ratio $\xi$ can also be calculated. The energy dissipated in each cycle is

$$W = 4G\gamma_f^2\left(\frac{\gamma_m}{\gamma_f} - 1\right) \tag{106}$$

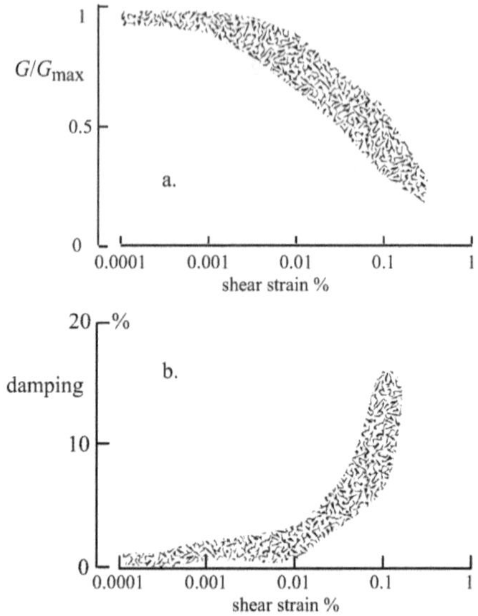


**Figure 48.** Range of (a) secant shear stiffness degradation data (normalised with shear stiffness at very small strain) $G_{\max}$ and (b) damping ratio for Quiou sand from resonant column and torsional shear tests (after LoPresti et al., 1997).

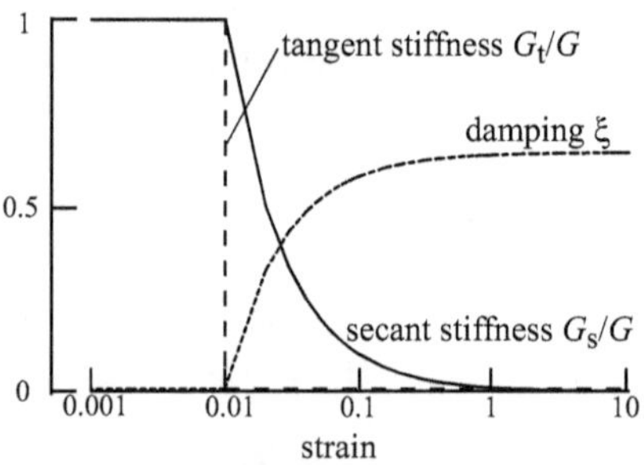


**Figure 49.** Elastic-perfectly plastic Mohr-Coulomb model: stiffness and damping variation with strain (yield strain =0.01).

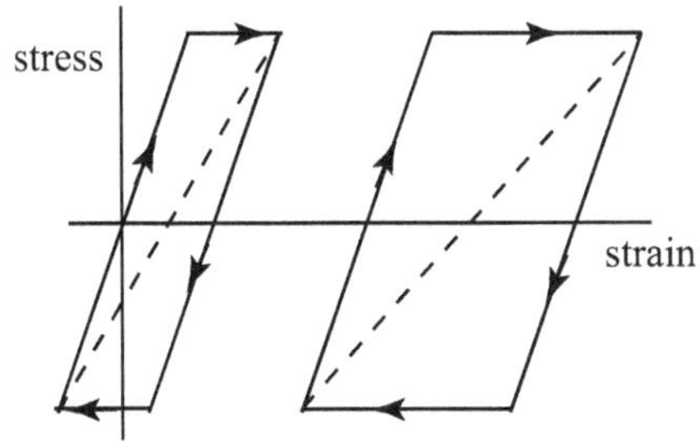


**Figure 50.** Variation of secant stiffness with strain amplitude for cycles of loading of elastic-perfectly plastic material.

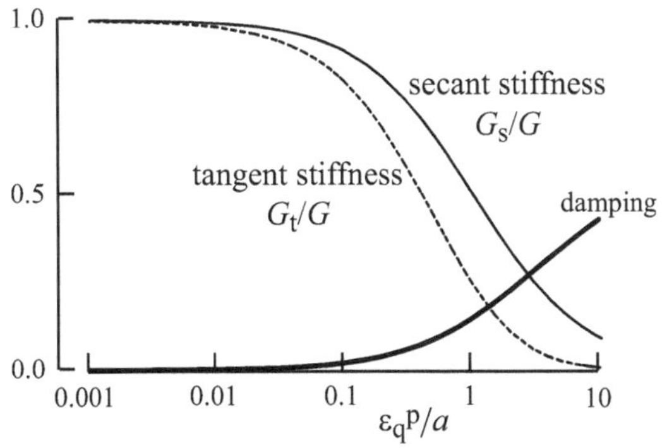


**Figure 51.** Elastic-hardening plastic Mohr-Coulomb model: variation of secant and tangent shear stiffness and damping ratio with strain.

and the maximum elastic energy stored in each cycle is

$$\Delta W = \frac{1}{2} G \gamma_f^2 \frac{\gamma_m}{\gamma_f} \tag{107}$$

so that

$$\xi = \frac{2}{\pi} \left( 1 - \frac{\gamma_f}{\gamma_m} \right) \tag{108}$$

and this is also plotted in Fig. 49.

### 11.2 Elastic-Hardening Plastic Model

The variation of stiffness with strain in monotonic straining and the variation of damping ratio in cyclic shearing are shown in Fig. 51. Obviously

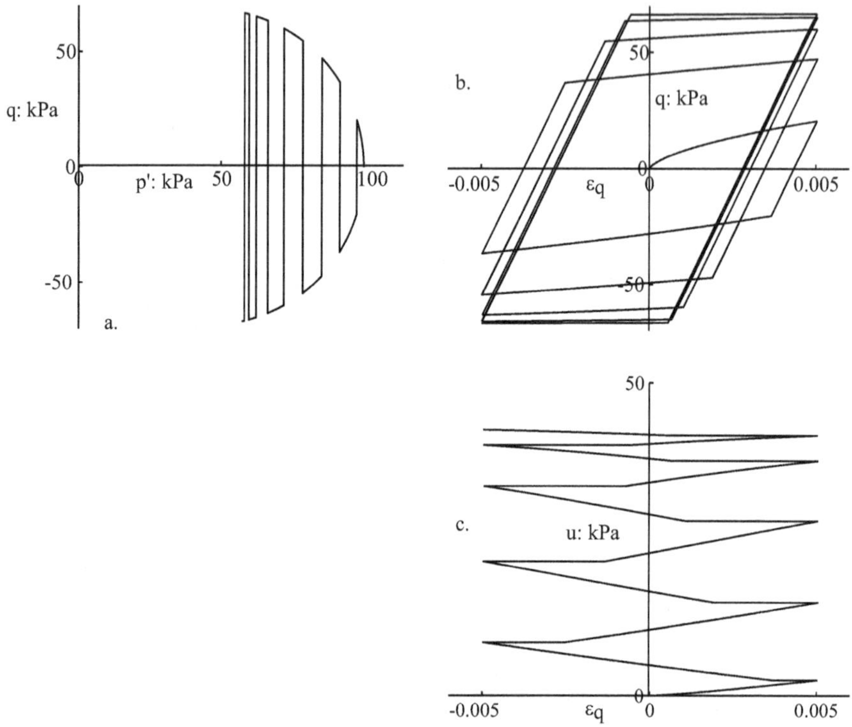


**Figure 52.** Cam clay: cyclic undrained triaxial compression tests conducted with constant total mean stress ($\delta p = 0$) and constant shear strain amplitude: (a) effective stress paths; (b) stress:strain response; (c) development of pore pressure.

the tangent stiffness now falls more gradually than for the elastic-perfectly plastic model (Fig. 49).

### 11.3 Cam Clay

From Fig. 47 we can readily deduce the response of Cam clay to cyclic undrained loading. Under constant amplitude of deviator stress $q$ the model will predict yielding in the first loading but thereafter purely elastic response. Under cycles of constant amplitude of distortional strain $\varepsilon_q$, yielding occurs in each reversal path in order to generate sufficient plastic strain (Fig. 52). The effective stress path moves progressively to lower mean ef-

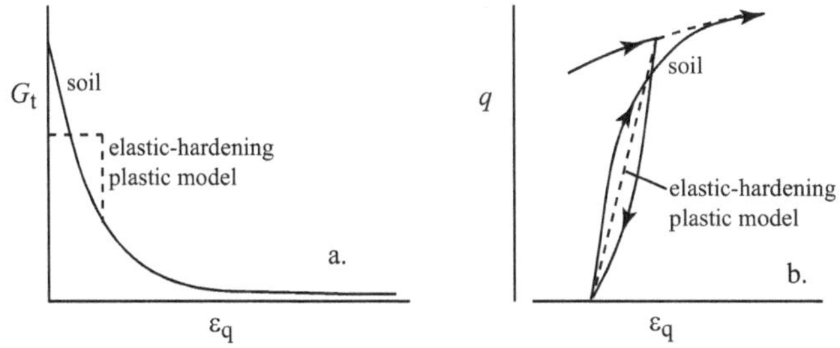


**Figure 53.** Comparison of typical capability of elastic-hardening plastic model with actual soil behaviour (a) variation of tangent stiffness with monotonic shearing; (b) unload-reload response.

fective stress and settles beneath the critical state point $\eta = M$: at this mean effective stress the perfect plasticity of the critical state is able to generate sufficient plastic distortional strain without any further change in mean effective stress.

## 12 Kinematic Hardening Plasticity

In section §2 we described some of the kinematic aspects of stiffness of soils and we showed a schematic variation of stiffness with non-monotonic loading in Fig. 4. How far have we progressed towards being able to reproduce this character of response?

Although our elastic-hardening plastic models are an evident improvement on the elastic-perfectly plastic models in that they provide for a steady decrease of tangent stiffness after yield occurs, rather than an immediate drop to zero (Fig. 10), there is still the dramatic fall in stiffness as the stress path crosses the yield surface (Fig. 53a) whereas real soils tend to show much more gradual stiffness changes. There is also a significant difference on unloading (Fig. 53b). The elastic-plastic models described here predict that the yield surface will expand as the stress state pushes it outwards—and the more it expands the larger the elastic region that remains. In fact, our kinematic observation suggests that, though the elastic region may indeed change in size as the stresses push it around, it is the change in position that is possibly more significant. Unloading paths develop plasticity in a way that the Cam clay model cannot describe.

Such response can be described using a kinematic hardening extension of the hardening plasticity models. Our models are so far essentially isotropic hardening models: the Cam clay yield locus retains its shape and orien-

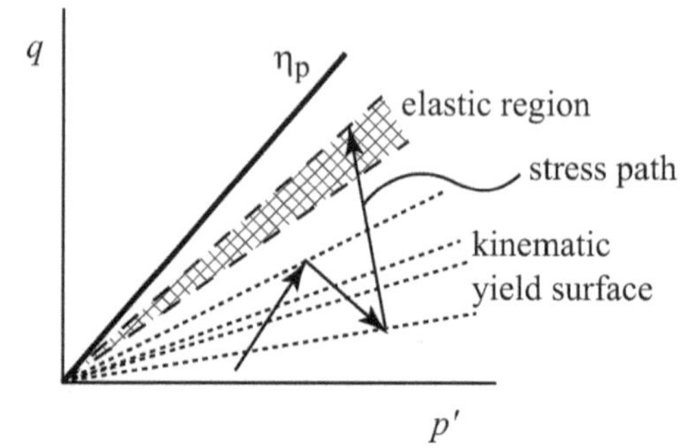


**Figure 54.** Kinematic hardening Mohr-Coulomb model.

tation, and always passes through the origin of stress space whatever the stress path that interacts with it; the yield locus of the Mohr-Coulomb model becomes a progressively more open cone.

### 12.1 Kinematic Hardening Mohr-Coulomb Model

Although we are describing some of the nonlinearity of stiffness and post-peak softening of sand we are not yet able to describe the nonlinearity and reversed plasticity that is observed when the direction of loading is reversed. Our models tell us that the behaviour will be purely elastic for all stress ratios lower than the previous maximum stress ratio. A simple way to overcome this shortcoming is to add kinematic hardening to the model so that the elastic region of high stiffness is carried round with the recent stress history (Fig. 54). The boundary of this elastic region is of course *the* yield surface but in order to retain a hierarchical link with the previous model we describe the previous yield surface (or, in this model, the current strength surface) as a *bounding* surface and use the idea of bounding surface plasticity (Dafalias and Popov, 1975) to permit the plastic hardening stiffness to depend on the separation of the yield surface and bounding surface. By careful choice of the stiffness interpolation function we can make the stiffness vary continuously and smoothly from the elastic value to the fully plastic value as the stress state moves from within the kinematic yield surface towards the outer surface.

This model, named Severn-Trent sand in its complete form (Gajo and Muir Wood, 1999), has been successfully calibrated against triaxial test data for Hostun sand (Fig. 55). The effects of different initial density or stress level are automatically described using a single set of soil parameters. We

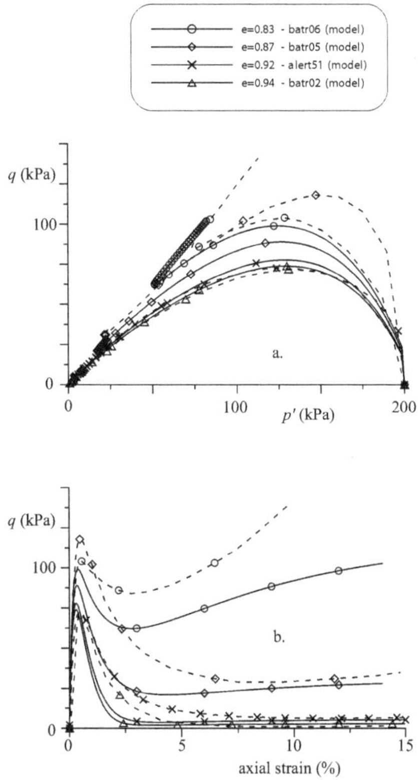


**Figure 55.** Comparison of simulation and experiment: effects of initial density on triaxial undrained compression tests on Hostun sand with initial mean effective stress $p'_i = 200$ kPa (Gajo and Muir Wood, 1999).

ignore the practical problem of maintaining complete homogeneity within a sand sample which is undergoing softening and merely propose that the model is demonstrably able to describe the sort of softening that might develop within a localised shear band. More detailed analysis of the effects of localisation in sand samples is given by Gajo et al. (2004) again based on a model equivalent to Severn-Trent sand.

The use of the model to simulate cyclic undrained loading leading to eventual liquefaction is shown in Fig. 56. The model fails after 25 cycles; the actual soil sample fails after 89 cycles. Number of cycles to liquefaction is not a particularly reliable parameter to use for model calibration—though there is an obvious significant difference between samples which liquefy in

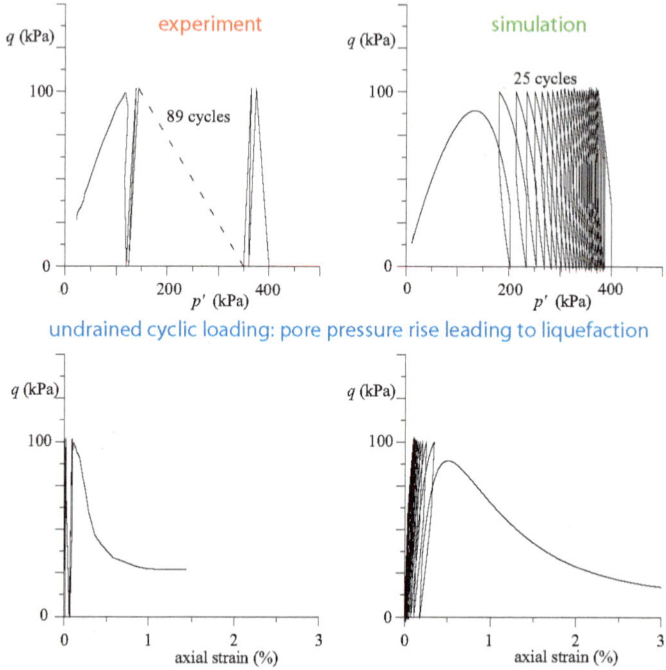


**Figure 56.** Comparison of simulation and experiment: undrained cyclic loading of loose Hostun sand.

one or two cycles and those which, like the one shown in Fig. 56, survive for many cycles. The important conclusion is that this model is able to reproduce an important aspect of soil response which is linked to the volume change characteristics of the soil.

## 12.2 Kinematic Hardening 'Bubble' Cam Clay Model

A kinematic hardening extension of a Cam clay-like model is illustrated in Fig. 57b (Al-Tabbaa and Muir Wood, 1989). The elastic region is now confined to an elastic 'bubble' which floats around in stress space with the current stress state. Plastic strains occur whenever the 'bubble' moves but the plastic stiffness is controlled by the separation, $b$, of the 'bubble' and some outer *bounding* surface and falls as the 'bubble' approaches this bounding surface. A translation rule is introduced to describe the way in which the 'bubble' decides how much to change in size and how much to

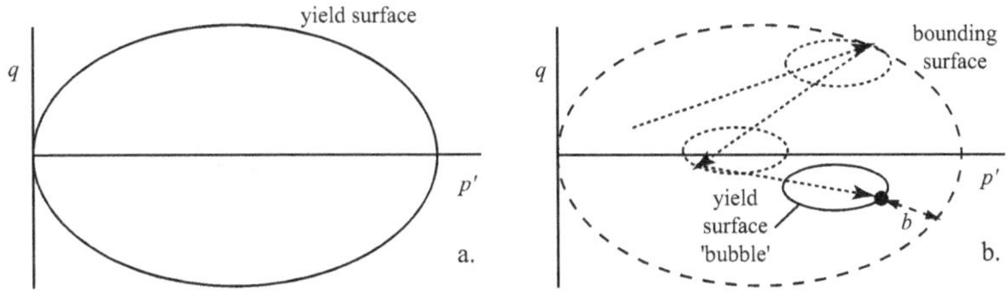


**Figure 57.** (a) Cam clay; (b) kinematic hardening extension of Cam clay: 'bubble' bounds elastic region and moves with stress history.

change in position as the stress engages with it.

The 'bubble' has the same shape as the bounding surface and a size which is a fixed proportion $R$ of the size of the bounding surface. With appropriate formulation this model can be made to behave identically to Cam clay when the soil is being loaded with the 'bubble' in contact with the bounding surface (which then looks rather like the Cam clay yield surface—but is not actually a yield surface because it does not control the onset of development of plastic strains) (Fig. 57a, b). There is thus a hierarchical development of the model, adding desirable features (smooth variation of stiffness, plasticity on stress reversal) to an already somewhat familiar model, Cam clay. (Setting the size ratio of the 'bubble' $R = 1$ the model is identical to Cam clay.)

In conformity with Cam clay, an assumption of normality is made for the definition of the plastic flow mechanism and plastic hardening is driven by plastic volumetric strain only. Since the sizes of the 'bubble' and outer surface are tied together in a constant ratio, plastic volumetric hardening produces a change in size of both surfaces.

The concept of the model is shown graphically in Fig. 57b in a conventional triaxial stress plane. The 'bubble' and the outer surface have the same elliptical shape. The outer surface passes though the stress origin. The 'bubble', which encloses the elastic domain moves around within the outer surface following a kinematic hardening rule.

As in Cam-clay, the scalar hardening variable $p_o$, which controls isotropic hardening, is dependent only on plastic volumetric strain increments.

Kinematic hardening is used to define the translation rule of the 'bubble'. For any stress point, $\boldsymbol{\sigma}$, on the 'bubble' $f$=0 there is a conjugate point $\boldsymbol{\sigma}_c$ on the outer surface $F = 0$ having the same direction of the outward normal as shown in Fig. 57. The smooth transition between the 'bubble' and the outer surface requires the centre of the 'bubble' to translate relative to the centre of the outer surface in a direction parallel to the line joining the current stress and the conjugate point. This geometric requirement can be combined with the consistency condition on the 'bubble' (the stress state must lie inside or on the 'bubble' at all times) to deduce the relative amounts of hardening and translation required to accommodate the new stress state. The starting point in describing a smooth variation of stiffness as the 'bubble' approaches the outer surface is to define a normalised distance $b$ between the current stress and the conjugate stress. This distance vanishes when the two surfaces are in contact; it is normalised with its maximum value, obtained when the 'bubble' is touching the outer surface at a point diametrically opposite the conjugate point $\boldsymbol{\sigma}_c$ (with the current stress on the 'bubble' diametrically opposite this point of tangency). The hardening modulus is made a function of this normalised separation of the surfaces and allows a steady fall of stiffness with strain towards the Cam-clay value as the 'bubble' approaches the outer surface.

Typical response of this 'bubble' model for a cyclic triaxial test is shown in Fig. 58. There is a combination of hysteretic cyclic stress:strain response together with development of permanent strains.

## 13 Viscoelastic Model

The models that we have described are rate-independent. The dominant rate effects seen in most soils are the consequence of delayed dissipation of pore pressures (consolidation). However, we know that in clays secondary consolidation or creep occurs and even for sands there are observable continuing strains at constant effective stress. We have also assumed that the primary source of energy dissipation for soils comes from soil plasticity, permanent rearrangements of particles. However, an alternative route to the dissipation of energy comes from the addition of viscosity as an element of the constitutive response dependent on rate of strain.

A simple parallel model (Fig. 59a) would then say that:

$$\sigma = k\varepsilon + \lambda\dot{\varepsilon} \tag{109}$$

where $\dot{\varepsilon}$ is the strain rate. For a sinusoidal variation of strain with time $\varepsilon = \alpha \sin \omega t$ the stress will be:

$$\sigma = k\alpha \sin \omega t + \lambda\omega\alpha \cos \omega t = \alpha\sqrt{k^2 + \lambda^2\omega^2} \sin \omega t + \phi \tag{110}$$

where $\tan\phi = \lambda\omega/k$ is a phase angle representing a lag between strain and stress. The typical stress strain response follows an ellipse as shown in Fig. 59b. We can contrast this with the typical response seen in cyclic

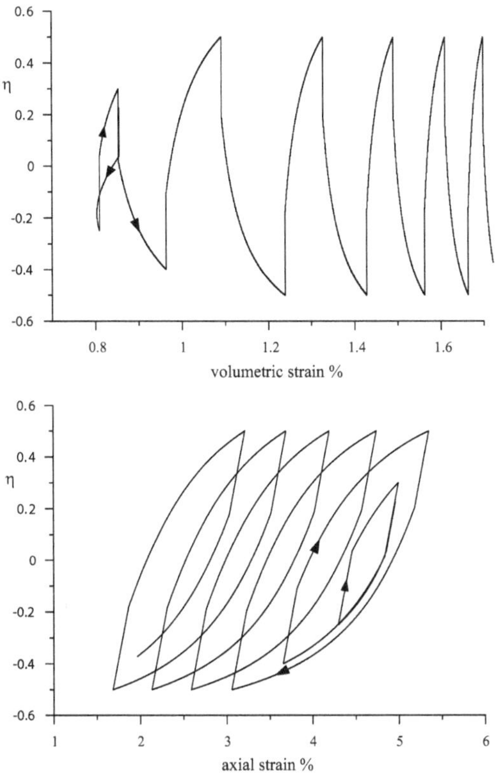


**Figure 58.** Kinematic hardening extension of Cam clay: response in cyclic triaxial test with constant mean effective stress $p'$.

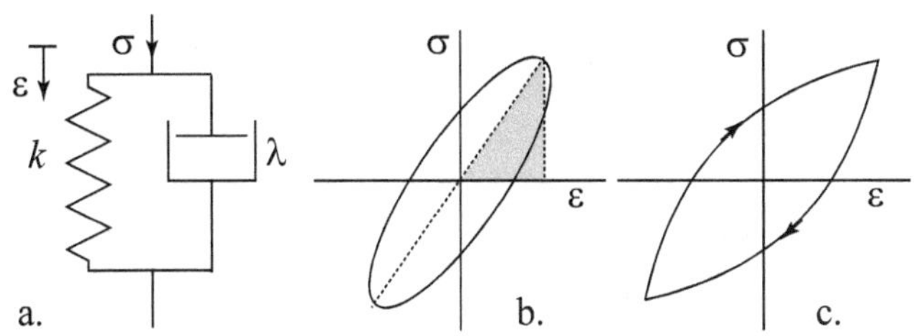


**Figure 59.** Viscoelastic model.

loading of soils (Fig. 59c) which shows a sharp change in stiffness on reversal of the direction of straining—a character of response that we can more easily match within an elastic-plastic modelling framework.

We can calculate a damping ratio from the ratio of the energy dissipated in each cycle (the area of the ellipse) to $4\pi$ times the maximum elastic energy stored (the area of the shaded triangle). The energy dissipated through viscosity is:

$$\int_0^{2\pi/\omega} \sigma \frac{\mathrm{d}\varepsilon}{\mathrm{d}t}\mathrm{d}t = \pi\lambda\omega\alpha^2 \tag{111}$$

The maximum elastic energy stored is

$$\frac{1}{2}\sigma\varepsilon = \frac{1}{2}k\alpha^2 \tag{112}$$

so that the damping ratio is:

$$\xi = \frac{\lambda\omega}{2k} \tag{113}$$

and depends on angular frequency $\omega$.

## 14 Strain-Based Models

The results of cyclic tests on soils are typically presented in terms of curves showing the degradation of secant modulus with cyclic strain amplitude and the increase of damping with cyclic strain amplitude (Fig. 48). We can use these curves—which should implicitly be mutually consistent—to generate a strain-space model of deviatoric soil behaviour for static and irregular cyclic loading. This section is inspired by the model presented in Itasca (2000) .

Given information about secant stiffness, $G_s(\gamma)$, we can convert this to the more useful information about tangent stiffness:

$$G_s = \frac{\tau}{\gamma} \qquad G_t = \frac{\mathrm{d}\tau}{\mathrm{d}\gamma} = G_s + \gamma\frac{\mathrm{d}G_s}{\mathrm{d}\gamma} \tag{114}$$

We will assume that this relation will apply along any strain path from a point of strain reversal. The initial loading is seen as a special case in which the stress:strain relation is scaled down by a factor of 2 from the general version described by (114): this is the so-called Masing rule. The general character of the response that such a model will describe is shown in Fig. 60a.

This implied modelling procedure can be seen as a generalisation of the Mroz-Iwan-Prevost approach to the modelling of the undrained—deviatoric—response of soils using fields of hardening moduli. Fig. 61 shows deviatoric

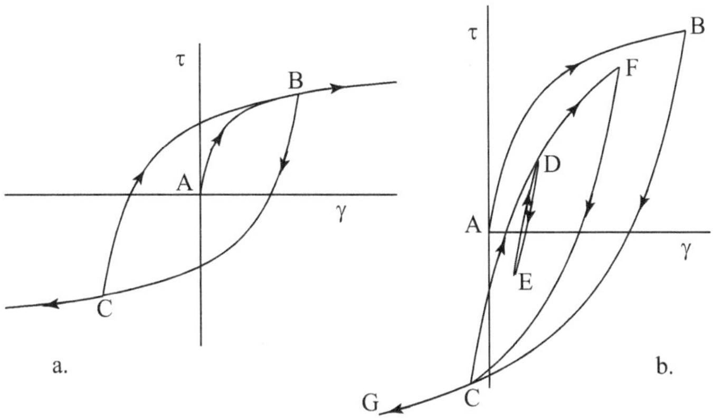


**Figure 60.** (a) Masing rule for stress:strain response: geometry of AB is identical to geometry of BC but both stress and strain are scaled by a factor of 2; (b) cycles within cycles.

views of principal stress space (but it could equally be some other deviatoric view of general stress space) with three different sets of initial 'yield' surfaces—these are drawn as circles for illustrative purposes. It is supposed that there is a certain distortional stiffness associated with the space within the smallest 'yield' surface and between each subsequent pair of surfaces. The stress state must always lie within all the 'yield' surfaces and the surfaces move with the current stress in order to ensure that this rule is not broken. The right hand diagram of each pair shows the stress:strain response, with the stiffness changing step by step as the successive surfaces are crossed.

Thus Fig. 61a shows symmetrical paths AB and AC which pick up identical stress:strain response. Fig. 61b shows paths DE and DF starting from a 'negative' initial stress D, perhaps the consequence of an episode of overconsolidation. The initial stiffness for the reversal of direction of straining DE is higher than that for the continuation of the 'unloading' straining DF. Finally, Fig. 61c shows paths GH and GI starting from a much more anisotropic initial state of stress where the 'yield' surfaces have become bunched together as a result of the loading AB in Fig. 61a. The stress:strain response now contrasts the continued loading GH, which is identical to the corresponding section of path AB with the effect of strain reversal GI which is not only initially much stiffer but also reproduces the response seen in the initial loading AB but at twice the scale. And the geometry of the shifting

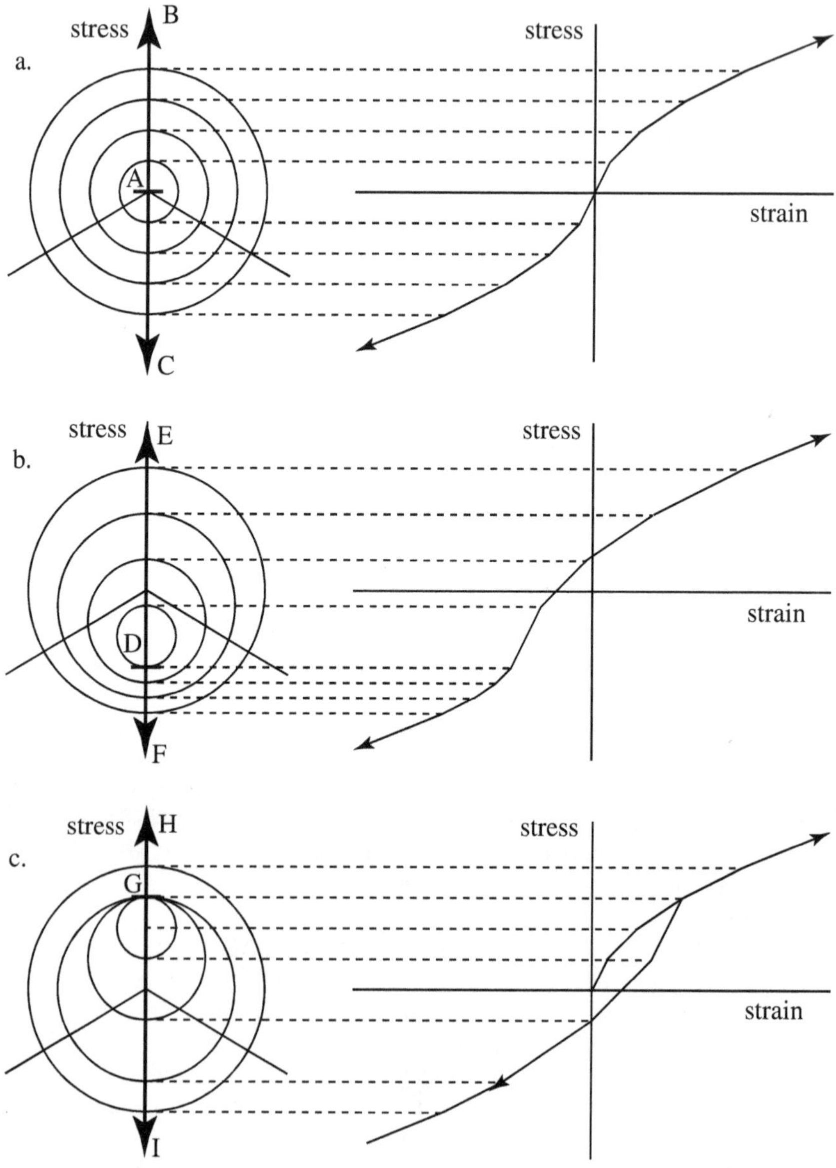


**Figure 61.** Fields of hardening moduli for deviatoric modelling.

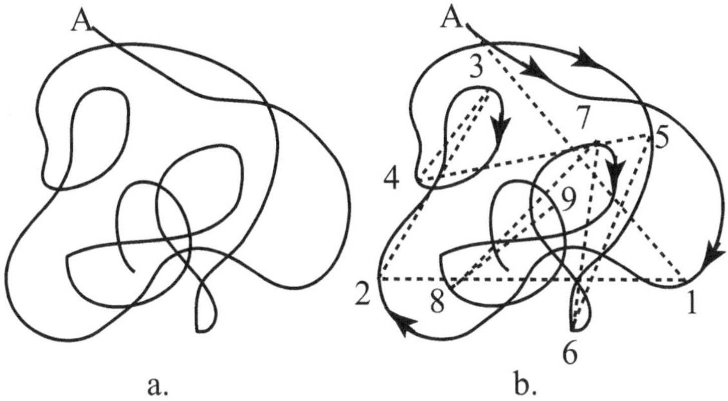


**Figure 62.** Detection of points of strain reversal $(1, 2, \ldots 9)$ in complex strain path.

'yield' surfaces shows very clearly why this should be so. The Masing rule is demonstrated rather effectively.

With such a series of independent 'yield' surfaces the geometry of the surfaces will all the time indicate whether a particular increment implies a change in stiffness of not. In generalising the continuous description of Fig. 60 we could use some implicit shape of loading surface which would enable us to distinguish between unloading and continuing loading. Alternatively we require some means of deciding whether we are continuing to strain the soil, with a consequent steadily reducing shear stiffness, or whether we have a point of strain reversal, in which case we have an immediate jump of stiffness to the initial fully elastic value. Fig. 60b illustrates the general scheme for a single degree of strain freedom. Initial loading AB is followed by unloading/reversal BC with an immediate jump in tangent stiffness. Reversal at C gives rise to a further jump. The unloading/reloading cycle DED shows high stiffness each time the direction of straining changes, and the stiffness reverts to the value that it had on path CF at the end of the cycle. Similarly, reversal at F rejoins the original unloading path BC at C and continues to G with a lower stiffness.

Given a rather random strain path as shown in Fig. 62, we can follow Itasca (2000) and adopt the pragmatic rule that, so long as the distance from the previous point of strain reversal is still increasing then we have continuing modulus degradation. Once we detect a maximum in this distance then we start afresh. We might propose that once the distance from reversal point $i$ exceeds the distance between reversal points $i$ and $i-1$ then the stiffness should revert to the value that it would have had for continued straining beyond point $i-1$ from the previous reversal point $i-2$, provided that was a longer path. The principles are rather straightforward when the straining has just a single component. Their application becomes a little more subtle when there are several degrees of deviatoric freedom (Fig. 62).

Itasca (2000) suggest several possible functions that might be used to describe the modulus degradation curve.

A simple hyperbolic model:

$$\frac{G_s}{G_{\max}} = \frac{1}{1+\gamma/\gamma_1} \tag{115}$$

where the reference strain $\gamma_1$ is chosen by fitting the observed strain amplitude when the secant stiffness has fallen to half of the small strain value.

The cubic expression:

$$\frac{G_s}{G_{\max}} = s^2(3-2s) \tag{116}$$

where

$$s = \frac{\ln(\gamma_2/\gamma)}{\ln(\gamma_2/\gamma_1)} \tag{117}$$

has zero slope for $s=0$, $\gamma=\gamma_2$ and $s=1$, $\gamma=\gamma_1$ and can be used to describe the modulus variation between these limits $\gamma_1<\gamma<\gamma_2$. Beyond these limits, for $\gamma<\gamma_1$ and $s>1$, $G=G_{\max}$ and for $\gamma>\gamma_2$ and $s<0$, $G=0$. Within the operational range:

$$\frac{G_t}{G_{\max}} = s^2(3-2s) - \frac{6s(1-s)}{\ln(\gamma_2/\gamma_1)} \tag{118}$$

In fact there is an additional limit if strain softening is to be avoided. The value of $s=s_{lim}$ for which the tangent modulus falls to zero is:

$$s_{lim} = \frac{1}{4}\left\{\left(3+\frac{6}{\ln(\gamma_2/\gamma_1)}\right)+\sqrt{\left[\left(3+\frac{6}{\ln(\gamma_2/\gamma_1)}\right)^2-\frac{48}{\ln(\gamma_2/\gamma_1)}\right]}\right\} \tag{119}$$

The corresponding value of strain is:

$$\gamma_{lim} = \frac{\gamma_2}{(\gamma_2/\gamma_1)^{s_{lim}}} \tag{120}$$

and secant stiffness $G_{slim}$. Then for $s<s_{lim}$, $G_t/G_{\max}=0$ and $G_s = G_{slim}\gamma_{lim}/\gamma$.

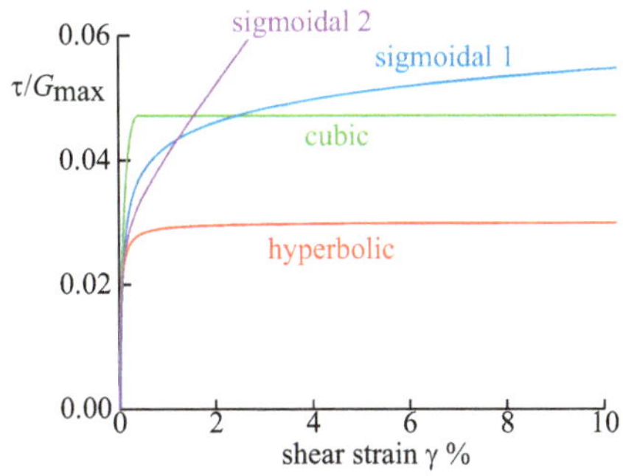


**Figure 63.** Stress:strain response predicted by different models.

Alternatively a sigmoidal model:

$$\frac{G_s}{G_{\max}} = \frac{a}{1+\exp[-\frac{\ln(\gamma/\gamma_1)}{\ln(\gamma_2/\gamma_1)}]} = \frac{a}{1+\exp[s-1]} \tag{121}$$

where $\gamma_1$ and $\gamma_2$ are (different) reference strains and $a$ is a soil constant. Or a slightly more elaborate model:

$$\frac{G_s}{G_{\max}} = b + \frac{a}{1+\exp[-\frac{\ln(\gamma/\gamma_1)}{\ln(\gamma_2/\gamma_1)}]} = b + \frac{a}{1+\exp[s-1]} \tag{122}$$

where $b$ is a further soil constant and the optimum values of the reference strains will no doubt differ in each case. Evidently if $b \neq 0$ then the secant modulus never falls to zero and the shear stress is (unrealistically) predicted to continue to increase indefinitely with continuing strain (Fig. 63). For $b \neq 0$ the damping ratio peaks and then falls as the strain amplitude increases which also feels physically uncomfortable. These may be acceptable disadvantages if this particular model enables a better fit to the available data for an appropriate range of strains.

For each of these models we can fit a modulus degradation curve available from the literature or from a programme of tests on our soil and calculate the corresponding variation of damping ratio (Fig. 64). The values of soil constants used are given in Tab. 1. It is interesting to note that the typical damping suggested in published curves is lower than that emerging from any of these models indicating some apparent incompatibility between modulus degradation and damping findings. It is also interesting to compare the stress:strain responses that emerge from the assumed link between secant

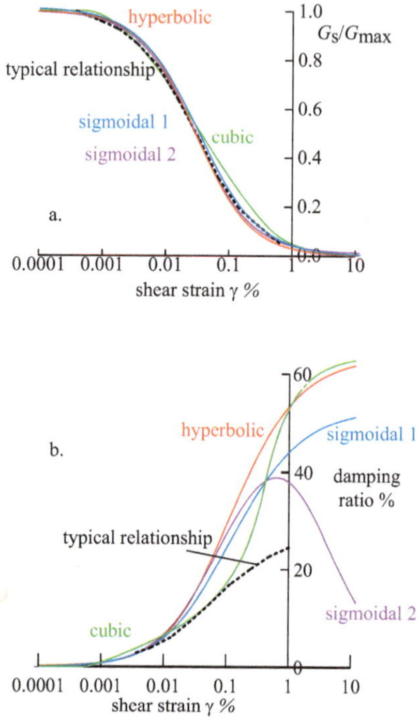


**Figure 64.** (a) Modulus degradation with cyclic strain amplitude; (b) variation of damping ratio with cyclic strain amplitude. (Typical relationships for non-plastic soils taken from Kramer (1996)).

**Table 1.** Values of soil constants

| soil constant | models | | | |
|---|---|---|---|---|
| | hyperbolic | cubic | sigmoidal 1 | sigmoidal 2 |
| $\gamma_1$ | 0.03 | 0.000473 | 0.03162 | 0.02818 |
| $\gamma_2$ | - | 2.51189 | 0.01049 | 0.01 |
| $a$ | - | - | 1.014 | 0.99 |
| $b$ | - | - | - | 0.01 |

**Table 2.** Values of soil constants for simulations in Figs. 65-68

| | |
|---|---|
| *elastic* | |
| $\chi$ | 100 |
| $\alpha$ | 0.6 |
| *critical state line* | |
| $\breve{v}$ | 1.1 |
| $\Delta v$ | 0.6 |
| $\beta$ | 0.6 |
| *strength and state parameter* | |
| $R_c$ | 1.2 |
| $\zeta_R$ | 1.1 |
| *distortional hardening* | |
| $\zeta_S$ | 0.0005 |
| *dilatancy* | |
| $\zeta_D$ | 0.5 |
| *reference stress* | |
| $\sigma_{rc}$ | 100 kPa |

stiffness and strain (Fig. 63). The need for self-consistency between the different aspects of the response is very evident.

The simple model described in §1.2 corresponds directly to the hyperbolic model of (115). Applying the Masing convention of Fig. 60 we can discover how it behaves under conditions of cyclic loading. Given the intimate links between volume change and shearing—and between state parameter and current strength—it is of interest to look at the response under undrained conditions and to discover how this response changes with the initial density of the soil. The results are shown in Figs. 65-68. The values of soil constants used in these simulations are given in Tab. 2. In each group of figures the initial specific volume rises (initial density falls) from bottom to top.

The figures show the expected dependence of response on initial density. Loose soil liquefies—it loses all its strength with the build up of pore pressure during the cycles of loading. Dense soil generates significant negative pore pressures which stabilise the soil and maintain its strength. The hyperbolic model gives rather large stress:strain cycles (Figs. 65 and 66) with rather steep gradients and very rapid fall in incremental stiffness as the top of the cycle is reached. This is a consequence of the rather low value of $\zeta_S$ (73) or $\gamma_1$ (115) used in the model: the higher the value of this soil constant the faster the shear stress climbs (the shear stress has half its ultimate value for a strain $\gamma = \gamma_1$) and the more the plastic strain near failure contributes to the imposed amplitude of cyclic strain. There is nothing sacred about the hyperbolic formulation apart from its simplicity—other functions could be chosen which show a more steady fall in stiffness with strain.

Looking at the shear stress generation (Fig. 66) the loss of strength is apparent as the pore pressure builds up. This is hidden when we look at the generation of stress ratio (Fig. 65) because the available stress ratio does not change hugely as the effective stress level falls (or rises).

The results are shown here in terms of shear stress $\tau$ and effective normal stress $\sigma'_z$ for the simple model but exactly the same results could be obtained working with distortional stress $q$ and effective mean stress $p'$.

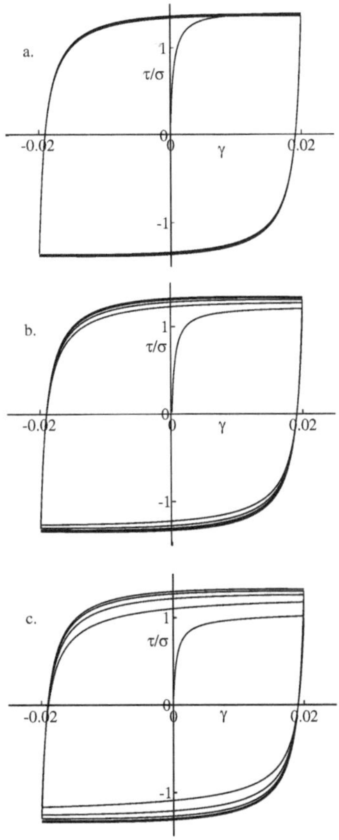


**Figure 65.** Cyclic undrained tests on hyperbolic model: stress ratio and shear strain: initial specific volume (a) 1.1, (b) 1.3, (c) 1.5.

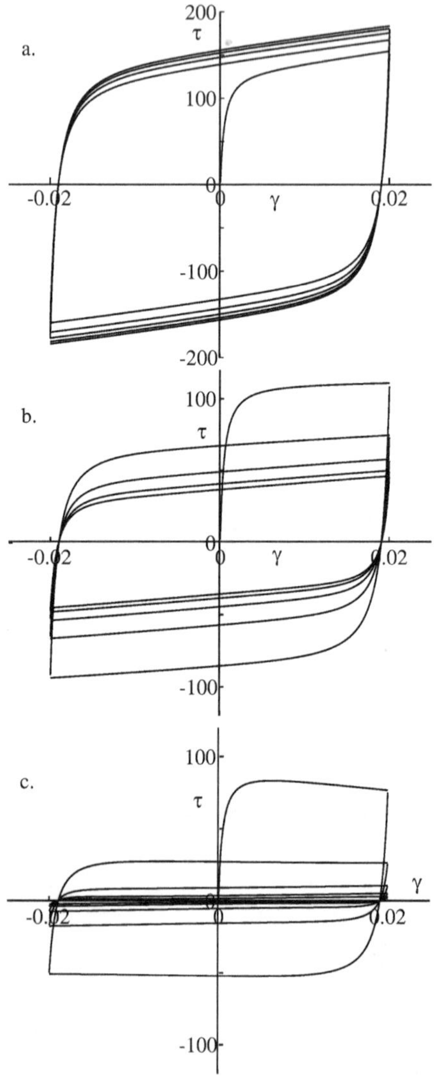


**Figure 66.** Cyclic undrained tests on hyperbolic model: shear stress and shear strain: initial specific volume (a) 1.1, (b) 1.3, (c) 1.5.

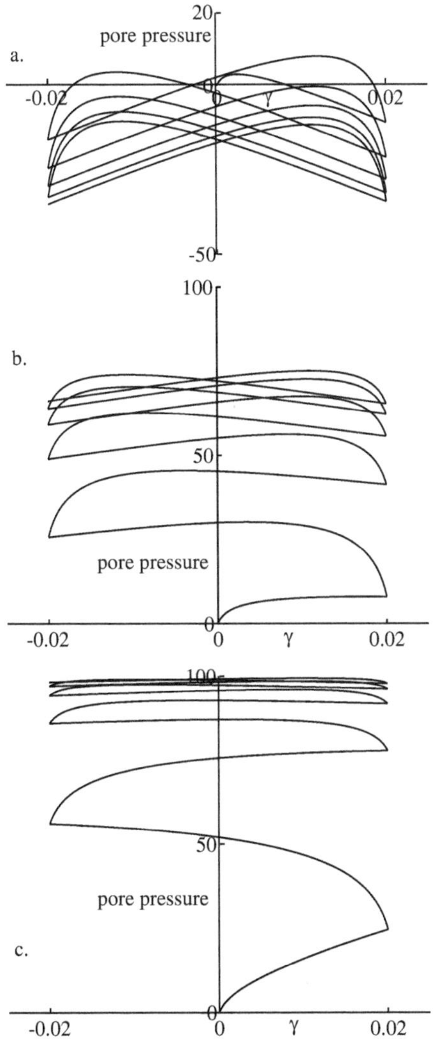


**Figure 67.** Cyclic undrained tests on hyperbolic model: pore pressure and shear strain: initial specific volume (a) 1.1, (b) 1.3, (c) 1.5.

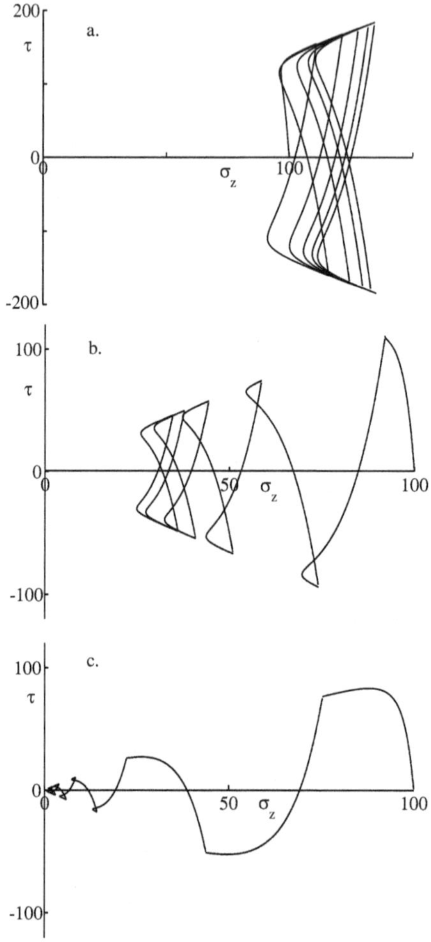


**Figure 68.** Cyclic undrained tests on hyperbolic model: effective stress path: initial specific volume (a) 1.1, (b) 1.3, (c) 1.5.

# List of Symbols

| | | |
|---|---|---|
| $^{e}$ | elastic component | §4 |
| $_{nc}$ | normally consolidated | §2 |
| $_{oc}$ | overconsolidated | §2 |
| $^{p}$ | plastic component | §4 |
| $a$ | soil parameter for strain degradation model | §14 |
| $A$ | area of sample | §1.1 |
| $b$ | separation of bubble and bounding surface | §12.2 |
| $b$ | soil parameter for strain degradation model | §14 |
| $c_u$ | undrained strength | §10 |
| $C_c$ | compression index | §10 |
| $C_s$ | swelling index | §10 |
| $d_{\max}$ | maximum particle size | §6 |
| $d_{\min}$ | minimum particle size | §6 |
| $d_{50}$ | median particle size | §6 |
| $\boldsymbol{D}$ | elastic stiffness matrix | §4 |
| $\boldsymbol{D}^{ep}$ | elastic-plastic stiffness matrix | §4 |
| $e$ | void ratio | §2 |
| $E_h$ | horizontal Young's modulus | §6 |
| $E_o$ | one-dimensional stiffness | §2 |
| $E_v$ | vertical Young's modulus | §6 |
| $f$ | yield function | §4 |
| $F$ | bounding surface | §12.2 |
| $F$ | axial force in triaxial apparatus | §1.1 |
| $g$ | plastic potential function | §4 |
| $G$ | shear modulus | §4.1 |
| $G_{\max}$ | shear stiffness at very small strain | §11.1 |
| $G_s$ | $=\tau/\gamma$ secant stiffness | §2 |
| $G_t$ | $=\mathrm{d}\tau/\mathrm{d}\gamma$ tangent stiffness | §2 |
| $H$ | plastic modulus | §9 |
| $I_i$ | $i=1,2,3$ invariants of the stress tensor | §3 |

| | | |
|---|---|---|
| $J_2$ | second invariant of deviatoric stress | §3 |
| $k$ | dummy constant | §4.1 |
| $k$ | stiffness | §13 |
| $K$ | bulk modulus | §4.1 |
| $M$ | strength parameter | §4.1 |
| $M$ | $= q/p'$ at critical state | §10 |
| $M^*$ | dilatancy parameter | §4.1 |
| $n$ | $= \sigma'_{z_{max}}/\sigma'_z$ overconsolidation ratio | §2 |
| $N$ | reference volume for normal compression line | §10 |
| $p'$ | $= (\sigma'_a + 2\sigma'_r)/3$ volumetric stress (mean effective stress) | §1.1 |
| $p'_f$ | mean effective stress at failure | §10 |
| $p'_i$ | initial mean effective stress | §9.1 |
| $p'_o$ | size of yield locus | §10 |
| $p'_r$ | reference stress for plastic potential | §9.1 |
| $P$ | normal force | §5 |
| $q$ | $= \sigma_a - \sigma_r$ distortional stress | §1.1 |
| $q_f$ | distortional stress at failure | §10 |
| $Q$ | shear force | §5 |
| $R$ | $= \tau/\sigma'_z$ current mobilised strength in shear box | §5 |
| $R$ | ratio of sizes of bubble and bounding surface | §12.2 |
| $R_c$ | critical state stress ratio | §5, §9.2 |
| $R_D$ | $= d_{\max}/d_{\min}$ | §6 |
| $R_p$ | current peak strength of the soil | §8, §9.2 |
| $s$ | parameter for strain degradation models | §14 |
| $s_{lim}$ | limiting value of $s$ | §14 |
| $u$ | pore pressure | §1.1 |
| $v$ | $= 1 + e = \rho_s/\rho_d$ specific volume | §2, §7 |
| $v_\kappa$ | reference value of specific volume for unload-reload line | §10 |
| $\Delta v$ | range of specific volume for critical state line | §7 |
| $\breve{v}$ | minimum specific volume for critical state line | §7 |
| $v_f$ | critical state specific volume | §7 |

| | | |
|---|---|---|
| $\delta W$ | work done in a small increment of strain | §1.1 |
| $W$ | energy dissipated in cycle | §11.1 |
| $\Delta W$ | elastic energy stored in cycle | §11.1 |
| $\delta W_d$ | distortional work | §1.1 |
| $W^p$ | plastic work | §4.1 |
| $\delta W_v$ | volumetric work | §1.1 |
| $x$ | horizontal displacement | §5, §9.2 |
| $z$ | vertical displacement | §5, §9.2 |
| $\alpha$ | exponent for stiffness | §2 |
| $\alpha$ | strain amplitude | §13 |
| $\beta$ | soil parameter governing slope of critical state line | §7 |
| $\gamma$ | shear strain | §2 |
| $\gamma_f$ | shear strain to failure | §11.1 |
| $\gamma_{lim}$ | limiting value of $\gamma$ | §14 |
| $\gamma_m$ | maximum shear strain | §11.1 |
| $\gamma_1$ | reference strain | §14 |
| $\gamma_2$ | reference strain | §14 |
| $\Gamma$ | intercept on critical state line | §10 |
| $\delta_{ij}$ | Kronecker delta | §1.1 |
| $\varepsilon$ | generalised distortional strain | §2 |
| $\dot{\varepsilon}$ | strain rate | §13 |
| $\delta\boldsymbol{\varepsilon}$ | vector of strain increments | §4 |
| $\delta\varepsilon_a$ | axial strain increment | §1.1 |
| $\delta\varepsilon_p$ | $= \delta\varepsilon_a + 2\delta\varepsilon_r$ volumetric strain increment | §1.1 |
| $\delta\varepsilon_q$ | $= 2(\delta\varepsilon_a - \delta\varepsilon_r)/3$ distortional strain increment | §1.1 |
| $\delta\varepsilon_r$ | radial strain increment | §1.1 |
| $\varepsilon_s$ | shear strain | §1.2, §5 |
| $\varepsilon_x$, $\varepsilon_y$, $\varepsilon_z$ | principal strains | §2 |
| $\varepsilon_z$ | vertical strain | §1.2 |
| $\varepsilon_z^\sigma$ | vertical strain from compression | §2, §9.2, §5 |
| $\varepsilon_z^\tau$ | vertical strain from shearing (dilatancy) | §9.2, §5 |

| | | |
|---|---|---|
| $\zeta_D$ | dilatancy parameter | §5, §9.2 |
| $\zeta_R$ | parameter linking strength and state parameter | §8, §9.2 |
| $\zeta_S$ | soil parameter controlling plastic stiffness | §9.1, §9.2 |
| $\eta$ | $= q/p'$ stress ratio | §1.1 |
| $\eta_i$ | initial stress ratio | §9.1 |
| $\eta_p$ | limiting value of stress ratio | §9.1 |
| $\eta_y$ | hardening variable: yield stress ratio | §9.1 |
| $\kappa$ | slope of unload-reload line | §10 |
| $\lambda$ | viscous damping | §13 |
| $\lambda$ | slope of critical state line and normal compression line | §10 |
| $\dot{\Lambda}$ | plastic multiplier | §4 |
| $\Lambda$ | $\dfrac{\lambda - \kappa}{\lambda}$ | §10 |
| $\nu$ | Poisson's ratio | §10 |
| $\nu_{vh}$ | Poisson's ratio linking the horizontal and vertical strain increments | §6 |
| $\xi$ | damping ratio | §11.1,§13 |
| $\rho_d$ | dry density | §2 |
| $\rho_s$ | density of soil mineral | §2 |
| $\delta\boldsymbol{\sigma}$ | vector of stress increments | §4 |
| $\sigma'_a$ | axial effective stress | §1.1 |
| $\boldsymbol{\sigma}_c$ | conjugate stress | §12.2 |
| $\sigma_{ij}$ | total stress tensor | §1.1 |
| $\sigma'_{ij}$ | effective stress tensor | §1.1 |
| $\sigma'_r$ | radial effective stress | §1.1 |
| $\sigma_{rc}$ | reference stress for critical state line | §7, §14 |
| $\sigma'_{ref}$ | reference stress for stiffness | §2 |
| $\sigma_x, \sigma_y, \sigma_z$ | principal stresses | §2, §3 |
| $\sigma'_z$ | vertical effective stress | §1.2, §5 |
| $\tau$ | shear stress | §1.2, §2, §5 |
| $\phi'$ | angle of shearing resistance | §3,§4.1 |
| $\phi$ | phase lag between strain and stress | §13 |

| | | |
|---|---|---|
| $\phi_c$ | angle of shearing resistance at critical state | §10 |
| $\phi'_m$ | mobilised friction | §1.1 |
| $\phi'_p$ | peak angle of shearing resistance | §8 |
| $\chi$ | hardening parameter | §9 |
| $\chi$ | soil parameter for stiffness magnitude | §2 |
| $\psi$ | $= v - v_f$ state parameter | §7 |
| $\Psi_1$ | composite soil parameter | §9.2 |
| $\Psi_2$ | composite soil parameter | §9.2 |
| $\Psi_3$ | composite soil parameter | §9.2 |
| $\omega$ | angular frequency of harmonic strain variation | §13 |

## Bibliography

A. Al-Tabbaa and D. Muir Wood. An experimentally based 'bubble' model for clay. In S. Pietruszczak and G. N. Pande, editors, *Numerical Models in Geomechanics NUMOG III*, pages 91–99. Elsevier Applied Science, London, 1989.

R. H. Bassett. *The behaviour of granular materials in the simple shear apparatus.* PhD thesis, University of Cambridge, Cambridge, 1967.

K. Been and M. G. Jefferies. Discussion: A state parameter for sands. *Géotechnique*, 36(1):449–461, 1986.

M. Budhu. *Simple shear deformation of sands.* PhD thesis, University of Cambridge, Cambridge, 1979.

A. Casagrande. Characteristics of cohesionless soils affecting the stability of slopes and earth fills. *J. Boston Soc. Civil Engineers*, 23(1):13–32, 1936.

Y. F. Dafalias and E. P. Popov. A model of nonlinear hardening materials for complex loading. *Acta Mechanica*, 21:173–192, 1975.

A. Gajo and D. Muir Wood. Severn-Trent sand: a kinematic hardening constitutive model for sands: the $q-p$ formulation. *Géotechnique*, 49 (5):595–614, 1999.

A. Gajo, D. Bigoni, and D. Muir Wood. Multiple shear band development and related instabilities in granular materials. *Journal of Mathematics and Physics of Solids*, 52:2683–2724, 2004.

A. Gajo, D. Muir Wood, and D. Bigoni. On certain critical material and testing characteristics affecting shear band development in sand. *Géotechnique*, 57(5):449–461, 2007.

J. Graham and G. T. Houlsby. Elastic anisotropy of a natural clay. *Géotechnique*, 33(2):165–180, 1983.

G. Gudehus. A comparison of some constitutive laws for soils under radially symmetric loading and unloading. In W. Wittke, editor, *Proc. 3rd Conf. Numerical methods in geomechanics*, pages 1309–1323, Aachen, Rotterdam, 1979. A. A. Balkema.

S. Hall, D. Muir Wood, E. Ibraim, and C. Viggiani. Localised deformation patterning in 2D granular materials revealed by digital image correlation. *Granular matter*, 12(1):1–14, 2010.

Itasca. *FLAC: Fast Lagrangian Analysis of Continua.* Itasca Consulting Group, Minneapolis, 2000.

S. L. Kramer. *Geotechnical earthquake engineering.* Prentice Hall, 1996.

M. R. Kuhn. Scaling in granular materials. In R. Garcia-Rojo, H. J. Herrmann, S. McNamara, editor, *Powders and Grains*, volume 1, pages 115–122, Leiden, 2005. A. A. Balkema.

P. V. Lade and J. M. Duncan. Elasto-plastic stress-strain theory for cohesionless soil. *ASCE, Journal of the Geotechnical Engineering Division*, 101(GT10):1037–1053, 1975.

D. C. F. LoPresti, M. Jamiolkowski, O. Pallara, A. Cavallaro, and S. Pedroni. Shear modulus and damping of soils. *Géotechnique*, 47(3):603–617, 1997.

H. Matsuoka and T. Nakai. A new failure criterion for soils in three-dimensional stresses. In PA Vermeer and HJ Luger, editor, *Proc. IUTAM Symp. on Deformation and failure of granular materials*, pages 253–263, Rotterdam, 2001. A. A. Balkema.

D. Muir Wood. *Soil behaviour and critical state soil mechanics.* Cambridge University Press, Cambridge, 1990.

D. Muir Wood. Some observations of volumetric instabilities in soils. *International Journal of Solids and Structures*, 39(13-14):3429–3449, 2002.

D. Muir Wood. Experimental inspiration for kinematic hardening models of soil. *Journal of Engineering Mechanics*, 130(6):656–664, 2004.

D. Muir Wood. Keynote address, I. S. Yamaguchi: Soil: discontinuum-continuum. Geomechanics and geotechnics of particulate media. ed. M. Hyodo, H. Murata & Y. Nakata, 2006.

D. Muir Wood. Modelling granular materials: discontinuum-continuum. In *Structures and granular solids: from scientific principles to engineering applications*, pages 21–40. CRC Press/Balkema (eds. J. F. Chen, J. Y. Ooi, J. G. Teng), Edinburgh, 2008.

D. Muir Wood and K. Maeda. Changing grading of soil: effect on critical states. *Acta Geotechnica*, 3(1):3–14, 2008.

D. Muir Wood, K. Belkheir, and D.-F. Liu. Strain-softening and state parameter for sand modelling. *Géotechnique*, 44(2):335–339, 1994.

D. Muir Wood, M. L. Lings, D. F. T. Nash, and A. Gajo. Anisotropy of soils: laboratory measurements and constitutive implementation. In Publications Committee of the 15th ICSMGE, editor, *Proc. 15th ICSMGE*, volume 1, pages 321–324, Istanbul, 2001. Balkema.

M. Oda and H. Kazama. Microstructure of shear bands and its relation to the mechanisms of dilatancy and failure of dense granular soils. *Géotechnique*, 48(4):483–494, 1998.

K. H. Roscoe, A. N. Schofield, and C. P. Wroth. On the yielding of soils. *Géotechnique*, 8(1):22–52, 1958.

A. N. Schofield and C. P. Wroth. *Critical state soil mechanics.* McGraw-Hill, London, 1968.

M. A. Stroud. *The behaviour of sand at low stress levels in the simple shear apparatus.* PhD thesis, University of Cambridge, Cambridge, 1971.

S. Sture, M. M. Alawi, and H.-Y. Ko. True triaxial and directional shear cell experiments on dry sand, 1988. Final report GL-88-1 to Department of the Army, US Army Corps of Engineers, Vicksburg.

D. W. Taylor. *Fundamentals of soil mechanics.* John Wiley, New York, 1948.

I. Vardoulakis. Equilibrium bifurcation of granular earth bodies. In *Advances in analysis of geotechnical instabilities*, volume SM study 13, paper 3, pages 65–119. University of Waterloo Press, Waterloo, Ontario, 1978.

C. P. Wroth. Soil behaviour during shear—existence of critical voids ratios. *Engineering*, 186:409–413, 1958.

M. Zytynski, M. F. Randolph, R. Nova, and C. P. Wroth. On modelling the unloading-reloading behaviour of soils. *International Journal for Numerical and Analythical Methods in Geomechanics*, 2:87–94, 1978.

# Creep Versus Transient Loading Effects in Geotechnical Problems

Claudio di Prisco [*]

[*] Dipartimento di Ingegneria Strutturale, Politecnico di Milano, Milano, Italy

## 1 Introduction

As already observed, when cyclic environmental loading is taken into consideration, it becomes crucial to account for both huge numbers of cycles and very long periods of time. Very often in geotechnics, distinguishing between displacements/settlements due to cyclic loading and viscous effects becomes quite difficult but in many problems, if the main issue is to quantify the accumulation of strains with time, this is essential. Moreover, in some cases, when frequencies characterizing the loading are quite high, the two phenomena are also strictly coupled and can influence each other. Therefore, in this chapter, a very practical problem (the evaluation of displacements along an unstable slope) is first broached. Subsequently the theory of standard visco-plasticity is briefly presented and an approach based on multiple mechanism visco-plasticity is discussed.

As already specified in the introduction to this volume, with term environmentally induced cyclic loading a large variety of loads are referred to. Among these, to stress the importance of the time factor in evaluating the response of geotechnical systems to this type of load, two very common problems will be tackled: (a) the interpretation of inclinometer data collected along a potentially unstable slope and (b) the evaluation of the settlements of an offshore structure due to wave loading. In the first case, the response of partially activated landslides to the cyclic variation in the water table due to intense seasonal rainfalls is employed to highlight the role of time and of delayed plasticity when failure conditions are reached, or better when additional resistant resources are absent. In the second case, with term, environmentally induced loading, the sea waves are intended and, in contrast, the object consists of evaluating settlements of an offshore structures due to cyclic loads of very small amplitudes under loading conditions very far from the ultimate ones. In this last case the number of cycles is enormous whereas in the previous one it is not; in this case the frequency of loading is quite high, whilst in the previous, frequencies are very low.

## 2 From Elasto-Plasticity to Visco-Plasticity: Perzyna's Approach

As can be easily inferred, the fabric rearrangement of the micro-structure of soils takes place with time and is generally due either to physical agents, chemical reactions or to the variation of loading conditions; in this latter case, the time period during which such an evolution occurs depends severely on the type of material considered: some minutes/hours when granular soils are concerned (di Prisco and Imposimato, 1996), a few weeks/months for clays and rocks. When a material is loaded, a stress wave passes through the continuum. If the load increment induces a micro-structural rearrangement within the continuum, at the micro-scale the stress distribution changes with time and the micro-structural fabric continuously modifies. The two effects are strictly linked and it is difficult defining which is the cause and which is the consequence. In granular soils, for instance, the final micro-structural configuration is reached passing throughout a huge number of intermediate not equilibrated configurations: the final result is not unique but statistically determined. The time period necessary to obtain the final configuration (several minutes) is the required amount to the micro stresses in order to reach a final equilibrated distribution with the external applied macro stresses. The internal fabric rearrangement is caused by the micro accelerations, which develop in order to balance the externally imposed macro-load increments. When a new equilibrated micro-configuration is reached, the kinetic process may be assumed to be concluded. This physical time may be interpreted as the time period during which plastic strains take place. For this reason the term delayed plasticity may be considered to be more appropriate than visco-plasticity. In clayey soils the viscous processes occurring at the micro-scale are strictly related to the micro-structural rearrangement and the time scale dilates abruptly. Analogously, viscous effects in granular materials are more evident when high confining pressures are taken into account (Lade, 1994). Indeed, when confining pressures are sufficiently severe, grains are subjected to dramatic stress concentrations, which may cause local grain crushes and the grains cease to be interpretable as inert particles. Since the development of elastic strains does not require a micro-structural rearrangement, the elastic response, which is necessarily linked to the current internal micro-structure, can be assumed to be quasi-instantaneous. The most simple way of experimentally analysing the time dependence of the mechanical behaviour of materials is to perform load controlled tests.

Standard creep tests are performed by imposing finite load increments followed by variable time periods during which the boundary conditions are kept constant. In soil mechanics, the most common tests are the oedometric tests on saturated clays. As is well known, in this case, because of the consolidation phenomenon, it is sometimes quite delicate separating the effects due to viscosity from those induced by the change with time in the pore water pressure. Only by using theoretical tools such as that introduced by Terzaghi with reference to one dimensional consolidation, may such a goal be achieved. For many and various reasons, strain controlled test results obtained by imposing different strain rates or jump tests during which the strain rate is varied instantaneously are more difficult of interpretation (Tatsuoka et al., 2001).

As is well known, from an engineering point of view, the time factor plays a key role in the following cases:

- The overconsolidation induced by viscous strains which develop with time (ageing) influences dramatically the mechanical response of in situ soils and consequently common engineering predictions. With time, even if effective stress is kept constant, void ratio decreases. When the material is subsequently loaded, it behaves like an overconsolidated material, but only for limited increments of load.
- During undrained load controlled tests, even though loads are kept constant, both when clays (Singh and Mitchell, 1968) and loose sands (di Prisco and Imposimato, 1999) are considered, unstable phenomena may take place with time. In this case the phenomenon is not a typical tertiary creep such as that experimentally observed when rocklike materials are tested. The kinematic constraint on the volumetric strain rate ($\dot{\varepsilon}_{\mathrm{vol}} = 0$), as well as the time dependency of the material mechanical behaviour, induce a variation in the effective state of stress: in particular a decrease in the effective mean pressure. When the initial effective stress level is sufficiently high, unexpected specimen collapses may occur.
- Also under drained conditions, when the system is statically redundant, viscous effects may induce relevant changes in the stress state (earth forces against retaining structures may change with time, deep anchorage forces in rocks and fine soils may decrease severely with time, slope movements may develop with time even without any loading perturbation).
- When rocklike materials are considered, one dimensional creep tests are usually performed. In this case, even though the load is kept constant, unstable responses may occur: after a previous stable phase

during which a continuous decrease in strain rate takes place (primary creep), the mechanical response becomes unstable, i.e. the strain rate increases (tertiary creep). In this case the phenomenon is very complex because damage evolution and fracture propagation with time should be taken into account at the same time.

## 2.1 Constitutive Modelling

Obviously, standard elasto-plasticity cannot be employed to simulate any one of the previous time dependent mechanical phenomena. In fact, as was stressed in the previous chapter, elasto-plasticity disregards a priori the time factor and it assumes the mechanical response of the material to any load increment to be instantaneous. For this reason, the elasto-plastic constitutive relationship can be written as it follows:

$$\dot{\varepsilon}_{ij} = \dot{\varepsilon}^{\mathrm{el}}_{ij} + \dot{\varepsilon}^{\mathrm{pl}}_{ij} = C^{\mathrm{ep}}_{ijhk}\dot{\sigma}'_{hk} \tag{1}$$

where $C^{\mathrm{ep}}_{ijhk}$ stands for the elasto-plastic compliance tensor, $\dot{\varepsilon}^{\mathrm{el}}_{ij}$ is the elastic strain increment tensor, $\dot{\varepsilon}^{\mathrm{pl}}_{ij}$ is the plastic one and the dot stands for the time-derivative, $\dot{\varepsilon}_{ij}$ and $\dot{\sigma}'_{hk}$ for the strain and the effective stress rate tensor, respectively. As is evident, (1) implies that strains cannot evolve without any effective stress perturbation.

To make a plastic constitutive model time-dependent, many different approaches have been conceived. Here, the two most common ones will be briefly cited. The first one is more suitable for capturing the effects of physical/chemical agents (causing for instance damage processes) on the mechanical behaviour of materials, the second, which is referred to as *delayed plasticity*, and is suitable for describing the time dependent mechanical response to mechanical perturbations.

**Approach 1** - Within the framework of standard single potential elasto-plasticity, two sets of hardening parameters are defined: $\alpha_{ij}$ and $\psi_i$: the former depending, as usual, on irreversible strains $\varepsilon^{\mathrm{pl}}_{rs}$, and the latter depending directly on time $t$. Therefore, the yield function $f$ (Olszak and Perzyna, 1966; Nova, 1982; Sekiguchi, 1984 and Kamei and Hirai, 1990) can be symbolically written as follows:

$$f = f\left(\sigma'_{ij},\ \alpha_{ij},\ \psi_i\right) \tag{2}$$

The hardening rules for the two sets of state variables are

$$\begin{aligned} \dot{\alpha}_{ij} &= \frac{\partial \alpha_{ij}}{\partial \varepsilon_{rs}^{\text{pl}}} \dot{\varepsilon}_{rs}^{\text{pl}} \\ \dot{\psi}_i &= \frac{\partial \psi_i}{\partial t} \dot{t}, \end{aligned} \tag{3}$$

where $\partial$ stands for the partial derivative, whereas $t$ and $\dot{t}$ stand, respectively, for the time and the time rate. By imposing both flow and consistency rules:

$$\begin{aligned} \dot{\varepsilon}_{ij}^{\text{pl}} &= \Lambda \frac{\partial g}{\partial \sigma'_{ij}} \\ \dot{f} &= \frac{\partial f}{\partial \sigma'_{ij}} \dot{\sigma}'_{ij} + \frac{\partial f}{\partial \alpha_{ij}} \dot{\alpha}_{ij} + \frac{\partial f}{\partial \psi_{ij}} \dot{\psi}_{ij} \end{aligned} \tag{4}$$

where $g$ stands for the plastic potential, the plastic multiplier $\Lambda$ is obtained:

$$\Lambda = -\frac{\dfrac{\partial f}{\partial \sigma'_{ij}} \dot{\sigma}'_{ij} + \dfrac{\partial f}{\partial \psi i} \dot{\psi}_i}{\dfrac{\partial f}{\partial \alpha_{ij}} \dfrac{\partial \alpha_{ij}}{\partial \varepsilon_{rs}^{\text{pl}}} \dfrac{\partial g}{\partial \sigma'_{rs}}} \tag{5}$$

Plastic strains are easily distinguished from the viscous ones since in (5) two distinct terms can be easily separated and the relative constitutive parameters easily calibrated. Nevertheless, the coupling between the two mechanisms is imposed by the denominator in (5).

**Approach 2** - According to the delayed plasticity theory introduced by Perzyna (Perzyna, 1963), the mechanical response once more is defined by the superimposition of an instantaneous elastic and a delayed plastic strain rate tensor, but the visco-plastic strain increment $\dot{\varepsilon}_{ij}^{\text{vp}}$ substituting $\dot{\varepsilon}_{ij}^{\text{pl}}$ in (1) is defined, in this case, as follows:

$$\dot{\varepsilon}_{ij}^{\text{vp}} = \tilde{\gamma}\ \Phi(f) \frac{\partial g}{\partial \sigma'_{ij}} \tag{6}$$

where $\Phi(f) \geq 0$ is the viscous nucleus and $\tilde{\gamma}$ is a positive constitutive parameter. This means that the time dependence is introduced by modifying the flow rule and by abolishing the consistency rule. The plastic potential defines the direction of the visco-plastic strain rate tensor while the yield function influences its modulus by means of the viscous nucleus $\Phi$. As consistency rule is abolished, the yield function may be either positive or negative, without any constraint, i.e. the stress state may be either external or internal with respect to the yield locus. The most general approach of such a visco-plastic theory assumes a viscous nucleus depending on the over-stresses (Eisenberg and Yen, 1981and Tsakmakis, 1996), which may be interpreted from a micro-structural point of view as a measure of the unbalanced forces. Phillips and Wu (1973) suggest in (6) $f$ to be substituted by the distance between the current state of stress and a conjugated state of stress on the current yield locus obtained by means of a radial mapping rule.

Implicitly, (6) assumes that the yield function may be used as a scalar measure (and from this the great success of this simplified approach derives) of the probability of occurrence with time of irreversible strains.

From (6) one derives that, even if the effective state of stress is kept constant, delayed plastic strains may develop with time. If strain hardening is assumed, also state variables will change with time, and, consequently, the yield locus, too (Michael and Katona, 1984, Adachi et al., 1987, di Prisco and Imposimato, 1996).

The viscous nucleus is generally assumed to be a monotonic function of $f$ and, in particular:

$$\frac{d\Phi}{df} \geq 0 \tag{7}$$

(6) and (7) imply that, if the effective stress state is kept constant and, for instance, the plastic potential is assumed to be exclusively dependent on the effective state of stress, in case of positive hardening, strain rates continuously decrease with time. Indeed, the strain acceleration tensor is defined as follows:

$$\ddot{\varepsilon}_{ij}^{\mathrm{vp}} = \tilde{\gamma}\, \dot{\Phi}\,(f)\, \frac{\partial g}{\partial \sigma'_{ij}} + \gamma\, \Phi\,(f)\, \frac{\partial}{\partial t}\left(\frac{\partial g}{\partial \sigma'_{ij}}\right) \tag{8}$$

where

$$\dot{\Phi} = \frac{\partial \Phi}{\partial f}\frac{\partial f}{\partial t} = \frac{\partial \Phi}{\partial f}\left(\frac{\partial f}{\partial \sigma'_{ij}}\dot{\sigma}'_{ij} - H\, \Phi\,(f)\right) \tag{9}$$

$$H = -\frac{\partial f}{\partial \alpha_{ij}}\frac{\partial \alpha_{ij}}{\partial \varepsilon_{rs}^{\mathrm{vp}}}\frac{\partial g}{\partial \sigma'_{rs}} \tag{10}$$

being $H$ the hardening modulus. Therefore, from (7), (9) and (10), since $\Phi\,(f) \geq 0$ and $\dot{\sigma}'_{ij} = 0$, when $H > 0$, $\dot{\Phi} < 0$ and, as a consequence, if $\frac{\partial}{\partial t}\left(\frac{\partial g}{\partial \sigma'_{rs}}\right) = 0$ it derives that $\left\|\dot{\varepsilon}_{ij}^{\mathrm{vp}}\right\|$ always reduces with time. On the contrary, when $H < 0$, tertiary creep can be reproduced, since in this case a continuous increase in the visco-plastic strain rate modulus is obtained.

In case of two different plastic mechanisms are accounted for, even the

transition from primary to secondary and tertiary creep can be simulated (Fig. 1).
When limit state is reached, state variable evolution stops. As a consequence, during load controlled tests, if the load imposed overpasses material strength and the mechanical response is assumed to be ductile, strain rate assumes a positive constant value Indeed, under these conditions $H = 0$ and $\frac{\partial}{\partial t}\left(\frac{\partial g}{\partial \sigma'_{ij}}\right) = 0$ and, as a consequence, during creep tests, $\left\|\ddot{\varepsilon}_{ij}^{\mathrm{vp}}\right\| = 0$.

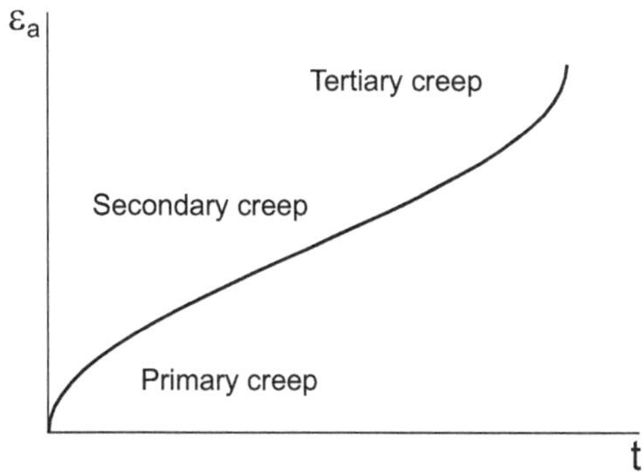


**Figure 1.** Schematic representation of the unstable mechanical response of a rocklike material during a standard compression triaxial creep test ($\varepsilon_a$ stands for axial strain).

The positive constitutive parameter $\tilde{\gamma}$ in (6) influences the strain rate and consequently the rapidity with which the asymptotic strain value is reached. When $\tilde{\gamma} \to \infty$ the mechanical response becomes instantaneous and standard elasto-plasticity is recovered. To describe the material's mechanical response during the evolution of time, even the choice of the viscous nucleus is quite important. For instance, when either both $f$ is linearly dependent on $\sigma'_{ij}$ and $\Phi(f)$ on $f$ or, more generally, when $\frac{\partial \Phi}{\partial \sigma'_{ij}}$ is constant, this visco-plastic approach reduces to the standard Maxwell viscous model.
By contrast, a standard definition for $\Phi(f)$ is to assume:

$$\Phi(f) = \left\langle f^{\tilde{\alpha}} \right\rangle \tag{11}$$

where $\tilde{\alpha}$ is a constitutive parameter and brackets impose to consider only the positive branch of $f$, whereas for negative values of $f$ the viscous nucleus is assumed to be nil. This implies a strong non-linearity in the definition of the viscous nucleus, reinterpreting, in the light of visco-plasticity, the standard Kotter constitutive modelling constraints of plasticity ($f\Lambda = 0$, $\dot{f}\Lambda = 0$).

An alternative highly non-linear definition for $\Phi(f)$ is given here below:

$$\Phi(f) = e^{\tilde{\alpha} f}; \tag{12}$$

in this case, the viscous nucleus is always positive. Even in case of $f < 0$, that is within the yield locus, plastic strains may develop and yield locus may evolve. Owing to this hypothesis, ageing phenomenon can be easily simulated.

When this definition for the viscous nucleus is employed, it is possible to demonstrate that:

$$\lim_{t \to +\infty} \int_t^{0^+} \dot{\varepsilon}_{ij}^{\text{vp}}(\tau)\, d\tau = \lim_{t \to +\infty} \int_t^{0^+} \tilde{\gamma}\, \Phi(f) \frac{\partial g}{\partial \sigma'_{ij}} d\tau = +\infty \tag{13}$$

i.e., when the time period becomes unlimited, although the strain rate progressively tends to decrease in size, plastic strains do not reach a finite value. From an engineering viewpoint, this is not a strong shortcoming, since with time the accumulation rate of irreversible strains tends to zero.

Generally, the calibration procedure for viscous parameters $\tilde{\gamma}$ and $\tilde{\alpha}$ is totally uncoupled with respect to the calibration of the plastic parameters. At a first approximation, strain controlled test results are usually considered to be time-independent and are used to calibrate the plastic parameters. By contrast, $\tilde{\gamma}$ and $\tilde{\alpha}$ are easily calibrated on the experimental data derived from standard creep tests.

During creep tests, when a load increment is instantaneously applied, the image point of the effective state of stress can stay outside the yield locus, (Fig. 2a) and $\Phi(f)$ can assume large positive values. Subsequently, it rapidly decreases. When a very long time period between two subsequent incremental load steps elapses, if the viscous nucleus is chosen according to (12), the effective state of stress, at the end of the time period, will stay within the yield locus, the $\Phi(f)$ value remains small and the initial slope of the axial strain versus time curve is not steep. On the contrary, when the specimen is rapidly loaded, i.e. the load increments are very close to each other, the effective state of stress continuously remains outside the yield locus, the $\Phi(f)$ value is quite large and the strain increment is quantitatively relevant.

To discuss the viscous nucleus definition (12) and the Perzyna's assumption (6), in the following both undrained and q-constant experimental triaxial test results will be briefly considered, respectively.

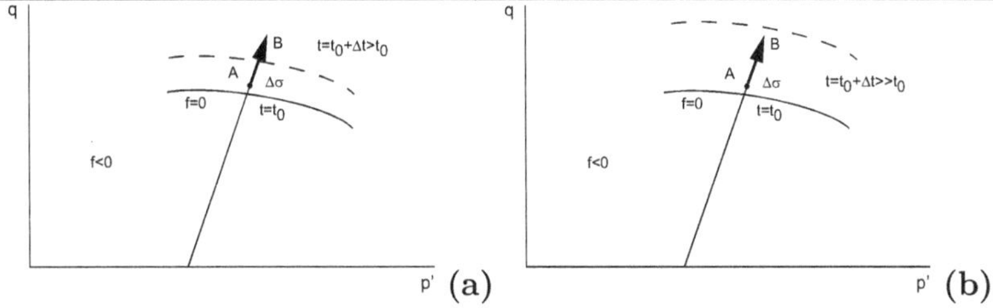


**Figure 2.** Yield function in the triaxial plane ($q - p'$, where $q$ is the difference between the axial and the radial stress, while $p'$ is the effective mean pressure) and effective stress state: (a) high stress rate; (b) low stress rate. $\Delta t$ is the time period elapsing between two different stress increments.

1. Under undrained triaxial conditions, even if the total state of stress is kept constant, delayed strains may cause an increase in pore pressure. The increase in pore pressure is associated with a decrease in the effective mean pressure and an increase in the effective stress level. In turn this may cause a further increase in pore pressure. This process may converge or not, it depends on the load time history imposed and on the stress level at which the drainage valve is closed. In Fig. 3 the effective stress path corresponding with a collapsing response relative to a loose sand specimen, as well as the axial strain and the pore pressure experimental and numerical curves, are shown. As is suggested in Fig. 3c, during the undrained creep test, the effective stress path enters into the yield locus and (at the beginning) the axial strain rate decreases rapidly. When the yield locus is reached again (the corresponding stress level (point B) is much higher), the axial strain rate increases and collapse takes place. In Fig. 3 experimental data are compared with numerical simulations obtained by employing a single potential strain hardening visco-plastic model with a viscous nucleus definition coincident with that expressed by (12) (di Prisco and Imposimato, 1999). Viscous models, characterized by a single plastic potential and a standard definitions for the viscous nucleus, that is allowing the development of irreversible strains only outside the yield locus, would not be capable of reproducing these experimental results. In fact, within the yield locus the elastic strains, associated with the decrease in the effective mean pressure would cause a volumetric dilation and the effective stress path would have stopped.
2. When drained triaxial q-constant creep tests are performed on loose sand specimens, and, in particular, when unloading steps are im-

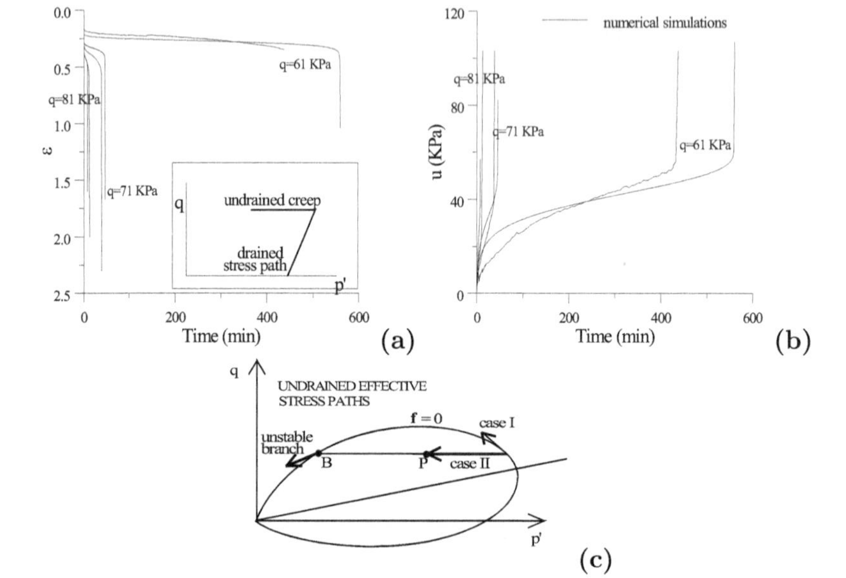


**Figure 3.** Loose sand specimens; experimental results and numerical simulations concerning undrained creep tests at different initial stress levels ($p'_0 = 100kPa$): (a) axial strain versus time, (b) pore pressure $u$ versus time, (c) schematic representation of the effective stress path for undrained creep tests (di Prisco and Imposimato, 1999).

posed ($\Delta p' < 0$), the material suffers an instantaneous dilatative volumetric response, followed by a continuous decrease in volume (Fig. 4): capital letters refer to the instants of time corresponding with cell pressure decrements. This mechanical response seems to confirm the distinction introduced by Perzyna (Perzyna, 1963) between reversible/immediate and plastic/delayed strain increments. Indeed, the decrement in the effective mean pressure necessarily causes a negative (dilation) volumetric elastic strain increment, while the delayed plastic part is positive. By increasing the stress level, the instantaneous elastic effect seems not to change, but the following plastic evolution becomes more quantitatively relevant. In Fig. 4c the relative numerical simulations obtained by means of the previously cited constitutive model are illustrated. The quantitative difference between experimental and numerical results is mainly due to the elastic numerical volumetric strains which are too small with respect to the experimental ones (Figs. 4a, b and c).

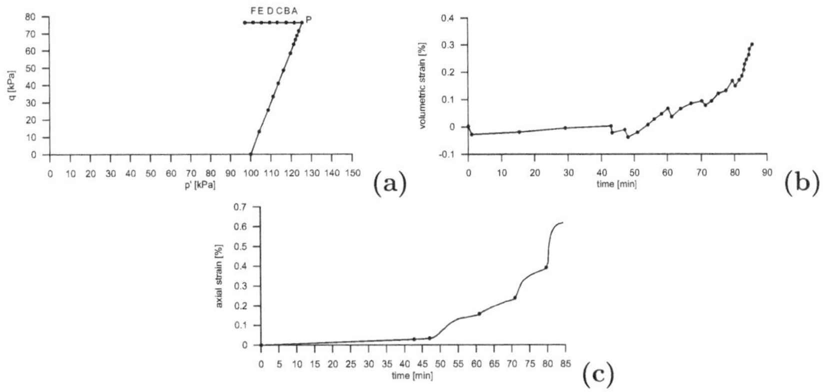


**Figure 4.** Triaxial q constant standard compression test on a loose Hostun RF sand specimen: (seven finite load decrements, after di Prisco and Imposimato (1996)). a) Effective stress path, b) volumetric experimental response, c) volumetric numerical response.

From a numerical point of view, the Perzyna's approach (Perzyna, 1963) allows us not to define a loading-unloading criterion. This makes simple the numerical implementation of this kind of constitutive models into F.E.M. codes (Hueghes and Taylor, 1978). Unfortunately, the numerical convergence to the solution is deeply linked to the chosen time step. The suitable time step depends on many factors: (i) the stress rate, (ii) the current stress level and (iii) the stress increment intensity, i.e. the current value of the viscous nucleus $\Phi(f)$. The time step must be controlled by the $\Phi(f)$ value: when $\Phi(f)$ is small, $\Delta t$ may be suitably increased.

Moreover, if the consistency rule is abolished, evolving rules for model constitutive parameters on state variable can be easily introduced. This is particularly useful when large strains take place, as during strain localization or during particular cyclic stress paths (di Prisco et al., 1993), causing large volumetric strains. For instance, when sands are considered, a dependency of the constitutive parameters on the relative density may be assumed and a unified approach for any initial density may be employed (di Prisco et al., 2002).

## 2.2 Impulsive and Cyclic Perturbations

An interesting application of visco-plastic theory consists in the analysis of the mechanical response of materials to impulsive loads. In this case, the load $F$ is increased for a fixed time period (Fig. 5). To describe the mechanical response of the system not only the load increment amplitude $\bar{F}$ but even the impulse $I$ defined as:

$$I = \int_{t_0}^{t_0+\Delta t} F(\tau)\, d\tau = \bar{F}\Delta t \tag{14}$$

must be assigned. As is known, an elasto-plastic response is totally inde-

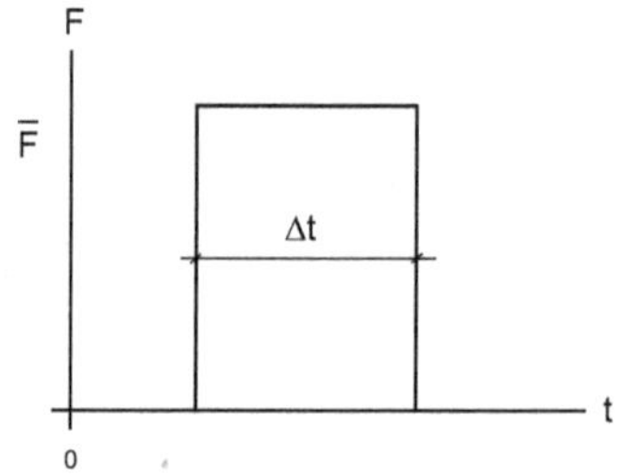


**Figure 5.** Schematic representation of an impulsive load.

pendent of impulse $I$, at least when macro-inertial effects are disregarded; on the contrary when a visco-plastic approach is employed, the response becomes severely dependent on the impulse intensity $I$ and in particular when the material response becomes potentially unstable. When, for instance, $H < 0$, collapse may occur or not according not only to $\bar{F}$ but more markedly to $I$. To stress this point, di Prisco et al. (2007) have numerically analysed, by means of a SEM (Spectral Element Method) numerical code within which a visco-plastic strain hardening constitutive model is implemented, the response of a biaxial granular soil specimen subjected to a very rapidly oscillating vertical load. By changing the frequency and the constitutive parameters defining the viscous nucleus, the accumulation rate in irreversible strains changes abruptly (Fig. 6): in some cases the collapse occurs in some others it does not. This dependence, here discussed with respect to very rapid impulsive loads, when the characteristic time is changed, that is when silty sands as well as silts or clayey sands are considered, may become quite important even in the case of environmentally induced loading characterized by very low frequencies. Analogously, under symmetric cyclic loading, the damping ratio, defined as the ratio between the area of

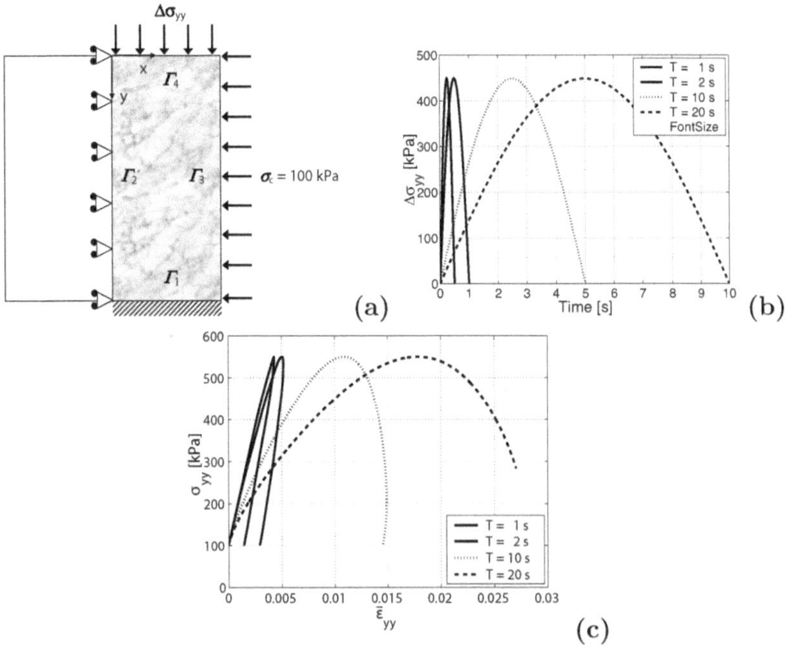


**Figure 6.** Bidimensional dry dense sand specimens: (a) isotropically consolidated under a confining pressure $\sigma_c$ of 100 kPa and axially loaded according to (b) four different loading time histories, (c) numerical vertical stress-strain curves (visco-plastic non-local version of the model after di Prisco et al. (2007)).

the cycle $W_D$ and the elastic stored energy $W_S$ (Fig. 7a), severely depends on the loading frequency. In particular, in contrast with what is assumed by standard elasto-plasticity, according to which the damping ratio depends on the strain amplitude but not on the loading frequency, visco-plasticity implies a reduction in the damping ratio at increasing values of frequency (Fig. 7b). The experimental data of Fig. 7b clearly show, however, that for high frequencies both plasticity and elasto–visco-plasticity fail to reproduce the experimental evidence. Nevertheless, here in the following, high frequencies and dynamic conditions will be disregarded. As a consequence, the visco-plastic approach will be accepted as a suitable tool for simulating the mechanical response of the material.

To stress such a point, let take now into consideration, as already anticipated in the introduction, the case of a landslide with a geometry that,

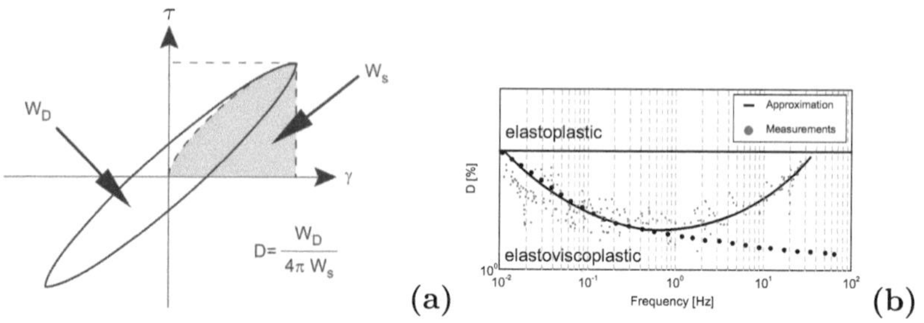


**Figure 7.** Influence of frequency on the damping ratio $D$ (experimental data concerning torsional tests on a silty sand - where $\tau$ and $\gamma$ stand for the shear stress and strain, respectively - after Meng (2003)):(a) schematic representation of the method used for the evaluation of $D$; (b) damping ratio $D$ versus loading frequency (experimental data vs elasto-plastic and elastovisco-plastic trends).

for the sake of simplicity, can be modelled by an infinitely long slope. In particular, the response of the landslide to intense rainfalls is analysed under the hypothesis that the more superficial partially saturated layers are sufficiently permeable to allow the infiltration of the water. After the raining event, an increase in the water table is therefore expected. The time lag between the meteorological event and the increase in the water table is mainly a function of the hydraulic characteristics of soils under partially saturated conditions but this problem is not faced here below. When an increase in the water table level $z_w$ is recorded by piezometers (Fig. 8a), an increase in the displacement rate is also recorded by the inclinometers positioned in the unstable area (Fig. 8b). Subsequently, with time, owing both to the evapotranspiration process and seepage along the slope that takes place constantly with time, the water table level reduces again and the landslide stops evolving. By putting in order a long seasonal succession of events and by plotting the data in a semi-logarithmic plane, in case of a progressive stabilization of the process a curve similar to that plotted in (Fig. 8c) results. The inclination $\beta$ is thus a function of the amplitude of variation in $z_w$, but in particular both of the frequencies of cycles and of their duration. The limit value for $\beta$, when the maximum amplitude of cycles tends toward zero, is $\beta_0$ (the viscous parameter for the system) that, in case of an elasto perfectly visco-plastic approach employing a viscous nucleus definition of the type defined in (11), is nil, but it is nil even in the case of cycles characterized by small amplitudes, not sufficiently large for the fulfilment of condition $f > 0$, coinciding with the activation of the plastic condition.

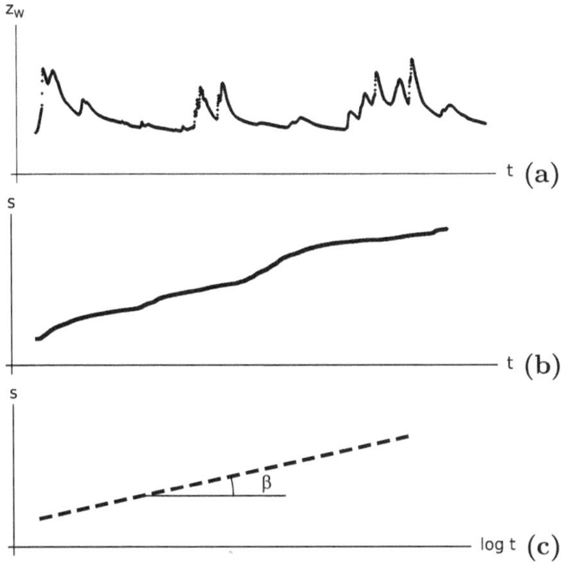


**Figure 8.** Schematic representation of the water table oscillation $z_w$ versus time (a), displacements $s$ versus time (b) and displacement versus time in semi-logarithmic scale.

To put in evidence in particular this last item, hereafter the response of the landslide to a sequence of cyclic variations in the water table is numerically analysed by employing the one dimensional scheme illustrated in (Fig. 9), where $\bar{H}$ stands for the projection of the stratum thickness along the vertical direction, $h$ for the thickness of the saturated zone of the stratum, $\alpha$ for the slope inclination, whereas, as usual, $N$ and $T$ for the forces acting normally and parallely to the slope, equilibrated with respect to the material weight and seepage forces.
By imposing under dynamic conditions the balance of momentum along the interface plane between the soil deposit and the underlying slope:

$$T = T_S - m\ddot{x}, \tag{15}$$

where $\ddot{x}$ stands for the stratum acceleration along the slope, $m$ for the soil mass of one unit wide soil deposit element $m = W/\text{g}$, with:

$$W = \gamma_d \left(\bar{H} - h\right) L + (\gamma' + \gamma_w)\, hL \tag{16}$$

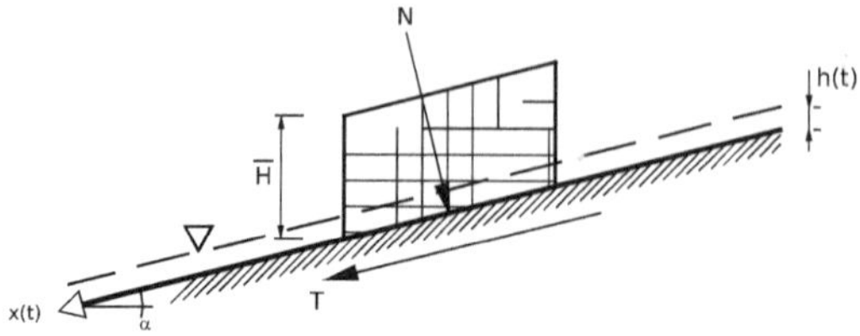


**Figure 9.** Schematic representation of the soil stratum on an inclined plane.

where g stands for the acceleration due to gravity, $\gamma_w$ for the water unit volume weight, $L$ for the landslide length, $\gamma_d$ and $\gamma'$ for the dry and buoyant unit volume soil weight, respectively;

$$T_S = W' \sin\alpha + \gamma_w h L \sin\alpha \tag{17}$$

whereas:

$$W' = \left[\gamma_d \left(\bar{H} - h\right) + \gamma' h\right] L \tag{18}$$

To determine $T$ of (15), a very approximate mechanical scheme has been employed: the shear band thickness $\Delta s$ during the process has been assumed to be constant, a priori known and within it both shear strain $\gamma$ and shear strain rate $\dot{\gamma}$ to be constant, that is:

$$\dot{\gamma} = \frac{\dot{x}}{\Delta s}. \tag{19}$$

Thus, by assuming within the band a rigid–perfectly-plastic visco-plastic one dimensional simple shear constitutive relationship, it follows that:

$$\dot{\gamma} = \frac{\tilde{\gamma}}{\sigma'_n} \left\langle \tau - \tau_f \right\rangle \tag{20}$$

where $\tilde{\gamma}$ is the viscous parameter, whereas $\tau$ and $\sigma'_n$ stand for shear and normal stresses, respectively. Since, by imposing the balance of momentum for a unit wide soil element along the direction orthogonal to the slope:

$$\begin{aligned} \sigma'_n &= \frac{W' \cos\alpha}{L} \\ \tau &= \frac{T}{L} \end{aligned} \tag{21}$$

by assuming a Coulomb-like yield locus:

$$\tau_f = \frac{W' \cos\alpha \tan\phi'}{L} \tag{22}$$

where $\phi'$ is the material internal friction angle and by substituting (18), (19), (21) and (22) into (20):

$$\frac{\dot{x}}{\Delta s \tilde{\gamma}} = \left\langle \tan\alpha + \frac{\gamma_w h \sin\alpha - m\ddot{x}}{\left(\gamma_d\left(\bar{H} - h\right) + \gamma' h\right)\cos\alpha} - \tan\phi' \right\rangle \tag{23}$$

Once the excursion of the water table is imposed, throughout the integration of (23), the evolution of the system with time can be evaluated. In Fig. 10 the response of the system to the sinusoidal trend of $h$ illustrated in Fig. 10a is represented. The integration of the non-linear differential equation (23) is quite simple. Cyclically, the accumulation of irreversible displacements will take place when the first two terms within brackets are greater than term $\tan\phi'$. By increasing either the shear band thickness or the viscous material parameter $\tilde{\gamma}$ (i.e. by decreasing the characteristic time), the displacement rate increases and, as a consequence, even the magnitude of the irreversible displacements accumulated with time, in correspondence with the same hydraulic perturbation, will increase. Since an elasto–perfectly visco-plastic mechanical behaviour is assumed, ratcheting will take place without any self feeding evolution. Indeed, since the accumulation of irreversible strains cannot induce any evolution of the constitutive relationship, the system will react always at the same manner to the same hydraulic perturbation. In contrast, if either hardening or softening are assumed, either stabilization processes or collapse will be attained, respectively. This example clearly

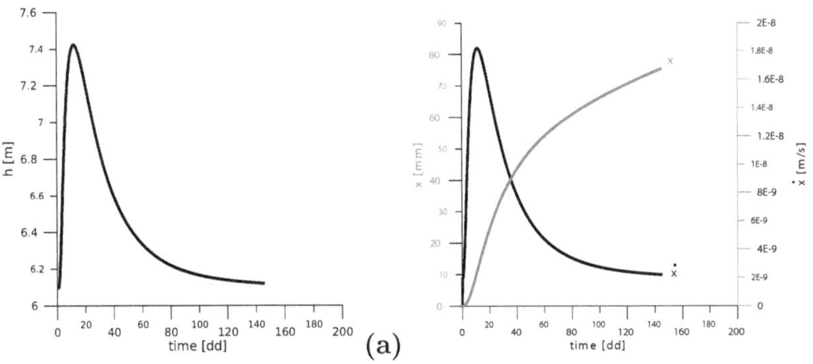


(a) (b)

**Figure 10.** Schematic representation of the mechanical response of the system during the evolution of time: (a) water table excursion; (b) displacement $x$ and displacement rate $\dot{x}$ versus time ($\alpha = 28$, $\phi' = 34.5$, $\gamma_d = 15.8\ kN/m^3$, $\gamma' = 9.2\ kN/m^3$, $\bar{H} = 40\ m$, $\Delta s = 0.5\ m$ and $\tilde{\gamma} = 7.1 \cdot 10^{-6}\ kPa/s$).

shows once more that when a visco-plastic constitutive relationship is assumed and the perturbation applied to the system is essentially asymmetric (in this case, because of the slope inclination, even the initial state of stress is severely asymmetric) the ratcheting is due to the absence of stabilizing factors.

## 3 Evaluation of Ratcheting Displacements in Granular Materials

In spite of the large number of constitutive models available in the literature based for instance on bounding plasticity theory, it is quite difficult finding constitutive approaches capable of quantitatively simulating the response of granular materials under repeated loading processes in particular when the stress state, around which the cyclic perturbation takes place, does not belong to the hydrostatic axis. In this section a theoretical framework based on multi-mechanism visco-plasticity is outlined. This is based on some theoretical assumptions that can be experimentally justified.

According to the approach introduced here below, that is designed to be quite general and not necessarily related to a certain model but adaptable to different visco-plastic models, (1) can be modified by adding two further terms and by substituting $\dot{\varepsilon}_{ij}^{\mathrm{pl}}$ with $\dot{\varepsilon}_{ij}^{\mathrm{vp}}$:

$$\dot{\varepsilon}_{ij} = \dot{\varepsilon}_{ij}^{\mathrm{el}} + \dot{\varepsilon}_{ij}^{\mathrm{vp}} + \dot{\varepsilon}_{ij}^{c} + \dot{\varepsilon}_{ij}^{r} \tag{24}$$

where $\dot{\varepsilon}_{ij}^{\mathrm{el}}$ refers to the elastic/reversible strain rates describing the response of the material at very small strains (that is $\gamma < 10^{-5}$), $\dot{\varepsilon}_{ij}^{\mathrm{vp}}$ to the mechanical response of the material at large strains (that is $\gamma > 10^{-3}$), whereas $\dot{\varepsilon}_{ij}^{c}$ to the intermediate response. Essentially, this term is hereafter introduced to simulate both the dissipated energy during the single cycle and the reduction in the material stiffness due to the increase in the loop size. Finally, $\dot{\varepsilon}_{ij}^{r}$ is added to reproduce the progressive accumulation of irreversible strains due to repeated loading (the superscript index $r$ stands indeed for ratcheting). The approach is based on the idea of calibrating the constitutive parameters relative to the four different mechanisms separately and by means of different experimental test results.

In particular, as is schematically suggested in Fig. 11, the first terms (the elasto strain-hardening visco-plastic ones) are conceived in order to simulate under standard load controlled cyclic tests a typical shake down response. This implies that, when the amplitude of the cycle is sufficiently small, irreversible strains do not develop (Fig. 11a). The third visco-plastic mechanism is conceived to simulate a response dominated by an ideal plastic adaptation where any accumulation in irreversible strains during the single cycle is missing: ratcheting is not contemplated. This implies a dissipating response of the system not associated with any evolution of the internal micro-structure (Fig. 11b).

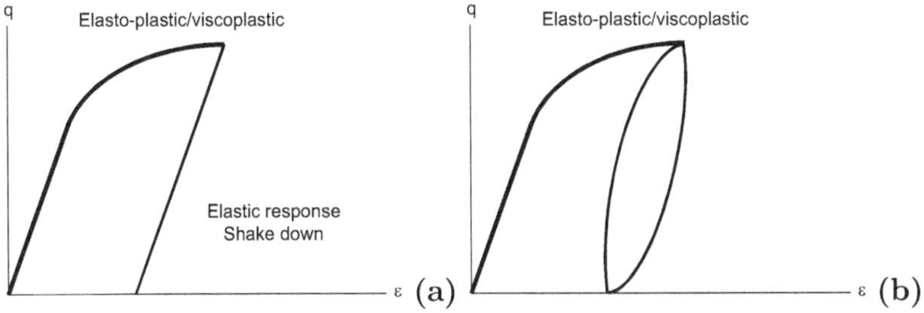


**Figure 11.** Schematic representation of the mechanical response of the system in case of shake down (a) and ideal-plastic adaptation (b).

In contrast, the last plastic mechanism is devoted to accounting for the ratcheting phenomenon. According to its definition, this can be suitable for reproducing (i) a ratcheting process characterized by a constant accumulation rate (curve a of Fig. 12 and response of Fig. 13a) - that, in the following, will be named perfect ratcheting - (ii) a ratcheting process characterized by a progressive stabilization of the response, that is a sort of progressive plastic adaptation (i.e. by a progressive reduction in the accumulation rate) (curve b of Fig. 12 and response of Fig. 13b), (iii) an unstable ratcheting process characterized by a continuous increase in the accumulation rate (curve c of Fig. 12 and response of Fig. 13c). As will be clarified below, in the case of no coupling between the principal plastic mechanism and that introduced to capture the ratcheting phenomenon, case (a) is necessarily obtained, whereas responses (b) and (c) can be simulated only by coupling the two mechanisms. To highlight this concept, in Fig. 14 the evolution of the yield function associated with the principal plastic mechanism, due to the accumulation of irreversible strains, is schematically drawn. In case (a) the yield locus does not evolve as a function of the irreversible strains due to ratcheting $\varepsilon_{ij}^{r}$; in case (b) these cause a positive hardening, i.e. in case of isotropic hardening, an increase in size of the yield locus. By contrast, in case (c), the evolution is negative (strain softening), i.e. a yield locus shrinkage. As far as the principal plastic/visco-plastic mechanism is concerned, the choice can be quite general, but with reference for instance to granular materials, it is quite suitable employing an anisotropic hardening and a non- associated flow rule. Indeed, the anisotropic hardening is essential for

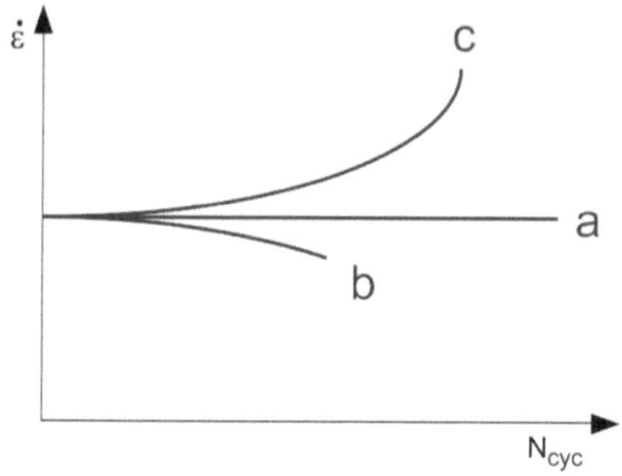


**Figure 12.** Schematic representation of the evolution of medium strain accumulation rate versus number of cycles $N_{\text{cyc}}$ for the three different responses schematically illustrated in Fig. 13.

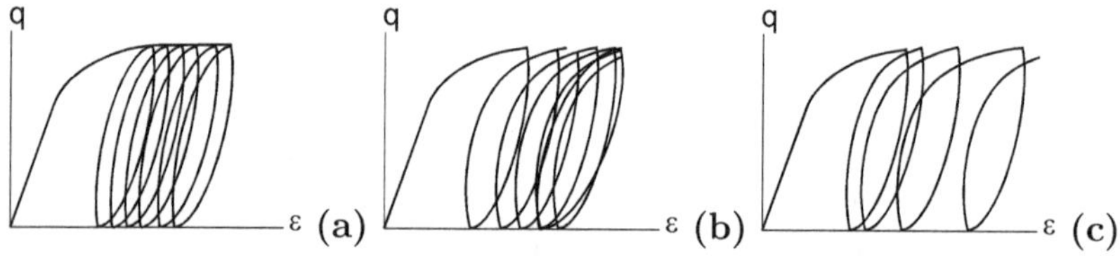


**Figure 13.** Schematic representation of the mechanical response of the system in case of perfect ratcheting (a), progressive stabilization toward the ideal plastic adaptation (b) and unstable cyclic response (c).

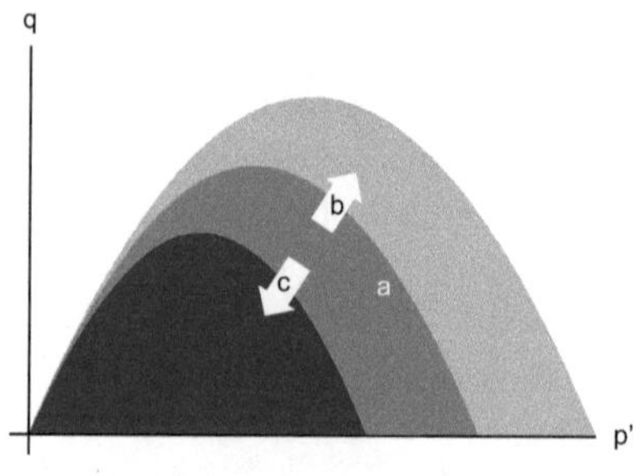


**Figure 14.** Schematic representation of the coupling between the ratcheting plastic mechanism and the principal one.

capturing the mechanical response of granular materials in case of cycles of large amplitudes (which are associated with large strains) either in the deviatoric plane or with a continuous rotation of the stress tensor principal axes. To exemplify it, in Fig. 15 experimental data obtained by employing a true triaxial apparatus (Lanier et al., 1991) are compared with numerical simulations retrieved by means of single potential anisotropic strain hardening elasto-plastic constitutive model (di Prisco et al., 1993). It is worth noting that if an isotropic strain hardening was used to simulate these experimental results, a sort of "infinitesimal" circle in Fig. 15c would be obtained. Indeed, in this case the mechanical response of the soil specimen would be, after the first monotonic loading, totally elastic. Additional observations are useless, since this subject was already discussed in the previous chapter.

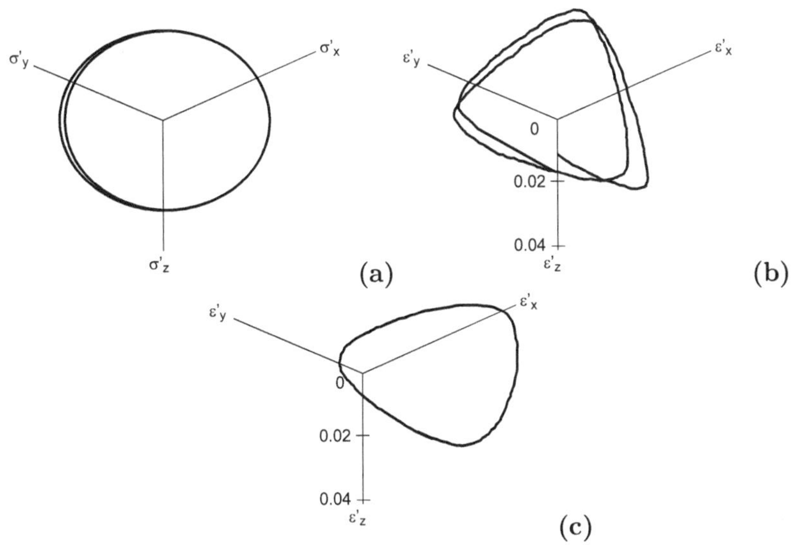


**Figure 15.** Large amplitude cyclic tests in the deviatoric plane (where $\sigma'_x$, $\sigma'_y$, $\sigma'_z$ are the effective principal stresses and $\varepsilon'_x$, $\varepsilon'_y$, $\varepsilon'_z$ the principal strains) performed on an RF Hostun sand specimen (after Lanier et al. (1991)): imposed stress path (a), experimental data (b), numerical simulations obtained by employing a isotropic strain hardening and an anisotropic strain hardening elasto-plastic constitutive model (c).

### 3.1 The Multimechanism Model

The hierarchical approach here discussed assumes a complete uncoupling between the principal plastic mechanism and the nesting multimechanism illustrated hereafter. For the sake of simplicity, even the single nesting visco-plastic mechanism are uncoupled from each other. This assumption allows us to define the hardening of the $n^{th}$ yield locus as a function uniquely of the irreversible strains governed by the corresponding $n^{th}$ plastic potential. Multi-mechanism plastic models, introduced in the 1960s, Fig. 16a, by Mroz (1967) and applied to soils for instance by Prevost (1985), Prevost and Keane (1990) and Prevost and Popescu (1993), are usually characterized by a variable number $N^*$ of plastic mechanisms, controlling the quality of the numerical simulations. A simple visco-plastic version is here described (di Prisco and Zambelli, 2009), according to which:

$$\dot{\varepsilon}^c_{ij} = \sum_{n=1}^{N^*} \dot{\varepsilon}^n_{ij} \tag{25}$$

where

$$\dot{\varepsilon}^n_{ij} = \phi^*_n \left(f_n\right) \frac{\partial g_n}{\partial \sigma'_{ij}}, \tag{26}$$

whereas $f_n$ and $g_n$ are the yield function and the plastic potential associated with the $n^{th}$ plastic mechanism, respectively, whereas the $n^{th}$ viscous nucleus for the sake of simplicity is defined as below:

$$\phi^*_n = \alpha^* \left\langle f_n \right\rangle \tag{27}$$

where, as stated above, the brackets imply a nil value for $\phi^*_n$ when $f_n < 0$ whereas $\alpha^*$ is a constitutive parameter. Each yield locus $f_n$ rotates in the effective stress space, keeping its opening constant, in accordance with the evolution of its axis $\chi^n_{ij}$ (Figs. 16b and c):

$$f_n = \beta^n I^{*n}_1 + \sqrt{J^{*n}_2} \tag{28}$$

with $I^{*n}_1 = \sqrt{3}\sigma'_{ij}\chi^n_{ij}$, $J^{*n}_2 = s^{*n}_{ij} s^{*n}_{ij}$, $s^{*n}_{ij} = \sigma'_{ij} - \frac{1}{\sqrt{3}} I^{*n}_1 \chi^n_{ij}$. By substituting in (28) $\beta^n = 0$, $g_n$ is also obtained. The evolution of $\chi^n_{ij}$ is described by a hardening rule of Prager's type (Prager and Druker, 1952):

$$\dot{\chi}^n_{ij} = w^n \dot{e}^{*n}_{ij} \tag{29}$$

where $w^n$ are constitutive parameters, while:

$$\dot{e}^{*n}_{ij} = \dot{\varepsilon}^n_{ij} - \frac{1}{\sqrt{3}} J^{*n}_1 \chi^n_{ij} \quad \text{with} \quad J^{*n}_1 = \sqrt{3}\dot{\varepsilon}^n_{ij}\chi^n_{ij} \tag{30}$$

This model is capable of describing (i) the non-linear response of the mate-

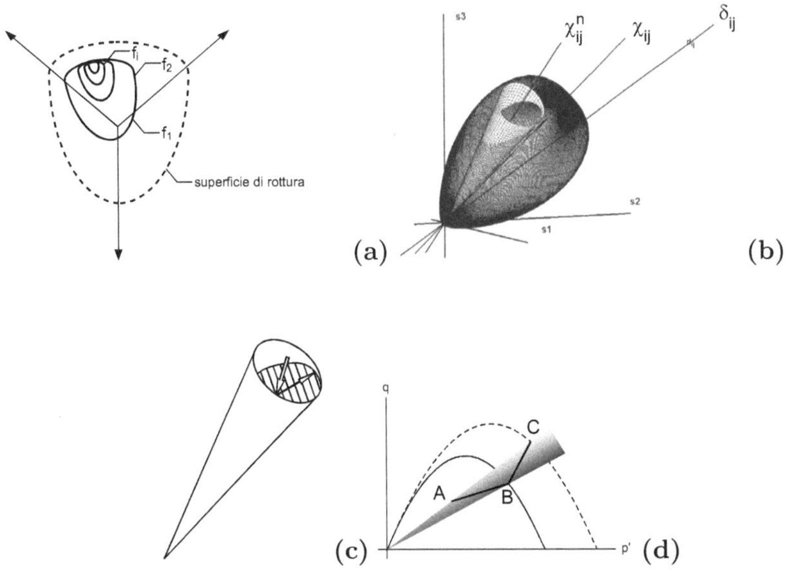


**Figure 16.** Schematic representation of nesting mechanisms: in the deviatoric plane (a), in the principal stresses space (b), nesting cones (c), in the triaxial plane (d).

rial even during small size cyclic tests, (ii) the associated energy dissipation and (iii) the decrease in the material stiffness when the loop size is increased. In other words, it is capable of simulating the previously defined ideal 'plastic adaptation' reponse. The model so far presented essentially requires the number $N^*$ of additional plastic mechanisms to be declared and the calibration of two series of constitutive parameters: $\beta^n$ and $w^n$. In Fig. 17a the $w^n$ and $\beta^n$ values employed to obtain the numerical curves of Fig. 17b are reported. The experimental curves plotted in Fig. 17b refer to a series of cyclic torsional shear test results (Zambelli et al., 2007) performed on dense Toyoura sand specimens at a constant confining pressure of 200 kPa. The points plotted refer to different cycles obtained by progressively increasing the size of each cycle.

During the unloading, the progressive activation of the different plastic mechanisms allows of always obtaining a plastic adaptative response and owing to the chosen plastic potential a perfect closed loop is always obtained and ratcheting is missing. In the example described above, the different yield loci are defined as open cones (Fig. 16c) of fixed opening angle.

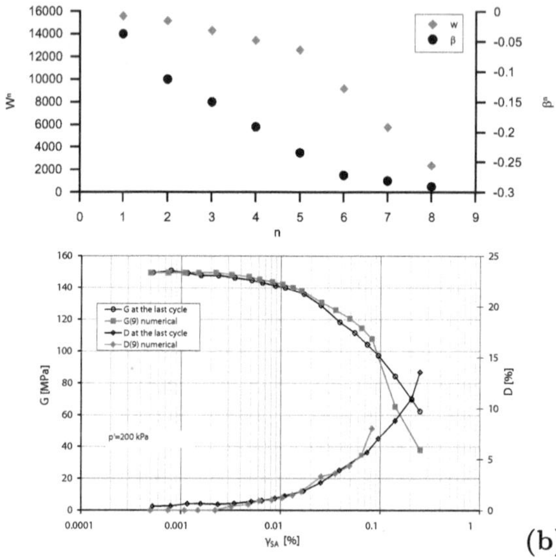


(a)

(b)

**Figure 17.** The model constitutive parameters calibration: (a) trends of the two series of constitutive parameters for each $n^{th}$ plastic mechanism ($N^* = 8$), (b) comparison between cyclic torsional shear test results after Zambelli et al. (2007) performed on dense Toyoura sand specimens and relative numerical simulations in term of the equivalent secant shear modulus $G$ versus the shear strain amplitude of the cycle $\gamma_{\mathrm{SA}}$.

### 3.2 The Ratcheting Mechanism

As was stated in the introduction to this paragraph, the principal visco-plastic mechanism provides the development of irreversible strains even if the effective state of stress is not altered. The ratcheting mechanism, in contrast, accounts for the development of irreversible strains associated with the cyclic perturbation of the effective state of stress.

The experimental data already illustrated in a previous chapter have clearly shown that

1. The irreversible strains accumulating during cyclic tests are mainly governed by the plastic potential associated with the principal plastic mechanism.
2. The intensity of the irreversible strains progressively accumulating (in case of cycles of constant amplitude) is governed by the amplitude of the cycle, by the medium stress level ($\eta$ in the triaxial plane) at which cycles are performed, and by the previous loading history. This

can be schematically summarized by assuming that, for instance, when standard triaxial compression tests are performed, $\varepsilon_a = \hat{\beta} \log N$ where $\varepsilon_a$ stands for the axial strain, $N_{\text{cyc}}$ for the number of cycles, whereas parameter $\hat{\beta}$ linearly depends on both $\Delta\eta$ (where $\eta = q/p'$) and $\eta_m$ the medium value of $\eta$ in the cycle.

Within the framework of elasto-visco-plasticty, a constitutive modelling approach capable of numerically reproducing the previously listed main features of the mechanical response of granular soils under cyclic loading can be therefore the following:

$$\dot{\varepsilon}_{ij}^{r} = \left( \sum_{n=1}^{N^*} \Phi_n^* \left( f_n \right) \right) \hat{\Phi} \left( f \right) \frac{\partial g}{\partial \sigma'_{ij}} \tag{31}$$

where $\hat{\Phi}(f)$ is an additional viscous nucleus characterized by an expression similar to that suggested in (12) but by different constitutive parameters. (31) is capable of accounting for all the items stressed above. Indeed:

- The presence of the plastic potential associated with the principal plastic mechanism for defining the direction of the ratcheting strain rate tensor derives from the assumption that small cycles work as perturbations but the evolution of the material micro-structure is essentially related once more only to its current configuration.
- The dependence on the amplitude of the cycle is provided by the term in between parentheses. When the soil specimen is not perturbed (that is when creep tests are performed), it becomes nil.
- The dependence on the current state of stress and on the loading history is provided by term $\hat{\Phi}(f)$.

Finally, the coupling between the principal and the ratcheting plastic mechanisms allow us to simulate both the progressive reduction and the progressive increase in $\dot{\varepsilon}$ (case b and c of Fig. 12). Coupling, in this case, implies irreversible strains, here indicated with the symbol $\varepsilon_{ij}^{r}$, not to be distinguished from those indicated with the symbol $\varepsilon_{ij}^{\text{vp}}$ associated with the principal plastic mechanism. Therefore, when cyclic tests of small amplitude are performed and ratcheting irreversible strains develop, the yield function $f$ progressively evolves. If this evolution causes, for instance, an increase in the size of $f$ (i.e. hardening) case b of Fig. 12 is derived. Otherwise, if a decrease in size results (typical of the mechanical response of cemented soils suffering a cyclic fatigue), case c of Fig. 12 is instead obtained. Fig. 18 (di Prisco and Zambelli, 2009) testifies the capability of this type of approach in quantitatively reproducing the mechanical behaviour of granular

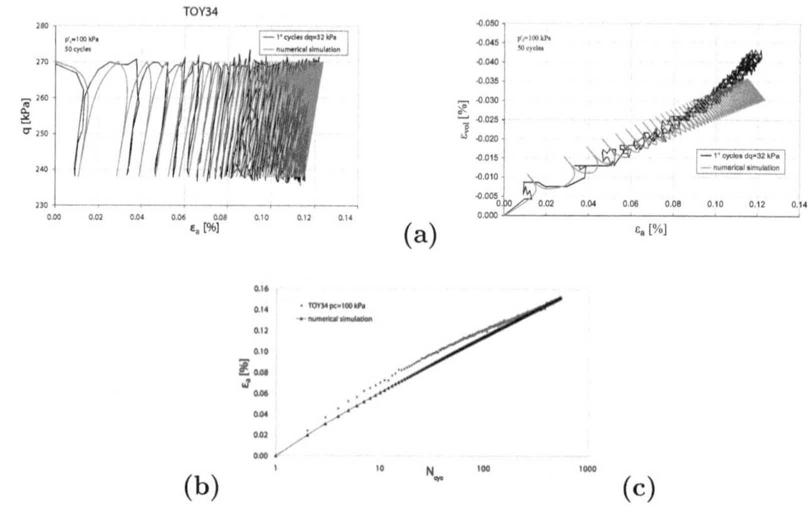


**Figure 18.** Comparison between cyclic triaxial experimental test results on Toyoura sand specimens after (Zambelli, 2006) and relative numerical simulations: a) vertical strain versus deviatoric stress, b) volumetric response, c) vertical strain accumulation as a function of the number of cycles $N_{\text{cyc}}$.

soils experimentally obtained by performing on Toyoura sand specimens cyclic drained compression triaxial tests. In particular, these data concern a test consisting of three phases. During the first phase the specimen is isotropically consolidate up to a cell pressure of 200 kPa; during the second phase the vertical stress $\sigma_a$ is increased to a value of 370 kPa whereas the radial one $\sigma_r$ is kept constant; finally, the specimen is cyclically loaded by imposing a $\Delta\sigma_a = 30\ kPa$ (even in this case $\sigma_r$ is kept constant).

### 3.3 A Numerical Simplified Application

In this paragraph, a simplified strategy, that can be considered as an extended version of the stress-path method, for evaluating settlements due to environmentally induced cyclic loading characterized by an enormous number of cycles is briefly suggested. Under the main hypothesis that during the evolution of time the local state of stress does not evolve (at any rate this could be updated by artificially adding the irreversible strains induced by cyclic loading histories), the method consists of four steps:

1. The subsoil domain is discretized and the local state of stress $\sigma'^0_{ij}$ under permanent loads is evaluated by employing a standard FEM

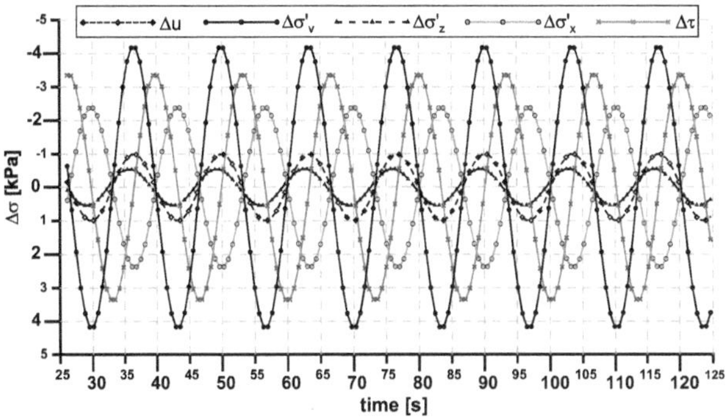


**Figure 19.** Local cyclic loading history induced by a heavy sea with a 4 meter high wave.

code, within which an elasto-plastic constitutive relationship is implemented. The results reported hereafter concern an exercise recently developed by the author (di Prisco et al., 2008) devoted to the evaluation of settlements induced by heavy sea on a very particular breakwater structure.

2. The local cyclic amplitude $\Delta\sigma'_{ij}$ is evaluated by using the same numerical code and, at least, an elastic constitutive relationship. This numerical step, to be solved, requires (i) the geometrical and stratigraphic definition of the geotechnical problem, (ii) the description of the synthetic loading history applied to the boundaries, (iii) the evaluation of both the previous and future local loading histories. In the case of the particular breakwater taken into consideration, at a certain depth under the foundation of the structure the expected loading history is illustrated, for instance, in Fig. 19. This local perturbation is larger than that calculated at the same depth without the structure, which, for the sake of brevity, is not here reported.
3. The constitutive model parameters are calibrated on standard laboratory experimental test results. Once the model parameters are calibrated, different load paths can be investigated and many observations can be retrieved from the analysis of the relative numerical results. If for instance the hierarchical model previously cited is employed, the data plotted in Fig. 20 are obtained. In particular, in Fig. 20 the influence for a normally consolidated silty loose sand both of the type of the cyclic perturbation and of its amplitude are taken into consideration. In Fig. 20a the cyclic stress paths accounted for are

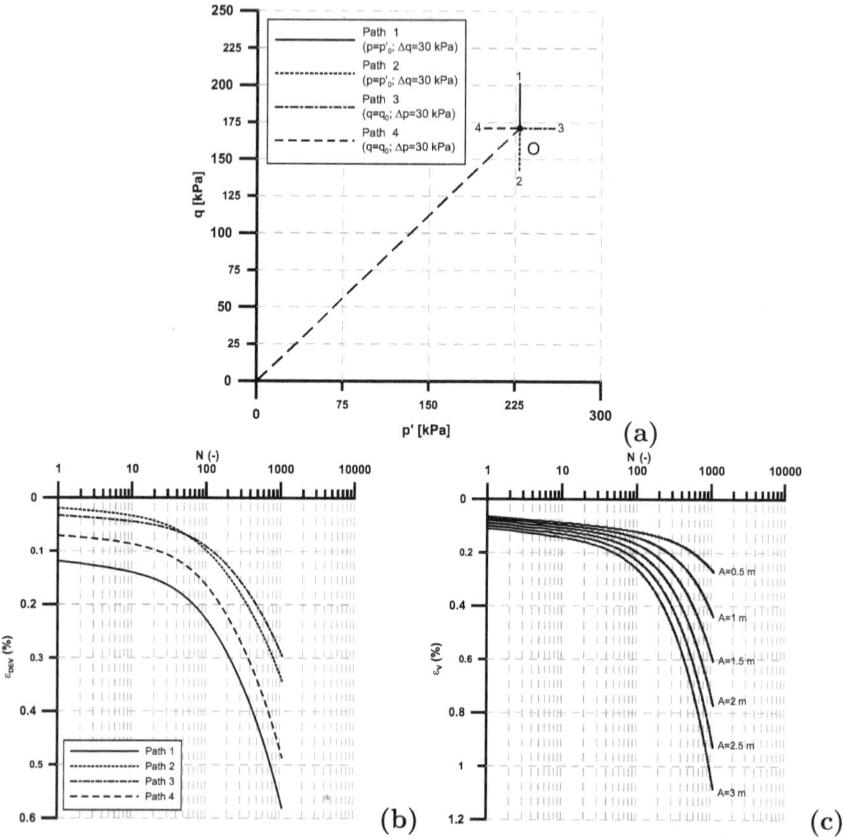


**Figure 20.** Heuristic use of the previously cited constitutive model: cyclic stress path definition (a) and deviatoric strain versus number of cycles (b), influence of the perturbation amplitude(c).

illustrated, whereas in Fig. 20b the accumulation of deviatoric strains is plotted versus the number of cycles for the four different stress paths of Fig. 20a. Finally, in Fig. 20c, for stress path 1 of Fig. 20a a comparison of the numerical simulations relative to different amplitudes is reported.

4. The accumulation of irreversible strains is locally evaluated by imposing on the REV (the Representative Elementary Volume) the past and future computed stress path (even undrained loading in case of offshore structures has to be accounted for) and by employing the

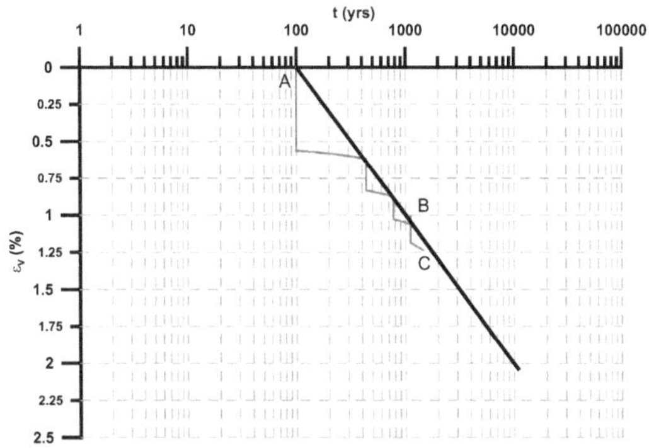


**Figure 21.** Exemplifying evolution with time of vertical deformation $\varepsilon_v$ relative to a given soil element: segment AB refers to the past, segment BC to the future.

constitutive model previously calibrated. With reference to the exercise here discussed, the perturbation is locally assumed to be applied under undrained conditions and only extreme events are accounted for. The applied local loading history is schematically given by the succession of three phases: phase 1, during which 1074 cycles under undrained conditions are applied, phase 2, during which dissipation of excess pore pressure, at constant total stresses, is permitted, phase 3, during which a sort of creep under drained conditions take place. The duration of this period is dominated by the return period of the considered event. In the case here analysed this is equal to 333 years. In 1000 years (representing the past of the stratum accounted for) three extreme events are assumed to have occurred. The last step, by contrast, is relative to the designed future for the structure, since within the next 50 years only one extreme event is expected. In Fig. 21 the inclined lines among the pseudo-vertical steps are due to viscous effects. The interpolating straight line describes the local effect of extreme events in the past: as is evident, the straight line inclination severely depends on (i) the return period of the event, on (ii) its intensity and (iii) duration.

5. The settlements are computed by integrating on the finite volume representing the spatial domain of interest the irreversible strains calculated at point 4.

## List of Symbols

| | | |
|---|---|---|
| el | elastic component | §2, §3 |
| ep | elasto-plastic component | §2, §3 |
| pl | plastic component | §2, §3 |
| vp | visco-plastic component | §2, §3 |
| $C^{\mathrm{ep}}_{ijhk}$ | elasto-plastic compliance tensor | §2.1 |
| $D$ | damping ratio | §2.2 |
| $\dot{e}^{*n}_{ij}$ | $= \dot{\varepsilon}^{n}_{ij} - \frac{1}{\sqrt{3}} J^{*n}_{1} \chi^{n}_{ij}$ | §3.1 |
| $f$ | yield function | §2, §3 |
| $f_n$ | yield function associated to the $n^{th}$ plastic mechanism | §3.1 |
| $F$ | load increment amplitude | §2.2 |
| g | acceleration due to gravity | §2.2 |
| $g$ | plastic potential | §2.1 |
| $g_n$ | plastic potential associated to the $n^{th}$ plastic mechanism | §3.1 |
| $G$ | equivalent secant shear modulus | §3.1 |
| $h$ | thickness of the saturated zone of the stratum | §2.2 |
| $H$ | hardening modulus | §2.1 |
| $\bar{H}$ | projection of the stratum thickness along the vertical direction | §2.2 |
| $I$ | impulse | §2.2 |
| $I^{*n}_{1}$ | $= \sqrt{3}\sigma'_{ij}\chi^{n}_{ij}$ | §3.1 |
| $J^{*n}_{1}$ | $= \sqrt{3}\dot{\varepsilon}^{n}_{ij}\chi^{n}_{ij}$ | §3.1 |
| $J^{*n}_{2}$ | $= s^{*n}_{ij} s^{*n}_{ij}$ | §3.1 |
| $L$ | landslide length | §2.2 |
| $N$ | force acting perpendicular to the slope | §2.2 |
| $N_{\mathrm{cyc}}$ | number of cycles | §3 |
| $N^{*}$ | number of plastic mechanisms | §3.1 |
| $m$ | soil mass of one unit wide/long soil deposit element | §2.2 |
| $p'$ | effective mean stress | §2.1 |
| $q$ | distortional stress | §2.1 |
| $s$ | displacement | §2.2 |

| | | |
|---|---|---|
| $s_{ij}^{*n}$ | $= \sigma'_{ij} - \dfrac{1}{\sqrt{3}} I_1^{*n} \chi_{ij}^n$ | §3.1 |
| $t$ | time | §2.1 |
| $T$ | force acting parallely to the slope | §2.2 |
| $T_s$ | $= W' \sin\alpha + \gamma_w hL \cdot \sin\alpha$ | §2.2 |
| $u$ | pore pressure | §2.1 |
| $w^n$ | constitutive parameters | §3.1 |
| $W$ | $= \gamma_d \left(\bar{H} - h\right) + \left(\gamma' + \gamma_w\right) hL$ | §2.2 |
| $W'$ | $= \left[\gamma_d \left(\bar{H} - h\right) + \gamma' h\right] L$ | §2.2 |
| $W_D$ | area of the cycle | §2.2 |
| $W_S$ | elastic stored energy | §2.2 |
| $x$ | coordinate | §2.2 |
| $\dot{x}$ | stratum displacement rate along the slope | §2.2 |
| $\ddot{x}$ | stratum acceleration along the slope | §2.2 |
| $z_w$ | water table level | §2.2 |
| $\alpha$ | slope inclination | §2.1 |
| $\tilde{\alpha}$ | constitutive parameter | §2.1 |
| $\alpha^*$ | constitutive parameter | §3.1 |
| $\alpha_{ij}$ | hardening parameter depending on irreversible strains | §2.1 |
| $\beta$ | inclination | §2.2 |
| $\hat{\beta}$ | parameter | §3.2 |
| $\beta_0$ | limit value for $\beta$ when the maximum amplitude of cycles tends towards zero | §2.2 |
| $\beta^n$ | constitutive parameter | §3.1 |
| $\gamma$ | shear strain | §2.2 |
| $\dot{\gamma}$ | shear strain rate | §2.2 |
| $\gamma'$ | buoyant unit volume soil weight | §2.2 |
| $\tilde{\gamma}$ | viscous constitutive parameter | §2 |
| $\gamma_d$ | dry unit volume soil weight | §2.2 |
| $\gamma_{\mathrm{SA}}$ | shear strain amplitude of the cycle | §3.1 |
| $\gamma_w$ | water unit volume weight | §2.2 |
| $\Gamma$ | plastic multiplier | §2.1 |

| | | |
|---|---|---|
| $\Delta s$ | shear band thickness | §2.2 |
| $\varepsilon_{ij}$ | strain tensor | §2, §3 |
| $\dot{\varepsilon}_{ij}$ | strain rate tensor | §2, §3 |
| $\ddot{\varepsilon}_{ij}$ | strain acceleration tensor | §2.1 |
| $\dot{\varepsilon}^{n}_{ij}$ | $= \phi^*_n (f_n) \dfrac{\partial g_n}{\partial \sigma'_{ij}}$ | §3.1 |
| $\varepsilon_a$ | axial strain | §2.1, §3.2 |
| $\dot{\varepsilon}^c$ | strain rate describing the intermediate response | §3 |
| $\dot{\varepsilon}^r$ | strain rate due to repeated loading (racheting) | §3 |
| $\varepsilon_{\mathrm{v}}$ | vertical deformation | §2.1, §3.2 |
| $\dot{\varepsilon}_{\mathrm{vol}}$ | volumetric strain accumulation | §2 |
| $\varepsilon_x$, $\varepsilon_y$, $\varepsilon_z$ | principal strains | §3 |
| $\eta$ | $= q/p'$ | §3.2 |
| $\eta_m$ | medium value of $\eta$ in the cycle | §3.2 |
| $\sigma_a$ | vertical stress | §3.2 |
| $\sigma_c$ | confining pressure | §2.2 |
| $\dot{\sigma}'_{ij}$ | effective stress rate tensor | §2, §3 |
| $\sigma_r$ | radial stress | §3.2 |
| $\sigma'_n$ | effective normal stress | §2.2 |
| $\sigma'_x$, $\sigma'_y$, $\sigma'_z$ | effective principal stresses | §3 |
| $\tau$ | shear stress | §2.2 |
| $\tau_f$ | shear stress at the yield locus | §2.2 |
| $\phi'$ | material internal friction angle | §2.2 |
| $\phi^*_n$ | viscous nucleus associated to the $n^{th}$ plastic mechanism | §3.1 |
| $\Phi$ | viscous nucleus | §2.1 |
| $\hat{\Phi}$ | additional viscous nucleus | §3.2 |
| $\chi^n_{ij}$ | axis of the $n^{th}$ yield locus | §3.1 |
| $\psi_i$ | hardening parameter depending on time | §2.1 |

## Bibliography


T. Adachi, F. Oka, and M. Mimura. Mathematical structure of an overstress elasto-viscoplastic model for clay. *Soils and Foundations*, 27(3):31–42, 1987.

C. di Prisco and S. Imposimato. Sand specimen undrained mechanical response to istantaneous load increments. In A. Khan, editor, *Proc. 7th Symposium Plasticity 1999 on Constitutive and Damage Modelling of Inelastic Deformation and Phase Transformation*, pages 557–560. Neat Press, Cancun, January 1999.

C. di Prisco and S. Imposimato. Time dependent mechanical behaviour of loose sand. *Mechanics of Cohesive-Frictional Materials*, 17(1):45–73, 1996.

C. di Prisco and C. Zambelli. A multi-mechanism visco-plastic constitutive model for simulating the phenomenon of ratcheting in granular soils. In *Proc. X International Conference on Computational Plasticity COMPLAS X*. E. Oñate and D. R. J. Owen, Barcelona, 2009.

C. di Prisco, R. Nova, and J. Lanier. A mixed isotropic kinematic hardening constitutive law for sand. In D. Kolymbas, editor, *Modern approach to plasticity*, pages 83–124. Balkema, 1993.

C. di Prisco, S. Imposimato, and E. C. Aifantis. A visco-plastic constitutive model for granular soils modified according to non-local and gradient approaches. *Int. Jour. Num. and Anal. Meth. in Geomechanics*, 26(2): 121–138, 2002.

C. di Prisco, M. Stupazzini, and C. Zambelli. Nonlinear SEM numerical analyses of dry dense sand specimens under rapid and dynamic loading. *Int. Jour. Num. and Anal. Meth. in Geomechanics*, 31(6):757–788, 2007.

C. di Prisco, C. Zambelli, and M. Secondi. Valutazione dei cedimenti indotti da onde e maree nei terreni di fondazione delle opere di sbarramento alle Bocche di Porto della Laguna Veneta. Technical Report 3 (in Italian), Politecnico di Milano and Technital S.p.A., Milan, 2008.

M. A. Eisenberg and C. F. Yen. A theory of multiaxial anisotropic viscoplasticity. *ASME J. Appl. Mech.*, 48:276–284, 1981.

T. R. Hueghes and R. L. Taylor. Unconditionally stable algorithms for quasi-static elasto/viscoplastic finite element analysis. *Computers and Structures*, 8:169–173, 1978.

T. Kamei and H. Hirai. An elasto-viscoplastic model with combined hardening of anisotropically consolidated cohesive soils. *Soils and Foundations*, 30(2):89–102, 1990.

P. V. Lade. Creep effects on satic an cyclic instability of granular soils. *ASCE J. of Geot. Engrg.*, 120(2):404–419, 1994.

J. Lanier, C. di Prisco, and R. Nova. Etude experimental et analyse theorique de l'anisotropie induite du sable d'Hostun. *Revue Française de Géotechnique*, 57:59–74, 1991.

J. Meng. *Simultaneous determination of shear modulus and damping ratio.* PhD thesis, Atlanta Georgia Institute, Atlanta, 2003.

G. Michael and M. G. Katona. Evaluation of visoplastic cap model. *ASCE J. of Geot. Engrg.*, 110(8):1106–1125, 1984.

Z. Mroz. On the description of anisotropic work-hardening. *Journal of the Mechanics and Physics of Solids*, 15:163–175, 1967.

R. Nova. A viscoplastic constitutive model for normally consolidated clay. In *Proceedings IUTAM Conf. Deformation and Failure of Granular Materials*, pages 287–295, Delft, 1982.

W. Olszak and P. Perzyna. The constitutive equations of teh flow theory for a non-stationary yield condition. In *Proceedings 11th Cong. Applied Mechanics*, pages 545–553, 1966.

P. Perzyna. The constitutive equations for rate sensitive plastic materials. *The Quarterly of Applied Mathematics*, 20:321–332, 1963.

A. Phillips and H. C. Wu. A theory of viscoplasticity. *Int. J. of Solids and Structures*, 9:15–30, 1973.

W. Prager and D. C. Druker. Soil mechanics and plastic analysis or limit design. *The Quarterly of Applied Mathematics*, 10:157–165, 1952.

J. H. Prevost. A simple plasticity theory for frictional cohesionless soils. *Int. J. Soil Dyn. Earth. Eng.*, 4(1):9–17, 1985.

J. H. Prevost and C. M. Keane. Shear stress-strain curve generation from simple material parameters. *ASCE J. of Geot. Engrg.*, 116:1255–1263, 1990.

J. H. Prevost and B. Popescu. Centrifuge validation of a numerical model for dynamic soil liquefaction. *Int. J. Soil Dyn. Earth. Eng.*, 12:73–90, 1993.

H. Sekiguchi. Theory of undrained creep rupture of normally consolidated clay based on elasto-viscoplasticity. *Soils and Foundations*, 24(1):129–147, 1984.

A. Singh and J. K. Mitchell. General stress-strain-time function for soils. *Jour. of ASCE SMFD*, 94(SM1):21–46, 1968.

F. Tatsuoka, F. Santucci de Magistris, K. Hayano, Y. Momoya, and J. Koseki. Some new aspects of time effects on the stress-strain behaviour of stiff geomaterials. Keynote Lecture. In Evangelista and Picarelli, editors, *The Geotechnics of Hard Soils-Soft Rocks, Proc. of Second Int. Conf. on Hard Soils and Soft Rocks (1998)*, volume 2, pages 1285–1371. Balkema, Naples, 2001.

C. Tsakmakis. Formulation of viscoplasticity laws using overstresses. *Acta Mechanica*, 115:179–202, 1996.

C. Zambelli. *Experimental and theoretical analysis of the mechanical behaviour of cohesionless soils under cyclic-dynamic loading.* PhD thesis, Politecnico di Milano, Milan, Italy, 2006.

C. Zambelli, C. di Prisco, A. d'Onofrio, and F. Santucci de Magistris. Cyclic mechanical behaviour of granular soils: experimental results and constitutive modelling. In *Proceedings 4th International Conference on Earthquake Geotechnical Engineering*, Thessaloniki, Greece, June 2007.

# Mathematical Models for Transient, Dynamic and Cyclic Problems in Geotechnical Engineering

M. Pastor *‡, P. Mira *‡ and J. A. Fernández Merodo ‡†

* Centro de Estudios y Experimentación de Obras Públicas CEDEX Madrid, Spain

‡ Grupo M2i, Dept. Applied Mathematics, ETS Ing. Caminos, Universidad Politécnica de Madrid, Spain

† Instituto Geológico y Minero de España IGME, Madrid, Spain

## 1 Introduction

Depending on their dependence on time, loading can be classified as static or transient. Two special cases of transient loading are (i) cyclic loading and (ii) dynamic loading. While accelerations are basic in the latter, they can be neglected in some problems in the former. It is also important to notice that there exist dynamic problems which are not cyclic (propagation of a fast landslide), and cyclic problems which are not dynamic (response of a marine foundation to forces induced by tides). In the case of geomaterials, and more specifically soils, there are cases where the problem is linear, such as the small amplitude vibrations of foundations, while in others, such as liquefaction of an earth dam during an earthquake, non linearity is a key factor.

Cyclic and dynamic problems are found in roads, railways, seismic engineering, coastal, harbour and ocean engineering, impact and impulsive wave propagation, and avalanches and fast catastrophic landslides foundations, just to mention a few.

Modelling of cyclic and dynamic problems in geomaterials is an area where much work has still to be done, improving (i) the mathematical models describing coupling with pore fluids, (ii) constitutive models describing material behaviour under cyclic loading, and (iii) numerical modelling, especially when failure has taken place and material has fluidized.

Concerning mathematical modelling, the dynamic response of geostructures is usually described by a set of PDEs which are of hyperbolic nature. Fundamental solutions are called waves, and in 1D undamped dry soil the solution consists on two waves which propagate in both directions without changing their shape or their amplitude. This is perhaps the main property of hyperbolic problems.

This Chapter is devoted to present some important fundamental concepts in solid dynamics, such as the two alternative formulations of the problem (first and second order hyperbolic partial differential equations or PDEs). Most of the existing models in solid dynamics are based on the second order PDE, using as a main variable the displacements. A very interesting alternative consists of using as variables the velocity and the stress fields. The discrete models which arise from them present important advantages, such as better propagation properties, and the possibility of using linear elements such as triangles in 2D and tetrahedral in 3D.

This Chapter is organized as follows. After introducing the two main alternative forms in which a problem can be cast in solid dynamics (first or second order hyperbolic equations), we will devote a first part to describe the former. Here we will introduce the Riemann invariants which can be used to develop suitable transparent boundary conditions. Then, in a second part, we will study the formulations based on the second order equations, in the frequency and the time domain.

## 2 Solid Dynamics: Alternative Formulations

The purpose of this Section is to describe the two alternative formulations which can be applied in Soil Dynamics in a simple manner, using the case of a column of soil -or a one dimensional elastic bar-. The field equations are:

(i) The balance of momentum equation

$$\frac{\partial \sigma}{\partial x} = \rho \frac{\partial^2 u}{\partial t^2} = \rho \frac{\partial v}{\partial t} \tag{1}$$

where $v$ is the velocity, $u$ the displacement, $\sigma$ the stress and $\rho$ the density.

(ii) The constitutive equation, which in the case of an elastic material is

$$\sigma = E \frac{\partial u}{\partial x} \tag{2}$$

where $E$ is the modulus of elasticity. Both equations are usually combined eliminating the stress, arriving to:

$$\frac{\partial^2 u}{\partial t^2} = c^2 \frac{\partial^2 u}{\partial x^2} \tag{3}$$

In above, $c$ is the wave velocity given by:

$$c^2 = \frac{E}{\rho} \tag{4}$$

Alternatively, we can write the system of first order equations:

$$\frac{\partial \mathbf{\Phi}}{\partial t} + \mathbf{A}\frac{\partial \mathbf{\Phi}}{\partial x} = \mathbf{0} \tag{5}$$

where the vector $\mathbf{\Phi}$ and the matrix $\mathbf{A}$ are given by:

$$\mathbf{\Phi} = \begin{pmatrix} \sigma \\ v \end{pmatrix} \quad \mathbf{A} = \begin{pmatrix} 0 & -E \\ -\rho & 0 \end{pmatrix} \tag{6}$$

Sometimes, it is preferred the so-called conservation law

$$\frac{\partial \mathbf{\Phi}}{\partial t} + \frac{\partial \mathbf{F}}{\partial x} = \mathbf{0} \tag{7}$$

where $\mathbf{F}$ is the flux vector $\mathbf{F} = (-Ev, -\sigma/\rho)^T$

# 3 First Order Hyperbolic Equations

## 3.1 1D Scalar Equation: Characteristics

The simplest equation with the structure of (6) and (7) is the 1D scalar convective transport equation

$$\frac{\partial \phi}{\partial t} + u\frac{\partial \phi}{\partial x} = 0 \quad x \in (0, L) \quad t \in (0, T) \tag{8}$$

which represents the 1D convection of a magnitude $\phi$ by a current of constant velocity $u$. The equation is obtained from the conservation of $\phi$ in a control volume of length $dx$ and normal section $A$ as:

$$Adx\frac{\partial \phi}{\partial t} = Au\phi - Au\left(\phi + \frac{\partial \phi}{\partial x}\right)$$

where $A$ and $u$ have been assumed to be constants. Its conservative form is written as:

$$\frac{\partial \phi}{\partial t} + \frac{\partial F}{\partial x} = 0$$

where $F = u\phi$. Above equation has to be complemented with suitable initial and boundary conditions for the problem to be well posed. The 1D scalar equation of convective transport is a particular case of the more general problem

$$a\frac{\partial \phi}{\partial t} + b\frac{\partial \phi}{\partial x} = c \tag{9}$$

Depending on the coefficients $a$ and $b$ the equation can be classified as:

(i) Linear: $a$ and $b$ are functions of $x, t$

(ii) Quasi linear, if they depend also on the variable $\phi$

(iii) Non linear, if they depend on partial derivatives of $\phi$

To gain insight in the properties of this equation, we will consider the variation of $\phi(x,t)$ along a line $\Gamma$ on the plane $(x,t)$

$$d\phi = \frac{\partial \phi}{\partial x}dx + \frac{\partial \phi}{\partial t}dt \tag{10}$$

Next, we will obtain the time derivative from the PDE as:

$$\frac{\partial \phi}{\partial t} = \frac{1}{a}\left(c - b\frac{\partial \phi}{\partial x}\right)$$

and we will substitute it into (10):

$$\frac{dt}{a}\left(c - b\frac{\partial \phi}{\partial x}\right) + dx\frac{\partial \phi}{\partial x} = d\phi$$

from where we obtain

$$d\phi = \frac{c}{a}dt + \frac{\partial \phi}{\partial x}\left(dx - \frac{b}{a}dt\right) \tag{11}$$

If we choose the curve $\Gamma$ such that along it

$$dx - \frac{b}{a}dt = 0 \tag{12}$$

Eq. (11) simplifies to:

$$d\phi = \frac{c}{a}dt \tag{13}$$

The solution of the problem can be obtained in two steps

(i) Find the characteristics $\Gamma$ integrating (12)

(ii) Integrate (13) along the characteristics, for which we need the value of $\phi$ at a point in each characteristic. In the case illustrated in Fig. 1, the characteristics cut the $x$-axis and the $t$-axis on segments $C_1$ and $C_2$ To solve the problem, we need to know $\phi$ on both $C_1$ and $C_2$, which are respectively, the boundary and initial conditions. It is important to note that boundary conditions are only to be applied at points from where characteristics enter the domain.

The boundary condition reads $\phi\,(x=0,t)=g(t)\ \ 0\le t\le T$ and the initial condition is $\phi\,(x,t=0)=h(x)\ \ 0\le x\le L$.

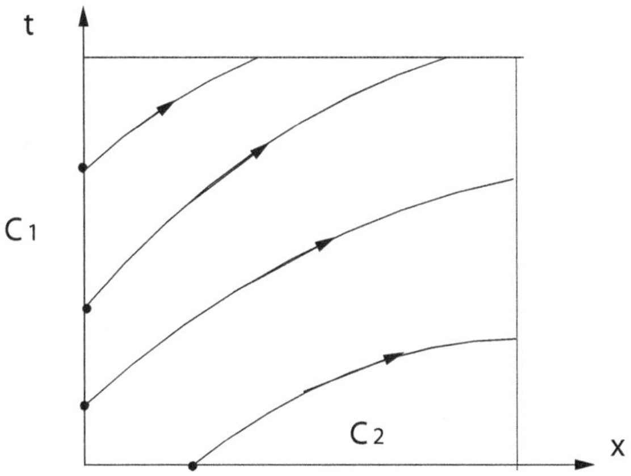


**Figure 1.** Characteristics of the first order scalar hyperbolic equation.

In the case of the scalar 1D convective transport equation, the characteristics are the straight lines $x-ut=C_t$ and along them $d\phi=0$, i.e., the convected magnitude $\phi$ is constant. The solution of the problem

$$
\begin{aligned}
&\frac{\partial\phi}{\partial t}+u\frac{\partial\phi}{\partial x}=0 \qquad x\in(0,L) \qquad t\in(0,T)\\
&\phi\,(x=0,t)=g(t) \quad 0\le t\le T\\
&\phi\,(x,t=0)=h(x) \quad 0\le x\le L
\end{aligned}
\tag{14}
$$

is:

$$\begin{aligned} \phi(x,t) &= h(x_0) = h(x-ut) \quad \text{if } x-ut \geq 0 \\ \phi(x,t) &= g(t_0) = g(x-ut) \quad \text{if } x-ut \leq 0 \end{aligned} \tag{15}$$

which has been obtained as illustrated in Fig. 2.

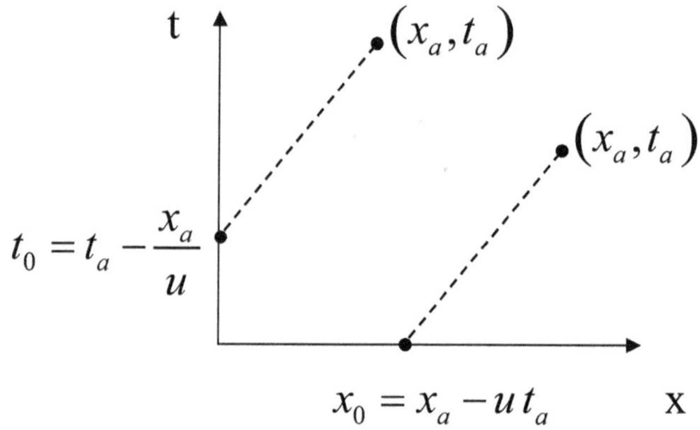


**Figure 2.** Characteristics and solution of the 1D advective transport problem.

In some cases, there exist source terms $S$ in the PDE,

$$\frac{\partial \phi}{\partial t} + u\frac{\partial \phi}{\partial x} = S \tag{16}$$

where we will assume that the source term is $S = G\phi$. In this case, the characteristics are the same than in the purely convective term, but the solution is obtained now by integrating:

$$d\phi = G\phi dt$$

and the solution is now:

$$\begin{aligned} \phi(x,t) &= h(x_0) = h(x-ut)\exp(Gt) \quad \text{if } x-ut \geq 0 \\ \phi(x,t) &= g(t_0) = g(x-ut)\exp(Gt) \quad \text{if } x-ut \leq 0 \end{aligned} \tag{17}$$

### 3.2 Systems of $1^{st}$ Order PDEs. Riemann Invariants

The structure of the systems of PDEs in 1D is very similar to that of the scalar equation. Indeed, as we have already seen, the propagation of an

elastic wave in a bar can be written as in (5)-(7).
We will assume that $\mathbf{A}$ has 2 eigenvalues $\{\lambda_1, \lambda_2\}$ with the two right eigenvectors $\{\mathbf{x}^{(1)}, \mathbf{x}^{(2)}\}$ satisfying the condition $\mathbf{A}\mathbf{x}^{(k)} = \lambda_k \mathbf{x}^{(k)},\ k = 1, 2$. We will introduce the eigenvectors matrix $\mathbf{P}$:

$$\mathbf{P} = \left\{\mathbf{x}^{(1)}, \mathbf{x}^{(2)}\right\}$$

where the $(i,j)$ term is $P_{ij} = x_i^{(j)}$. The matrix $\mathbf{A}$ can be diagonalized as

$$\mathbf{P}^{-1}\mathbf{A}\mathbf{P} = \mathbf{\Lambda} \quad \text{where} \quad \Lambda_{ii} = \lambda_i$$

If we multiply by $\mathbf{P}^{-1}$ the PDE, we obtain

$$\mathbf{P}^{-1}\frac{\partial \mathbf{\Phi}}{\partial t} + \mathbf{P}^{-1}\mathbf{A}\frac{\partial \mathbf{\Phi}}{\partial x} = \mathbf{0}$$

and

$$\begin{aligned} \mathbf{P}^{-1}\frac{\partial \mathbf{\Phi}}{\partial t} + \left(\mathbf{P}^{-1}\mathbf{A}\mathbf{P}\right)\mathbf{P}^{-1}\frac{\partial \mathbf{\Phi}}{\partial x} &= \mathbf{0} \\ \mathbf{P}^{-1}\frac{\partial \mathbf{\Phi}}{\partial t} + \mathbf{\Lambda}\mathbf{P}^{-1}\frac{\partial \mathbf{\Phi}}{\partial x} &= \mathbf{0} \end{aligned} \tag{18}$$

We will introduce a new vector $\mathbf{\Psi}$ defined as

$$d\mathbf{\Psi} = \mathbf{P}^{-1} d\mathbf{\Phi} \tag{19}$$

and we assume it to be differentiable, so:

$$\frac{\partial \mathbf{\Psi}}{\partial t} = \mathbf{P}^{-1}\frac{\partial \mathbf{\Phi}}{\partial t} \quad \frac{\partial \mathbf{\Psi}}{\partial x} = \mathbf{P}^{-1}\frac{\partial \mathbf{\Phi}}{\partial x} \tag{20}$$

From where, using (20) in (19) we arrive to:

$$\frac{\partial \mathbf{\Psi}}{\partial t} + \mathbf{\Lambda}\frac{\partial \mathbf{\Psi}}{\partial x} = \mathbf{0} \tag{21}$$

which is a system of uncoupled equations as $\mathbf{\Lambda}$ is diagonal:

$$\frac{\partial \psi^{(i)}}{\partial t} + \lambda_i \frac{\partial \psi^{(i)}}{\partial x} = 0 \tag{22}$$

with solutions of the form

$$\psi^{(i)} = F^{(i)}\left(x - \lambda_i t\right) \tag{23}$$

These results can be directly extrapolated to systems of $n$ $1^{st}$ order PDEs. The uncoupled problem has, therefore, $n$ characteristics which can be obtained following the procedure described for the scalar case. The solution consists, therefore, on $n$ magnitudes $\psi_i$ propagating along the $n$ characteristics. These magnitudes are often referred to as "Riemann invariants" of the problem. As in the case of the scalar equation, we will need $n$ initial conditions and $n$ boundary conditions to be applied where characteristics enter the domain.

In the simple case we are considering, the two eigenvalues $\lambda_k$ and eigenvectors $\mathbf{x}^{(k)}$ of matrix $\mathbf{A}$ can be obtained from:

$$\mathbf{A}\mathbf{x}^{(k)} = \lambda_k \mathbf{x}^{(k)} \quad k = 1,2 \tag{24}$$

as:

$$\lambda_1 = +\sqrt{E/\rho} \quad \lambda_2 = -\sqrt{E/\rho} \tag{25}$$

$$\mathbf{x}^{(1)} = \left(-\sqrt{E\rho}, 1\right)^T \quad \mathbf{x}^{(2)} = \left(+\sqrt{E\rho}, 1\right)^T \tag{26}$$

The matrix of eigenvectors $\mathbf{P}$ and its inverse are

$$\mathbf{P} = \begin{pmatrix} -\sqrt{E\rho} & 1 \\ +\sqrt{E\rho} & 1 \end{pmatrix} \quad \mathbf{P}^{-1} = \frac{1}{2} \begin{pmatrix} -\dfrac{1}{\sqrt{E\rho}} & 1 \\ +\dfrac{1}{\sqrt{E\rho}} & 1 \end{pmatrix} \tag{27}$$

and $d\mathbf{\Psi} = \mathbf{P}^{-1} d\mathbf{\Phi}$ can be integrated to give

$$\mathbf{\Phi} = \mathbf{P}\mathbf{\Psi} \tag{28}$$

where

$$\mathbf{\Psi} = \frac{1}{2\sqrt{E\rho}} \begin{pmatrix} -\sigma + v\sqrt{E\rho} \\ \sigma + v\sqrt{E\rho} \end{pmatrix} \tag{29}$$

Substituting (23) in (15), and taking into account that

$$\mathbf{P}^{-1}\mathbf{A}\mathbf{P} = \mathbf{\Lambda} = \begin{pmatrix} c & 0 \\ 0 & -c \end{pmatrix} \tag{30}$$

we arrive to the new system

$$\begin{aligned} \frac{\partial \psi^{(1)}}{\partial t} + c\frac{\partial \psi^{(1)}}{\partial x} = 0 \\ \frac{\partial \psi^{(2)}}{\partial t} - c\frac{\partial \psi^{(2)}}{\partial x} = 0 \end{aligned} \tag{31}$$

which is uncoupled. These equations describe the convective transport of the two magnitudes $\psi^{(1)}$ and $\psi^{(2)}$ with velocities $+c$ and $-c$ which are the Riemann invariants of the problem. Here, for convenience, we will multiply them by $2\sqrt{E\rho}$.

Riemann invariants can be applied to derive suitable boundary conditions of the "incoming wave" and "transparent boundary" types. If we consider a bar of length $L$ extending from $x = 0$ to $x = L$, the transparent boundary condition at $x = L$ consists on making zero the Riemann invariant entering the domain from the exterior, i.e.

$$\begin{gathered} \psi^{(2)} = \sigma + v\sqrt{E\rho}\Big|_{x=L} = 0 \\ \text{or} \\ \psi^{(2)} = \sigma + \rho c v|_{x=L} = 0 \end{gathered} \tag{32}$$

while the incoming wave boundary condition is written as

$$\psi^{(2)} = \sigma + v\sqrt{E\rho}\Big|_{x=L} = \psi^{(2)}_{\text{Inc}} \tag{33}$$

where $\psi^{(2)}_{\text{Inc}}$ characterizes the wave entering the domain from the left.

### 3.3 Generalization to Solid Dynamics Problems. Conservative and Primitive Equations Formulation

**Conservative Formulation** Concerning two and three dimensional problems in isotropic elastic materials, the method described above can be generalized in a straightforward manner. The constitutive equation can be written as:

$$\sigma_{ij} = D^e_{ijkl}\varepsilon_{kl} \tag{34}$$

where the constitutive tensor is written in terms of Lamé's constants $\lambda$ and $\mu$ as:

$$D^e_{ijkl} = \lambda\delta_{ij}\delta_{kl} + \mu\left(\delta_{ik}\delta_{jl} + \delta_{il}\delta_{jk}\right) \tag{35}$$

with

$$\lambda = \frac{\nu E}{(1+\nu)(1-2\nu)} \quad \mu = \frac{\nu E}{2(1+\nu)}$$

where $\nu$ is the Poisson ratio. Substituting (35) into (34) we get

$$\sigma_{ij} = \lambda \varepsilon_v \delta_{ij} + 2\mu\varepsilon_{ij}$$

where $\varepsilon_v$ is the volumetric strain: $\varepsilon_v = \mathrm{tr}(\varepsilon_{ij})$.

From here, we obtain

$$\frac{\partial}{\partial t}\begin{pmatrix}\sigma_{11}\\ \sigma_{22}\\ \sigma_{33}\\ \sigma_{12}\\ \sigma_{23}\\ \sigma_{31}\end{pmatrix} = \frac{\partial}{\partial x_1}\begin{pmatrix}(\lambda+2\mu)\,v_1\\ \lambda v_1\\ \lambda v_1\\ \mu v_2\\ 0\\ \mu v_3\end{pmatrix} + \frac{\partial}{\partial x_2}\begin{pmatrix}\lambda v_2\\ (\lambda+2\mu)\,v_2\\ \lambda v_2\\ \mu v_1\\ \mu v_3\\ 0\end{pmatrix} + \frac{\partial}{\partial x_3}\begin{pmatrix}\lambda v_3\\ \lambda v_3\\ (\lambda+2\mu)\,v_3\\ 0\\ \mu v_2\\ \mu v_1\end{pmatrix} \tag{36}$$

The balance of momentum is:

$$\rho\frac{\partial \mathbf{v}}{\partial t} = \mathrm{div}\boldsymbol{\sigma} + \mathbf{b} \tag{37}$$

where $\mathbf{b}$ is the vector of body forces. Expanding this equation:

$$\frac{\partial}{\partial t}\begin{pmatrix}v_1\\ v_2\\ v_3\end{pmatrix} = \frac{1}{\rho}\frac{\partial}{\partial x_1}\begin{pmatrix}\sigma_{11}\\ \sigma_{12}\\ \sigma_{13}\end{pmatrix} + \frac{1}{\rho}\frac{\partial}{\partial x_2}\begin{pmatrix}\sigma_{21}\\ \sigma_{22}\\ \sigma_{23}\end{pmatrix} + \frac{1}{\rho}\frac{\partial}{\partial x_3}\begin{pmatrix}\sigma_{31}\\ \sigma_{32}\\ \sigma_{33}\end{pmatrix} + \begin{pmatrix}b_1/\rho\\ b_2/\rho\\ b_3/\rho\end{pmatrix} \tag{38}$$

Equations (36) and (38) can be combined, resulting on:

$$\frac{\partial \boldsymbol{\Phi}}{\partial t} + \frac{\partial \mathbf{F}_k}{\partial x_k} = \mathbf{S} \tag{39}$$

where

$$\boldsymbol{\Phi} = (\sigma_{11}, \sigma_{22}, \sigma_{33}, \sigma_{12}, \sigma_{23}, \sigma_{31}, v_1, v_2, v_3)^T \tag{40}$$

$$
\begin{aligned}
\mathbf{F}_1 &= -\left\{(\lambda+2\mu)\,v_1, \lambda v_1, \lambda v_1, \mu v_2, 0, \mu v_3, \frac{\sigma_{11}}{\rho}, \frac{\sigma_{21}}{\rho}, \frac{\sigma_{31}}{\rho}\right\}^T \\
\mathbf{F}_2 &= -\left\{\lambda v_2, (\lambda+2\mu)\,v_2, \lambda v_2, \mu v_1, \mu v_3, 0, \frac{\sigma_{21}}{\rho}, \frac{\sigma_{22}}{\rho}, \frac{\sigma_{23}}{\rho}\right\}^T \\
\mathbf{F}_3 &= -\left\{\lambda v_3, \lambda v_3, (\lambda+2\mu)\,v_3, 0, \mu v_2, \mu v_1, \frac{\sigma_{31}}{\rho}, \frac{\sigma_{32}}{\rho}, \frac{\sigma_{33}}{\rho}\right\}^T
\end{aligned}
\tag{41}
$$

and

$$
\mathbf{S} = \left(0, 0, 0, 0, 0, 0, \frac{b_1}{\rho}, \frac{b_2}{\rho}, \frac{b_3}{\rho}\right)^T \tag{42}
$$

$k$ ranges from 1 to the dimensions of the problem, $\mathbf{\Phi}$ is the vector of unknowns, $\mathbf{F}_k$ is the vector of fluxes along axis $x_k$ and $\mathbf{S}$ is the source terms vector. This is referred to as the "conservative form" of the solid dynamics problem.

So far, we have considered elastic materials. In the case of viscoplastic materials, the constitutive equation is:

$$
\frac{\partial \boldsymbol{\sigma}}{\partial t} = \mathbf{D}^e \left(\dot{\boldsymbol{\varepsilon}} - \dot{\boldsymbol{\varepsilon}}^{\mathrm{vp}}\right) \tag{43}
$$

where the viscoplastic strain rate $\dot{\boldsymbol{\varepsilon}}^{\mathrm{vp}}$ is given by a suitable constitutive law. Above equation can be written as:

$$
\frac{\partial \boldsymbol{\sigma}}{\partial t} - \mathbf{D}^e \dot{\boldsymbol{\varepsilon}} = \mathbf{D}^e \dot{\boldsymbol{\varepsilon}}^{\mathrm{vp}} \tag{44}
$$

which shows that the fluxes are the same than those of the elastic case, but the source term will include the contributions from $\mathbf{D}^e \dot{\boldsymbol{\varepsilon}}^{\mathrm{vp}}$.

Finally, in the case of an elastoplastic problem, the structure of the equations is of the type (39), but the fluxes will depend also on the components of the elastoplastic tensor.

**Primitive Variables Formulation** The equation of an elastic wave propagating along a direction characterized by unit vector $\mathbf{n}$ is written as

$$
\mathbf{u} = \mathbf{g} f\left(\mathbf{n} \cdot \mathbf{x} \pm ct\right) = \mathbf{g} f\left(s \pm ct\right) \tag{45}
$$

where $c$ is the velocity of propagation , $\mathbf{g}$ is a polarization vector describing the direction of vibration of particles and $s$ the abscissa along the direction of propagation. The negative sign corresponds to waves travelling in the direction of $\mathbf{n}$. Particle velocities are found differentiating with respect to time:

$$\mathbf{v} = \pm \mathbf{g} c f' \tag{46}$$

The constitutive equation for a rate independent material can be written as:

$$\frac{\partial \sigma_{ij}}{\partial t} - D_{ijkl} \frac{\partial v_k}{\partial x_l} = 0 \tag{47}$$

where, taking into account (46),

$$\frac{\partial v_k}{\partial x_l} = g_k n_l c f'' \tag{48}$$

it results in:

$$\frac{\partial \sigma_{ij}}{\partial t} - D_{ijkl} g_k n_l c f'' = 0 \tag{49}$$

If we now multiply by $n_j$

$$\frac{\partial (\sigma_{ij} n_j)}{\partial t} - (n_j D_{ijkl} n_l)\, g_k c f'' = 0$$

introduce the acoustic tensor $\mathbf{Q}$,

$$Q_{ik} = n_j D_{ijkl} n_l$$

and substitute

$$g_k c f'' = \frac{\partial v_k}{\partial s}$$

we finally arrive to:

$$\frac{\partial (\sigma_{ij} n_j)}{\partial t} - Q_{ik} \frac{\partial v_k}{\partial s} = 0 \tag{50}$$

The balance of momentum equation expressed in terms of the velocity is:

$$\frac{\partial v_i}{\partial t} - \frac{1}{\rho} \frac{\partial \sigma_{il}}{\partial x_l} = \frac{1}{\rho} b_i \tag{51}$$

Taking into account that we are considering plane waves, and no variation exists in directions perpendicular to $\mathbf{n}$, above equation can be written as:

$$\frac{\partial v_i}{\partial t} - \frac{1}{\rho}\frac{\partial \sigma_{ij}}{\partial s}\frac{\partial s}{\partial x_j} = \frac{\partial v_i}{\partial t} - \frac{1}{\rho}\frac{\partial \sigma_{ij}}{\partial s} n_j$$

from which we obtain

$$\frac{\partial v_i}{\partial t} - \frac{1}{\rho}\frac{\partial}{\partial s}(\sigma_{ij} n_j) = \frac{1}{\rho} b_i \tag{52}$$

Equations (50) and (52) describe the propagation of a plane wave in an unbounded continuum. They can be written in a more compact way as:

$$\frac{\partial}{\partial t}\begin{pmatrix} t_i \\ v_i \end{pmatrix} + \begin{pmatrix} 0 & -Q_{ik} \\ -\delta_{ik}/\rho & 0 \end{pmatrix} \frac{\partial}{\partial s}\begin{pmatrix} t_k \\ v_k \end{pmatrix} = \begin{pmatrix} 0 \\ b_i/\rho \end{pmatrix} \tag{53}$$

or

$$\frac{\partial}{\partial t}\begin{pmatrix} \mathbf{t} \\ \mathbf{v} \end{pmatrix} + \begin{pmatrix} \mathbf{0} & -\mathbf{Q} \\ -\mathbf{I}/\rho & \mathbf{0} \end{pmatrix} \frac{\partial}{\partial s}\begin{pmatrix} \mathbf{t} \\ \mathbf{v} \end{pmatrix} = \begin{pmatrix} \mathbf{0} \\ \mathbf{b}/\rho \end{pmatrix} \tag{54}$$

where $\mathbf{t}$ is the traction per unit area in a plane normal to the direction of propagation.

It has to be stressed that above equations can be applied to rate independent materials which behaviour can be described by (47) and therefore they are not restricted to linear elastic materials.

In the case of isotropic elastic materials, the acoustic tensor can be easily found to be:

$$\mathbf{Q} = \begin{pmatrix} \lambda + 2\mu & 0 & 0 \\ 0 & \mu & 0 \\ 0 & 0 & \mu \end{pmatrix} \tag{55}$$

and the wave propagation equations can be written as:

$$\frac{\partial}{\partial t}\begin{pmatrix} t_x \\ t_y \\ t_z \\ v_x \\ v_y \\ v_z \end{pmatrix} - \begin{pmatrix} 0 & 0 & 0 & -M & 0 & 0 \\ 0 & 0 & 0 & 0 & -G & 0 \\ 0 & 0 & 0 & 0 & 0 & -G \\ -1/\rho & 0 & 0 & 0 & 0 & 0 \\ 0 & -1/\rho & 0 & 0 & 0 & 0 \\ 0 & 0 & -1/\rho & 0 & 0 & 0 \end{pmatrix} \frac{\partial}{\partial s}\begin{pmatrix} t_x \\ t_y \\ t_z \\ v_x \\ v_y \\ v_z \end{pmatrix} = \begin{pmatrix} 0 \\ 0 \\ 0 \\ b_x/\rho \\ b_y/\rho \\ b_z/\rho \end{pmatrix} \tag{56}$$

where $M = \lambda + 2\mu$ and $G = \mu$. In above, the reference system has been chosen so that the $x$ axis coincides with the direction of propagation. If the problem being analyzed is two-dimensional, the system can be written as:

$$\frac{\partial}{\partial t}\begin{pmatrix} t_x \\ t_z \\ v_x \\ v_z \end{pmatrix} - \begin{pmatrix} 0 & 0 & -M & 0 \\ 0 & 0 & 0 & -G \\ -\rho & 0 & 0 & 0 \\ 0 & -\rho & 0 & 0 \end{pmatrix} \frac{\partial}{\partial s}\begin{pmatrix} t_x \\ t_z \\ v_x \\ v_z \end{pmatrix} = \begin{pmatrix} 0 \\ 0 \\ b_x/\rho \\ b_z/\rho \end{pmatrix} \tag{57}$$

Above equations can be rearranged in order to show the uncoupling of two waves with speeds of propagation $c_P = \sqrt{M/\rho}$ and $c_S = \sqrt{G/\rho}$ corresponding to the $P$ wave and $SV$ wave:

$$\begin{aligned} \frac{\partial}{\partial t}\begin{pmatrix} t_x \\ v_x \end{pmatrix} + \begin{pmatrix} 0 & -M \\ -1/\rho & 0 \end{pmatrix} \frac{\partial}{\partial s}\begin{pmatrix} t_x \\ v_x \end{pmatrix} = \begin{pmatrix} 0 \\ b_x/\rho \end{pmatrix} \\ \frac{\partial}{\partial t}\begin{pmatrix} t_z \\ v_z \end{pmatrix} + \begin{pmatrix} 0 & -M \\ -1/\rho & 0 \end{pmatrix} \frac{\partial}{\partial s}\begin{pmatrix} t_z \\ v_z \end{pmatrix} = \begin{pmatrix} 0 \\ b_z/\rho \end{pmatrix} \end{aligned} \tag{58}$$

If the material is of elasto-viscoplastic type, the results would be very similar to those obtained for the elastic case, the left hand side of (56)-(58) being identical, with the diference of an extra source term depending on the viscoplastic strain rate.

**Absorbing and Incoming Wave Boundary Conditions** The absorbing boundary conditions can be easily obtained when the solution consists on plane waves approaching the boundary along its normal. In a general case, a first order approximation can be obtain by assuming this condition. In the case of elastic and viscoplastic materials, the solution can be described as uncoupled waves of the form (58), for which the Riemann invariants can be easily obtained following the method described in Section 3.2. Using a local coordinate system with axes normal to the boundary and tangent to it, the equations are projected onto this coordinate system, and the conditions to be applied read:

$$\begin{aligned} \psi_P^{(2)} &= t_n + \rho c_P v_n = 0 \\ \psi_S^{(2)} &= t_t + \rho c_S v_t = 0 \end{aligned} \tag{59}$$

where sub indexes $n$ and $t$ refer to the tangential and normal components, $\mathbf{t}$ being the normal unit tractions on the surface.

In the case of a wave entering the domain, the boundary conditions are

$$\begin{aligned} \psi_P^{(2)} &= t_n + \rho c_P v_n = \psi_P^{(\mathrm{Inc})} \\ \psi_S^{(2)} &= t_t + \rho c_S v_t = \psi_S^{(\mathrm{Inc})} \end{aligned} \tag{60}$$

where $\psi_P^{(\mathrm{Inc})}$ and $\psi_S^{(\mathrm{Inc})}$ can be obtained from the information available for the incoming wave and the definition of both invariants. A variant of these boundary conditions consist on projecting the equations onto the direction of the incoming wave, instead of onto the normal to the domain.
It is interesting to notice that the absorbing wave condition is a particular case of the incident wave, obtained by making the Riemann invariant zero.

# 4 Second Order Equations

## 4.1 Second Order Equation: 2 and 3D Problems

The second order equation is the most widely used formulation. It consists on:

(1) The balance of momentum equation

$$\mathrm{div}\boldsymbol{\sigma} + \mathbf{b} = \rho \frac{d^2\mathbf{u}}{dt^2} \tag{61}$$

where $\mathbf{b}$ is the vector of body forces, $\boldsymbol{\sigma}$ the stress tensor, and $\mathbf{u}$ the displacement field. In some cases, it is written in terms of velocities as:

$$\mathrm{div}\boldsymbol{\sigma} + \mathbf{b} = \rho \frac{d\mathbf{v}}{dt} \tag{62}$$

(2) A suitable constitutive relation relating stress and strain increments. In the particular case of an elastoplastic material it is given by

$$d\boldsymbol{\sigma} = \mathbf{D}^{\mathrm{ep}} d\boldsymbol{\varepsilon} \tag{63}$$

For linear elastic materials, the constitutive equation is

$$d\boldsymbol{\sigma} = \mathbf{D}^{\mathrm{e}} d\boldsymbol{\varepsilon} \tag{64}$$

and, for elasto viscoplastic materials,

$$d\boldsymbol{\sigma} = \mathbf{D}^{\mathrm{e}} \left( d\boldsymbol{\varepsilon} - d\boldsymbol{\varepsilon}^{\mathrm{vp}} \right) \tag{65}$$

(3) A kinematic relation between the strain and displacements

$$d\varepsilon_{ij} = \frac{1}{2} \left( \frac{\partial u_i}{\partial x_j} + \frac{\partial u_j}{\partial x_i} \right) \tag{66}$$

In above we have assumed small strains, and no rotations have been taken into account.

Above equations have to be complemented, for the problem to be well posed, by boundary and initial conditions. Concerning the latter, both displacements and velocities at time zero has to be given.

In the case of linear elastic materials, the constitutive tensor is

$$D^e_{ijkl} = \lambda \delta_{ij} \delta_{kl} + \mu \left( \delta_{il} \delta_{jk} + \delta_{ik} \delta_{jl} \right) \tag{67}$$

from where we obtain the stress as:

$$\sigma_{ij} = \lambda \delta_{ij} u_{k,k} + \mu \left( u_{i,j} + u_{j,i} \right) \tag{68}$$

After substitution in the balance of linear momentum equation (61) we arrive to Navier equation

$$(\lambda + \mu)\, u_{j,ji} + \mu u_{i,jj} + \rho b_i = \rho \frac{\partial^2 u_i}{\partial t^2} \tag{69}$$

of which a particular case is the 1D bar

$$\frac{\partial^2 u}{\partial t^2} = c^2 \frac{\partial^2 u}{\partial x^2}$$

So far, we have used tensor notation, but it is worth considering an alternative vector notation which is frequently used in Finite Elements context. The stress and strain are represented as

$$\begin{aligned} \boldsymbol{\sigma}^* &= \left( \sigma_{xx}, \sigma_{yy}, \sigma_{zz}, \sigma_{xy}, \sigma_{yz}, \sigma_{zx} \right)^T \\ \boldsymbol{\varepsilon}^* &= \left( \varepsilon_{xx}, \varepsilon_{yy}, \varepsilon_{zz}, \gamma_{xy}, \gamma_{yz}, \gamma_{zx} \right)^T \end{aligned} \tag{70}$$

respectively. For the sake of simplicity we will drop the star superindexes in what follows, understanding that whenever the vector form is used, we refer to the star representation. Notice that in order to keep the work, we have used the engineering shear strains $\gamma_{ij}$ which are defined as $\gamma_{ij} = 2\varepsilon_{ij}$. A convenient way to relate strains to displacements is by means of an operator matrix $\mathbf{S}$ defined as

$$\begin{pmatrix} \varepsilon_{xx} \\ \varepsilon_{yy} \\ \varepsilon_{zz} \\ \gamma_{xy} \\ \gamma_{yz} \\ \gamma_{zx} \end{pmatrix} = \begin{pmatrix} \partial_x & 0 & 0 \\ 0 & \partial_y & 0 \\ 0 & 0 & \partial_z \\ \partial_y & \partial_x & 0 \\ 0 & \partial_z & \partial_y \\ \partial_z & 0 & \partial_x \end{pmatrix} \begin{pmatrix} u_x \\ u_y \\ u_z \end{pmatrix} = \mathbf{Su} \tag{71}$$

The divergence of the stress tensor can be written as:

$$\mathrm{div}\boldsymbol{\sigma} = \sigma_{ij,j} = \begin{pmatrix} \partial_x\sigma_{xx} + \partial_y\sigma_{xy} + \partial_z\sigma_{xz} \\ \partial_x\sigma_{xy} + \partial_y\sigma_{yy} + \partial_z\sigma_{yz} \\ \partial_x\sigma_{xz} + \partial_y\sigma_{yz} + \partial_z\sigma_{zz} \end{pmatrix}$$

$$\sigma_{ij,j} = \begin{pmatrix} \partial_x & 0 & 0 & \partial_y & 0 & \partial_z \\ 0 & \partial_y & 0 & \partial_x & \partial_z & 0 \\ 0 & 0 & \partial_z & 0 & \partial_y & \partial_x \end{pmatrix} \begin{pmatrix} \sigma_{xx} \\ \sigma_{yy} \\ \sigma_{zz} \\ \sigma_{xy} \\ \sigma_{yz} \\ \sigma_{zx} \end{pmatrix}$$

$$\mathrm{div}\boldsymbol{\sigma} = \mathbf{S}^T\boldsymbol{\sigma} \tag{72}$$

The balance of momentum is written in vector form as:

$$\mathbf{S}^T\boldsymbol{\sigma} + \mathbf{b} = \rho\frac{\partial^2\mathbf{u}}{\partial t^2} \tag{73}$$

Finally, the constitutive equation is written in vector form as:

$$d\sigma_i = D_{ij}^{\mathrm{ep}} d\varepsilon_j \tag{74}$$

In the case of linear elastic materials, the equations are:

$$\begin{aligned} &\mathbf{S}^T\boldsymbol{\sigma} + \mathbf{b} = \rho\frac{\partial^2\mathbf{u}}{\partial t^2} \\ &\boldsymbol{\sigma} = \mathbf{D}^e\boldsymbol{\varepsilon} \\ &\boldsymbol{\varepsilon} = \mathbf{Su} \end{aligned} \tag{75}$$

The equation can be written in terms of displacements by substituting the stress and the strain:

$$\left(\mathbf{S}^T\mathbf{D}^e\mathbf{S}\right)\mathbf{u} + \mathbf{b} = \rho\frac{\partial^2\mathbf{u}}{\partial t^2} \tag{76}$$

Above equations have to be complemented by suitable initial and boundary conditions. We will assume that the problem is defined in a domain $\mathbf{x} \in \Omega$ with a boundary $\partial\Omega$

(i) Initial conditions

$$\mathbf{u}(\mathbf{x}, t_0) = \mathbf{u}_0(\mathbf{x}) \quad \mathbf{v}(\mathbf{x}, t_0) = \mathbf{v}_0(\mathbf{x}) \tag{77}$$

(ii) Boundary conditions, which can be of several types

(ii.a) Prescribed displacements on $\mathbf{x} \in \partial_u\Omega : \mathbf{u}(\mathbf{x}, t) - \bar{\mathbf{u}}(\mathbf{x}, t) = \mathbf{0}$

(ii.b) Prescribed tractions on $\mathbf{x} \in \partial_t\Omega : \boldsymbol{\sigma}\mathbf{n} - \mathbf{t} = \mathbf{0}$, where $\mathbf{n}$ is the outer normal to the boundary

(iii) Absorbing boundaries, where the boundary conditions are those described in (59)-(60).

These conditions can be easily implemented in displacement or velocity based finite elements.

### 4.2 1D Scalar Problems. Characteristics and Boundary Conditions

To illustrate the concepts introduced in Section 4.1, we will consider here the simple case of wave propagation in a 1D elastic bar. We recall the governing equations, which are:

$$\left.\begin{aligned} \rho\frac{\partial^2 u}{\partial t^2} &= \frac{\partial \sigma}{\partial x} \\ \sigma &= E\varepsilon \\ \varepsilon &= \frac{\partial u}{\partial x} \end{aligned}\right\} \quad \frac{\partial^2 u}{\partial t^2} = c^2\frac{\partial^2 u}{\partial t^2} \tag{78}$$

where $c = \sqrt{E/\rho}$. The characteristic lines can be obtained as:

$$\frac{dx}{dt} = \pm c$$

and along them,

$$\begin{aligned} d\left(\frac{\partial u}{\partial t}\right) \pm cd\left(\frac{\partial u}{\partial x}\right) &= 0 \\ d\left(\frac{\partial u}{\partial t} \pm c\frac{\partial u}{\partial x}\right) &= 0 \end{aligned}$$

From where, integrating along the two sets of characteristics, we arrive at:

$$R^{(1)} = \frac{\partial u}{\partial t} - c\frac{\partial u}{\partial x} \qquad R^{(2)} = \frac{\partial u}{\partial t} + c\frac{\partial u}{\partial x} \tag{79}$$

where $R^{(1)}$ and $R^{(2)}$ are the Riemann invariants of waves travelling towards the right and the left, respectively.

The problem requires initial conditions on displacements and velocities and boundary conditions at both extremes of the bar to be well posed. They are a particular 1D case of those given in previous section. The initial conditions are:

$$u(x, t_0) = u_0(x) \quad v(x, t_0) = v_0(x) \quad 0 \leq x \leq L \tag{80}$$

The boundary conditions are applied at both ends, $x = 0$ and $x = L$. They can be (i) prescribed displacements (ii) prescribed stress, (iii) prescribed incoming wave, and (iv) absorbing boundary type.

Concerning the latter, at the right end, the absorbing boundary condition is:

$$R^{(2)} = \frac{\partial u}{\partial t} + c\frac{\partial u}{\partial x} = 0 \tag{81}$$

while the incoming wave condition at $x = L$ is

$$R^{(2)} = \frac{\partial u}{\partial t} + c\frac{\partial u}{\partial x} = R^{(\mathrm{Inc})} \tag{82}$$

In general, it can be seen that in both left and right extremes, above conditions can be written as:

$$\frac{\partial u}{\partial t} + c\frac{\partial u}{\partial n} = 0 \tag{83a}$$

$$\frac{\partial u}{\partial t} + c\frac{\partial u}{\partial n} = \left(\frac{\partial u}{\partial t} + c\frac{\partial u}{\partial n}\right)^{\mathrm{Inc}} \tag{83b}$$

It is interesting to notice that this equation is just another form of the one found for first order equations (59)-(60). Indeed, from (82), taking into account that $v = \frac{\partial u}{\partial t}, \frac{\partial u}{\partial x} = \varepsilon$ and $\sigma = E\varepsilon$, we arrive to

$$R^{(2)} = v + c\frac{\sigma}{E} = R^{(\mathrm{Inc})}$$

which is equivalent to (59)-(60).

As a first example we will consider the case of an infinite bar where a wave is propagating to the right. We will assume that the travelling wave is $u(x,t) = u_0(x - ct)$. For computational reasons, the domain which will be studied is $x \in (0, L)$. The boundary conditions to be applied are:

(i) an incident wave condition at $x = 0$

(ii) an absorbent boundary condition at $x = L$

The incident wave is characterized by :

$$\left(\frac{\partial u}{\partial t} + c\frac{\partial u}{\partial n}\right)^{\text{Inc}}$$

with

$$\frac{\partial u}{\partial n} = -\frac{\partial u}{\partial x}$$

Therefore, we obtain

$$\left(\frac{\partial u}{\partial t} - c\frac{\partial u}{\partial x}\right)^{\text{Inc}} = cu_0u' + cu_0u' = 2cu_0u'$$

The incident wave boundary condition results on

$$\frac{\partial u}{\partial t} + c\frac{\partial u}{\partial n} = 2cu_0u'$$

which is verified at $x = 0$.

The absorbent boundary condition is written at $x = L$ as

$$\frac{\partial u}{\partial t} + c\frac{\partial u}{\partial x} = 0$$

which is fulfilled by the solution $u(x,t) = u_0(x - ct)$.

## 4.3 2D Scalar Problems

In some 2D cases, such as membrane structures, the equation describing the wave propagation can be written as

$$\frac{\partial^2 u}{\partial t^2} = \text{div}\left(c^2 \text{grad} u\right) \tag{84}$$

where $u$ is the displacement normal to the membrane, and $c$ the wave propagation. Other cases of interest are the propagation of Rayleigh waves in the surface of the soil (Das, 1983), and the surface waves in the sea (Mei, 1983).

Initial conditions consist on providing both the displacement $u$ and the velocity $v$ at the initial time $t_0$, and boundary conditions can be of the four types described above, i.e., prescribed displacement, prescribed boundary tractions (related to normal derivatives of the displacement), absorbing boundary and incident wave boundary.

As above, a simple approximation consists of assuming a planar wave approaching the boundary, resulting on:

$$\frac{\partial u}{\partial t} + c\frac{\partial u}{\partial n} = 0 \quad \text{(absorbing boundary)} \tag{85}$$

and

$$\frac{\partial u}{\partial t} + c\frac{\partial u}{\partial n} = r^{(\mathrm{Inc})} \tag{86}$$

for the known incident wave condition. Alternatively, if we project onto the incident wave direction, using the relation

$$\frac{\partial u}{\partial n} = \frac{\partial u}{\partial \xi}\cos\beta$$

where $\xi$ is the abscissa along the incoming wave direction, we obtain:

$$\frac{\partial u}{\partial t} + c\frac{1}{\cos\beta}\frac{\partial u}{\partial n} = r^{(\mathrm{Inc})} \tag{87}$$

## 5 Coupled Problems in Soil Dynamics

One fundamental aspect of geomaterials is the coupling between soil skeleton and the pore fluids (water, air, oil,...). Much effort has been devoted in the past to understand coupling between soil skeleton and pore water. The first model was proposed by Biot (1941, 1955) for linear elastic materials. Further development was produced at Swansea University, where Zienkiewicz and co-workers (Zienkiewicz et al., 1980, 1999; Zienkiewicz and Shiomi, 1984) extended the theory to non linear materials and large deformation problems. It is worth mentioning the contributions to this field of Lewis and Schrefler (1998), Coussy (1995) and de Boer (2000). The purpose of this Section is to present the Biot-Zienkiewicz equations and the simplified forms which are used nowadays in the study of soil dynamics problems.

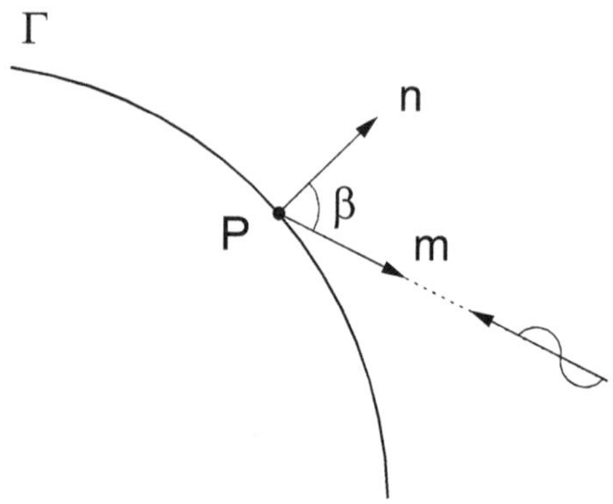


**Figure 3.** Incident wave conditions.

### 5.1 Biot-Zienkiewicz Equations

We will consider the case of a saturated soil consisting of a solid skeleton and a fluid phase, water, which fills completely the voids. We will consider a lagrangian formulation for the soil skeleton, and the movement of the fluid phase will be described relative to it. Indeed, we will introduce the Darcy velocity as

$$\mathbf{v}_w = \mathbf{v} + \mathbf{w}/n \tag{88}$$

where $\mathbf{v}_w$ is the absolute velocity of the pore water, $\mathbf{v}$ the velocity of the soil skeleton, $\mathbf{w}$ the Darcy velocity and $n$ the porosity. The total stress tensor $\boldsymbol{\sigma}$ acting on the mixture can be decomposed into a hydrostatic pore pressure term $p_w\mathbf{I}$ and an effective stress tensor $\boldsymbol{\sigma}'$ acting on soil skeleton as:

$$\boldsymbol{\sigma} = \boldsymbol{\sigma}' - p_w\mathbf{I} \tag{89}$$

where $\mathbf{I}$ is the second order identity tensor.

The First Biot Zienkiewicz equation is the balance of momentum for the mixture,

$$\operatorname{div}(\boldsymbol{\sigma}' - p_w\mathbf{I}) + \rho\mathbf{b} = \rho\frac{d\mathbf{v}}{dt} + \rho_w n\frac{d\mathbf{w}}{dt} \tag{90}$$

where $\rho$ is the density of the mixture and $\mathbf{b}$ the vector of body forces.

(i) soil skeleton

$$d\theta_1 = \mathrm{tr}\,(d\varepsilon) = d\varepsilon_v \tag{91}$$

where tr denotes the trace operator and $d\varepsilon_v$ is the increment of volumetric strain

(ii) deformation of soil grains caused by the pore pressure

$$d\theta_2 = \frac{1-n}{K_s} dp_w \tag{92}$$

where $K_s$ is the volumetric stiffness of soil particles

(iii) deformation of pore water caused by pore pressure

$$d\theta_3 = \frac{n}{K_w} dp_w \tag{93}$$

where $K_w$ is the volumetric stiffness of water.

From here, the balance of pore fluid mass is written as:

$$\mathrm{div}\mathbf{w} + \left(\frac{1-n}{K_s} + \frac{n}{K_w}\right)\frac{dp_w}{dt} + \mathrm{tr}\left(\frac{d\varepsilon}{dt}\right) = 0 \tag{94}$$

or

$$\mathrm{div}\mathbf{w} + \frac{1}{Q}\frac{dp_w}{dt} + \mathrm{div}\mathbf{v} = 0 \tag{95}$$

where we have introduced the mixed volumetric stiffness of water and solid particles

$$\frac{1}{Q} = \left(\frac{1-n}{K_s} + \frac{n}{K_w}\right) \tag{96}$$

The balance of momentum of the pore fluid is:

$$-\mathrm{grad}p_w + \rho_w \mathbf{b} - \frac{1}{k_w}\mathbf{w} = \rho_w \frac{d\mathbf{v}}{dt} + \rho_w \frac{d}{dt}\left(\frac{\mathbf{w}}{n}\right) \tag{97}$$

where $\rho_w$ is the density of the pore water, and $k_w$ the permeability. We will assume that Darcy law can be used to describe the interaction between pore water and soil skeleton, although other alternatives can be chosen.
The mathematical model is completed with a suitable constitutive equation and a kinematic relation between strain and displacement. The Biot-Zienkiewicz equations are, therefore:

(B1) the balance of momentum for the mixture

(B2) the balance of mass of the pore water

(B3) the balance of momentum of the pore fluid

(B4) the constitutive equation

(B5) the kinematic relation between strain and displacement.

These 5 equations involve 5 unknowns: displacements, Darcy velocity, effective stress, strain, and pore pressure. In practical applications the discretization is based on displacements, Darcy velocities, and pore pressures. As an alternative, the relative displacement of the pore water relative to the skeleton $\mathbf{U}$ can be used in the analysis instead of $\mathbf{w}$.

### 5.2 The Displacement-Pore Pressure Swansea Model

The most widely used form of Biot-Zienkiewicz equations is the so called "displacement- pore pressure" model, which is obtained in the cases where fluid accelerations are small by eliminating the Darcy velocity between equations (B2) and (B3). The result is:

$$\frac{1}{Q}\frac{dp_w}{dt} + \mathrm{div}\mathbf{v} - \mathrm{div}\left(k_w \mathrm{grad} p_w\right) = 0 \tag{98}$$

The equations are 4, involving 4 unknowns, and the analysis is based on displacements and pore pressures, from where the name of the model comes. The displacement-pore pressure Swansea model is usually formulated in terms of displacements instead of velocities, which is the version presented above. The main equations are:

$$\mathrm{div}\left(\boldsymbol{\sigma}' - p_w \mathbf{I}\right) + \rho \mathbf{b} = \rho \frac{d^2\mathbf{u}}{dt^2} \tag{99}$$

$$\frac{1}{Q}\frac{dp_w}{dt} + \mathrm{div}\left(\frac{d\mathbf{u}}{dt}\right) - \mathrm{div}\left(k_w \mathrm{grad} p_w\right) = 0 \tag{100}$$

together with the constitutive and kinematics relations.

## List of Symbols

| | | |
|---|---|---|
| $^e$ | elastic | §3.3, §4.1 |
| $^{ep}$ | elastoplastic | §4.1 |
| $^{Inc}$ | incoming | §3, §4 |
| $_n$ | component normal to the boundary | §3.3 |
| $_t$ | component tangent to the boundary | §3.3 |
| $^{vp}$ | viscoplastic | §3.3, §4.1 |
| $a$ | coefficient | §3.1 |
| $\mathbf{A}$ | $\begin{pmatrix} 0 & -E \\ -\rho & 0 \end{pmatrix}$ | §2, §3.2 |
| $A$ | normal section of control volume | §3.1 |
| $b$ | coefficient | §3.1 |
| $\mathbf{b}$ | vector of body forces | §2, §3.3, §4.1, §5 |
| $c$ | wave velocity | §2, §3, §4 |
| $c$ | coefficient | §3.1 |
| $c_P$ | $=\sqrt{M/\rho}$ | §3.3 |
| $c_S$ | $=\sqrt{G/\rho}$ | §3.3 |
| $\mathbf{D}$ | constitutive tensor | §3.3, §4.1 |
| $E$ | modulus of elasticity | §2, §3, §4.2 |
| $F$ | $= u\phi$ | §3.1 |
| $\mathbf{F}$ | flux vector | §2 |
| $\mathbf{F}_k$ | vector of fluxes along axis $x_k$ | §3.3 |
| $g$ | boundary condition | §3.1 |
| $\mathbf{g}$ | polarization vector | §3.3 |
| $G$ | $= \mu$ | §3.3 |
| $h$ | initial condition | §3.1 |
| $\mathbf{I}$ | identity tensor | §3.3, §5 |
| $k_w$ | permeability | §5 |
| $K_s$ | volumetric stiffness of soil particles | §5.1 |
| $K_w$ | volumetric stiffness of water | §5.1 |
| $L$ | bar length | §3, §4.2 |

| | | |
|---|---|---|
| $M$ | $= \lambda + 2\mu$ | §3.3 |
| $n$ | porosity | §5.1 |
| $\mathbf{n}$ | unit vector | §3.3 |
| $p_w$ | pore pressure | §5 |
| $\mathbf{P}$ | eigenvectors matrix of $\mathbf{A}$ | §3.2 |
| $Q$ | mixed volumetric stiffness of water and solid particles | §5 |
| $\mathbf{Q}$ | acoustic tensor | §3.3 |
| $s$ | abscissa along the direction of propagation | §3.3 |
| $S$ | source term | §3.1 |
| $\mathbf{S}$ | source terms vector | §3.3 |
| $\mathbf{S}$ | operator matrix | §4.1 |
| $R^{(1)}$ | Riemann invariant of waves travelling towards the right | §4.2 |
| $R^{(2)}$ | Riemann invariant of waves travelling towards the left | §4.2 |
| $t$ | time | §2, §3, §4, §5 |
| $\mathbf{t}$ | traction per unit area in a plane normal to the direction of propagation | §3.3 |
| $T$ | time period | §3 |
| $u$ | displacement | §2, §4 |
| $u$ | velocity | §3.1 |
| $\mathbf{u}$ | displacement vector | §3.3, §4.1, §5 |
| $\mathbf{U}$ | displacement of the pore water relative to the skeleton | §5.1 |
| $v$ | velocity | §2, §3.2 |
| $\mathbf{v}$ | velocity vector | §3.3, §4.1 |
| $\mathbf{v}$ | velocity of soil skeleton | §5 |
| $\mathbf{v}_w$ | absolute velocity of pore water | §5.1 |
| $\mathbf{w}$ | Darcy velocity | §5.1 |
| $x$ | horizontal coordinate | §2, §3, §4 |
| $\mathbf{x}$ | vector of coordinates | §3.3, §4.1, §5 |
| $\mathbf{x}^{(i)}$ | $i = 1, 2$ right eigenvector of $\mathbf{A}$ | §3.2 |
| $dx$ | length of control volume | §3.1 |
| $\gamma_{ij}$ | $2\varepsilon_{ij}$ engineering shear strain | §4.1 |
| $\delta_{ij}$ | Kronecker's delta | §3.3, §4.1 |

| | | |
|---|---|---|
| $\varepsilon$ | strain | §4.2 |
| $\boldsymbol{\varepsilon}$ | strain tensor | §3.3, §4.1 |
| $\boldsymbol{\varepsilon}^*$ | strain tensor in vector form | §4.1 |
| $\dot{\boldsymbol{\varepsilon}}$ | strain tensor rate | §3.3 |
| $\varepsilon_v$ | volumetric strain | §3.3, §5.1 |
| $d\theta_1$ | volume change of soil skeleton | §5.1 |
| $d\theta_2$ | volume change due to deformation of soil grains | §5.1 |
| $d\theta_3$ | volume change due to deformation of pore water | §5.1 |
| $\lambda$ | Lamé's constant | §3.3, §4.1 |
| $\lambda_i$ | $i = 1, 2$ eigenvalue of $\mathbf{A}$ | §3.2 |
| $\boldsymbol{\Lambda}$ | $\mathbf{P}^{-1}\mathbf{AP}$ | §3.2 |
| $\mu$ | Lamé's constant | §3.3, §4.1 |
| $\nu$ | Poisson's coefficient | §3.3 |
| $\xi$ | abscissa along the incoming wave direction | §4.3 |
| $\rho$ | density | §2, §3, §4 |
| $\rho$ | density of the mixture | §5 |
| $\rho_w$ | density of the water | §5.1 |
| $\sigma$ | stress | §2, §3.2, §4.2 |
| $\boldsymbol{\sigma}$ | stress tensor | §3.3, §4.1, §5.1 |
| $\boldsymbol{\sigma}'$ | effective stress tensor | §5 |
| $\boldsymbol{\sigma}^*$ | stress tensor in vector form | §4.1 |
| $\phi$ | scalar variable | §3.1 |
| $\boldsymbol{\Phi}$ | vector of unknowns | §2, §3 |
| $\boldsymbol{\Psi}$ | vector of Riemann invariants | §3 |

## Bibliography

M. A. Biot. General theory of three-dimensional consolidation. *Journal of Applied Physics*, 12:155–164, 1941.

M. A. Biot. Theory of elasticity and consolidation for a porous anisotropic solid. *Journal of Applied Physics*, 26:182–185, 1955.

O. Coussy. *Mechanics of Porous Continua.* J. Wiley and Sons, Chichester, 1995.

B. M. Das. *Fundamental of Soil Dynamics.* Elsevier, New York, 1983.

R. de Boer. *Theory of porous media.* Springer-Verlag, Berlin, 2000.

R. L. Lewis and B. A. Schrefler. *The Finite Element Method in the Static and Dynamic Deformation and Consolidation of Porous Media.* J. Wiley and Sons, Chichester, 1998.

C. C. Mei. *The applied dynamics of surface waves.* J. Wiley and Sons, Chichester, 1983.

O. C. Zienkiewicz and T. Shiomi. Dynamic behaviour of saturated porous media: The generalised biot formulation and its numerical solution. *Int. J. Num. Anal. Meth. Geomech.*, 8(1):71–96, 1984.

O. C. Zienkiewicz, C. T. Chang, and P. Bettess. Drained, undrained, consolidating dynamic behaviour assumptions in soils. *Géotechnique*, 30: 385–395, 1980.

O. C. Zienkiewicz, A. H. C. Chan, M. Pastor, B. A. Shrefler, and T. Shiomi. *Computational Geomechanics.* J. Wiley and Sons, Chichester, 1999.

# Discretization Techniques for Transient, Dynamic and Cyclic Problems in Geotechnical Engineering: First Order Hyperbolic Partial Differential Equations

Manuel Pastor [*‡]

[*] Centro de Estudios y Experimentación de Obras Públicas CEDEX Madrid, Spain

[‡] Grupo M2i, Dept. Applied Mathematics, ETS Ing. Caminos, Universidad Politécnica de Madrid, Spain

## 1 Introduction

In the Chapter devoted to mathematical modelling of transient, cyclic and dynamic problems we have studied two alternative ways to describe these problems, (i) as a system of first order, or (ii) as second order partial differential equations. We will devote this Chapter to Discretization techniques which can be applied for the former.

The Chapter is structured as follows:

First of all, we will present a fundamental difficulty found in standard Boubnov Galerkin Finite Element methods, which do not work here. Then, we will introduce the Finite Differences Method where we will consider their stability, numerical diffusion and dispersion. In this way, we will gain insight into the problem found in standard FEM. In particular, we will describe the Lax Wendroff scheme, which is the basis of the Taylor Galerkin method in FE described in last Section of the Chapter.

## 2 Model Equation: the 1D Convective Transport Problem

We will recall some fundamental aspects of this problem which were presented in previous Chapter. Let us consider the transport of a magnitude having a concentration $\phi$ by a steady, uniform current of velocity $v$ (Fig. 1).

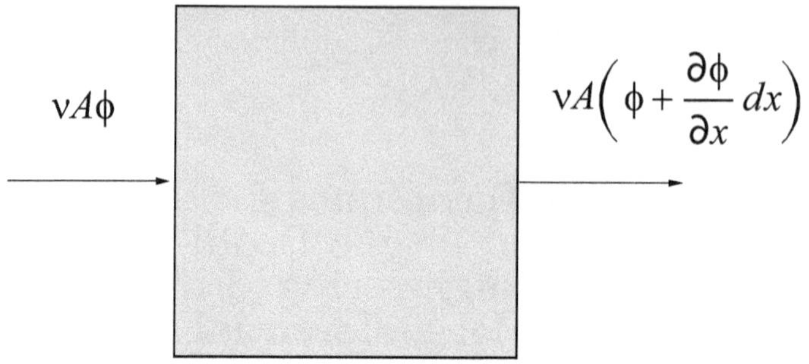


**Figure 1.** Convection of a specie by an uniform current.

The equation of balance of mass for the specie being convected is

$$Adx\frac{\partial \phi}{\partial t} = vA\phi - vA\left(\phi + \frac{\partial \phi}{\partial x}dx\right)$$

from where we obtain

$$\frac{\partial \phi}{\partial t} + v\frac{\partial \phi}{\partial x} = 0 \tag{1}$$

This Partial Differential Equation has to be complemented by suitable initial and boundary conditions:

$$\phi\left(x, t=0\right) = f_0\left(x\right) \tag{2a}$$

$$\phi\left(x=0, t\right) = g\left(t\right) \tag{2b}$$

It is important to notice that the boundary condition is only given at the part of the domain where the current enters the domain, or, better, where the information concerning the concentration enters the domain. This is an important difference with parabolic or elliptic problems, where we should have provided information in both sides of the channel.

The general solution of the problem is of the type:

$$\phi\left(x, t\right) = \phi\left(x - vt\right) \tag{3}$$

which is a "wave" propagating to the right with the velocity of the current. This can be easily checked by substituting in the PDE (1) and verifying it is satisfied.

## 3 Classical Galerkin Approximation: a Fundamental Problem

The solution of (1) can be approximated using finite elements. We will describe the classical Galerkin method using a simple one dimensional problem, consisting on the convection by a fluid of constant velocity of a pollutant which concentration is specified at the left boundary. We will assume that the velocity $v$ is equal to 1, the length of the domain is $L = 5$, and at time zero the concentration within the domain is zero.

First of all, we will introduce the simple mesh consisting of 5 nodes and 4 elements which can be seen in Fig. 2 below.

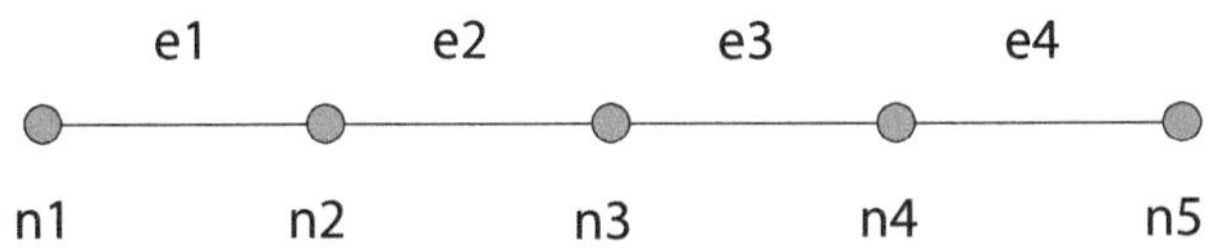


**Figure 2.** One dimensional finite element mesh for the convective transport problem.

We will define the global shape functions $N_j(x) \quad j = 1, \ldots, 5$, and introduce the nodal variables $\hat{\Phi}_j(t) \quad j = 1, \ldots, 5$ which will be used to approximate the solution as:

$$\phi(x,t) \approx \hat{\phi}(x,t) = \sum_{j=1}^{5} N_j(x)\,\hat{\Phi}_j(t) \tag{4}$$

or, in a more compact manner:

$$\hat{\phi}(x,t) = \mathbf{N} \cdot \hat{\mathbf{\Phi}} \tag{5}$$

where we have introduced the vectors of shape functions $\mathbf{N}$ and of nodal unknowns $\hat{\mathbf{\Phi}}$ which depends, respectively, on $x$ and $t$. The boundary condition of Dirichlet type (prescribed concentration) is

$$\phi(x=0,t) = g(t)$$

In this case, the concentration on the left boundary is constant:

$$\phi\left(x=0,t\right)=1$$

This boundary condition is immediately satisfied if we choose $\hat{\Phi}_1\left(t\right)=1$, i.e. we prescribe the value at node 1.

If we substitute the approximation in the PDE, it will not satisfy it. We will define the error or residual $R_\Omega$ as:

$$R_\Omega=\frac{\partial\hat{\phi}}{\partial t}+v\frac{\partial\hat{\phi}}{\partial x} \tag{6}$$

Next, we will substitute (4) in above:

$$R_\Omega=\frac{\partial}{\partial t}\left(\sum_{j=1}^{5}N_j\left(x\right)\hat{\Phi}_j\left(t\right)\right)+v\frac{\partial}{\partial x}\left(\sum_{j=1}^{5}N_j\left(x\right)\hat{\Phi}_j\left(t\right)\right)$$

from where we obtain:

$$R_\Omega=\sum_{j=1}^{5}N_j\left(x\right)\frac{d\hat{\Phi}_j}{dt}+\sum_{j=1}^{5}v\hat{\Phi}_j\frac{dN_j}{dx} \tag{7}$$

The Galerkin method consists of obtaining the unknowns $\hat{\Phi}_j$ using the equations:

$$\int_\Omega N_i R_\Omega d\Omega=0 \tag{8}$$

which provides us with the same number of equations than unknowns, taking into account that in node 1 the nodal value is known.

Eq. (8) can be interpreted as making the error orthogonal to the subspace where we are building the approximation. When approximating functions, the best approximation to a given function using a certain basis is such that the residual or error is orthogonal to all vectors of the subspace.

If we further develop (8) using (7), we obtain:

$$\int_\Omega N_i\left(\sum_{j=1}^{5}N_j\left(x\right)\frac{d\hat{\Phi}_j}{dt}\right)d\Omega=-\int_\Omega N_i\left(\sum_{j=1}^{5}v\hat{\Phi}_j\frac{dN_j}{dx}\right)d\Omega$$

from where:

$$\left(\int_{\Omega} N_i N_j d\Omega\right) \frac{d\hat{\Phi}_j}{dt} = -v \left(\int_{\Omega} N_i \frac{dN_j}{dx} d\Omega\right) \hat{\Phi}_j \tag{9}$$

where we have used the implicit summation convention of Einstein.

This equation can be written in matrix form as:

$$\mathbf{M}\frac{d\hat{\mathbf{\Phi}}}{dt} = -v\mathbf{H}\hat{\mathbf{\Phi}} \tag{10}$$

The mass matrix $\mathbf{M}$ and the discrete convective matrix $\mathbf{H}$ are obtained by assembling the contributions of all elements in the mesh. All element matrices are equal:

$$\begin{aligned} \mathbf{M}^{(e)} &= \frac{h}{6}\begin{pmatrix} 2 & 1 \\ 1 & 2 \end{pmatrix} = \frac{1}{6}\begin{pmatrix} 2 & 1 \\ 1 & 2 \end{pmatrix} \\ \mathbf{H}^{(e)} &= \frac{1}{2}\begin{pmatrix} -1 & 1 \\ -1 & 1 \end{pmatrix} \end{aligned} \tag{11}$$

where $h$ is the length of the elements, which is unity in this case. The assembled matrices are:

$$\mathbf{M} = \frac{h}{6}\begin{pmatrix} 2 & 1 & 0 & 0 & 0 \\ 1 & 4 & 1 & 0 & 0 \\ 0 & 1 & 4 & 1 & 0 \\ 0 & 0 & 1 & 4 & 1 \\ 0 & 0 & 0 & 1 & 2 \end{pmatrix}$$

and

$$\mathbf{H} = \frac{1}{2}\begin{pmatrix} -1 & 1 & 0 & 0 & 0 \\ -1 & 0 & 1 & 0 & 0 \\ 0 & -1 & 0 & 1 & 0 \\ 0 & 0 & -1 & 0 & 1 \\ 0 & 0 & 0 & -1 & 1 \end{pmatrix} \tag{12}$$

The Eq. (10) is a ODE which has to be integrated with respect to time. We will introduce a time discretization, considering a series of time stations $t_0, t_1, \ldots, t_n$ with

$$t_n = t_0 + n\,\Delta t$$

The value of the vector of unknowns at time $n$ will be denoted $\hat{\mathbf{\Phi}}^n$. We will introduce a simple forwards approximation of the time derivative:

$$\frac{d\hat{\mathbf{\Phi}}^n}{dt} = \frac{\hat{\mathbf{\Phi}}^{n+1} - \hat{\mathbf{\Phi}}^n}{\Delta t} \tag{13}$$

which will be substituted in (10) to yield:

$$\hat{\mathbf{\Phi}}^{n+1} = \hat{\mathbf{\Phi}}^n - \Delta t v \mathbf{M}^{-1}\mathbf{H}\hat{\mathbf{\Phi}}^n \tag{14}$$

or

$$\begin{aligned} \hat{\mathbf{\Phi}}^{n+1} &= \left(\mathbf{I} - \Delta t v \mathbf{M}^{-1}\mathbf{H}\right)\hat{\mathbf{\Phi}}^n \\ \hat{\mathbf{\Phi}}^{n+1} &= \mathbf{A}\hat{\mathbf{\Phi}}^n \end{aligned} \tag{15}$$

where $\mathbf{I}$ is the identity matrix of order 5 and $\mathbf{A}$ the iteration matrix. The scheme is said to be explicit as the matrix of coefficients in (10) is the mass matrix. Indeed, this problem can be solved using a Jacobi iteration scheme. Usually, a reasonably accurate solution is obtained with 3-5 iterations.

We will start with an initial solution $\hat{\mathbf{\Phi}}^0 = (1,0,0,0,0)^T$, and choose an increment of time $\Delta t = 0$.

We will obtain:

$$\begin{aligned} \hat{\mathbf{\Phi}}^1 &= \mathbf{A}\hat{\mathbf{\Phi}}^0 \\ \hat{\mathbf{\Phi}}^2 &= \mathbf{A}\hat{\mathbf{\Phi}}^1 \\ \hat{\mathbf{\Phi}}^3 &= \mathbf{A}\hat{\mathbf{\Phi}}^2 \\ &\dots \\ \hat{\mathbf{\Phi}}^{n+1} &= \mathbf{A}\hat{\mathbf{\Phi}}^n \end{aligned} \tag{16}$$

The results present important oscillations which grow up with time which can be seen in Figs. 3 and 4.

The former shows how the solution evolves with time at two control nodes, while the latter gives the concentration in the domain at two different times. The analytical solution is a step function, located at $x_s = n\Delta t$. This type of behaviour will be obtained no matter the increment of time used.

The reason is that the proposed scheme is unconditionally unstable, i.e., it will not converge for any value of $\Delta t$.

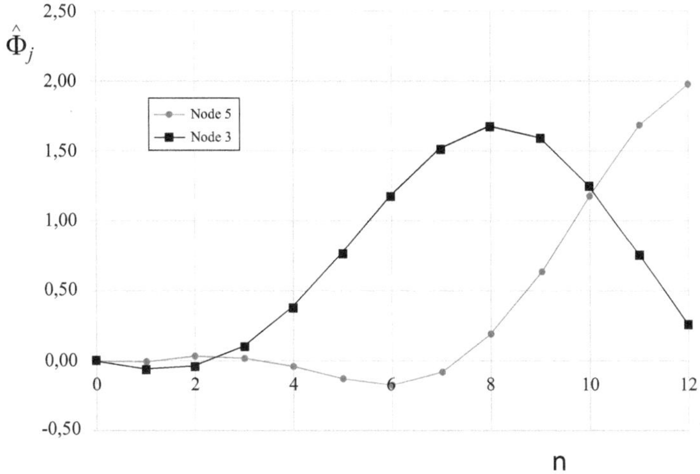


**Figure 3.** Evolution of concentration at nodes 3 and 5 as a function of time step.

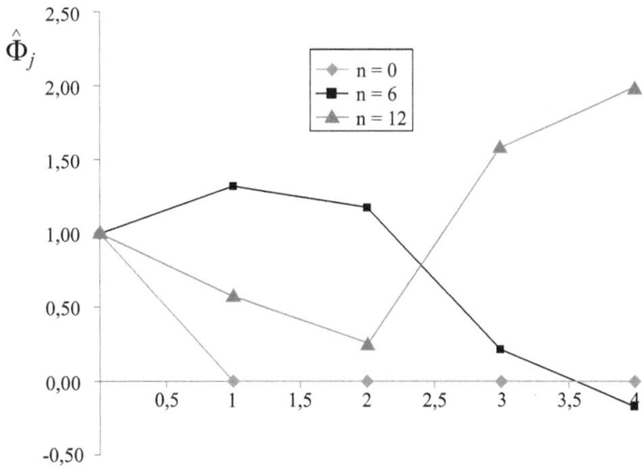


**Figure 4.** Concentration in the mesh at time steps 0, 6 and 12.

A simple proof of why the error is growing can be obtained by considering the scheme given in (15). As the exact solution will fulfil this equation, we can write:

$$\bar{\bar{\mathbf{\Phi}}}^{n+1} = \mathbf{A}\bar{\bar{\mathbf{\Phi}}}^{n}$$

where $\bar{\bar{\mathbf{\Phi}}}^{n}$ is the exact value at time $n$. If we substract from this (15), we find that the error at times $n+1$ and $n$ are related by the same numerical scheme we are using to obtain the solution:

$$\boldsymbol{\varepsilon}^{n+1} = \mathbf{A}\boldsymbol{\varepsilon}^{n}$$

It can be shown that the necessary and sufficient condition for the error not to grow is that the moduli of all the eigenvalues have to be smaller than unity. In the case we are considering here, there are complex eigenvalues with their modulus larger than one.

Sometimes, in order to save computer time, the consistent mass matrix $\mathbf{M}$ is approximated by a diagonal matrix with diagonal terms which are the sum of all the coefficients in the same row. This diagonal matrix is referred to as "lumped mass matrix" or $\mathbf{M}_L$.

In our case, $\mathbf{M}_L$ is obtained immediately from $\mathbf{M}$

$$\mathbf{M} = \frac{h}{6}\begin{pmatrix} 2 & 1 & 0 & 0 & 0 \\ 1 & 4 & 1 & 0 & 0 \\ 0 & 1 & 4 & 1 & 0 \\ 0 & 0 & 1 & 4 & 1 \\ 0 & 0 & 0 & 1 & 2 \end{pmatrix}$$

as:

$$\mathbf{M}_L = \frac{h}{6}\begin{pmatrix} 3 & 0 & 0 & 0 & 0 \\ 0 & 6 & 0 & 0 & 0 \\ 0 & 0 & 6 & 0 & 0 \\ 0 & 0 & 0 & 6 & 0 \\ 0 & 0 & 0 & 0 & 3 \end{pmatrix}$$

The iteration matrix $\mathbf{A}$ has the eigenvalues: $\{1 \pm 0.3536,\ 1.0(\text{triple})\}$ and it is given by:

$$\mathbf{A} = (\mathbf{I} - \Delta t v \mathbf{M}_L^{-1}\mathbf{H}) =$$

$$= \begin{pmatrix} 1 & 0 & 0 & 0 & 0 \\ 0 & 1 & 0 & 0 & 0 \\ 0 & 0 & 1 & 0 & 0 \\ 0 & 0 & 0 & 1 & 0 \\ 0 & 0 & 0 & 0 & 0 \end{pmatrix} - \frac{v\Delta t}{h} \begin{pmatrix} 2 & 0 & 0 & 0 & 0 \\ 0 & 1 & 0 & 0 & 0 \\ 0 & 0 & 1 & 0 & 0 \\ 0 & 0 & 0 & 1 & 0 \\ 0 & 0 & 0 & 0 & 2 \end{pmatrix} \cdot \frac{1}{2} \begin{pmatrix} -1 & 1 & 0 & 0 & 0 \\ -1 & 0 & 1 & 0 & 0 \\ 0 & -1 & 0 & 1 & 0 \\ 0 & 0 & -1 & 0 & 1 \\ 0 & 0 & 0 & -1 & 1 \end{pmatrix}$$

$$= \begin{pmatrix} 1 & 0 & 0 & 0 & 0 \\ 0 & 1 & 0 & 0 & 0 \\ 0 & 0 & 1 & 0 & 0 \\ 0 & 0 & 0 & 1 & 0 \\ 0 & 0 & 0 & 0 & 0 \end{pmatrix} - \frac{v\Delta t}{2h} \begin{pmatrix} -2 & 2 & 0 & 0 & 0 \\ -1 & 0 & 1 & 0 & 0 \\ 0 & -1 & 0 & 1 & 0 \\ 0 & 0 & -1 & 0 & 1 \\ 0 & 0 & 0 & -2 & 2 \end{pmatrix}$$

The iteration equation for node 3 is:

$$\hat{\Phi}_j^{n+1} = \hat{\Phi}_j^n - \frac{v\Delta t}{h}\left(\frac{\hat{\Phi}_{j+1}^n - \hat{\Phi}_{j-1}^n}{2}\right) \quad \text{with } j = 2 \tag{17}$$

In order to gain more insight of what the solution to this problem could be, we will consider some basic finite difference schemes and try to export the results to the finite element world.

## 4 Finite Difference Schemes

### 4.1 Finite Difference Grids. Notation

The Finite Difference Method is based on constructing a grid in the space-time domain. If we consider a one dimensional domain of length $L$, $0 \leq x \leq L$ and a time interval $0 \leq t \leq T$, a finite difference mesh could be the one depicted in Fig. 5.

We have divided the space and the time axes in intervals of lengths $\Delta x$ and $\Delta t$ respectively. The points at which vertical and horizontal lines coincide are called "nodes" of the finite difference mesh. A particular node in the grid at $x_j = j\Delta x \;\; t_n = t_0 + n\Delta t$ will be characterized by the pair of values $(j, n)$. The values of the unknown at this point will be denoted as

$$\phi(x_j, t_n) = \phi_j^n$$

### 4.2 Approximation of Derivatives: Difference Formulas

Partial derivatives with respect to time and space can be approximated as combinations of the values of the unknown at the nodes. For instance, the

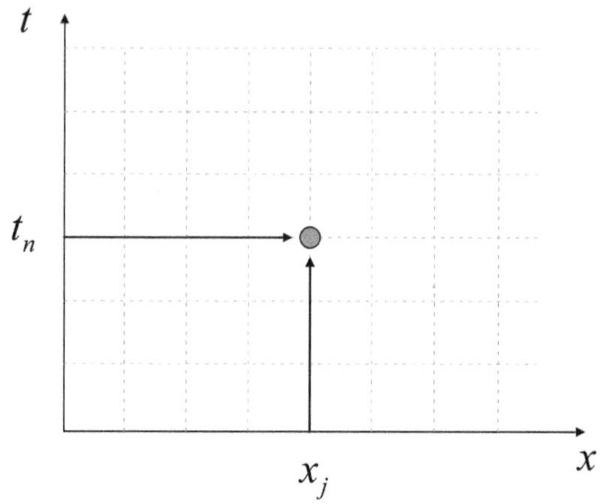


**Figure 5.** Finite difference mesh.

partial derivative with respect to time at $(j, n)$ can be approximated as:

$$\left.\frac{\partial \phi}{\partial t}\right|_j^n = \frac{\phi_j^{n+1} - \phi_j^n}{\Delta t} + \mathcal{O}\left(\Delta t\right) \tag{18}$$

where we have used the notation $\mathcal{O}\left(\Delta t\right)$ to indicate the derivative is first order accurate. This derivative is "forwards in time" or FT for short.

Concerning the first order space derivative, there are several alternatives. We can approximate it as:

$$\left.\frac{\partial \phi}{\partial x}\right|_j^n = \frac{\phi_j^n - \phi_{j-1}^n}{\Delta x} + \mathcal{O}\left(\Delta x\right) \tag{19}$$

which is "backwards in space" or BS and first order accurate.

Alternatively, we can obtain higher accuracy if we approximate the special derivative using a central difference formula:

$$\left.\frac{\partial \phi}{\partial x}\right|_j^n = \frac{\phi_{j+1}^n - \phi_{j-1}^n}{2\Delta x} + \mathcal{O}\left(\Delta x^2\right) \tag{20}$$

which is "centered in space" or CS for short, and second order accurate. It would seem reasonable at first sight to assume that we could obtain a more accurate result using the CS approximation. We will compare next both approximations.

### 4.3 Alternative FD Schemes: FTBS, FTCS and Lax Wendroff

A finite difference scheme is obtained by substituting the partial derivatives with respect to time and space by their discrete approximations. If we use a forwards in time and centred in space formulas, and substitute them in our model PDE (1), we obtain:

$$\frac{\phi_j^{n+1} - \phi_j^n}{\Delta t} + v\frac{\phi_{j+1}^n - \phi_{j-1}^n}{2\Delta x} = 0$$

from where we obtain the FTCS scheme:

$$\phi_j^{n+1} = \phi_j^n - \frac{v\Delta t}{\Delta x}\left(\frac{\phi_{j+1}^n - \phi_{j-1}^n}{2}\right) \tag{21}$$

It is very important to notice that this expression is exactly the same expression obtained for the classical Galerkin FE method presented in the preceding Section.

The non dimensional number

$$C = \frac{v\Delta t}{\Delta x}$$

represents the ratio between the physical velocity and a numerical velocity, and it is referred to as Courant number. It plays a paramount role in numerical analysis of first order hyperbolic PDEs. The **FTCS** scheme can be written as:

$$\boxed{\phi_j^{n+1} = \phi_j^n - \frac{C}{2}\left(\phi_{j+1}^n - \phi_{j-1}^n\right)} \tag{22}$$

The **FTBS** scheme can be obtained in a similar way, resulting on:

$$\boxed{\phi_j^{n+1} = \phi_j^n - C\left(\phi_j^n - \phi_{j-1}^n\right)} \tag{23}$$

Both schemes are said to be explicit, as the values at time $t_{n+1}$ can be obtained directly from the known values at time $t_n$ without having to solve any system of equations.

Fig. 6 gives the "stencil" of both schemes, which describe the way in which the information is obtained in both schemes.

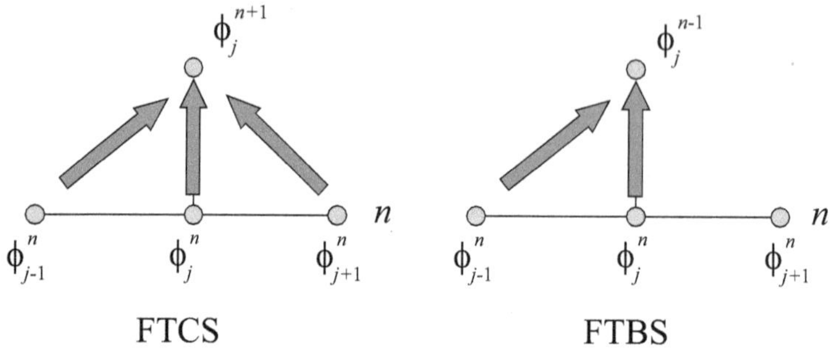


**Figure 6.** Stencil of FTCS and FTBS finite difference schemes.

We will use both schemes to solve the following problem:

Study the transport by a uniform current of constant velocity $v$ of a pollutant in a channel of length 2 units with a concentration at time zero which is given by:

$$\begin{array}{ll} 0 \leq x < 0.8 & \phi(x,0) = 0 \\ 0.8 \leq x < 1 & \phi(x,0) = (x - 0.8)/2 \\ 1 \leq x < 1.2 & \phi(x,0) = (1.2 - x)/2 \\ 1.2 \leq x < 1 & \phi(x,0) = 0 \end{array}$$

The boundary condition is:

$$\phi(x = 0, t) = 0$$

i.e., no pollutant enters the domain. We will use $\Delta x = 0.1$ and study the cases $C = \{0.5, 1, 1.5\}$.

This problem can be easily programmed using a worksheet. The results are given in Figs. 7 (a) to (c) and 8 (a) to (c).

The analytical solution of this problem consists on a translation of the triangular wave profile with the current velocity. We can observe in Fig. 7 that we get the exact solution if Courant number is equal to one. In the case $C = 0.5$, the profile is damped as if the problem we are solving would be one including diffusion. This diffusion has been introduced by the algorithm and is called "numerical diffusion". We can observe that the choice of $C = 1.5$ results on numerical oscillations which grow up as time increases:

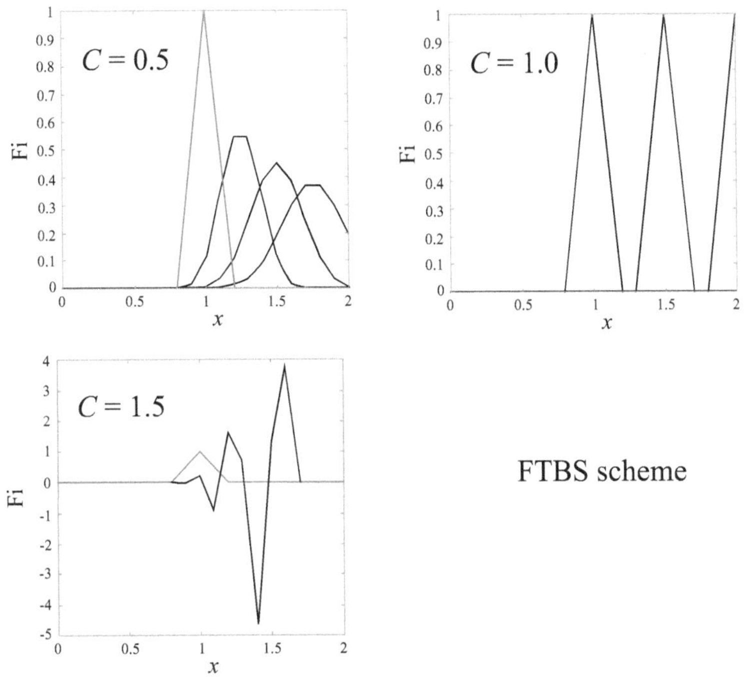


**Figure 7.** Convection of a triangular wave with FTBS scheme: (a) $C = 0.5$; (b) $C = 1$; (c) $C = 1.5$.

the scheme is said to be unstable. It will be shown that stability is achieved if the Courant number is smaller than one: the scheme is said to be "conditionally stable".

Next, we present in Fig. 8 the results obtained for the same case with the FTCS algorithm. In principle, the accuracy should be higher, as we are using a more accurate description of the space derivative, which in FTCS is second order.

In spite of the higher accuracy of the space difference formula chosen, we can see how the scheme is always unstable, with oscillations which grow with time. The scheme is said to be "unconditionally unstable".

Finally, we will present the Lax Wendroff scheme, which is obtained by (i) performing a Taylor series expansion in time, followed by (ii) a discretization

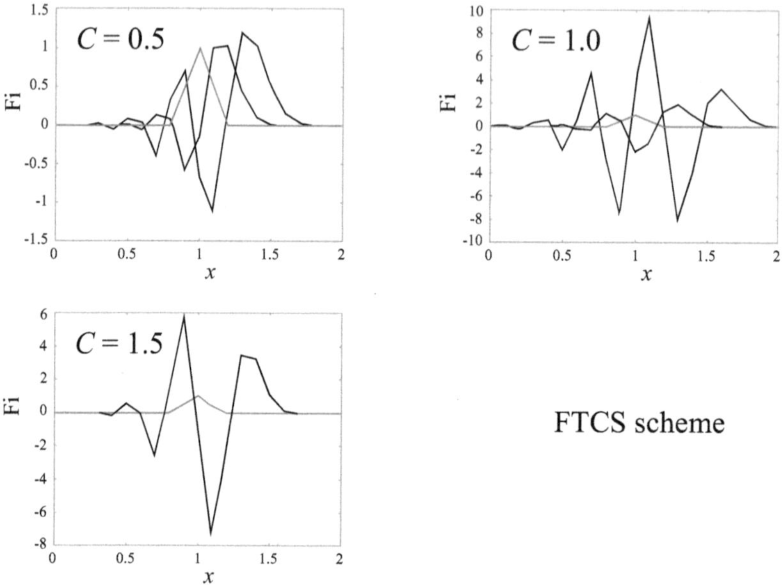


**Figure 8.** Convection of a triangular wave with FTCS scheme: (a) $C = 0.5$; (b) $C = 1$; (c) $C = 1.5$.

in space. We will begin by writing the unknown at time $t_{n+1}$ as:

$$\phi^{n+1} = \phi^n + \Delta t \frac{\partial \phi}{\partial t}\bigg|^n + \frac{1}{2}\Delta t^2 \frac{\partial^2 \phi}{\partial t^2}\bigg|^n \tag{24}$$

Next, we use the PDE to obtain the partial derivatives with respect to time:

$$\frac{\partial \phi}{\partial t}\bigg|^n = -v\,\frac{\partial \phi}{\partial x}\bigg|^n \tag{25}$$

And, differentiating again with respect to time:

$$\left.\frac{\partial^2\phi}{\partial t^2}\right|^n = \frac{\partial}{\partial t}\left(-v\left.\frac{\partial\phi}{\partial x}\right|^n\right) = -v\frac{\partial}{\partial x}\left(\left.\frac{\partial\phi}{\partial t}\right|^n\right)$$

$$= -v\frac{\partial}{\partial x}\left(-v\left.\frac{\partial\phi}{\partial x}\right|^n\right) \tag{26}$$

$$= v^2\left.\frac{\partial^2\phi}{\partial x^2}\right|^n$$

We will substitute now these expressions in the Taylor series expansion:

$$\phi^{n+1} = \phi^n - v\,\Delta t\left.\frac{\partial\phi}{\partial x}\right|^n + \frac{1}{2}v^2\Delta t^2\left.\frac{\partial^2\phi}{\partial x^2}\right|^n \tag{27}$$

And now, we are in position to approximate the space derivatives by difference formulas. As the Taylor expansion is second order accurate in time, we will use second order accurate approximations for the space:

$$\left.\frac{\partial\phi}{\partial x}\right|^n = \frac{\phi_{j+1}^n - \phi_{j-1}^n}{\Delta x} + \mathcal{O}\left(\Delta x^2\right)$$

$$\left.\frac{\partial^2\phi}{\partial x^2}\right|^n = \frac{\phi_{j+1}^n - 2\phi_j^n + \phi_{j-1}^n}{\Delta x^2} + \mathcal{O}\left(\Delta x^2\right) \tag{28}$$

After substituting in the PDE and rearranging, we obtain the **Lax Wendroff scheme**:

$$\boxed{\phi_j^{n+1} = \frac{1}{2}C\left(1+C\right)\phi_{j-1}^n + \left(1-C^2\right)\phi_j^n - \frac{1}{2}C\left(1-C\right)\phi_{j+1}^n} \tag{29}$$

which is an centred in space explicit scheme with the same stencil than the FTCS scheme discussed above.

If we solve the test example of the convection of a triangular wave by a uniform current of constant velocity, we would obtain the results depicted in Fig. 9.

We can see that the scheme is conditionally stable, the solution being exact for $C = 1$.

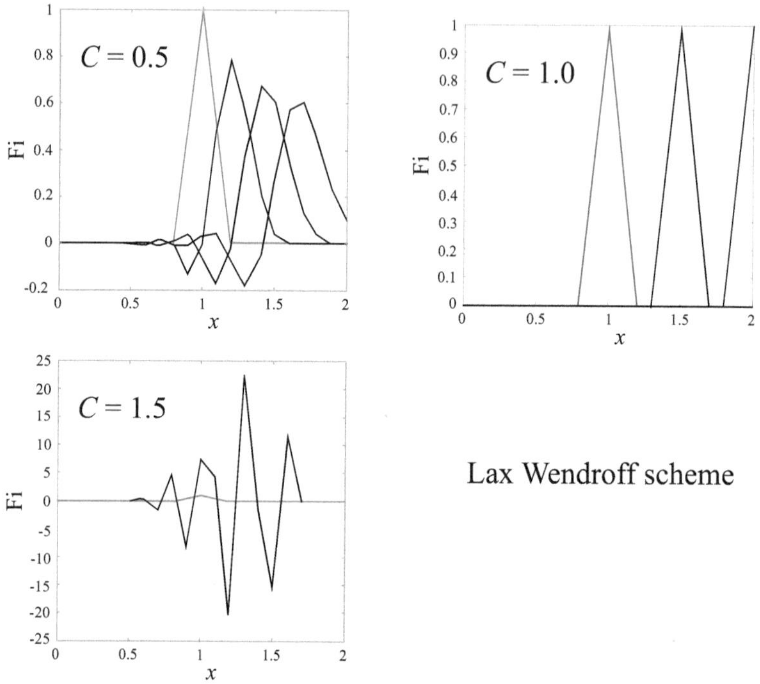


**Figure 9.** Convection of a triangular wave with Lax Wendroff scheme: (a) $C = 0.5$; (b) $C = 1$; (c) $C = 1.5$.

The three schemes presented in this Section have been selected for several reasons. First of all, we have introduced the concepts of stability and numerical diffusion. We have seen that the centred in space FTCS scheme is always unstable, while the FTBS and Lax Wendroff are conditionally stable. The second reason is that the FTCS algorithm coincides in 1D problems with the classical Galerkin FEM scheme described in this Chapter when the consistent mass matrix $\mathbf{M}$ is approximated by the lumped mass matrix $\mathbf{M}_L$, as we have seen. Therefore, a centred, explicit Finite Element scheme based on the classical Galerkin method will never work. Finite differences could provide us with some hints for developing FE schemes which work. There are two simple ways which can be derived from the FTBS and Lax Wendroff schemes: the upwind or Petrov Galerkin methods, where the weighting functions do not coincide with the shape functions, and the Taylor Galerkin scheme which is an extension in the FE realm of Lax Wendroff scheme.

## 4.4 Analysis of Stability: Von Neumann Method

We will assume that the FD scheme is linear, and can be cast in the form

$$L(\mathbf{\Phi}) = L\left(\phi_j^{n+1}, \phi_{j-1}^n, \phi_j^n, \ldots\right) = 0 \tag{30}$$

The solution of the scheme $\bar{\mathbf{\Phi}}$ will also satisfy the condition

$$L(\bar{\mathbf{\Phi}}) = L\left(\bar{\phi}_j^{n+1}, \bar{\phi}_{j-1}^n, \bar{\phi}_j^n, \ldots\right) = 0 \tag{31}$$

If we now introduce the error $\varepsilon_j^n$ at $x = x_0 + j\Delta x$ and $t = t_0 + n\Delta t$

$$\varepsilon_j^n = \phi_j^{n+1} - \bar{\phi}_j^{n+1} \tag{32}$$

we can easily see that

$$\begin{aligned} L\left(\mathbf{\Phi} - \bar{\mathbf{\Phi}}\right) &= 0 \\ L(\boldsymbol{\varepsilon}) &= 0 \end{aligned} \tag{33}$$

and therefore, the error satisfies the finite difference equation. We will assume that the error at time $t_n$ can be written as

$$\varepsilon_j^n = \sum_{l=-N}^{N} E_l^n \exp\left(I k_l j \Delta x\right) \quad I = \sqrt{-1} \tag{34}$$

Because of the linearity of the scheme, we can analyze the behaviour of a particular component $E^n \exp(Ikj\Delta x)$. It is interesting to note that the wave lengths which can be represented by the grid have a maximum value $\lambda_{\text{max}} = 2L = 2N\Delta x$, and a minimum value $\lambda_{\text{min}} = 2\Delta x$ where $N = L\Delta x$ is the number of divisions along $x$. Therefore, a particular wave length can be written as

$$\lambda_l = 2l\Delta x \quad l = 1 \ldots N \tag{35}$$

Concerning wave numbers, they can be obtained as

$$k_l = \frac{\pi}{l\Delta x} \quad l = 1 \ldots N \tag{36}$$

with

$$k_{\text{min}} = \frac{2\pi}{\lambda_{\text{max}}} = \frac{\pi}{N\Delta x} \quad k_{\text{max}} = \frac{2\pi}{\lambda_{\text{min}}} = \frac{\pi}{\Delta x} \tag{37}$$

An important magnitude is the phase lag between two consecutive grid points, which is:

$$\varphi = k\Delta x \tag{38}$$

Taking into account (36) we obtain

$$\varphi = \frac{\pi}{l} \quad l = 1 \dots N$$

The error can be written in terms of $\varphi$ as

$$E^n \exp\left(lkj\Delta x\right) = E^n \exp(Ij\varphi) \tag{39}$$

It is also useful to introduce the relative wave length

$$\Lambda = \frac{\lambda}{\Delta x} = 2l = \frac{2\pi}{\varphi} \tag{40}$$

which relates a particular wavelength to the grid spacing. It can vary between $\Lambda_{\text{min}} = 2$ and $\Lambda_{\text{max}} = 2$ as $\Lambda_l = 2l$. Taking all above definitions into account, we can write the error in some representative points of the finite difference grid as:

$$\begin{aligned} \varepsilon_j^{n+1} &= E^{n+1} \exp(Ij\varphi) \\ \varepsilon_{j+1}^{n} &= E^{n} \exp(I\left(j+1\right)\varphi) \\ \varepsilon_j^{n} &= E^{n} \exp(Ij\varphi) \\ \varepsilon_{j-1}^{n} &= E^{n} \exp(I\left(j-1\right)\varphi) \end{aligned} \tag{41}$$

We will analyze next the stability of the FD schemes FTCS, FTBS and Lax Wendroff.

In the case of the FTCS scheme (22), we know that the error satisfies the FD equation:

$$\varepsilon_j^{n+1} = \varepsilon_j^n - \frac{C}{2}\left(\varepsilon_{j+1}^n - \varepsilon_{j-1}^n\right) \tag{42}$$

We will substitute next (41) into (42), which results on:

$$E^{n+1} \exp\left(Ij\varphi\right) = E^n \exp\left(Ij\varphi\right) - \frac{C}{2}\left\{E^n \exp\left(I\left(j+1\right)\varphi\right) - E^n \exp\left(I\left(j-1\right)\varphi\right)\right\} \tag{43}$$

From here, we divide by $E^n \exp\left(Ij\varphi\right)$:

$$E^{n+1} = E^n - \frac{C}{2} E^n \left(\exp\left(I\varphi\right) - \exp\left(-I\varphi\right)\right)$$

We will introduce a measure of the error amplification

$$\xi^n = \frac{E^{n+1}}{E^n} = 1 - C\left(\frac{\exp\left(I\varphi\right) - \exp\left(-I\varphi\right)}{2}\right) = 1 - I\,C\sin\varphi \tag{44}$$

The condition which must be fulfilled by the FD scheme for being stable is:

$$|\xi^n| \leq 1 \tag{45}$$

In the case of the FTCS scheme, we have

$$|\xi^n| = \left(1 + C^2\sin^2\varphi\right)$$

which is larger than 1 for all values of $C$. Therefore, the scheme is said to be **unconditionally unstable**. The importance of this result lies in the fact that centred FDs are equivalent to classical FEM approximations.

In the case of the scheme FTBS we obtain:

$$\xi^n = \frac{E^{n+1}}{E^n} = \{1 - C + C\cos\varphi\} - I\,\sin\varphi \tag{46}$$

The stability condition is satisfied for Courant numbers $C \leq 1$. The scheme is said to be **conditionally stable**. This is why we have obtained in Fig. 7 (c) oscillations which destroy the accuracy of the computed solution, as $C = 1.5$. However, in the cases $C = 0.5$ and $C = 1$ the solution was stable.

The stability condition in this case can be obtained by means of geometrical considerations, as illustrated in Fig. 10.

In the case of the Lax-Wendroff scheme, we obtain

$$\xi^n = \frac{E^{n+1}}{E^n} = \left\{1 - C^2 + C^2\cos\varphi\right\} - IC\sin\varphi \tag{47}$$

and the stability condition is $C \leq 1$. The Lax-Wendroff scheme is, therefore **conditionally stable**. As in the previous case, this explain the results obtained in Fig. 9.

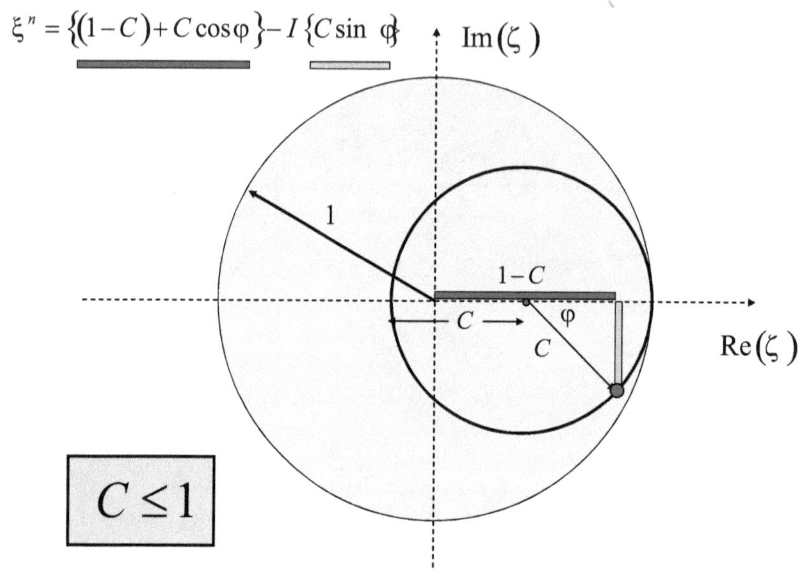


**Figure 10.** Stability of FTBS method.

### 4.5 Numerical Diffusion and Dispersion

If we look again at the examples presented in Figs. 7 and 9, where a wave of triangular shape was convected by a uniform current, we can observe that when the Courant number is 1, the numerical predictions coincide with the analytical solution, in which the shape of the wave does not change. However, at $C = 0.5$, we observe a diffusion which does not exist in the model. This spurious effect is referred to as "numerical diffusion" or "numerical damping" and causes a deterioration of the quality of the numerical predictions. The reason is that the numerical scheme introduces a diffusion which depends on the relative wave length and the Courant number for a considered scheme. In addition to this, waves of different lengths propagate with different velocities in a given mesh, instead of having a velocity given by the convective velocity $u$.

We will consider the case of the scalar convective transport 1D (1) and we will represent a fundamental solution as

$$\bar{\phi} = A \exp\{Ik(x - vt)\} \tag{48}$$

which can be easily verified to satisfy (1). The ratio between the analytical solutions at two consecutive grid point $j$ and $j+1$ at a certain time $n$ is:

$$\frac{\bar{\phi}_{j+1}^n}{\bar{\phi}_j^n} = \exp(-Ikv\Delta x) = \exp(-I\varphi) \tag{49}$$

where we have used the definition (38) $\varphi = k\Delta x$.

The ratio between the analytical solutions at two consecutive times $n$ and $n+1$ at a certain point $j$ is:

$$\frac{\bar{\phi}_j^{n+1}}{\bar{\phi}_j^n} = \exp(-Ikv\Delta t) = \exp(-IC\varphi) \tag{50}$$

where we have used the equalities $v\Delta t = C\Delta x$ which results from the definition of Courant number and $\varphi = k\Delta x$.

Therefore, if we compare the solutions at times $t_n$ and $t_{n+1}$, we can see that the amplitudes are the same, and the phase difference is $-IC\varphi$.

Concerning the computed solution, the ratio $\dfrac{\phi_j^{n+1}}{\phi_j^n}$ can be obtained following the procedure described in previous Section. Indeed, as both the error and the computed solution verify the FD equation, we have

$$\xi = \frac{\phi_j^{n+1}}{\phi_j^n} = \frac{\varepsilon_j^{n+1}}{\varepsilon_j^n} \tag{51}$$

This value has been obtained when studying the stability of FTBS and Lax Wendroff methods:

$$\begin{aligned} \xi|_{FTBS} &= \frac{E^{n+1}}{E^n} = \{1 - C + C\cos\varphi\} - I\sin\varphi \\ \xi|_{LW} = \xi^n &= \frac{E^{n+1}}{E^n} = \{1 - C^2 + C^2\cos\varphi\} - IC\sin\varphi \end{aligned} \tag{52}$$

The damping $D$ is obtained as the modulus of $\xi$. It is interesting to notice that $D$ depends on the Courant number and $\varphi$, or the relative wave number $\Lambda = \dfrac{\lambda}{\Delta x} = \dfrac{2\pi}{\varphi}$. Therefore, for a given initial condition, the components with different relative wave lengths present different damping. We illustrate this fact in Figs. 11 and 12, where we have represented for FTBS and Lax Wendroff schemes the damping at different Courant numbers as a function of the relative wavelength.

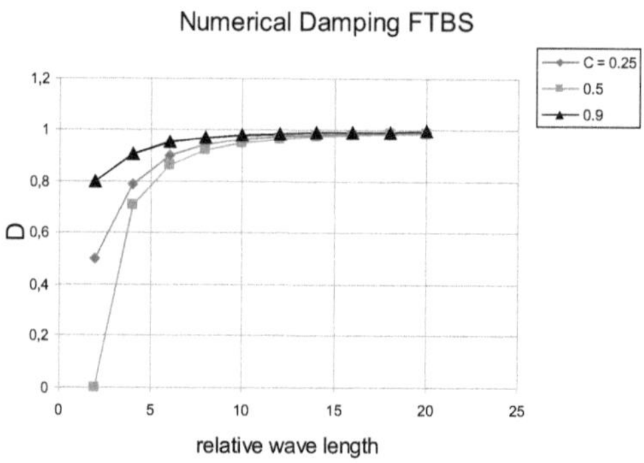


**Figure 11.** Diffusion and damping properties of FTBS scheme.

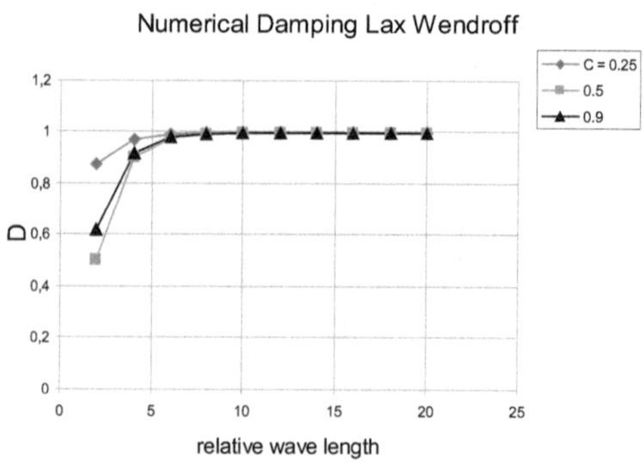


**Figure 12.** Diffusion and damping properties of Lax Wendroff scheme.

From above figures, we can conclude that the FTBS scheme is more dissipative than Lax Wendroff, specially for low numbers of the relative wave length. Moreover, in the former, the $\Lambda = 2\Delta x$ component disappears for $C = 0.5$, as $D = 0$. Highly dissipative schemes have to be avoided when modelling the propagation of waves over very long distances.

Concerning the numerical dispersion, the analysis consists of comparing the analytical phase difference $\exp(-IC\varphi)$ with that of the numerical scheme, which is given by:

$$\theta = \tan^{-1} \frac{\mathrm{Im}(\xi)}{\mathrm{Re}(\xi)} \tag{53}$$

The dispersion error is quantified usually by the quotient

$$\frac{\theta}{\overline{\theta}} = \frac{\tan^{-1} \frac{\mathrm{Im}(\xi)}{\mathrm{Re}(\xi)}}{-C\varphi} \tag{54}$$

We provide in Figs. 13 and 14 the dispersion error of FTBS and Lax Wendroff schemes. It is worth observing that the FTBS scheme at $C = 0.5$ has no dispersion error, i.e., the analytical and numerical phases coincide for all possible wave lengths. At $C = 0.9$, the computed short wavelengths travel faster than they should, which correspond to errors larger than unity. In all three cases considered the dispersion ratio is zero at relative wave lengths of 2, which means this wavelength has zero velocity. Due to the high damping, they disappear from the solution.

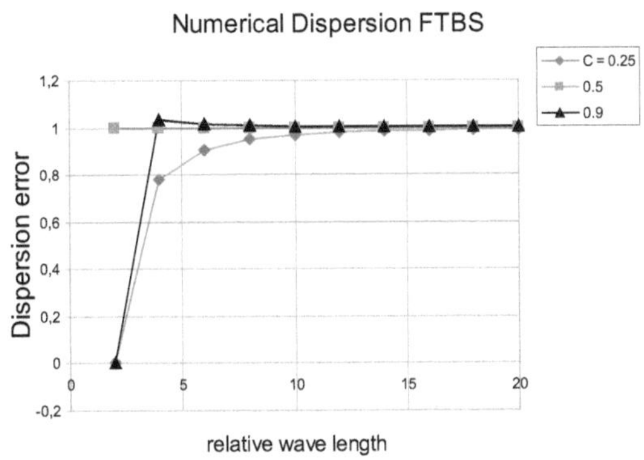


**Figure 13.** Numerical dispersion properties of FTBS scheme.

In the Lax Wendroff scheme, we observe in all cases a dispersion factor smaller than unity, smaller for the smaller wave lengths, which means that small wave lengths will be propagating slower. As the damping is not high, they appear as oscillations behind shocks. This is illustrated in Fig. 15, where we have depicted a discontinuity which propagates to the right.

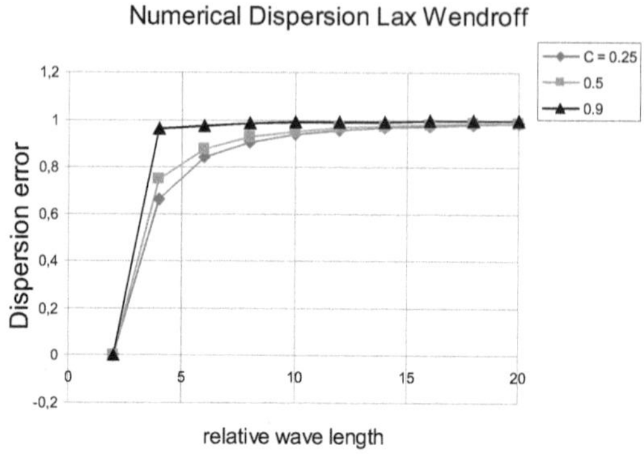


**Figure 14.** Numerical dispersion properties of Lax Wendroff scheme.

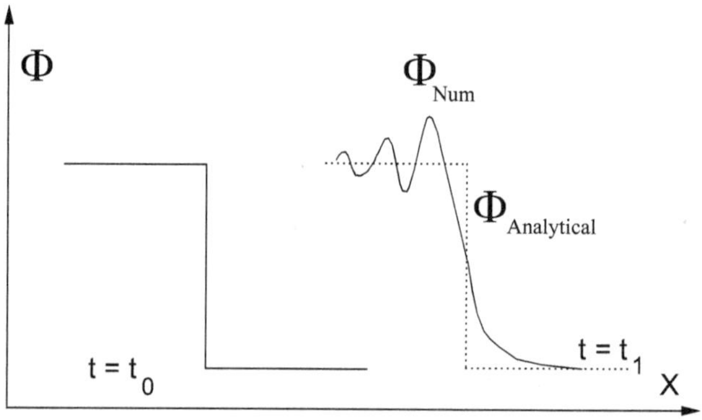


**Figure 15.** Trailing oscillations behind a propagating discontinuity.

## 5 Finite Elements That Work: the Taylor Galerkin Scheme

The Taylor Galerkin algorithm is obtained following the same procedure than in the Lax Wendroff scheme, the only difference being that the space discretization is performed using the Galerkin method.

The Taylor Galerkin method was introduced independently by Donea (1984) and Löhner et al. (1984), and applied to fluid dynamics problems by Peraire (1986), Peraire et al. (1986), Donea et al. (1987) and Quecedo and Pastor

(2002). The interested reader can find a detailed description in the text by Zienkiewicz and Taylor (2000).

This Section is devoted to present this simple yet accurate algorithm, starting from the very simple case of the convective transport equation in 1D, following with its implementation for quasi-linear equations and systems, where we will introduce the two-step variant of the Taylor Galerkin method. Finally, we will analyze the discretization of source and diffusive terms.

## 5.1 The Taylor Galerkin Algorithm for the 1D Convective Transport Equation

We will start with the 1D convective transport equation (1) and the Taylor series expansion (55)

$$\phi^{n+1} = \phi^n - v\ \Delta t \frac{\partial \phi}{\partial x}\bigg|^n + \frac{1}{2} v^2 \Delta t^2 \frac{\partial^2 \phi}{\partial x^2}\bigg|^n \tag{55}$$

Next, we will introduce the finite element approximation:

$$\phi(x,t) \approx \hat{\phi}(x,t) = \sum_{j=1}^{5} N_j(x)\, \hat{\Phi}_j(t) \tag{56}$$

or,

$$\hat{\phi}(x,t) = \mathbf{N} \cdot \hat{\boldsymbol{\Phi}}$$

where we have introduced the global shape functions $N_j(x)\ j = 1, \ldots, 5$, and nodal variables $\hat{\Phi}_j(t)\ j = 1, \ldots, 5$.

If we substitute the approximation (56) in (55), multiply both sides by $N_i(x)$ and integrate in our domain, we obtain:

$$\int_\Omega N_i \hat{\phi}^{n+1} d\Omega = \int_\Omega N_i \hat{\phi}^n d\Omega - v\Delta t \int_\Omega N_i \frac{\partial \hat{\phi}}{\partial x}\bigg|^n d\Omega + v^2 \frac{\Delta t^2}{2} \int_\Omega N_i \frac{\partial^2 \hat{\phi}}{\partial x^2}\bigg|^n d\Omega \tag{57}$$

We will study each term separately:

$$\int_\Omega N_i \hat{\phi}^{n+1} d\Omega = \left( \int_\Omega N_i N_j d\Omega \right) \hat{\Phi}_j^{n+1} = M_{ij} \hat{\Phi}_j^{n+1}$$

$$\int_\Omega N_i \hat{\phi}^n d\Omega = \left( \int_\Omega N_i N_j d\Omega \right) \hat{\Phi}_j^n = M_{ij} \hat{\Phi}_j^n$$

where $\mathbf{M}$ is the mass matrix.

$$\int_\Omega N_i \frac{\partial \hat{\phi}}{\partial x}\bigg|^n d\Omega = \left(\int_\Omega N_i \frac{\partial N_j}{\partial x} d\Omega\right) \hat{\Phi}_j^n = H_{ij}\hat{\Phi}_j^n$$

where we have introduced the discrete convective operator matrix $\mathbf{H}$ as:

$$H_{ij} = \int_\Omega N_i \frac{\partial N_j}{\partial x} d\Omega$$

Finally, the term involving second derivatives of $\hat{\phi}$ cannot be obtained as the approximation is continuous of class $\mathcal{C}^0$, with discontinuous first derivatives and second derivatives not defined at nodes between elements. We will integrate by parts this term to get:

$$\int_\Omega N_i \frac{\partial^2 \hat{\phi}}{\partial x^2}\bigg|^n d\Omega = -\int_\Omega \frac{\partial N_i}{\partial x}\frac{\partial \hat{\phi}}{\partial x}\bigg|^n d\Omega + \int_{\partial\Omega} N_i \frac{\partial \hat{\phi}}{\partial n}\bigg|^n d\Gamma$$

where $\partial\Omega$ is the boundary of the domain (in 1D problems the two points limiting the interval) and $n$ the unit normal to the boundary.

This expression can be further elaborated, arriving to:

$$\begin{aligned} &-\left(\int_\Omega \frac{\partial N_i}{\partial x}\frac{\partial N_j}{\partial x} d\Omega\right)\hat{\Phi}_j^n + \int_{\partial\Omega} N_i \frac{\partial \hat{\phi}}{\partial n}\bigg|^n d\Gamma \\ &= -K_{ij}\hat{\Phi}_j^n + \int_{\partial\Omega} N_i \frac{\partial \hat{\phi}}{\partial n}\bigg|^n d\Gamma \end{aligned} \tag{58}$$

where $\mathbf{K}$ is a matrix which is usually referred to as "stiffness matrix" in structural analysis. Therefore, after collecting all the terms we obtain:

$$\mathbf{M}\hat{\mathbf{\Phi}}^{n+1} = \mathbf{M}\hat{\mathbf{\Phi}}^n - v\Delta t\mathbf{H}\hat{\mathbf{\Phi}}^n - v^2\frac{\Delta t^2}{2}\mathbf{K}\hat{\mathbf{\Phi}}^n + \mathbf{f}^n$$

where $\mathbf{f}^n$ is the vector arising from the integral along the boundary.

This equation can be cast as:

$$\mathbf{M}\Delta\hat{\mathbf{\Phi}}^{n+1} = \mathbf{rhs}|^n \tag{59}$$

where **rhs** means the right hand side vector and the vector of unknowns is:

$$\Delta\hat{\mathbf{\Phi}}^n = \hat{\mathbf{\Phi}}^{n+1} - \hat{\mathbf{\Phi}}^n$$

The algorithm is explicit, as the coefficient matrix is the mass matrix, and it can be solved using an iterative scheme of Jacobi type:

$$\Delta\hat{\mathbf{\Phi}}^n\Big|^{(k+1)} = \mathbf{M}_L^{-1}\left(\mathbf{rhs}|^n - \mathbf{M}\,\Delta\hat{\mathbf{\Phi}}^n\Big|^{(k)}\right) \tag{60}$$

which usually converges within 3 to 5 iterations.

## 5.2 Quasi Linear Equations and Systems. Conservation Forms

First order hyperbolic equations can be cast in an alternative form called conservative or conservation form. In the case of the convective transport equation, we introduce the flux $F$ as:

$$F(x,t) = v\phi(x,t)$$

and the equation can be written as:

$$\frac{\partial \phi}{\partial t} + \frac{\partial F}{\partial x} = 0 \tag{61}$$

Another example is the Burger's equation, which is the simplest equation incorporating a convective term similar to those found in the eulerian balance of momentum equations of fluid dynamics:

$$\frac{\partial \phi}{\partial t} + \phi\frac{\partial \phi}{\partial x} = 0 \tag{62}$$

where we can define the flux as:

$$F = \frac{1}{2}\phi^2 \tag{63}$$

This equation is an example of quasi linear hyperbolic PDEs.

To derive the Taylor Galerkin algorithm for a quasi linear equation written in conservation form, we use the Taylor series expansion:

$$\phi^{n+1} = \phi^n + \Delta t\frac{\partial \phi}{\partial t}\Big|^n + \frac{1}{2}\Delta t^2\frac{\partial^2 \phi}{\partial t^2}\Big|^n \tag{64}$$

From here, and using the PDE written in conservation form we obtain:

$$\frac{\partial \phi}{\partial t} = -\frac{\partial F}{\partial x}$$

and

$$\frac{\partial^2 \phi}{\partial t^2} = \frac{\partial}{\partial t}\left(-\frac{\partial F}{\partial x}\right) = -\frac{\partial}{\partial x}\left(\frac{\partial F}{\partial t}\right)$$

We will write the time derivative of the flux as:

$$\frac{\partial F}{\partial t} = \frac{\partial F}{\partial \phi}\frac{\partial \phi}{\partial t} = -A\frac{\partial F}{\partial x} \tag{65}$$

where

$$A = \frac{\partial F}{\partial \phi}$$

From here, we obtain:

$$\phi^{n+1} = \phi^n - \Delta t \frac{\partial F}{\partial x}\bigg|^n + \frac{1}{2}\Delta t^2 \frac{\partial}{\partial x}\left(A\frac{\partial F}{\partial x}\right)\bigg|^n \tag{66}$$

which is the one step Taylor Galerkin algorithm for quasi linear equations.

In the case of a one dimensional system of $1^{st}$ order PDEs, the unknowns and fluxes would be vectors of a certain dimension. The system could be written in a compact manner as:

$$\frac{\partial \phi}{\partial t} + \frac{\partial F}{\partial x} = 0 \tag{67}$$

and the Taylor Galerkin algorithm would be (65), the only difference being now that $A$ is a matrix.

Space Discretization of (65) is performed through the method of Galerkin. First, we substitute the finite element approximation of the unknowns in (65), and then the equation is multiplied by the shape functions and integrated in the domain:

$$\int_\Omega N_i \hat{\phi}^{n+1} d\Omega = \int_\Omega N_i \hat{\phi}^n d\Omega - \Delta t \int_\Omega N_i \left.\widehat{\frac{\partial F}{\partial x}}\right|^n d\Omega + \frac{1}{2}\Delta t^2 \int_\Omega N_i \frac{\partial}{\partial x}\left(\widehat{A\frac{\partial F}{\partial x}}\right)\bigg|^n d\Omega \tag{68}$$

The third term is obtained easily if we are using linear elements. The flux is obtained at nodes and interpolated using the shape functions as:

$$F = N_j \hat{F}_j$$

The partial derivative is constant within elements. The contribution of a particular element ($e$) to the integral is then obtained as:

$$\int_{\Omega_e} N_i \left.\widehat{\frac{\partial F}{\partial x}}\right|^n d\Omega_e = \left.\overline{\frac{\partial F}{\partial x}}\right|^n \int_{\Omega_e} N_i d\Omega \tag{69}$$

where the overbar refers to the constant value at the element. In the case of 1D linear elements considered here, the integral of a shape function over an element is:

$$\int_{\Omega} N_i d\Omega = \frac{1}{2} l_e$$

where $l_e$ is the length of the element.

Concerning the term involving second order derivatives, it is integrated by parts (in 2 and 3D problems we will use the Gauss theorem):

$$\int_{\Omega} N_i \frac{\partial}{\partial x}\left.\left(\widehat{A\frac{\partial F}{\partial x}}\right)\right|^n d\Omega = -\int_{\Omega} \frac{\partial N_i}{\partial x} \left.\widehat{A\frac{\partial F}{\partial x}}\right|^n d\Omega + \int_{\partial\Omega} N_i \left.\widehat{A\frac{\partial F}{\partial n}}\right|^n d\Gamma \tag{70}$$

The integral in the domain is easily obtained, as both partial derivatives are constant within each element. Depending on the nature of $A$ the integral could be directly computed or we will have to use numerical integration techniques. In the case of Burgers equation, $A = \phi$, and the integral is obtained as the product of the element length by the partial derivatives (constant) in the element by the average of $\phi$ in the element.

### 5.3 The Two Step Taylor Galerkin Algorithm

In the preceding Section we have presented a one step Taylor Galerkin algorithm for systems of quasi linear first order PDEs. The algorithm involves products of a jacobian matrix $A$ by a vector, which introduces an important computational load.

The two step Taylor Galerkin method is an improvement which circumvents this problem. It was introduced by Löhner et al. (1984). It consists on two steps:

In the **first step**, the solution is advanced in time $\Delta t/2$:

$$\phi^{n+\frac{1}{2}} = \phi^n - \frac{\Delta t}{2} \left.\frac{\partial F}{\partial x}\right|^n \tag{71}$$

It is important to notice that the flux derivative is obtained at elements and not at nodes. Therefore, this expression is reformulated using element averaged values as:

$$\bar{\phi}^{n+\frac{1}{2}} = \bar{\phi}^{n} - \frac{\Delta t}{2} \frac{\partial F}{\partial x}\bigg|^{n} \tag{72}$$

Once the values of the unknowns at the elements are known, we can obtain the fluxes $\bar{F}^{n+\frac{1}{2}}$ at time $n+\frac{1}{2}$ as they are functions of the unknowns.

In the **second step**, we obtain the values of the unknowns at time $n+1$ as:

$$\phi^{n+1} = \phi^{n} - \Delta t \, \frac{\partial F}{\partial x}\bigg|^{n+\frac{1}{2}} \tag{73}$$

We will use Galerkin method to obtain:

$$\int_\Omega N_i \hat{\phi}^{n+1} d\Omega = \int_\Omega N_i \hat{\phi}^{n} d\Omega - \Delta t \int_\Omega N_i \, \frac{\partial F}{\partial x}\bigg|^{n+\frac{1}{2}} d\Omega \tag{74}$$

The first two terms are:

$$\int_\Omega N_i \hat{\phi}^{n} d\Omega = M_{ij} \hat{\Phi}_j^{n}$$

$$\int_\Omega N_i \hat{\phi}^{n+1} d\Omega = M_{ij} \hat{\Phi}_j^{n+1}$$

The last term is obtained using integration by parts or Gauss theorem, as the fluxes at time $n+\frac{1}{2}$ are known at element level and not at nodes:

$$\int_\Omega N_i \, \frac{\partial F}{\partial x}\bigg|^{n+\frac{1}{2}} d\Omega = -\int_\Omega \frac{\partial N_i}{\partial x} \, \bar{F}\Big|^{n+\frac{1}{2}} d\Omega + \int_\Omega N_i \, \frac{\partial \bar{F}}{\partial n}\bigg|^{n+\frac{1}{2}} d\Omega \tag{75}$$

The second step can be written as:

$$\mathbf{M} \Delta \hat{\mathbf{\Phi}}^{n} = \mathbf{rhs}\big|^{n+\frac{1}{2}}$$

which can be solved with an iterative solver such as Jacobi or Preconditioned Conjugate Gradient.

### 5.4 Diffusive and Source Terms

The two step algorithm can be extended easily to the general case involving diffusive and source terms. Let us consider the equation:

$$\frac{\partial \phi}{\partial t} + \frac{\partial F}{\partial x} = S + D \tag{76}$$

where $S$ is the source and $D$ the diffusive term:

$$D = \frac{\partial}{\partial x}\left(k\frac{\partial \phi}{\partial x}\right) \tag{77}$$

The **first step** is:

$$\phi^{n+\frac{1}{2}} = \phi^n + \frac{\Delta t}{2}\left(S - \left.\frac{\partial F}{\partial x}\right|^n\right) \tag{78}$$

where we have not included the diffusive terms.

The **second step** is:

$$\phi^{n+1} = \phi^n + \Delta t\left(S + D + \left.\frac{\partial F}{\partial x}\right|^{n+\frac{1}{2}}\right) \tag{79}$$

We find here two additional terms, source and diffusive, to be discretized. Concerning the former, its contribution is:

$$\int_\Omega N_i \bar{S}^{n+\frac{1}{2}} d\Omega \tag{80}$$

This term can be approximated using the value of the source term at the element.

The diffusive term is approximated by its value at time $t_n$:

$$\int_\Omega N_i \frac{\partial}{\partial x}\left(k\frac{\partial \phi}{\partial x}\right)^n d\Omega$$

We will integrate it by parts, obtaining:

$$-\int_\Omega \frac{\partial N_i}{\partial x} k \frac{\partial \hat{\phi}}{\partial x} d\Omega + \int_\Omega N_i k \frac{\partial \hat{\phi}}{\partial n} d\Omega$$

which can be written as:

$$-K_{ij}\hat{\Phi}_j^n + f_{Di} \tag{81}$$

The algorithm described above performs well provided that the intensity of the source term is not very high. Otherwise, we will have to introduce special splitting algorithms to deal with this term. This situation happens quite frequently in the analysis of fast catastrophic landslides, where source terms are due to the slope of the terrain and the basal friction.

### 5.5 A $4^{th}$ Order Runge Kutta Splitting Method for Strong Sources

We will describe here a technique which has been used by the authors (Quecedo et al., 2004; Mabssout et al., 2006) to deal with source terms within the framework of the two steps Taylor Galerkin algorithm. The original idea is described by Hirsch (1988).

The system of hyperbolic equations will be discretized using a splitting operator technique in which each of the two operators, convective transport and sources, will be treated separately. Therefore, we will consider two problems:

(i) A pure convection problem of the type

$$\left.\begin{aligned} \frac{\partial \phi}{\partial t} + v\frac{\partial \phi}{\partial x} &= 0 \\ \phi\left(x, t^n\right) &= \phi^n\left(x\right) \end{aligned}\right\} \Rightarrow \phi^{\text{adv}} \tag{82}$$

(ii) The source problem, which is a ordinary differential equation:

$$\begin{aligned} \frac{d\phi}{dt} &= S(\phi, t) \\ \phi\left(x, t^n\right) &= \phi^{\text{adv}} \end{aligned} \tag{83}$$

Above decomposition can be written in a compact manner as:

$$\phi^{n+1} = S\left(\Delta t\right) \text{Adv}\left(\Delta t\right) \phi^n \tag{84}$$

where $S$ and Adv are the differential operators for the source and convective transport parts.

The second part of the splitting concerns the source term. The solution which we will obtain is:

$$\phi_i^{n+1} = \phi_i^{\text{adv}} + \Delta t S(\phi_i^s) \tag{85}$$

Runge-Kutta algorithms provide a high accuracy in the evaluation of ODEs. A detailed description is provided by Hirsch (1988). The general form of a RK algorithm is

$$
\begin{aligned}
\phi^1 &= \phi^n \\
\phi^2 &= \phi^n + \Delta t \alpha_2 H(\phi^1) \\
\phi^3 &= \phi^n + \Delta t \alpha_3 H(\phi^2) \\
&\cdots\cdots \\
\phi^k &= \phi^n + \Delta t \alpha_k H(\phi^{k-1}) \\
\phi^{n+1} &= \phi^n + \Delta t \sum_{k=1}^{k} \beta_k H(\phi^k)
\end{aligned} \tag{86}
$$

where $H$ is the time derivative of the unknown, in our case $S$, and the consistency condition $\sum_{k=1}^{k} \beta_k = 1$ has to be fulfilled.

Here we have chosen a $4^{th}$ order RK algorithm, as it provides an excellent combination of accuracy and computational effort. The coefficients are given by:

$$
\begin{aligned}
\alpha_2 &= \frac{1}{2}; \quad \alpha_3 = \frac{1}{2}; \quad \alpha_4 = 1; \\
\beta_1 &= \frac{1}{6}; \quad \beta_2 = \beta_3 = \frac{1}{3}; \quad \beta_4 = \frac{1}{6};
\end{aligned} \tag{87}
$$

from where we obtain

$$
\begin{aligned}
\phi^1 &= \phi^n \\
\phi^2 &= \phi^n + \frac{1}{2} \Delta t H(\phi^1) \\
\phi^3 &= \phi^n + \frac{1}{2} \Delta t H(\phi^2) \\
\phi^4 &= \phi^n + \frac{1}{2} \Delta t H(\phi^3) \\
\phi^{n+1} &= \phi^n + \frac{\Delta t}{6} \left[ H(\phi^1) + 2H(\phi^2) + 2H(\phi^3) + H(\phi^4) \right]
\end{aligned} \tag{88}
$$

## List of Symbols

| | | |
|---|---|---|
| $\square^{n+\frac{1}{2}}$ | element averaged value | §5.3 |
| $A$ | normal section of control volume | §2 |
| $A$ | $= \dfrac{\partial F}{\partial \phi}$ | §5 |
| $\mathbf{A}$ | iteration matrix | §3 |
| Adv | differential operator for the convective transport part | §5.5 |
| $C$ | $= \dfrac{v\Delta t}{\Delta x}$ Courant number | §4 |
| $D$ | damping | §4.5 |
| $D$ | diffusive term | §5.4 |
| $\mathbf{f}^n$ | vector arising from the integral along the boundary | §5 |
| $F$ | flux | §5 |
| $f_0$ | initial condition | §2 |
| $g$ | boundary condition | §2, §3 |
| $h$ | length of the elements | §3 |
| $H$ | time derivative of the unknown | §5.5 |
| $\mathbf{H}$ | discrete convective matrix | §3, §5 |
| $I$ | $= \sqrt{-1}$ | §4 |
| $\mathbf{I}$ | identity matrix | §3 |
| $k_l$ | $l = 1, \ldots, N$ wave number | §4 |
| $k_{\max}$ | $= \dfrac{2\pi}{\lambda_{\min}}$ | §4.4 |
| $k_{\min}$ | $= \dfrac{2\pi}{\lambda_{\max}}$ | §4.4 |
| $\mathbf{K}$ | stiffness matrix | §5 |
| $l_e$ | length of the element | §5 |
| $L$ | length of the domain | §3, §4 |
| $\mathbf{M}$ | mass matrix | §3, §5 |
| $\mathbf{M}_L$ | lumped mass matrix | §3, §5 |
| $n$ | unit normal to the boundary | §5 |
| $N$ | $= L\Delta x$ number of divisions along $x$ | §4 |
| $\mathbf{N}$ | vector of shape functions | §3, §5 |

| | | |
|---|---|---|
| **rhs** | right hand side vector | §5 |
| $R_\Omega$ | residual | §3 |
| $S$ | source term | §5.4, §5.5 |
| $t$ | time | §2, §3, §4, §5 |
| $t_n$ | $= t_0 + n\Delta t$ time stations | §3, §4 |
| $\Delta t$ | length of the time interval | §3, §4, §5 |
| $T$ | time period | §4 |
| $v$ | velocity | §2, §3, §4, §5 |
| $x$ | horizontal coordinate | §2, §3, §4, §5 |
| $x_j$ | $= j\Delta x$ | §4 |
| $\Delta x$ | length of the space interval | §4 |
| $\alpha_k$ | coefficient | §5.5 |
| $\beta_k$ | coefficient | §5.5 |
| $\varepsilon_j^n$ | error at time $t_n$ on node $j$ | §3, §4 |
| $\theta$ | $= \tan^{-1} \dfrac{\mathrm{Im}\,(\xi)}{\mathrm{Re}\,(\xi)}$ | §4.5 |
| $\lambda_l$ | $l = 1, \ldots, N$ wave length | §4 |
| $\lambda_{\max}$ | maximum wave length | §4.4 |
| $\lambda_{\min}$ | minimum wave length | §4.4 |
| $\Lambda$ | $= \dfrac{2\pi}{\varphi}$ relative wave length | §4 |
| $\xi$ | $= \dfrac{\phi_j^{n+1}}{\phi_j^n}$ | §4.5 |
| $\xi^n$ | error amplification | §4 |
| $\phi$ | concentration | §2, §4, §5 |
| $\hat{\phi}$ | approximated solution | §3, §5 |
| $\phi^{\mathrm{adv}}$ | solution of a pure convection problem | §5.5 |
| $\phi_j^n$ | $= \phi\,(x_j, t_n)$ | §4 |
| $\hat{\Phi}_j$ | nodal variable | §3, §5 |
| $\hat{\Phi}_j^n$ | $= \hat{\Phi}_j(t_n)$ | §3, §5 |
| $\bar{\Phi}_j^n$ | solution of the scheme at time $t_n$ on node $j$ | §4 |
| $\bar{\hat{\Phi}}_j^n$ | exact solution at time $t_n$ on node $j$ | §3 |
| $\varphi$ | $= k\Delta x$ phase lag between two consecutive grid points | §4 |

| | | |
|---|---|---|
| $\Omega$ | domain | §5 |
| $\partial\Omega$ | boundary of the domain | §5 |

## Bibliography


J. Donea. A Taylor-Galerkin method for convective transport problems. *Int. J. Num. Meth. Engng.*, 20(1):101–109, 1984.

J. Donea, S. Giuliani, H. Laval, and L. Quartapelle. Time-accurate solution of advection-diffusion problems by finite elements. *Comp. Meth. Appl. Mech. Engng.*, 45(1-3):123–145, 1984b.

J. Donea, L. Quartapelle, and V. Selmin. An analysis of time discretization in the finite element solution of hyperbolic problems. *J. Comp. Phys.*, 70(2):463–499, 1987.

C. Hirsch. *Numerical Computation of Internal and External Flows*, volume II. J. Wiley and Sons, 1988.

R. Löhner, K. Morgan, and O. C. Zienkiewicz. The solution of non-linear hyperbolic equation systems by the finite element method. *Int. J. Num. Meth. Fluids*, 4(11):1043–1063, 1984.

M. Mabssout, M. Pastor, M. I. Herreros, and M. Quecedo. A Runge-Kutta, Taylor-Galerkin scheme for hyperbolic systems with source terms. Application to shock wave propagation in viscoplastic geomaterials. *Int. J. Num. Anal. Meth. Geomech.*, 30(13):1337–1355, 2006.

J. Peraire. *A finite element method for convection dominated flows.* PhD thesis, University of Wales, Swansea, 1986.

J. Peraire, O. C. Zienkiewicz, and K. Morgan. Shallow water problems. A general explicit formulation. *Int. J. Num. Meth. Engng.*, 22(3):547–574, 1986.

M. Quecedo and M. Pastor. A reappraisal of Taylor-Galerkin algorithm for drying-wetting areas in shallow water computations. *Int. J. Num. Meth. Fluids*, 38(6):515–531, 2002.

M. Quecedo, M. Pastor, M. I. Herreros, and J. A. Fernández Merodo. Numerical modelling of the propagation of fast landslides using the finite element method. *Int. J. Num. Meth. Engng.*, 59(6):755–794, 2004.

O. C. Zienkiewicz and R. L. Taylor. *The Finite Element Method.* Butterworth-Heinmann, 5th edition, 2000.

# Discretization Techniques for Transient, Dynamic and Cyclic Problems in Geotechnical Engineering: Second Order Equation

M. Pastor [*‡], J. A. Fernández Merodo [‡†], P. Mira [*‡], S. López Querol [§] I. Herreros [‡†] and M. Mabssout [£]

[*] Centro de Estudios y Experimentación de Obras Públicas CEDEX Madrid, Spain

[‡] Grupo M2i, Dept. Applied Mathematics, ETS Ing. Caminos, Universidad Politécnica de Madrid, Spain

[†] Instituto Geológico y Minero de España IGME, Madrid, Spain

[§] Universidad de Castilla La Mancha, Spain

[£] Faculté des Sciences et Techniques, Tanger, Morocco

## 1 Introduction

The second order equations are the form more widely used both in solids and in soil dynamics. This Chapter is devoted to present the Discretization techniques which are used. The solution strategy consists on (i) perform a discretization in space from which a system of ordinary differential equations (ODE) is obtained, and (ii) discretize this system in time. An important aspect is that of imposing suitable boundary conditions. Here we will present a simple technique which can be used for elastic or viscoplastic materials assuming that the wave is planar in the neighbourhood of the boundary. Finally, we will present some applications to cyclic and dynamic problems.

## 2 Discretization of Second Order Equations of Solid Dynamics

### 2.1 General Problem

The second order equation is the most widely used formulation. It consists on:

(1) The balance of momentum equation

$$\text{div}\boldsymbol{\sigma} + \mathbf{b} = \rho \frac{d^2\mathbf{u}}{dt^2} \tag{1}$$

where $\mathbf{b}$ is the vector of body forces, $\boldsymbol{\sigma}$ the stress tensor, and $\mathbf{u}$ the displacement field.

The equivalent vector form is

$$\mathbf{S}^T\boldsymbol{\sigma}^* + \mathbf{b} = \rho \frac{\partial^2\mathbf{u}}{\partial t^2} \tag{2}$$

with

$$\mathbf{S}^T = \begin{pmatrix} \partial_x & 0 & 0 & \partial_y & 0 & \partial_z \\ 0 & \partial_y & 0 & \partial_x & \partial_z & 0 \\ 0 & 0 & \partial_z & 0 & \partial_y & \partial_x \end{pmatrix}$$

$$\boldsymbol{\sigma}^* = (\sigma_{xx}, \sigma_{yy}, \sigma_{zz}, \sigma_{xy}, \sigma_{yz}, \sigma_{zx})^T$$

For the sake of simplicity we will drop the star superindexes in what follows, understanding that whenever the vector form is used, we refer to the star representation.

(2) A suitable constitutive relation relating stress and strain increments. In the particular case of an elastoplastic material it is given by

$$d\boldsymbol{\sigma} = \mathbf{D}^{ep} d\boldsymbol{\varepsilon} \tag{3}$$

while for a linear elastic material it can be written as

$$\boldsymbol{\sigma} = \mathbf{D}^{e} \boldsymbol{\varepsilon} \tag{4}$$

(3) A kinematic relation between the strain and displacements

$$d\varepsilon_{ij} = \frac{1}{2}\left(\frac{\partial u_i}{\partial x_j} + \frac{\partial u_j}{\partial x_i}\right) \tag{5}$$

or, in vector form

$$\begin{pmatrix} \varepsilon_{xx} \\ \varepsilon_{yy} \\ \varepsilon_{zz} \\ \gamma_{xy} \\ \gamma_{yz} \\ \gamma_{zx} \end{pmatrix} = \begin{pmatrix} \partial_x & 0 & 0 \\ 0 & \partial_y & 0 \\ 0 & 0 & \partial_z \\ \partial_y & \partial_x & 0 \\ 0 & \partial_z & \partial_y \\ \partial_z & 0 & \partial_x \end{pmatrix} \begin{pmatrix} u_x \\ u_y \\ u_z \end{pmatrix} = \mathbf{S}\mathbf{u} \tag{6}$$

In above we have assumed small strains, and no rotations have been taken into account.

Above equations have to be complemented, for the problem to be well posed, by suitable boundary and initial conditions. Concerning the latter, both displacements and velocities at time zero has to be given.

Boundary conditions can be of different types:

(i) Dirichlet boundary conditions, where displacements or velocities are prescribed

$$\mathbf{u} - \bar{\mathbf{u}} = 0 \qquad \text{on } \Gamma_u \tag{7}$$

(ii) Neumann or prescribed tractions

$$\sigma_{ij} n_j - \bar{t}_i = 0 \qquad \text{on } \Gamma_q \tag{8}$$

where $n_j$ is the $j$-th component of the outwards unit vector normal to the boundary $\Gamma_q$ and $\bar{t}_i$ is the vector of forces per unit area acting on $\Gamma_q$

(iii) Absorbing wave conditions

$$\begin{aligned} \psi_P^{(2)} &= t_n + \rho c_P v_n = 0 \\ \psi_S^{(2)} &= t_t + \rho c_S v_t = 0 \end{aligned} \tag{9}$$

where the sub indexes $t$ and $n$ refer to the components along the normal and tangent plane to the boundary.

(iv) Incident wave conditions In the case of a wave entering the domain, the boundary conditions are

$$\begin{aligned} \psi_P^{(2)} &= t_n + \rho c_P v_n = \psi_P^{(\mathrm{Inc})} \\ \psi_S^{(2)} &= t_t + \rho c_S v_t = \psi_S^{(\mathrm{Inc})} \end{aligned} \tag{10}$$

where $\psi_P^{(\mathrm{Inc})}$ and $\psi_S^{(\mathrm{Inc})}$ can be obtained from the information available for the incoming wave.

It is important to notice that we have assumed that the wave is plane when approaching the boundary and that either elastic or elasto-viscoplastic behaviour in the absorbing and incident wave boundaries.

In the case of the second order equation, the equations can be discretized using finite differences, finite elements, or any other alternative techniques.

Most of the times, finite elements are preferred because of their flexibility. The discretization process consists on two steps, performed first in space and then in time. It can be found in texts like those of The system of partial differential equations can be discretized using standard Galerkin techniques, as described in Zienkiewicz and Taylor (2000) and Hughes (1987).

## 2.2 Discretization in Space

The components of the displacement vector are approximated within an element as

$$u_x \approx u_x^h = \sum_{J=1}^{nnode} N_J(\mathbf{x})\, \hat{u}_{Jx}(t) \tag{11}$$

where $\mathbf{u}^h(\mathbf{x}, t)$ is the approximation of the displacements, $N_J(\mathbf{x})$ are the element shape functions and $\hat{u}_{Jx}(t)$ the nodal values of the nodal components of the displacements, which depend on time and the sub indexes $x$, $y$ and $z$ refer to the space components along the axes.

If we introduce the notation

$$\mathbf{N}_J = \begin{pmatrix} N_J & 0 & 0 \\ 0 & N_J & 0 \\ 0 & 0 & N_J \end{pmatrix} \quad \hat{\mathbf{u}}_J = \begin{pmatrix} \hat{u}_{Jx} \\ \hat{u}_{Jy} \\ \hat{u}_{Jz} \end{pmatrix} \tag{12}$$

we can write the approximation as:

$$\begin{pmatrix} u_x \\ u_y \\ u_z \end{pmatrix} \approx \sum_{J=1}^{nnode} \begin{pmatrix} N_J & 0 & 0 \\ 0 & N_J & 0 \\ 0 & 0 & N_J \end{pmatrix} \begin{pmatrix} \hat{u}_{Jx} \\ \hat{u}_{Jy} \\ \hat{u}_{Jz} \end{pmatrix} \tag{13}$$

or

$$\mathbf{u}(\mathbf{x}, t) \approx \mathbf{u}^h(\mathbf{x}, t) = \sum_{J=1}^{nnode} \mathbf{N}_J(\mathbf{x})\, \hat{\mathbf{u}}_J(t) = \sum_{J=1}^{nnode} N_J(\mathbf{x})\, \hat{\mathbf{u}}_J(t) \tag{14}$$

Above expression can be generalized to global shape functions as:

$$\mathbf{u}^h(\mathbf{x}, t) = \mathbf{N}^{(g)T}(\mathbf{x})\, \hat{\mathbf{u}}^{(g)}(t)$$

where

$$\mathbf{N}^{(g)T} = \left\{ \begin{pmatrix} N_1^{(g)} & 0 & 0 \\ 0 & N_1^{(g)} & 0 \\ 0 & 0 & N_1^{(g)} \end{pmatrix} \cdots \begin{pmatrix} N_n^{(g)} & 0 & 0 \\ 0 & N_n^{(g)} & 0 \\ 0 & 0 & N_n^{(g)} \end{pmatrix} \right\} \tag{15a}$$
$$= \left\{ \mathbf{N}_1^{(g)}, \mathbf{N}_2^{(g)}, \ldots, \mathbf{N}_n^{(g)} \right\}$$

$$\begin{aligned} \hat{\mathbf{u}}^{(g)}(t) &= \left\{ \left( \hat{u}_{1x}^{(g)}, \hat{u}_{1y}^{(g)}, \hat{u}_{1z}^{(g)} \right) \cdots \left( \hat{u}_{nx}^{(g)}, \hat{u}_{ny}^{(g)}, \hat{u}_{nz}^{(g)} \right) \right\}^T \\ &= \left\{ \hat{\mathbf{u}}_1^{(g)T}, \hat{\mathbf{u}}_2^{(g)T} \ldots \hat{\mathbf{u}}_n^{(g)T} \right\}^T \end{aligned} \tag{15b}$$

are the global shape function matrix and displacement vector, $n$ being the total number of nodes in the finite element mesh. In what follows, for the sake of simplifying the notation, we will drop the $(g)$ super index.

When the approximated displacement field is substituted into the balance of momentum equation, we will obtain a residual

$$\mathbf{R}_\Omega = \rho \frac{\partial^2 \mathbf{u}^h}{\partial t^2} - \mathbf{b} - \mathbf{S}^T \boldsymbol{\sigma} \neq \mathbf{0} \tag{16}$$

The Galerkin method consists on finding an approximate solution which is orthogonal to all shape functions:

$$\int_\Omega N_I \mathbf{R}_\Omega d\Omega = \int_\Omega N_I \left( \rho \frac{\partial^2 \mathbf{u}^h}{\partial t^2} - \mathbf{b} - \mathbf{S}^T \boldsymbol{\sigma} \right) d\Omega = \mathbf{0} \tag{17}$$

We will analyze the three terms of above integral next.

Concerning the accelerations term, we have

$$\int_\Omega N_I \rho \frac{\partial^2 \mathbf{u}^h}{\partial t^2} d\Omega = \sum_{J=1}^{n} \left( \int_\Omega \rho N_I N_J d\Omega \right) \frac{d^2 \hat{\mathbf{u}}_J}{dt^2} \tag{18}$$

where we have used $\mathbf{u}^h(\mathbf{x}, t) = \mathbf{N}^T(\mathbf{x})\, \hat{\mathbf{u}}(t)$. We will define the mass matrix $\mathbf{M}$ as:

$$M_{IJ} = \int_\Omega \rho N_I N_J d\Omega \tag{19}$$

The contribution from the body forces term is:

$$\mathbf{F}_{bI} = \int_{\Omega} N_I \mathbf{b} d\Omega \tag{20}$$

Finally, we will analyze the contribution of the divergence of the stress tensor $-\int_{\Omega} N_I \mathbf{S}^T \boldsymbol{\sigma} d\Omega$. We will consider the tensor form as

$$-\int_{\Omega} N_I \frac{\partial \sigma_{ij}}{\partial x_j} d\Omega = 0 \tag{21}$$

where $N_I$ is a scalar shape function, and we will apply the Gauss theorem to this integral, obtaining

$$-\int_{\Omega} N_I \frac{\partial \sigma_{ij}}{\partial x_j} d\Omega = \int_{\Omega} \frac{\partial N_I}{\partial x_j} \sigma_{ij} d\Omega - \int_{\Gamma} N_I \sigma_{ij} n_j d\Gamma \tag{22}$$

The terms $\frac{\partial N_I}{\partial x_j} \sigma_{ij}$ can be collected in vector form, after some algebra, as the product

$$\mathbf{B}_I^T \boldsymbol{\sigma} \stackrel{def}{=} \begin{pmatrix} \partial_x N_I & 0 & 0 & \partial_y N_I & 0 & \partial_z N_I \\ 0 & \partial_y N_I & 0 & \partial_x N_I & \partial_z N_I & 0 \\ 0 & 0 & \partial_z N_I & 0 & \partial_y N_I & \partial_x N_I \end{pmatrix} \begin{pmatrix} \sigma_{xx} \\ \sigma_{yy} \\ \sigma_{zz} \\ \sigma_{xy} \\ \sigma_{yz} \\ \sigma_{zx} \end{pmatrix} \tag{23}$$

where we have introduced the matrix $\mathbf{B}_I$, which can be obtained as:

$$\mathbf{B}_I = \mathbf{S}^T \mathbf{N}_I \tag{24}$$

Concerning the boundary integral, we will assume that we have boundary conditions of the four types defined in (7) to (10) in $\Gamma_u$ ,$\Gamma_q$ ,$\Gamma_{\mathrm{Abs}}$ and $\Gamma_{\mathrm{Inc}}$ which fulfil the condition of $\Gamma_u \cap \Gamma_q \cap \Gamma_{\mathrm{Abs}} \cap \Gamma_{\mathrm{Inc}} = \{\emptyset\}$ being the null set the intersection of any pair of them.

Therefore, we can decompose the boundary integral into:

$$\begin{aligned} \int_{\Gamma} N_I \sigma_{ij} n_j d\Gamma = & \int_{\Gamma_u} N_I \sigma_{ij} n_j d\Gamma + \int_{\Gamma_q} N_I \sigma_{ij} n_j d\Gamma \\ & + \int_{\Gamma_{\mathrm{Abs}}} N_I \sigma_{ij} n_j d\Gamma + \int_{\Gamma_{\mathrm{Inc}}} N_I \sigma_{ij} n_j d\Gamma \end{aligned} \tag{25}$$

The first term corresponds to the reactions in all prescribed degrees of freedom,

$$\int_{\Gamma_u} N_I \sigma_{ij} n_j d\Gamma = R_{Ii} \tag{26}$$

The contribution from the boundary where tractions are prescribed is

$$\int_{\Gamma_q} N_I \sigma_{ij} n_j d\Gamma = \int_{\Gamma_q} N_I \bar{t}_i d\Gamma = \bar{t}_{Ii} \tag{27}$$

The contributions from absorbing and incident wave boundaries result in terms of the type

$$\begin{aligned}\int_{\Gamma_{\text{Abs}}} N_I \sigma_{ij} n_j d\Gamma &= -\int_{\Gamma_{\text{Abs}}} N_I \mathbf{A}_{\text{Abs}} \frac{\partial \mathbf{u}^h}{\partial t} d\Gamma \\ &= -\sum_{J=1}^{n} \left( \int_{\Gamma_{\text{Abs}}} N_I N_J \mathbf{A}_{\text{Abs}} d\Gamma \right) \frac{d\hat{\mathbf{u}}_J}{dt}\end{aligned} \tag{28}$$

and

$$\int_{\Gamma_{\text{Inc}}} N_I \sigma_{ij} n_j d\Gamma = -\int_{\Gamma_{\text{Inc}}} N_I \mathbf{A}_{\text{Inc}} \frac{\partial \mathbf{u}^h}{\partial t} d\Gamma + \mathbf{F}_I^{(\text{Inc})} \tag{29}$$

where the matrices $\mathbf{A}_{\text{Abs}}$ and $\mathbf{A}_{\text{Inc}}$, and the vector $\mathbf{F}_I^{(\text{Inc})}$ can be obtained from the Riemann invariants. In the case of 2D problems,

$$\mathbf{A}_{\text{Abs}} = \mathbf{T}^{-1}\mathbf{V}\mathbf{T}$$

where

$$\mathbf{T} = \begin{pmatrix} n_x & n_y \\ -n_y & n_x \end{pmatrix} \quad \mathbf{V} = \begin{pmatrix} \rho c_P & 0 \\ 0 & \rho c_S \end{pmatrix}$$

$n_x$ and $n_y$ being the components of the normal unit vector to the boundary, and $c_P$ and $c_S$ the $P$ and $S$ wave velocities.

Finally, the equation of balance of momentum discretized in space is:

$$\sum_{J=1}^{n} M_{IJ} \frac{d^2\hat{\mathbf{u}}_J}{dt^2} + \int_\Omega \mathbf{B}_I^T \boldsymbol{\sigma} d\Omega + \sum_{J=1}^{n} C_{IJ} \frac{d\hat{\mathbf{u}}_J}{dt} - \mathbf{F}_{bI} - \mathbf{R}_I - \bar{\mathbf{t}}_I + \mathbf{F}_I^{(\text{Inc})} = \mathbf{0} \tag{30}$$

where $C_{IJ}$ are given in (28). In a more compact manner

$$\mathbf{M}\frac{d^2\hat{\mathbf{u}}}{dt^2} + \mathbf{C}\frac{d\hat{\mathbf{u}}}{dt} + \int_\Omega \mathbf{B}^T \boldsymbol{\sigma} d\Omega - \mathbf{F} = \mathbf{0} \tag{31}$$

where we have introduced

$$\mathbf{F}_I = \mathbf{F}_{bI} + \mathbf{R}_I + \bar{\mathbf{t}}_I - \mathbf{F}_I^{(\mathrm{Inc})}$$

and the matrix $\mathbf{B}$ is the discrete divergence matrix, collecting all $\mathbf{B}_I$ terms. It is interesting to note that the contribution from incident and absorbing boundaries consists on a viscous damping term $\mathbf{C}\frac{d\hat{\mathbf{u}}}{dt}$ and a forcing term $\mathbf{F}$.

### 2.3 Discretization in Time

The second step is the discretization in time of the system of ordinary differential equation (31). Among the many alternatives suggested in the literature, the Generalized Newmark algorithm proposed by Katona and Zienkiewicz and described in Zienkiewicz and Taylor (2000) provides a good combination of computational cost and accuracy. It begins by writing (31) at time $n+1$:

$$\mathbf{M}\frac{d^2\hat{\mathbf{u}}}{dt^2} + \mathbf{C}\frac{d\hat{\mathbf{u}}}{dt} + \int_\Omega \mathbf{B}^T \boldsymbol{\sigma} d\Omega - \mathbf{F}\Bigg|^{n+1} = \mathbf{0} \tag{32}$$

and introducing the variable

$$\Delta\ddot{\hat{\mathbf{u}}} = \ddot{\hat{\mathbf{u}}}^{n+1} - \ddot{\hat{\mathbf{u}}}^{n} \tag{33}$$

where the upper dots refer to time derivatives, and approximating the velocities and displacements at time $n$ as

$$\begin{aligned} \dot{\hat{\mathbf{u}}}^{n+1} &= \left(\dot{\hat{\mathbf{u}}}^{n} + \Delta t \ddot{\hat{\mathbf{u}}}^{n}\right) + \beta_1 \Delta t \Delta\ddot{\hat{\mathbf{u}}} \\ \hat{\mathbf{u}}^{n+1} &= \left(\hat{\mathbf{u}}^{n} + \Delta t \dot{\hat{\mathbf{u}}}^{n} + \frac{1}{2}\Delta t^2 \ddot{\hat{\mathbf{u}}}^{n}\right) + \frac{1}{2}\beta_2 \Delta t^2 \Delta\ddot{\hat{\mathbf{u}}} \end{aligned} \tag{34}$$

where the terms within brackets are known at time $t^n$ and are often referred to as "predictors"

$$\begin{aligned} \dot{\hat{\mathbf{u}}}^{n+1} &= \dot{\hat{\mathbf{u}}}^{n+1(p)} + \beta_1 \Delta t \Delta\ddot{\hat{\mathbf{u}}} \\ \hat{\mathbf{u}}^{n+1} &= \hat{\mathbf{u}}^{n+1(p)} + \frac{1}{2}\beta_2 \Delta t^2 \Delta\ddot{\hat{\mathbf{u}}} \end{aligned} \tag{35}$$

Substituting these values on (31), we arrive at the discretized system:

$$(\mathbf{M} + \beta_1 \Delta t \mathbf{C})\,\Delta\ddot{\hat{\mathbf{u}}} + \int_\Omega \mathbf{B}^T \boldsymbol{\sigma}^{*n+1} d\Omega + \left\{\mathbf{M}\ddot{\hat{\mathbf{u}}}^{n} + \mathbf{C}\dot{\hat{\mathbf{u}}}^{n+1(p)} - \mathbf{F}\right\} = \mathbf{0} \tag{36}$$

# 3 Elastodynamics

In the case of elastic materials, (31) can be simplified. The term $\int_\Omega \mathbf{B}^T \boldsymbol{\sigma} d\Omega$ can be evaluated taking into account (i) the constitutive relation $\boldsymbol{\sigma} = \mathbf{D}^e \boldsymbol{\varepsilon}$, (ii) the relation between strain and displacements $\boldsymbol{\varepsilon} = \mathbf{S}\mathbf{u}^h$ and the relation between the approximated displacement field and the nodal displacements $\mathbf{u}^h = \mathbf{N}^T \hat{\mathbf{u}}$. The stress can be written as:

$$\boldsymbol{\sigma} = \mathbf{D}^e \mathbf{B} \hat{\mathbf{u}} \tag{37}$$

where we have used the definition $\mathbf{B}^T = \mathbf{NS}$. Substituting the stress in (31), we arrive to:

$$\mathbf{M}\frac{d^2\hat{\mathbf{u}}}{dt^2} + \mathbf{C}\frac{d\hat{\mathbf{u}}}{dt} + \mathbf{K}\hat{\mathbf{u}} - \mathbf{F} = \mathbf{0} \tag{38}$$

where we have introduced the elastic stiffness matrix $\mathbf{K}$ defined as:

$$\mathbf{K} = \int_\Omega \mathbf{B}^T \mathbf{D}^e \mathbf{B} d\Omega \tag{39}$$

As in the general case, the second step is the discretization in time. We will use again (33)-(35), arriving to the discretized system:

$$\left(\mathbf{M} + \beta_1 \Delta t \mathbf{C} + \frac{1}{2}\beta_2 \Delta t^2 \mathbf{K}\right) \Delta\ddot{\hat{\mathbf{u}}} + \left\{\mathbf{M}\ddot{\hat{\mathbf{u}}}^n + \mathbf{C}\dot{\hat{\mathbf{u}}}^{n+1(p)} + \mathbf{K}\hat{\mathbf{u}}^{n+1(p)} - \mathbf{F}\right\} = \mathbf{0} \tag{40}$$

which can be solved in $\Delta\ddot{\hat{\mathbf{u}}}$.

## 3.1 Analysis in the Frequency Domain

We will assume here that both the external forces and the system response are harmonic functions of the type:

$$\begin{aligned} \hat{\mathbf{u}}(\mathbf{x}, t) &= \mathrm{Re}\left(\hat{\mathbf{U}}(\mathbf{x}) \exp(-i\omega t)\right) \\ \mathbf{F}(\mathbf{x}, t) &= \mathrm{Re}\left(\hat{\mathbf{F}} \exp(-i\omega t)\right) \end{aligned} \tag{41}$$

Substituting into (38) we obtain:

$$\left\{\mathbf{K} - i\omega\mathbf{C} - \omega^2\mathbf{M}\right\} \hat{\mathbf{U}} = \hat{\mathbf{F}} \tag{42}$$

where $\hat{\mathbf{F}}$ includes (i) the reactions at the nodes where displacements have been prescribed, (ii) the body forces, (iii) the incident wave forcing.

From here, we can obtain some important properties of the system. We will consider first the eigenvalues and eigenmodes of the undamped system:

$$\left\{\mathbf{K} - \omega^2 \mathbf{M}\right\} \hat{\mathbf{U}} = \mathbf{0} \tag{43}$$

which corresponds to its free oscillations. We will denote by $\omega_k$ and $\hat{\mathbf{U}}^{(k)}$ its $k$-th eigenvalue and eigenvector (or natural frequency and normal mode of vibration) which are obtained from:

$$\det\left(\mathbf{K} - \omega^2 \mathbf{M}\right) = \mathbf{0} \tag{44}$$

In what follows we will assume that the normal modes have been normalized with respect to the mass matrix $\mathbf{M}$ as:

$$\hat{\mathbf{U}}^{(k)T} \mathbf{M} \hat{\mathbf{U}}^{(k)} = 1$$

An important property of the modes is their orthogonality. If we consider two different modes $(i)$ and $(j)$, we will have:

$$\begin{aligned} \mathbf{K}\hat{\mathbf{U}}^{(i)} &= \omega_i^2 \mathbf{M} \hat{\mathbf{U}}^{(i)} \\ \mathbf{K}\hat{\mathbf{U}}^{(j)} &= \omega_j^2 \mathbf{M} \hat{\mathbf{U}}^{(j)} \end{aligned} \tag{45}$$

We will premultiply them by $\hat{\mathbf{U}}^{(j)T}$ and $\hat{\mathbf{U}}^{(i)T}$ respectively, subtract both equations and take into account the symmetry of both stiffness and mass matrices, arriving to:

$$\hat{\mathbf{U}}^{(j)T} \mathbf{K} \hat{\mathbf{U}}^{(i)} - \hat{\mathbf{U}}^{(i)T} \mathbf{K} \hat{\mathbf{U}}^{(j)} = 0 = \left(\omega_i^2 - \omega_j^2\right) \hat{\mathbf{U}}^{(i)T} \mathbf{K} \hat{\mathbf{U}}^{(j)}$$

If the natural frequencies $(i)$ and $(j)$ are different, we will have:

$$\hat{\mathbf{U}}^{(i)T} \mathbf{K} \hat{\mathbf{U}}^{(j)} = 0 \tag{46}$$

The same method can be applied to obtain:

$$\hat{\mathbf{U}}^{(i)T} \mathbf{M} \hat{\mathbf{U}}^{(j)} = 0 \tag{47}$$

Knowledge of the natural modes and frequencies of a system is important, as it provides insight on its behaviour. For instance, when designing a dam on a rock foundation under earthquake action, we can compare the frequency spectrum of the earthquake with the natural modes of the dam, and change the design if they coincide.

The second type of analysis which can be done is the harmonic response, which consists on applying forcing with different angular frequencies and obtain the response of the system. The solution can be written as:

$$\hat{\mathbf{U}} = \left\{\mathbf{K} - i\omega\mathbf{C} - \omega^2\mathbf{M}\right\}^{-1}\hat{\mathbf{F}} \tag{48}$$

In the case of undamped systems, we have:

$$\hat{\mathbf{U}} = \left\{\mathbf{K} - \omega^2\mathbf{M}\right\}^{-1}\hat{\mathbf{F}}$$

In this case, when we excite the system with one of its natural frequencies, we obtain an infinite response, and we say that the system is in resonance. In real cases, this will not happen because as the response increases, the system turns non linear and some damping appears. However, failure of the structure can happen in this situation.
The analysis is usually performed by selecting a control degree of freedom (for instance, horizontal displacement at the top of the dam), and studying its variation when the frequency changes.

# 4 Coupled Problems

## 4.1 The Displacement-Pore Pressure Swansea Model

The most widely used form of Biot-Zienkiewicz equations is the so called "displacement- pore pressure" model, which is obtained in the cases where fluid accelerations are small by eliminating the Darcy velocity. The main equations are:

$$\mathrm{div}\left(\boldsymbol{\sigma}' - p_w\mathbf{I}\right) + \rho\mathbf{b} = \rho\frac{d^2\mathbf{u}}{dt^2} \tag{49a}$$

$$\frac{1}{Q^*}\frac{dp_w}{dt} + \mathrm{div}\left(\frac{d\mathbf{u}}{dt}\right) - \mathrm{div}\left(k_w \mathrm{grad} p_w\right) = 0 \tag{49b}$$

together with the constitutive and kinematics relations.

The system of partial differential equations can be discretized using standard Galerkin techniques, as described in Zienkiewicz and Taylor (2000) Zienkiewicz et al. (2000). The displacement and pressure are discretized as $\mathbf{u} = \mathbf{N}_u\hat{\mathbf{u}}$ and $\mathbf{p}_w = \mathbf{N}_p\hat{\mathbf{p}}_w$, resulting on the nonlinear system

$$\mathbf{M}\frac{d^2\hat{\mathbf{u}}}{dt^2} + \int_\Omega \mathbf{B}^T\boldsymbol{\sigma}'d\Omega - Q\hat{\mathbf{p}}_w - \mathbf{f}_u = \mathbf{0}$$

$$\mathbf{Q}^T\frac{d\hat{\mathbf{u}}}{dt} + \mathbf{H}\hat{\mathbf{p}}_w + \mathbf{C}\frac{d\hat{\mathbf{p}}_w}{dt} - \mathbf{f}_p = \mathbf{0} \tag{50}$$

where

$$\mathbf{M} = \int_\Omega \rho\mathbf{N}_u^T\mathbf{N}_u d\Omega \qquad \mathbf{C} = \int_\Omega \frac{1}{Q^*}\mathbf{N}_p^T\mathbf{N}_p d\Omega$$

$$\mathbf{Q} = \int_\Omega S_w\alpha\mathbf{B}^T\mathbf{m}\mathbf{N}_p d\Omega \quad \mathbf{H} = \int \nabla\mathbf{N}_p^T k_w \nabla\mathbf{N}_p d\Omega \tag{51}$$

and

$$\mathbf{f}_u = \int_\Omega \mathbf{N}_u^T\mathbf{b}d\Omega + \int_{\Gamma_t} \mathbf{N}_u^T\bar{\mathbf{t}}d\Gamma$$

$$\mathbf{f}_p = \int_{\Gamma_q} \mathbf{N}_p^T k_w \frac{\partial\mathbf{p}}{\partial\mathbf{N}}d\Gamma + \int_\Omega \nabla\mathbf{N}_p^T k_w\rho_w\mathbf{b}d\Omega - \left(\int_\Omega \nabla\mathbf{N}_p k_w\mathbf{N}_u d\Omega\right)\ddot{\mathbf{u}} - \int_\Omega \mathbf{N}_p^T\mathbf{s}_0 d\Omega \tag{52}$$

in (51) $\alpha$ is the Biot parameter, which in the case of soils is assumed to be 1, $Q^*$ the mixed volumetric stiffness of water and soil particles, $S_w$ the degree of saturation, $\mathbf{m}$ the vector representation of the identity matrix, which for 3D situations is $(1, 1, 1, 0, 0, 0)^T$; in (52) $\mathbf{s}_0$ represents volume changes in the soil caused by temperature and $\partial\mathbf{p}/\partial\mathbf{n}$ is the derivative of the pressure along the normal to the boundary.

Time derivatives of displacements and pressures are approximated in a typical step $(n)$ of computation using the Generalized Newmark scheme GN22 for displacements and GN11 for pore pressures, as proposed by Katona and Zienkiewicz (Zienkiewicz and Taylor, 2000), which uses as main variables $\Delta\ddot{\mathbf{u}}^n$ and $\Delta\dot{\mathbf{p}}_w^n$ at time step $n$. From here, we arrive at the following set of discretized equations:

$$\mathbf{M}\Delta\ddot{\mathbf{u}}^n + \int \mathbf{B}^T\boldsymbol{\sigma}'^{n+1} - \theta\Delta t\mathbf{Q}\Delta\dot{\mathbf{p}}_w^n - \mathbf{F}_u^{n+1} = \boldsymbol{\Phi}_u = \mathbf{0}$$

$$\beta_1\Delta t\mathbf{Q}^T\Delta\ddot{\mathbf{u}}^n + (\Delta t\theta\mathbf{H} + \mathbf{C})\Delta\dot{\mathbf{p}}_w^n - \mathbf{F}_p^{n+1} = \boldsymbol{\Phi}_p = \mathbf{0} \tag{53}$$

This non linear system can be solved using a Newton-Raphson scheme with a suitable Jacobian matrix.

## 4.2 The $u - W$ Model

One of the advantages of the displacement (velocity)-pressure model is the smaller number of unknowns in comparison to the full model. The model has been obtained assuming that the fluid accelerations are small, and therefore accuracy will be lost as pore fluid accelerations become important. The limits of validity of different approximations have been proposed by Zienkiewicz et al. (1980). Indeed, they propose three regions, denoted as (i) slow, (ii) moderate and (iii) fast phenomena.

It is possible to reformulate the full system in terms of displacements of the soil skeleton and the pore water relative to it (López Querol and Blázquez, 2006; López Querol et al., 2007). The model is formulated in terms of solid skeleton and pore water displacements relative to the skeleton. The final system of equations is:

$$
\begin{aligned}
&\operatorname{div}(d\boldsymbol{\sigma}) + \mathbf{Q}\operatorname{grad}\left[\operatorname{div}(d\mathbf{u} + d\mathbf{W})\right] + \rho d\mathbf{b} = \rho d\ddot{\mathbf{u}} + \rho_w d\ddot{\mathbf{W}} \\
&\mathbf{Q}\operatorname{grad}\left[\operatorname{div}(d\mathbf{u} + d\mathbf{W})\right] + \rho_w d\mathbf{b} - \mathbf{k}^{-1} d\dot{\mathbf{W}} = \rho_w d\ddot{\mathbf{u}} + \frac{\rho_w}{n} d\ddot{\mathbf{W}}
\end{aligned}
\tag{54}
$$

In above, we have introduced the variable $\mathbf{W}$ to denote the integral of the Darcy's velocity while $\mathbf{k}^{-1}$ represents the inverse of the permeability tensor ($1/k_w$ in the case it is a scalar). Pore pressures are computed using the equation of the balance of mass of the fluid.

## 4.3 A Note on Equal Order of Interpolation and Stabilization Techniques

Soil dynamics problems often involve a large number of time steps and meshes with a large number of nodes which results on large times of computation. Time can be saved by using explicit schemes when possible, and elements with a low order of interpolation, such as triangles in 2D an tetrahedra in 3D. If such elements are used, the displacement (or the velocity) field will be approximated using linear functions, and the pore pressures will have to be approximated using the same shape functions. Elements will be triangles with 3 nodes both for displacements and pore pressures, for instance.

Such equal order of interpolation elements present severe oscillations in the pore pressure field when the material has very small permeability and both the solid particles and the pore fluid have small compressibility. The situation is of similar nature to that found when using mixed formulations in Solid and Fluid Dynamics. The reason is that these mixed elements with equal interpolation do not satisfy the Babuska-Brezzi (Babuška, 1973; Brezzi, 1974) condition which is necessary to ensure stability or the much simpler patch test proposed by Zienkiewicz et al. (1986) for mixed formulations. It is important to note that the latter is a necessary but not a sufficient condition for stability. The interested reader can find in the text of Bathe (1996) a detailed description. Allowed elements are for instance, quadrilaterals Q8P4 with quadratic approximation of the displacements and nilinear for the pore pressure, or the T6P3 triangles, both in 2D.

In order to use the simple linear triangles or tetrahedras for computational efficiency circumventing the limitations imposed by the Babuska-Brezzi conditions, special stabilization techniques have to be used. The problem has attracted the attention of numerous researchers from Fluid, Solid and Soil Mechanics in the past years. In the context of Fluid Dynamics, it is worth mentioning the work of Brezzi and Pitkaranta (1984); Hughes et al. (1986); Hafez and Soliman (1991); Zienkiewicz and Wu (1991), just to mention a few.

These methods have been extended to Soil Dynamics problems by Zienkiewicz et al. (1993, 1994) and to Soil Mechanics problems by Pastor et al. (1996, 1997).

One of the most simple yet effective ways of stabilization is the fractional step algorithm, which was introduced by Chorin (1968) as a device to allow the use of standard time integration techniques in fluid dynamics problems. Among the several alternative formulations some forms were found later to provide the required stabilization for elements with equal order of interpolation of velocities and pressures. The discovery was first made by Schneider et al. (1978) and Kawahara and Ohmiya (1985) and later justified by Codina et al. (1995).

The method has been extended to Solid Dynamics by Zienkiewicz et al. (1998) and Quecedo et al. (2000), and to Soil Dynamics problems by Pastor et al. (1999a,b) and Mabssout et al. (2006). It is also worth noticing the work of Mira et al. (2003) for enhanced strain elements.

The $u - pw$ equations are written as:

$$\rho \frac{d\mathbf{v}}{dt} = \text{div}\boldsymbol{\sigma}' - \text{grad}p_w + \rho\mathbf{b} \tag{55a}$$

$$\frac{1}{Q^*}\frac{dp_w}{dt} = \operatorname{div}\left(k_w \operatorname{grad} p_w\right) - \operatorname{div}\mathbf{v} \tag{55b}$$

$$\boldsymbol{\sigma} = \boldsymbol{\sigma}' - p_w \mathbf{I} \tag{55c}$$

$$d\boldsymbol{\sigma}' = \mathbf{D}^{ep} d\boldsymbol{\varepsilon} \tag{55d}$$

The first equation is written in an incremental time using an intermediate velocity $\mathbf{v}^*$ as:

$$\rho\frac{\mathbf{v}^* - \mathbf{v}^n}{\Delta t} = \operatorname{div}\boldsymbol{\sigma}' + \rho\mathbf{b}|^n \tag{56a}$$

$$\rho\frac{\mathbf{v}^{n+1} - \mathbf{v}^*}{\Delta t} = -\operatorname{grad} p_w|^{n+\theta_2} = -\operatorname{grad}\left(p_w + \theta_2 \Delta p_w\right) \tag{56b}$$

where the increment $\Delta p_w$ is defined by

$$p_w^{n+1} = p_w^n + \Delta p_w$$

In above, $\theta_2 = 1/2$ is usually chosen. Eq. (55b) is then discretized in time as:

$$\frac{1}{Q^*}\frac{\Delta p_w}{\Delta t} = \operatorname{div}\left(k_w \operatorname{grad} p_w\right)|^n - \operatorname{div}\mathbf{v}^{n+\theta_1} \tag{57}$$

where we select $\theta_1 = 1$.

The velocity $\mathbf{v}^{n+1}$ can be obtained from (56b) as

$$\mathbf{v}^{n+1} = \mathbf{v}^* - \frac{\Delta t}{\rho}\operatorname{grad}\left(p_w + \theta_2 \Delta p_w\right) \tag{58}$$

After substitution in (57), we obtain

$$\left(\frac{1}{Q^*} - \frac{\Delta t^2 \theta_2}{\rho}\nabla^2\right)\frac{\Delta p_w}{\Delta t} = \operatorname{div}\left(k_w \operatorname{grad} p_w\right)|^n - \operatorname{div}\mathbf{v}^* + \Delta t \nabla^2 p_w^n \tag{59}$$

The fractional step method consist on performing the time step computations as follows:

(i) First of all, obtain the intermediate velocity field $\mathbf{v}^*$ using (56):

$$\mathbf{v}^* = \mathbf{v}^n + \frac{\Delta t}{\rho}\left(\operatorname{div}\boldsymbol{\sigma}' + \rho\mathbf{b}\right)^n \tag{60}$$

(ii) Then, solve implicitly (59) to obtain $\Delta p_w$ and $p_w^{n+1} = p_w^n + \Delta p_w$

$$\left( \frac{1}{Q^*} - \frac{\Delta t^2 \theta_2}{\rho} \nabla^2 \right) \frac{\Delta p_w}{\Delta t} = \text{div}\left( k_w \text{grad} p_w \right)\Big|^n - \text{div}\mathbf{v}^* + \Delta t \nabla^2 p_w^n \quad (61)$$

(iii) Finally, obtain the velocity field at time $n+1$ using (58)

$$\mathbf{v}^{n+1} = \mathbf{v}^* - \frac{\Delta t}{\rho} \text{grad}\left( p_w + \theta_2 \Delta p_w \right) \quad (62)$$

Equations (60)-(62) can be discretized using a standard Galerkin finite element technique, as shown in Zienkiewicz et al. (1999). It is important to notice that the time step is limited for the dry solid skeleton response, as the pore pressure laplacian equation has been solved implicitly.

## 5 Application Cases

The purpose of this Section is to provide a selection of representative examples of most of the alternative modelling techniques described in the preceding Sections.

### 5.1 Liquefaction of a Sand Layer Under Earthquake Loading

Fig. 1 shows a fully saturated sand layer subjected to a horizontal earthquake. The seismic input is the accelerogram given in Fig. 2 that correspond to the NS component accelerogram of the El Salvador earthquake 13/01/2001 in Santa Tecla. The sand layer is modelled by a sand column with both sides and the bottom assumed impermeable. Pore pressures are assumed to be zero at the surface of the sand layer. Repeated boundary conditions are assumed for the lateral nodes, which implies that the displacement of a right hand side node equals that of the corresponding left hand side node.

The finite element mesh is shown in Fig. 3 and consists of stabilized four node quadrilateral elements, where bilinear shape functions are used both for displacements and pressures. The material is assumed to be a very loose sand. Constitutive behaviour has been modelled using the Pastor-Zienkiewicz mode. Material properties and other relevant data used in the analysis are listed in Tab. 1 and 2.

The results can be seen in Fig. 3, which depicts the evolution of the pore pressure and the mean effective stress at stations A to E. It is interesting to

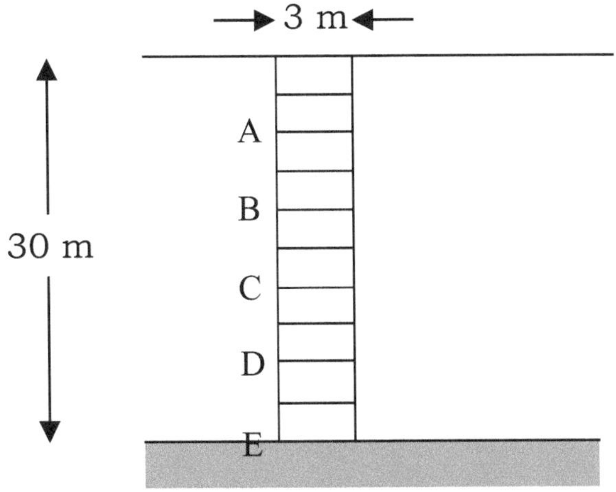


**Figure 1.** Soil layer problem: finite element mesh.

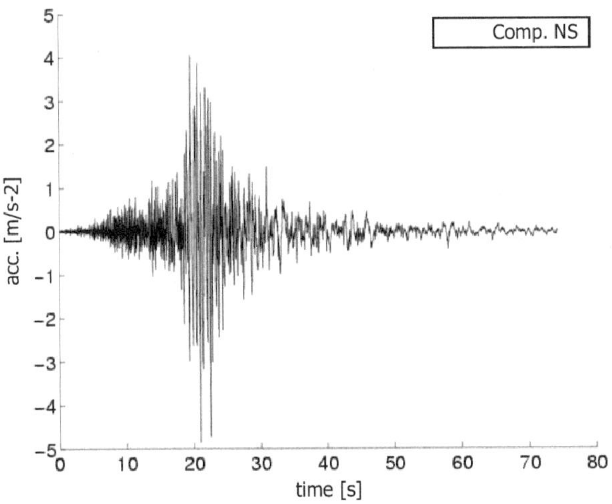


**Figure 2.** Seismic input: NS component accelerogram of the El Salvador earthquake 13/01/2001 in Santa Tecla.

note that the pore pressure increases and reaches a plateau between 7 and 17 seconds depending on the depth, and the mean effective stress decreases to zero.

**Table 1.** Data used in sand layer analysis.

| $\rho_s$ (kg/m$^3$) | $\rho_w$ (kg/m$^3$) | $K_s$ (Pa) | $K_w$ (Pa) | $n$ | $k$ (m/s) |
|---|---|---|---|---|---|
| 2720. | 980. | $1.e20$ | $1.092e9$ | 0.363 | $2.1e-3$ |

**Table 2.** Pastor-Zienkiewicz soil model parameters.

| $K_0$ (kPa) | $G_0$ (kPa) | $M_g$ | $M_f$ | $\alpha_g = \alpha_f$ | $H_0$ | $\beta_0$ | $\beta_1$ | $H_U$ (kPa) | $\gamma_{DM} = \gamma_U$ |
|---|---|---|---|---|---|---|---|---|---|
| 45000. | 22500. | 1-5 | 0.4 | 0.45 | 350. | 4.2 | 0.2 | $6.e3$ | 2 |

Fig. 4 provides further insight into the phenomenon of liquefaction. We have plotted profiles of pore pressure at several instants together with the vertical stress. In this way, when the excess pore pressure coincides with the vertical effective stress, it is possible to see how liquefaction extends from the surface. At 17 s. all the column is liquefied.

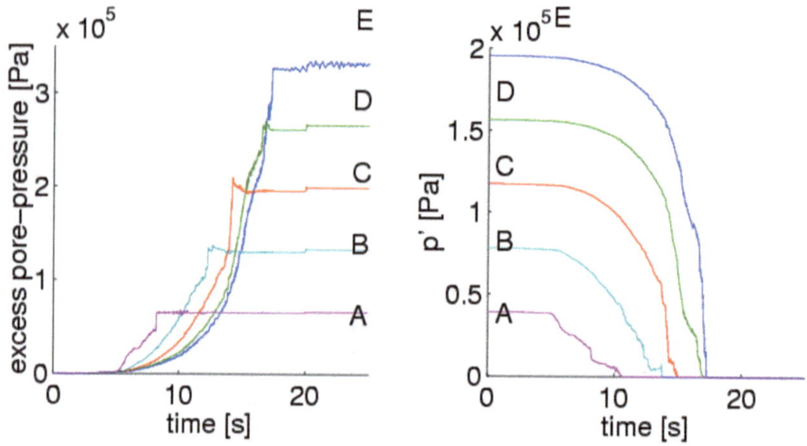


**Figure 3.** Evolution of excess pore pressure and mean effective stress at several stations.

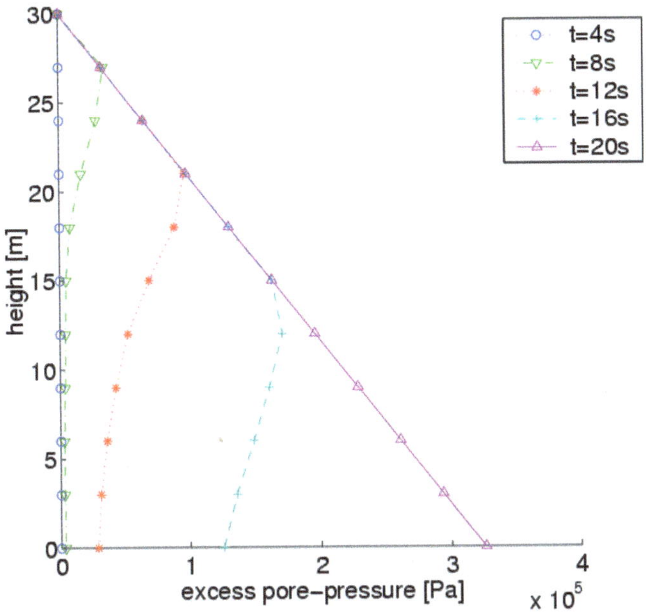


**Figure 4.** Vertical profiles of excess pore pressure at different times showing the extent of liquefaction.

### 5.2 Liquefaction Failure of an Embankment Under Earthquake Action

The case we will be considered next is that of an earthquake induced flowslide in very loose saturated sand. Here it is important to notice that proposed model is intended to describe the initiation of failure. In order to cope with the propagation of the flowslide, the authors have recently proposed a fluid-like, depth integrated model formulated in an Eulerian framework.

The problem consists of an embankment 10 m in height with slopes 2:1, founded on a sand layer which extends 10 m in depth and lies on a rigid rock bed. The material of both the embankment and the foundation is a very loose saturated sand. Initial conditions correspond to geostatic equilibrium under gravity forces. Suction at the surface has been assumed to be equal to -20 kPa. The finite element mesh can be seen in Fig. 5 and consists on 500 quadrilaterals with 8 nodes for displacements and 4 for pore pressures. A reduced integration rule has been used in the solid part to avoid locking. The number of nodes is 1611, with 3535 degrees of freedom.

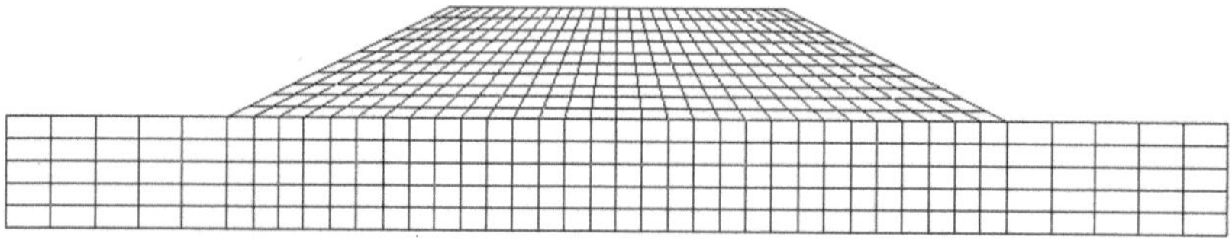

**Figure 5.** Finite element mesh of the embankment.

Loading is applied by prescribing horizontal accelerations at the base. We have used the input motion defined in Fig. 2. A simplified absorbing boundary condition has been applied at lateral boundaries. Concerning pore pressures, it has been assumed that no flux occurs at artificial boundaries, and the constant value of -20 kPa has been kept at the surface.

The behaviour of the loose material is represented using the Pastor-Zienkiewicz model for sand, Tab. 1 and 2. To better understand this behaviour we simulate a loading-unloading-reloading undrained triaxial test. The material loses progressively its resistant capacity, the effective stress decreases and the pore pressure increases, Fig. 6.

The results can be seen in Figs. 7-9, where the contours of pore pressure, $p'_w/p'_{w0}$ and displacements are given at different times. The ratio between the mean effective confining pressure and its initial value $p'_w/p'_{w0}$ has been used as an indicator of the extent of the liquefied zones. From these results, it can be concluded that failure of the embankment is caused by liquefaction of the outer liquefied zones.

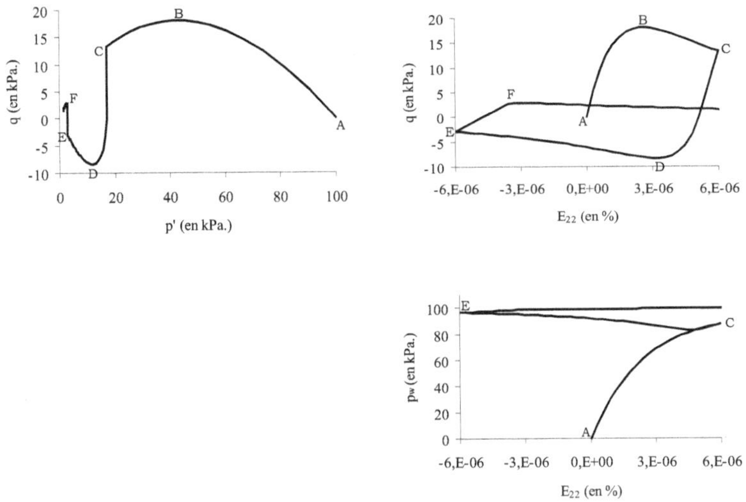


**Figure 6.** Undrained triaxial simulation of the loose sand: loading, unloading, reloading path.

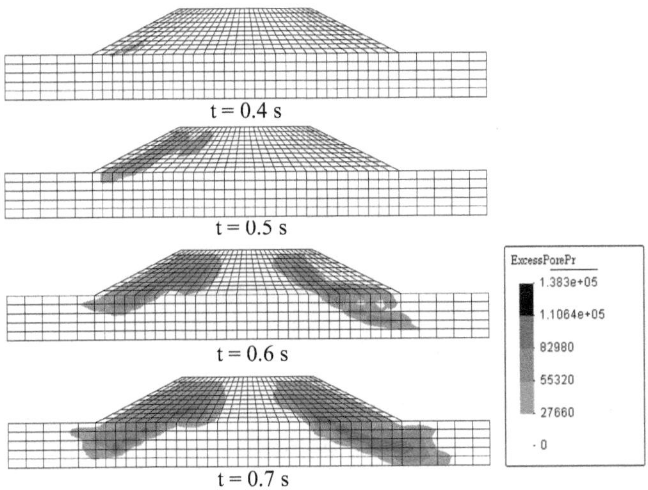


**Figure 7.** Evolution of excess pore pressure contour [Pa].

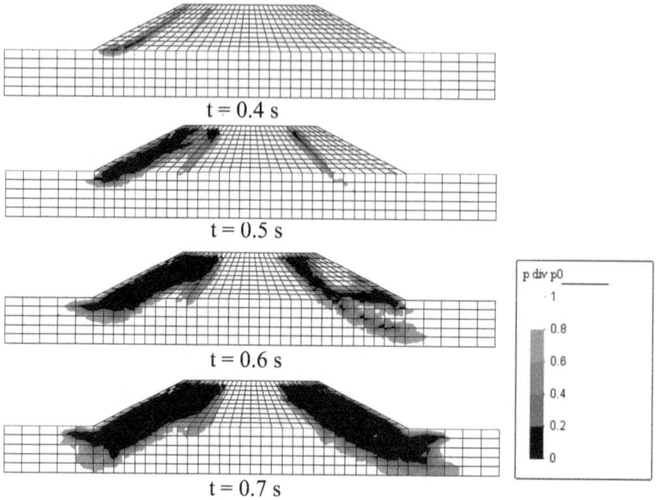


**Figure 8.** Evolution of $p'_w/p'_{w0}$.

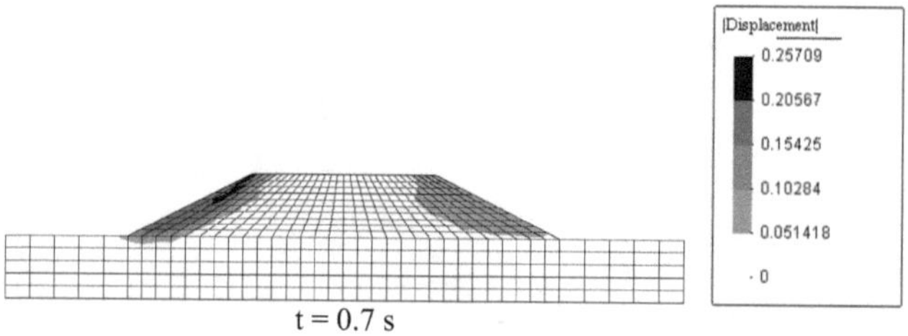


**Figure 9.** Deformation contour [m].

### 5.3 Liquefaction of Marine Foundations Under Wave Action

The proposed example concerns a breakwater structure of caisson type subjected to continued wave loading. The 24x24m concrete caisson rests on a rubble-mound bedding layer that transmits loading to the sand foundation. The finite element mesh is reproduced in Fig. 10.

Loading and unloading sand behavior is described using the Pastor-Zienkiewicz model. This constitutive model is able to reproduce the experimental cyclic

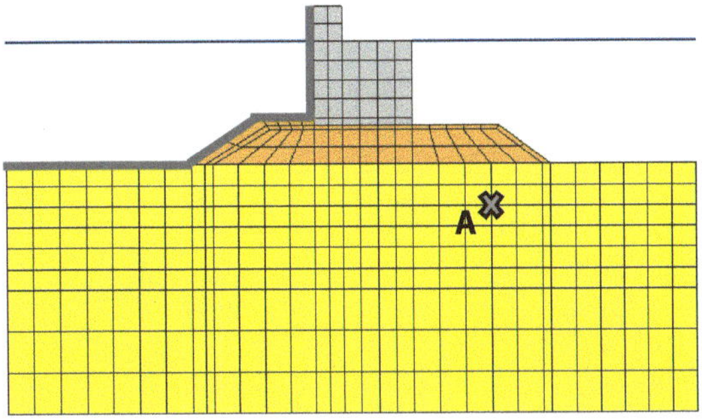


**Figure 10.** Finite element mesh of the marine structure.

non-drained triaxial test on the considered material with a relative density $D_r$ equal to 63% as shown in Fig. 11.

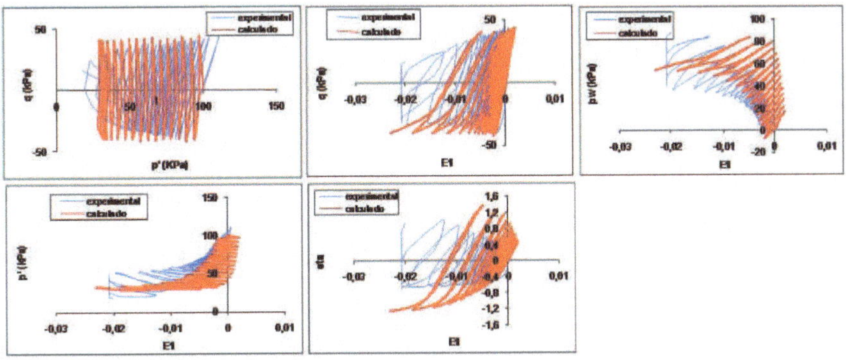

**Figure 11.** Cyclic non-drained triaxial test on the considered sand ($D_r$ = 63%): experimental and computed results.

Concerning the load, the wave loading is computed using an external program solving the Navier Stokes equations and the free surface problem with VOF techniques. Pore-pressures and hydrostatic loads histories are prescribed on the red line shown in Fig. 10. It corresponds to a high intensity storm with waves reaching 8 m height and 12 s period.

The excess pore water pressure and the mean effective stress contours after 20 minutes (approximately 100 wave cycles) are plotted in Fig. 12. It can be seen how the pore water pressure increases and how the stresses are reorganized under the caisson. Fig. 13 plots the time evolution of the excess pore pressure and the mean effective stress in point A defined in Fig. 10.

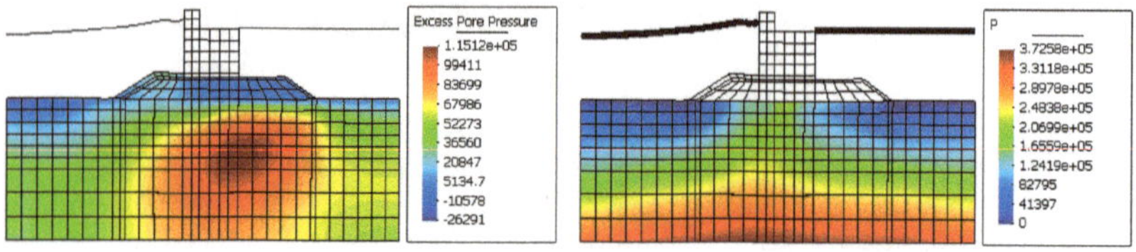


**Figure 12.** Excess pore water pressure and mean effective stress contours (in Pa) after 20 minutes.

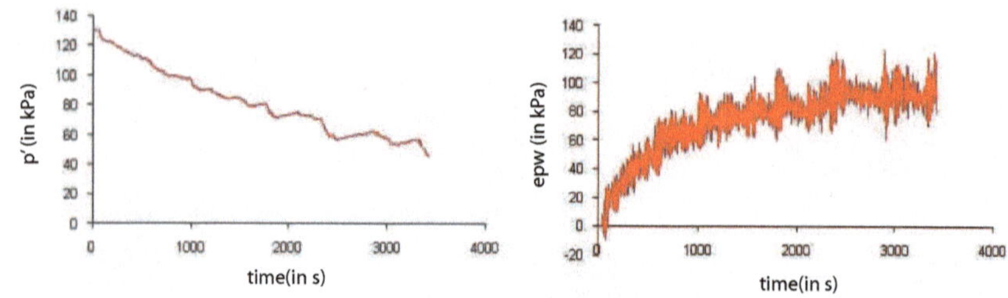


**Figure 13.** Time evolution of the excess pore pressure and the mean effective stress in point A.

The increase of pore pressure causes a dramatic reduction of soil strength. Indeed, the liquefied soil behaves as a viscous fluid where the structure can sink. This failure mechanism can be accelerated during earthquake events.

### 5.4 Application of $u-W$ Formulation to Dynamic Consolidation

As it was previously mentioned, the $u-W$ formulation, involves a higher number of degrees of freedom than the classic Swansea model. However, in addition of being more accurate when the fluid acceleration displays high values, this formulation, only based on displacements, performs other important advantages, like a greater stability when the permeability of the soil is very low or when the mixture formed by the solid skeleton and the fluid filling the voids is close to incompressible. Furthermore, a boundary condition of pressure is not necessary in this $u - W$ model, since the pore pressure is not an unknown. This circumstance allows us to obtain, as a result of the model, the displacements of the fluid throughout the analysed geometry, even at a saturated soil boundary. To illustrate this fact, the application of the $u - W$ formulation to a numerical example of dynamic consolidation is presented next.

Fig. 14 shows the deformed shape of a symmetric plane strain problem at two different times during the impact. The solid phase behaviour is modelled by means of the Pastor-Zienkiewicz law for sands. The loading zone is indicated in the figures, at the upper boundary, and the axis of symmetry is the left boundary. The arrows in the soil show the fluid velocity relative to the solid phase, and therefore, the flow directions within the whole geometry.

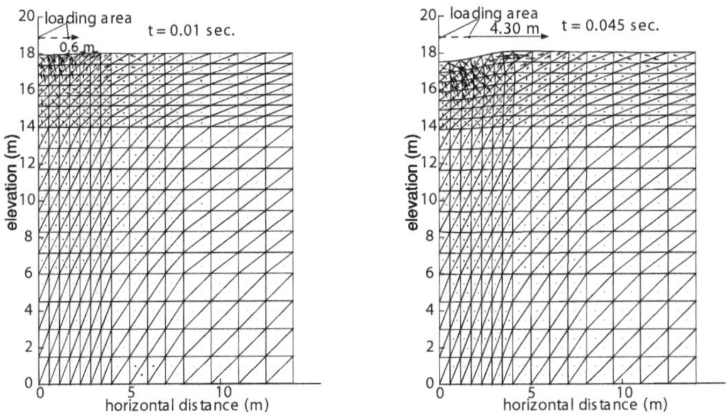


**Figure 14.** Deformed geometry of a plane strain numerical model of dynamic consolidation at 0.01 and 0.045 seconds (López Querol and Blázquez, 2006).

The impact due to a mass falling in the loading area is applied in the model by means of a set equal nodal forces, which added all together, in this particular case, follow the expression given by

$$F(t) = F_{\max} \cdot \sin(20\pi t) \quad t \leq 0.05 \text{ s}$$
$$F(t) = 0 \qquad t > 0.05 \text{ s}$$

with a maximum force, $F_{\max}$, equal to 20MN, equivalent to a mass of 20 T falling from 11.5 m (Pan and Selby, 2002). Initially, the water table is located at the upper boundary, where the arrows directions give the points of exit of the water during the impact. As it can be seen, this exit of water moves from the loading zone to the right boundary as the time passes. Thus, since at 0.01 seconds this point is located 0.6 m far from loading zone, after 0.045 seconds it appears at 4.30 m. The numerical results give a "wave of dryness" around the loading zone, which reproduces several field observations, like the field test at Algeciras (Spain), provided in Fig. 15.

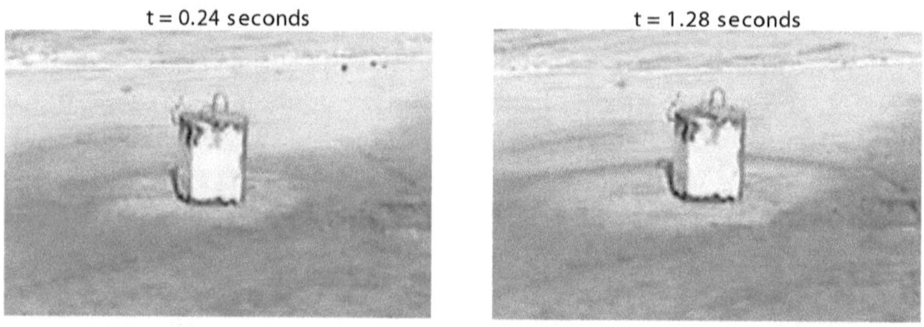


**Figure 15.** Pictures of an impact on a field test of dynamic consolidation on saturated sand in Algeciras (Spain) at instants 0.24 and 1.28 seconds (López Querol and Blázquez, 2006).

## List of Symbols

| | | |
|---|---|---|
| $\dot{\sqcup}$ | time derivative | §2.3, §3, §4 |
| $^{e}$ | elastic | §2, §3 |
| $^{\mathrm{ep}}$ | elastoplastic | §2, §4.3 |
| $^{(g)}$ | global | §2.2 |
| $_{\mathrm{Inc}}$ | incoming | §2.1 |
| $_{n}$ | component normal to the boundary | §2.1 |
| $^{n}$ | value at time step $n$ | §4 |
| $^{(p)}$ | predictors | §2.3, §3 |
| $_{t}$ | component tangent to the boundary | §2.1 |
| $\mathbf{A}_{\mathrm{Abs}}$ | matrix related to the Riemann invariants on $\Gamma_{\mathrm{Abs}}$ | §2.2 |
| $\mathbf{A}_{\mathrm{Inc}}$ | matrix related to the Riemann invariants on $\Gamma_{\mathrm{Inc}}$ | §2.2 |
| $\mathbf{b}$ | vector of body forces | §2, §4 |
| $\mathbf{B}$ | discrete divergence matrix (colletting all $\mathbf{B}_I$ terms) | §2, §3, §4.1 |
| $\mathbf{B}_I$ | $= \mathbf{S}^T\mathbf{N}_I$ | §2.2 |
| $\mathbf{C}$ | matrix related to the wave boundaries | §2, §3 |
| $\mathbf{C}$ | $= \int_\Omega \frac{1}{Q^*}\mathbf{N}_p^T\mathbf{N}_p d\Omega$ | §4.1 |
| $c_P$ | $P$ wave velocity | §2 |
| $c_S$ | $S$ wave velocity | §2 |
| $\mathbf{D}$ | constitutive tensor | §2, §3, §4.3 |
| $D_r$ | relative density | §5.3 |
| $\mathbf{f}_p$ | pressure forcing term | §4.1 |
| $\mathbf{f}_u$ | displacement forcing term | §4.1 |
| $F$ | force | §5.4 |
| $F_{\mathrm{max}}$ | maximum force | §5.4 |
| $\mathbf{F}$ | forcing term (colletting all $\mathbf{F}_I$ terms) | §2, §3 |
| $\hat{\mathbf{F}}$ | auxiliary forcing term | §3.1 |
| $\mathbf{F}_I$ | $= \mathbf{F}_{bI} + \mathbf{R}_I + \bar{\mathbf{t}}_I - \mathbf{F}_I^{(\mathrm{Inc})}$ | §2.2 |
| $\mathbf{F}_{bI}$ | contribution from the body forces term on node $I$ | §2.2 |
| $\mathbf{F}_{bI}^{(\mathrm{Inc})}$ | contribution from incident wave boundary on node $I$ | §2.2 |

| | | |
|---|---|---|
| $\mathbf{H}$ | $= \int \nabla \mathbf{N}_p^T k_w \nabla \mathbf{N}_p d\Omega$ | §4.1 |
| $\mathbf{I}$ | identity tensor | §4 |
| $\mathbf{k}$ | permeability tensor | §4.2 |
| $k_w$ | permeability | §4 |
| $\mathbf{K}$ | elastic stiffness matrix | §3 |
| $\mathbf{m}$ | vector representation of the identity matrix | §4.1 |
| $\mathbf{M}$ | mass matrix | §2, §3 |
| $\mathbf{M}$ | $= \int_\Omega \rho \mathbf{N}_u^T \mathbf{N}_u d\Omega$ | §4.1 |
| $n$ | total number of nodes in the finite element mesh | §2.2 |
| $n$ | time step | §2.3, §4.1 |
| $n$ | porosity | §4.2 |
| $\mathbf{n}$ | outwards unit vector normal to the boundary $\Gamma_q$ | §2, §4.1 |
| $\mathbf{N}$ | global shape function matrix | §2, §3 |
| $N_J$ | element shape functions | §2.2 |
| $\mathbf{N}_J$ | $= \begin{pmatrix} N_J & 0 & 0 \\ 0 & N_J & 0 \\ 0 & 0 & N_J \end{pmatrix}$ | §2.2 |
| $\mathbf{N}_p$ | pressure shape function matrix | §4.1 |
| $\mathbf{N}_u$ | displacement shape function matrix | §4.1 |
| $p_w$ | pore pressure | §4, §5.2 |
| $\hat{\mathbf{p}}_w$ | global vector of nodal value of the pore pressure | §4.1 |
| $\mathbf{Q}$ | $= \int_\Omega S_w \alpha \mathbf{B}^T \mathbf{m} \mathbf{N}_p d\Omega$ | §4.1 |
| $Q^*$ | mixed volumetric stiffness of water and solid particles | §4 |
| $\mathbf{R}_I$ | contribution due to the reactions in all prescribed degrees of freedom on node $I$ | §2.2 |
| $\mathbf{R}_\Omega$ | residual vector | §2.2 |
| $\mathbf{s}_0$ | vector representing volume changes in the soil caused by temperature | §4.1 |
| $\mathbf{S}$ | operator matrix | §2, §3 |
| $S_w$ | degree of saturation | §4.1 |
| $t$ | time | §2, §3, §4, §5 |

| | | |
|---|---|---|
| $\mathbf{t}$ | vector of forces per unit area acting on $\Gamma_q$ | §2, §4 |
| $\bar{\mathbf{t}}_I$ | $I = 1, \ldots, n$ contribution from the boundary where tractions are prescribed | §2.2 |
| $\mathbf{T}$ | $= \begin{pmatrix} n_x & n_y \\ -n_y & n_x \end{pmatrix}$ | §2.2 |
| $\mathbf{u}$ | displacement field | §2 §4 |
| $\hat{\mathbf{u}}$ | global vector of nodal value of the displacement | §2, §3, §4.1 |
| $\mathbf{u}^h$ | approximated displacement vector | §2, §3 |
| $\hat{\mathbf{u}}_J$ | vector of nodal value of the displacement | §2.2 |
| $\hat{\mathbf{U}}$ | auxiliary variable | §3.1 |
| $\hat{\mathbf{U}}^{(k)}$ | normal mode of vibration | §3.1 |
| $\mathbf{v}$ | velocity vector | §2, §4.3 |
| $\mathbf{v}^*$ | intermediate velocity | §4.3 |
| $\mathbf{V}$ | $= \begin{pmatrix} \rho c_P & 0 \\ 0 & \rho c_S \end{pmatrix}$ | §2.2 |
| $\mathbf{W}$ | integral of the Darcy's velocity | §4.2 |
| $\mathbf{x}$ | vector of coordinates | §2, §3, §4 |
| $\alpha$ | Biot parameter | §4.1 |
| $\beta_1$ | parameter | §2.3, §3, §4.1 |
| $\beta_2$ | parameter | §2.3, §3 |
| $\Gamma_{\text{Abs}}$ | absorbing wave boundary | §2.2 |
| $\Gamma_{\text{Inc}}$ | incident wave boundary | §2.2 |
| $\Gamma_q$ | Neumann boundary | §2 |
| $\Gamma_u$ | Dirichlet boundary | §2 |
| $\boldsymbol{\varepsilon}$ | strain tensor | §2, §3, §4 |
| $\theta$ | parameter | §4.1 |
| $\theta_1$ | parameter | §4.3 |
| $\theta_2$ | parameter | §4.3 |
| $\rho$ | density | §2, §3, §4 |
| $\rho_w$ | water density | §4.2 |
| $\boldsymbol{\sigma}$ | stress tensor | §2, §3, §4 |
| $\boldsymbol{\sigma}^*$ | stress tensor in vector form | §2, §3 |
| $\boldsymbol{\sigma}'$ | effective stress tensor | §4 |

| | | |
|---|---|---|
| $\psi_k^{(2)}$ | $k = P, S$ Riemann invariant | §2.1 |
| $\omega$ | angular frequence | §3.1 |
| $\omega_k$ | natural frequency of vibration | §3.1 |
| $\Omega$ | domain | §2, §3, §4 |

## Bibliography


I. Babuška. The finite element method with Lagrange multipliers. *Numerische Mathematik*, 20:179–192, 1973.

K. J. Bathe. *Finite Element Procedures.* Prentice Hall, 1996.

F. Brezzi. On the existence, uniqueness and approximations of saddle point problems arising from Lagrange multipliers. *RAIRO*, 8-R2:129–151, 1974.

F. Brezzi and J. Pitkaranta. On the stabilization of finite element approximations of the Stokes problem. In W. Hackbusch, editor, *Efficient solutions of elliptic problems, Notes on Numerical Fluid Mechanics*, volume 10, pages 11–19, Wiesbaden, 1984. Vieweg.

A. J. Chorin. *Numerical solution of incompressible flow problems. Studies in Numerical Analysis 2.* 1968.

R. Codina, M. Vázquez, and O. C. Zienkiewicz. A fractional step method for compressible flows: boundary conditions and incompressible limit. In M. Morandi Cecchi, K. Morgan, J. Periaux, B. A. Schrefler, and O. C. Zienkiewicz, editors, *Proc. Int. Conf. on Finite Elements in Fluids-New trends and applications*, pages 409–418, Venezia, Italy, October 1995.

M. Hafez and M. Soliman. Numerical solution of the incompressible Navier-Stokes equations in primitive variables on unstaggered grids. In *Proc. AIAA Conf.*, volume 91-1561-CP, pages 368–379, 1991.

T. J. R. Hughes. *The Finite Element Method. Linear static and Dynamic Finite Element Analysis.* Prentice Hall, London, 1987.

T. J. R. Hughes, L. P. Franca, and M. Balestra. A new finite element formulation for fluid dynamics. V. Circumventing the Babuska-Brezzi condition: A stable Petrov-Galerkin formulation of the Stokes problem accommodating equal order interpolation. *Comp. Meth. Appl. Mech. Eng.*, 59:85–99, 1986.

M. Kawahara and K. Ohmiya. Finite element analysis of density flow using velocity correction method. *Int. J. Num. Meth. Fluids*, 5:981–993, 1985.

S. López Querol and R. Blázquez. Liquefaction and cyclic mobility model for saturated granular media. *Int. J. Num. Anal. Meth. Geomech.*, 30: 413–439, 2006.

S. López Querol, J. A. Fernández Merodo, P. Mira, and M. Pastor. Numerical modelling of dynamic consolidation on granular soils. *Int. J. Num. Anal. Meth. Geomech.*, 32(12):1431–1457, 2007.

M. Mabssout, I. Herreros, and M. Pastor. Wave propagation and localization problems in saturated viscoplastic geomaterials. *Int. J. Num. Meth. Engng.*, 68(4):425–447, 2006.

P. Mira, M. Pastor, T. Li, and X. Liu. A new stabilized enhanced strain element with equal order of interpolation for soil consolidation problems. *Comp. Meth. Appl. Mech. Engrg.*, 192:4257–4277, 2003.

J. L. Pan and A. R. Selby. Simulation of dynamic compaction of loose granular soils. *Advances in Engineering Software*, 33:631–640, 2002.

M. Pastor, M. Quecedo, and O. C. Zienkiewicz. A mixed displacement-pressure formulation for numerical analysis of plastic failure. *Computers and Structures*, 62:13–23, 1996.

M. Pastor, T. Li, and J. A. Fernández Merodo. Stabilized finite elements for harmonic soil dynamics problems near the undrained-incompressible limit. *Soil Dynamics and Earthquake Engineering*, 16:161–171, 1997.

M. Pastor, T. Li, X. Liu, and O. C. Zienkiewicz. Stabilized low-order finite elements for failure and localization problems in undrained soils and foundations. *Comp. Meth. Appl. Mech. Engng.*, 174:219–234, 1999a.

M. Pastor, O. C. Zienkiewicz, T. Li, L. Xiaoqing, and M. Huang. Stabilized finite elements with equal order of interpolation for soil dynamics problems. *Archives of Computational Methods in Engineering*, 6(1):3–33, 1999b.

M. Quecedo, M. Pastor, and O. C. Zienkiewicz. Application of a fractional step method to localization problems. *Computers and Structures*, 74: 535–545, 2000.

G. E. Schneider, G. D. Raithby, and M. M. Yovanovich. Finite element analysis of incompressible flow incorporating equal order pressure and velocity interpolation. In K. Morgan C. Taylor and C. Brebbia, editors, *Numerical Methods for Laminar and Turbulent Flow*, Plymouth, 1978. Pentech Press.

O. C. Zienkiewicz and R. L. Taylor. *The Finite Element Method.* Butterworth-Heinmann, 5th edition, 2000.

O. C. Zienkiewicz and Wu. Incompressibility without tears! How to avoid restrictions of mixed formulations. *Int. J. Num. Meth. Engng.*, 32:1184–1203, 1991.

O. C. Zienkiewicz, C. T. Chang, and P. Bettess. Drained, undrained, consolidating dynamic behaviour assumptions in soils. *Géotechnique*, 30: 385–395, 1980.

O. C. Zienkiewicz, S. Qu, R. L. Taylor, and S. Nakazawa. The patch test for mixed formulations. *Int. J. Num. Meth. Engng.*, 23:1873–1883, 1986.

O. C. Zienkiewicz, M. Huang, Wu, and S. Wu. A new algorithm for the coupled soil-pore fluid problem. *Shock and Vibration*, 1:3–14, 1993.

O. C. Zienkiewicz, M. Huang, and M. Pastor. Computational soil dynamics - A new algorithm for drained and undrained conditions. In H. J. Siriwardne and M. M. Zaman, editors, *Comp. Meth. Adv. Geomechanics*, pages 47–59, Balkema, 1994.

O. C. Zienkiewicz, J. Rojek, R. L. Taylor, and M. Pastor. Triangles and tetrahedra in explicit dynamic codes for solids. *Int. J. Num. Meth. Engng.*, 43:565–583, 1998.

O. C. Zienkiewicz, A. H. C. Chan, M. Pastor, B. A. Shrefler, and T. Shiomi. *Computational Geomechanics*. J. Wiley and Sons, Chichester, 1999.

# Cyclic Mechanical Response of Rigid Bodies Interacting With Sand Strata

Claudio di Prisco [*]

[*] Dipartimento di Ingegneria Strutturale, Politecnico di Milano, Milano, Italy

## 1 Introduction

In this chapter, the cyclic/dynamic interaction of rigid bodies with homogeneous/heterogeneous sand strata is taken into consideration. For the sake of simplicity, the problem is discussed chiefly with reference to strip shallow footings, but many considerations concerning different systems like buried pipes will be also introduced. The objective of this chapter is mainly to outline the features of a sort of homogeneization theory called the 'macro-element approach' and to demonstrate its capabilities for describing the interaction between structures and soil strata even when complex cyclic loading is taken into account. As far as shallow foundations are concerned, in the recent past many authors have demonstrated that the use of the macroelement approach provides a useful tool for reproducing the structural response of the superstructures without totally disregarding geotechnical issues.

For instance, when the superstructure is highly redundant the loads transmitted to the foundation by the superstructure are a function of the irreversible settlements of the footing, the structural/geotechnical problems become more dramatically coupled and disregarding the coupling can sometimes become quite unsafe (di Prisco et al., 2004).

Indeed, in standard structural design, soil–structure interaction problems are approached by uncoupling geotechnical and structural analyses. For instance, the cyclic/dynamic response of foundation soil strata is either neglected or, once the structural problem is solved, addressed separately by employing 2D/3D numerical FEM codes. The soil–structure interaction problem is very rarely approached in a fully coupled manner; this is done by performing either small/large scale and centrifuge experimental tests (Paolucci et al., 2007; PWRI, 2005; Zeng and Steedman, 1998 and Gajan et al., 2005) or cyclic/dynamic finite element numerical analyses of large spatial domains, including the superstructure, the foundation and the surrounding soil. In principle, these numerical analyses, when a sufficiently sophisticated constitutive relationship is numerically implemented, allow us to take the soil–structure cyclic/dynamic interaction into account, but, unfortunately, since they are very time consuming, they cannot be used to perform extensive parametric numerical campaigns and to be employed as heuristic tools.

In contrast, cyclic soil–structure interaction problems can be successfully approached by employing the macro-element concept (Nova and Montrasio, 1991; Paolucci, 1997; Cremer et al., 2001; Cremer et al., 2002 and Pape and Sieffert, 2001) and this is particularly suitable in case of cyclic environmental loading (like windy or sea/ocean waves actions) for two reasons:

1. the foundation is in this case subjected to inclined and eccentric loads,
2. the number of cycles to be considered is enormous.

The main implicit assumption which the macroelement theory is based on is to subdivide the entire domain into three sub-structures: (i) the far field, (ii) the near field and (iii) the superstructure. The far field is the zone of the soil stratum which is not influenced by the presence of the structure; displacements can be there assumed to be known. The near-field consists of the soil stratum zone where soil–structure interaction irreversible mechanisms become dominant. In this perspective the definition of the spatial domain where plastic/irreversible strains develop is essential: for instance, in the case of rigid shallow strip footings, the size of the domain is imposed by the foundation width and embedment. Sometimes the definition of the boundaries of this domain is quite ambiguous since its geometry evolves with time. Regarding this point, in the following the case of shallow footings placed on loose sand strata will be discussed.

As is commonly done for standard plastic hinges developing in steel frames, the mechanical response of the entire substructure is interpreted as the addition of two terms: one associated with the elastic response of the entire infinite half-space and one to the plastic hinge.

## 2 Generalised Stress–Strain Variables: Shallow Foundations Under Monotonically Increasing Loading

As already observed by Nova and Montrasio (1991) with reference to shallow footings, when the interacting body with respect to the soil stratum can be assumed to be rigid, the mechanical interaction can be described by employing (at least in plane strain conditions) only three generalised stress (the vertical load component $V$, the horizontal load component $H$, and the overturning moment $M$) and three generalised strain variables (the vertical displacement $v$, the horizontal displacement $u$ and the foundation rotation $\vartheta$) (Fig. 1a). For convenience, the generalised stress variables (Prager, 1955) can be summarised in a three dimensional vector $\mathbf{Q}$, whereas the associated generalised strain variables can be collected in a corresponding three dimensional $\mathbf{q}$ vector. By disregarding, for the sake of simplicity, any time dependence of the system mechanical behaviour and by taking the developing of irreversible generalised strains (that is the analonomy of the constitutive relationship) into account, an incremental constitutive relationship can be written:

$$\dot{\mathbf{Q}} = \mathbf{D}(\mathbf{Q}, \dot{\mathbf{Q}}, \boldsymbol{\psi}(\mathbf{q}))\dot{\mathbf{q}} \quad (1)$$

where $\mathbf{D}$ stands for an incremental constitutive matrix depending only on the current state of stress, on the current stress rate $\dot{\mathbf{Q}}$, on the previous history of the system that in (1) is described by the vector of hidden variables $\boldsymbol{\psi}$ assumed to be dependent on the accumulated generalised plastic strain $\mathbf{q}^p$. The macro-element theory was initially conceived for rigid strip foundations

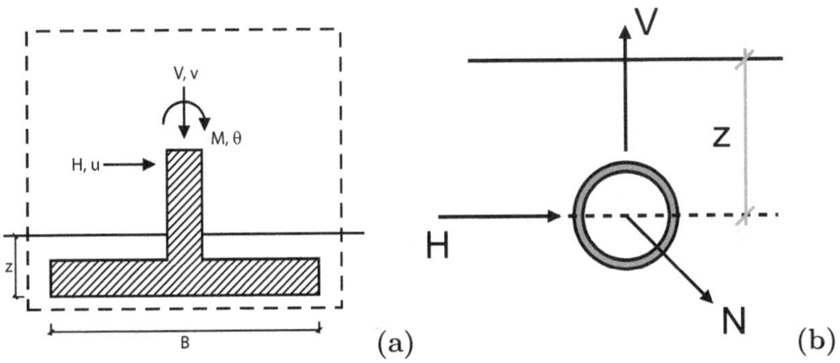


**Figure 1.** Generalised stresses and generalised strains for (a) a shallow strip footing of width $B$ embedded at a depth $Z$, (b) a pipe embedded in a soil stratum (where $Z$ stands for the depth of the embedment).

placed on homogeneous dry sand strata under monotonically increasing inclined and eccentric loading and more recently was extended to describe the system mechanical response under cyclic loading (di Prisco et al., 1998; Cremer et al., 2001; di Prisco et al., 2003a; di Prisco et al., 2003b; di Prisco et al., 2003c). The theory has been also extended to the case of rectangular shallow footings (Grange et al., 2008; Grange et al., 2009): in this case both vectors $\mathbf{Q}$ and $\mathbf{q}$ become six dimensional. Obviously, by taking into account another type of structure, both the dimensions and the single components of both vectors $\dot{\mathbf{Q}}$ and $\dot{\mathbf{q}}$ will change. For instance, in Fig. 1a the case of a buried pipe is considered. For each cross section, four different kinematic and static components will be assigned: the two components acting in the cross section plane $V$ and $H$, respectively, the longitudinal action $N$ and the torsional component $T$.

If the deformable half space interacting with the rigid body is assumed to behave elastically, the stiffness matrix in (1) becomes diagonal and its three terms can be evaluated by employing well known empirical or analytical solutions (Sieffert and Cevaer, 1991). Since the problem under consideration is elliptic, the domain influenced by the local interaction becomes unlimited. Moreover, as is well known, in this specific case (which is essentially 2D), if the soil is assumed to be elastic and homogeneous (that is elastic parameters to be constant within the unlimited spatial domain), the stiffness vanishes.

By contrast, when irreversible strains progressively develop and in particular a local failure mechanism activates, the process concentrates and, according to a well known process in the case of metal beams (Fig. 2), a sort of plastic hinge develops. Thus the stiffness matrix in (1) is intended to describe the response of the system during the entire process evolution, from the very beginning to the eventual final collapse. As is well known, if the steel beam schematically shown in Fig. 2a is considered, and a rigid-plastic constitutive relationship for the plastic hinge is assumed, the structural response of Fig. 2d is obtained, where $P_U$ is the ultimate load and $\delta$ is the maximum vertical displacement of the beam. Therefore, analogously, the inclination of the straight line passing through the origin of Fig. 2d, related to the elastic response of the beam, will correspond with the elastic response of the half space and the ultimate load $P_U$, associated with the development of the plastic hinge in the middle of the beam, will ideally correspond with the activation of the failure mechanism under the shallow footing.

If the mechanical local response of the cross section, where the plastic hinge develops, was more precisely described and the progressive plasticization of the cross section accounted for, the rigid-plastic scheme suggested in Fig. 2c would have to be abandoned and a sort of generalised strain hardening relationship to be chosen. Analogously, in the case of shallow footings, irreversible strains locally develop much earlier than the failure mechanism activation and more sophisticated relationships have to be introduced.

The previously defined analogy between the beam and the considered interaction problem becomes a bit more obscure when the local problem is tackled. Indeed, as is well known, in the case of the beam, the generalised stresses are, for instance, the bending moment, the axial force, the

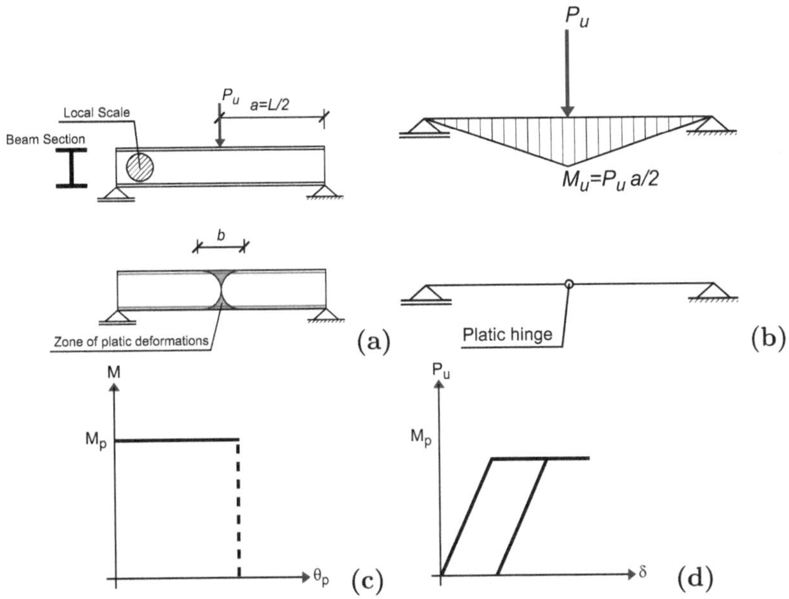


**Figure 2.** Structural analogy: (a) steel beam $L$ long under point loading, (b) development of the plastic hinge of characteristic length $b$, (c) plastic hinge constitutive relationship (rigid-plastic assumption) and (d) structural response.

torsional moment. By contrast, in the case of a shallow footing, the generalised stresses, as are indicated in (1), coincide with the external loads themselves. This lack of symmetry between the two mechanical systems is essentially due to the types of compared processes: in the case of the beam the structural response is analysed; in contrast, in the case of the footing, the local process of failure dominates the system response.

Before introducing the theoretical assumptions which the macro-element concept is based on, in the following section a series of experimental observations concerning the static interaction between shallow footings and dry sand strata are summarised.

## 2.1 Experimental Evidence

The main information, concerning the mechanical response of rigid shallow foundations under monotonically increasing loads, can be derived from the large amount of experimental test results published in literature in the last three decades and can be very schematically summarised as it follows:

1. Even during monotonically increasing loading tests, the response of the system is not linear from the very beginning of the test;
2. Coupling between the different generalised stress/strain variables is evident from very low generalised stress levels and dominates when failure is approached. For example, when a horizontal monotonically increasing load is applied vertical displacements develop although the vertical load is kept constant and the overturning moment nil;
3. Bearing capacity is severely affected both by the inclination and the eccentricity of the loads imposed; the so-called interaction domain describes this dependence. The interaction domain is a function of (i) the nature of the foundation soil (the relative density severely influences its size and shape), (ii) the roughness, shape and embedment of the footing itself. As is suggested by the schematic plots of Fig. 3, if the penetration mechanism was uncoupled with respect either to the sliding mechanism or to the toppling mechanism, as can be inferred when rigid and infinitely resistant underlying strata are considered, the interaction domains in the two planes $H-V$ and $M-V$, respectively, would be uncoupled, too. Indeed, sliding would always concern uniquely the interface zone and toppling would consist solely in the result of the detachment between the footing and the underlying stratum. On the contrary, the deformability and the limited strength of the soil necessarily introduce the previously cited coupling and interaction domains cannot be described by the four straight lines of Fig. 3.
4. Failure mechanism geometries severely depend on the applied combination of the generalised stress variables: to each point belonging to the interaction domain corresponds a unique failure mechanism. Two exemplifying pictures (after Nova and Montrasio (1991)) for a rigid strip shallow footing under a vertical centred load and an inclined centred load positioned on a homogeneous dense sand stratum respectively, are in Fig. 4.
5. The experimental results can be easily interpolated in a three dimensional space by employing expressions, quite common in the literature (Butterfield and Ticof, 1979; Georgiadis and Butterfield, 1988; Nova and Montrasio, 1991; Butterfield and Gottardi, 1994; Nova and Montrasio, 1997), like that reported here below introduced by Nova and Montrasio (1991):

$$F = \left(\frac{M}{\psi B}\right)^2 + \left(\frac{H}{\mu}\right)^2 - V^2\left(1 - \frac{V}{V_{\max}}\right)^{2\beta} \leq 0, \qquad (2)$$

where $\psi$, $\mu$, $\beta$ and $V_{\max}$ are parameters describing, respectively, the shape and size of the failure locus. A geometrical representation of the cxpression given in (2) is reported in an opportunely non-dimensional space in Fig. 5.

6. When overturning moments are applied (or better, under generalised strain controlled conditions, when tilting angles are imposed), local measures testify to a process of progressive concentration of the vertical stresses transmitted by the footing to the soil (and vice versa). As is schematically suggested in Fig. 6a, the uplift of the foundation dominates the response of the system. The $rB$ (Fig. 6a) zone is progressively detached with respect to the underlying soil. In Fig. 6b the vertical stress distributions are reported as a function of the tilting angle $\vartheta$ imposed during a cyclic test performed on a dense sand stratum (PWRI, 2005).
7. An interaction domain can be also defined to interpret the mechanical response at failure of pipes. In this case the rigid footing is substituted by the cross section of the pipe and tilting moment by a torque moment.

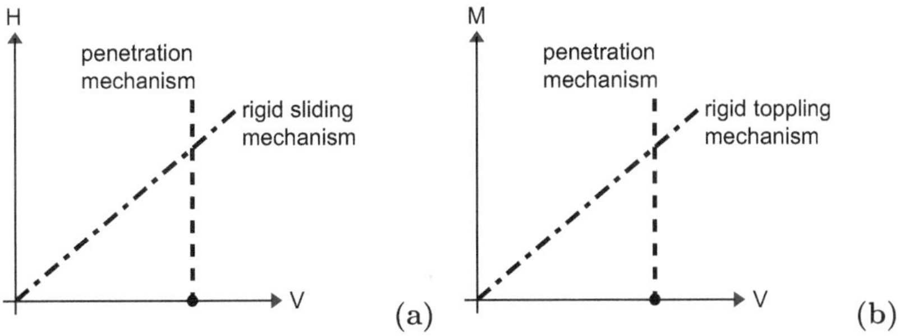


**Figure 3.** Schematic representation of uncoupled interaction domains for a shallow rigid strip footing, in the $H-V$ plane (a) and in the $M-V$ plane (b).

8. As is well known, when the soil is sufficiently rigid, standard general shear failure mechanisms develop; by contrast, when the soil stiffness is quite small (loose sand strata) a punching mechanism takes place and the corresponding bearing capacity cannot be evaluated. Indeed, owing to second order effects (the foundation sinks within the soil and large displacements must be accounted for) the corresponding generalised stress–strain curve is characterised by a continuous increase in the stress component: peaks and plateaus are absent. A plateau can

be envisaged solely if the foundation level is artificially maintained coincident with the ground level.

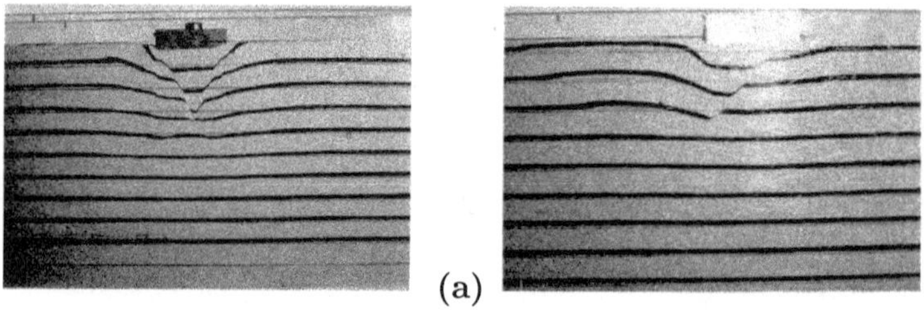

(a) (b)

**Figure 4.** Laboratory small scale experimental test results after (Nova and Montrasio, 1991): failure mechanisms concerning a rigid shallow strip footing under a vertical centred load (a) and an inclined centred load (b).

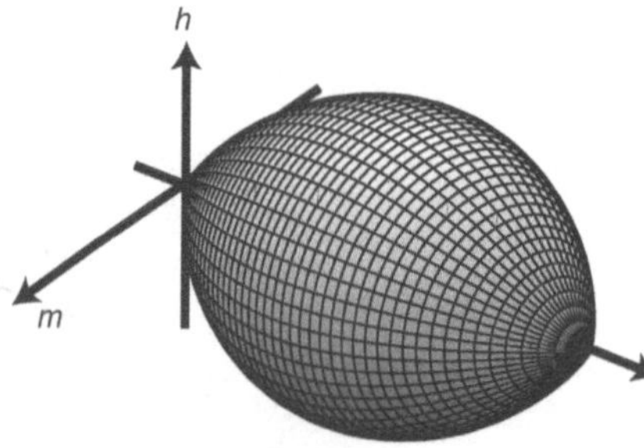


**Figure 5.** Interaction domain in the non-dimensional space ($m = M/\psi B V_{\max}$; $h = H/\mu V_{\max}$; $\xi = V/V_{\max}$).

9. If the stratum is not homogeneous, the shape of the interaction domain and, analogously, the shape of the corresponding failure mechanism, can change abruptly. For instance in Fig. 7, (these pictures refer to small scale experimental tests performed on rigid strip footings placed on loose sand strata (di Prisco et al., 2003a)), the evolution of the failure mechanism due to the presence of georeinforcements, placed within the underlying stratum according to the geometry of Fig. 7c, is illustrated. The failure mechanism is, indeed, much deeper than that obtained without geogrids (Fig. 7a).
10. The symmetry of the interaction domain with respect to the $V$ axis is lost in case of either inclined strata or anisotropy of soil.

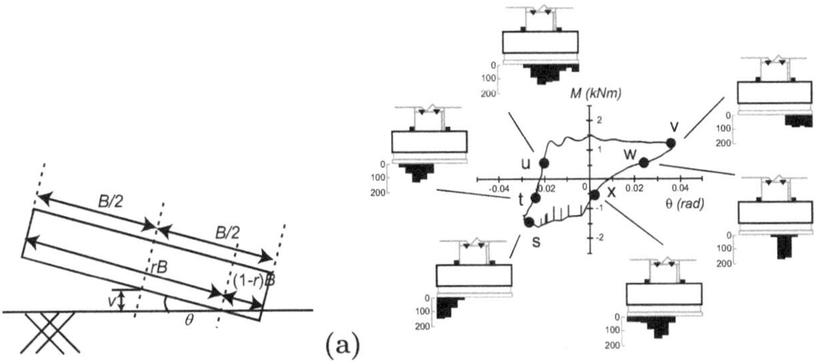


**Figure 6.** Laboratory small scale experimental test results after (Shirato et al., 2008): uplift of the rigid shallow foundation (a) and vertical stress distribution during a cyclic test at constant vertical load and variable tilting angle $\vartheta$ (b).

## 2.2 Numerical Homogeneization Analyses

As is well known, it is quite common today to investigate the macroscopic mechanical behaviour of the specimen, employing micro-structural approaches, in which the interaction among the single grains is numerically reproduced. Analogously, in the case of interface problems of large dimensions like those here accounted for, by changing abruptly the scale, to better understand the mechanical processes governing the problem, numerical analyses can be performed. These latter, owing to the change in scale, allow of being employed as heuristic tools but even to calibrate the constitutive parameters introduced for the macroelement constitutive relationships discussed in the following section. For the sake of completeness, four different approaches will be treated: (i) a very simple contact spring model to describe the uplift phenomenon, (ii) limit analysis to determine the interaction domain for the rigid cross section of either piles or pipes, (iii) finite and (iv) discrete element numerical analyses to describe the dependence of the interaction domain shape on geometrical and stratigraphic conditions.

Contact springs models are usually employed to demonstrate that the interaction problem (in case of a rigid shallow footing) is mainly dominated by the unilaterality of the local contact. Indeed, if bilateral elastic contact springs are assumed to locally describe the interaction between the rigid body and the underlying soil stratum, and an ideal loading test is performed, during which the vertical load is kept constant and the tilting angle

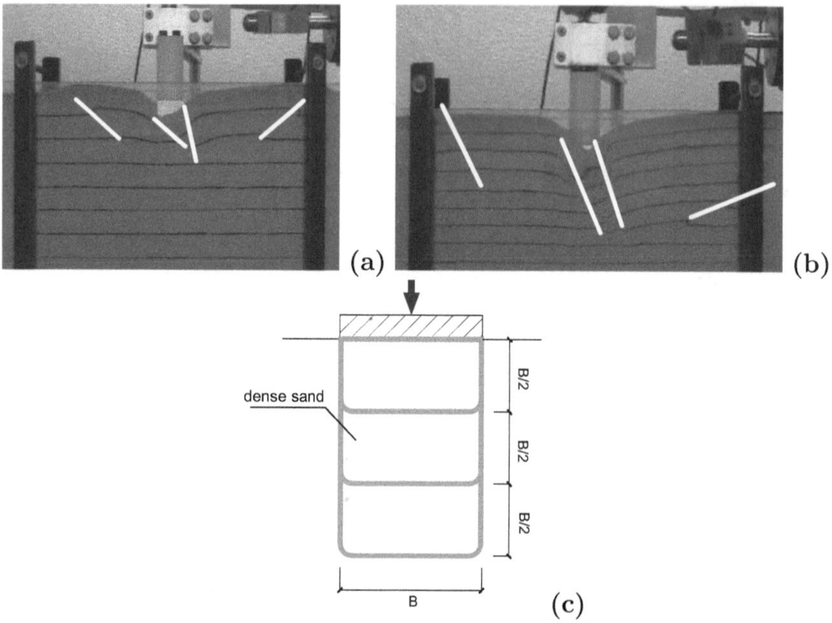


**Figure 7.** Laboratory small scale experimental test results after di Prisco et al. (2003a): failure mechanisms concerning a rigid shallow strip footing under a vertical centred load on an unreinforced dense sand stratum (a) and on a reinforced dense sand stratum (b) according to the geometry schematically outlined in Fig. 7c.

$\vartheta$ is monotonically increased, a linear relationship between the overturning moment $M$ and $\vartheta$ can be easily proven to be obtained (straight line $K_R$ inclined of Fig. 8a) and a maximum value for $M$ is not achieved. In contrast, if elastic unilateral springs are introduced to describe the local contact, the response is still linear until $M = M_0$ is reached (that is until all the springs are under compression - here $\vartheta_0$ stands for the tilting angle corresponding with the initiation of the detachment), whereas subsequently linearity is lost and an ultimate overturning moment is asymptotically approached. Finally, if the springs are also assumed to be elasto-plastic and an appropriate value to their ultimate load is assigned, asymptotically value $M_2$ of Fig. 8a is approached. By numerically performing different tests at different $V$ constant values, even the interaction domain in the $M-V$ plane can be obtained and this satisfactorily matches that experimentally determined and empirically described by means of the function given in (1). It is worth noting that in case of a rigid substratum the boundary of the interaction domain of Fig. 8b simply reduces to a straight line passing through the origin.

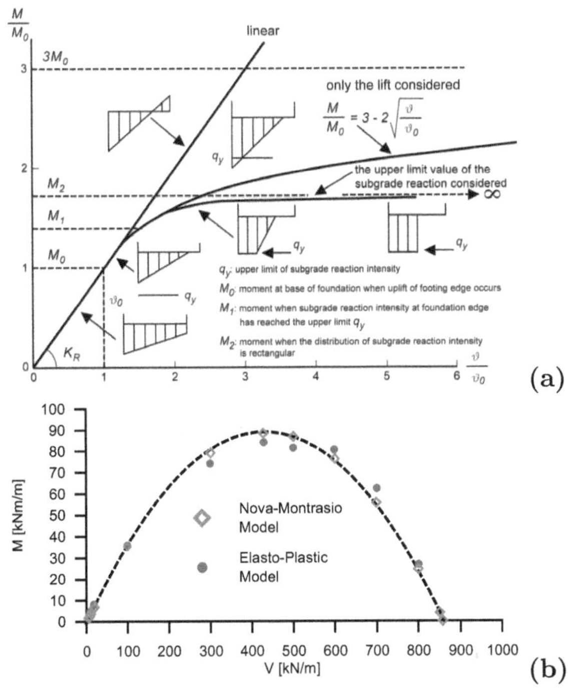


**Figure 8.** Simplified schematic interpretation of the mechanical response of the rigid footing subject to an overturning moment under constant vertical load: linear elastic, non-linear elastic and elastoplastic linear mechanical response (a), interaction domain in the $M - V$ plane obtained by using a generalised spring model of interaction and relative comparison with the locus defined by Nova-Montrasio (b).

To determine the interaction domain shape, limit analysis theory can be even employed, for instance, in particular in case of piles/pipes buried within clays. These approaches are commonly employed, for instance, (Randolph and Houlsby, 1984, di Prisco et al., 2003a), in case of piles horizontally loaded. The problem can be easily dealt with when undrained conditions are considered. For instance, in Fig. 9 the dependence of the interaction domain shape on the relative depth $Z/D$ of a pipe of diameter $D$ is plotted. At increasing values of $Z/D$ the interaction domain tends to become progressively more centred with respect to the $H$ axis (Fig. 1b) and progressively reduces to a circle.

The experimental results employed to determine the interaction domain cited above can be also numerically reproduced by performing finite element analyses in which both the mechanical behaviour of the soil and the local contact between the rigid footing and the underlying soil are satisfactorily simulated. Both these aspects are essential: the former naturally for conveniently reproducing the mechanical response of the system in the zone of the interaction domain where penetration mechanisms dominate, the latter when the values of the vertical loads are quite small and interface mechanisms govern the response of the system. As far as the penetration mechanisms are concerned, in particular, in case the analyst decides to implement an elasto-perfectly plastic constitutive relationship for the soil into the numerical code, the use of a non-associated flow rule is essential for correctly reproducing the system response, and not only its stiffness but even its strength. In case of loose sands, for instance, the use of non-associativeness, jointly to the large deformability of the material, inhibits the formation of a standard failure mechanism and allows the activation of a punching mechanism. As will be clarified in a following section and as was first suggested by Cremer et al. (2001) and Cremer et al. (2002), FEM analysis results can be therefore fruitfully employed to calibrate the constitutive parameters of constitutive relationships for the macro-element.

Analogously, even Distinct Element codes (Cundall and Strack, 1979) have been recently employed (Calvetti et al., 2004) to highlight the main features of very common soil–structure interaction problems. For instance, with reference to the cross section of a pipe embedded into a homogeneous loose sand stratum, the interaction domains for different values of $\zeta = Z/D$ are illustrated in Fig. 10. Once again, as $\zeta$ increases, the domain widens progressively and tends to translate. In Fig. 10b, the numerical prototype and the corresponding failure mechanisms are also reported, respectively.

### 2.3 Elasto-Plastic Approaches

Once the generalised stresses and strains are defined, in order to conceive a homogenised constitutive relationship, capable of accounting for the nonlinearities experimentally and numerically discussed above, and in particular the collapse of the system, the simplest choice is to employ an elasto-perfectly plastic relationship, that is to define an appropriate yield/failure locus $F(\mathbf{Q}, \boldsymbol{\alpha}_F) = 0$, and a plastic potential $G(\mathbf{Q}, \boldsymbol{\alpha}_G) = 0$, where vector $\boldsymbol{\alpha}_F$ and $\boldsymbol{\alpha}_G$ stand for two sets of parameters describing, respectively, the shape of the failure locus and of the plastic potential. Within the failure/yield

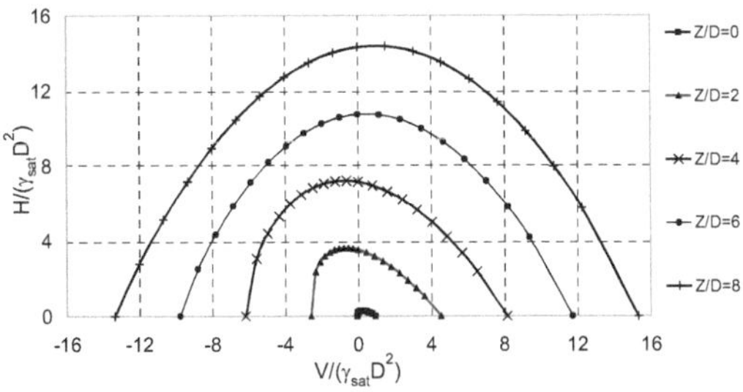


**Figure 9.** Dependence on relative depth $Z/D$ of the interaction domain relative to a rigid cross section of a pipe embedded within a homogeneous clay stratum under undrained conditions ($\gamma_{\text{sat}}$ stands for the saturated soil unit volume weight), evaluated by employing the kinematic method of the limit analysis, after di Prisco et al. (2004).

locus the mechanical behaviour is assumed to be elastic and uncoupled and coupling is assumed to be associated exclusively with ultimate conditions. Either DEM or FEM codes can thus be employed to evaluate the direction of the irreversible generalised strain vector at failure. An example of superimposition of generalised strain rate vectors at failure on the corresponding interaction domains is given in Fig. 10a. It is evident that for these systems associativeness cannot be accepted. Analogously, in case of shallow footings, if an associated flow rule was accepted when sliding mechanisms are activated, that is when in the plane $M = 0$ the straight line $H/V = \mu$ is approached, a mechanical/meaningless negative (i.e. upwards directed) unlimited vertical displacement would occurr.

An example of an elasto-perfectly plastic constitutive relationship conceived for describing the response of rigid footings under seismic actions is in Paolucci (1997), whereas in Cocchetti et al. (2009) an application to pipelines across landslides can be found. In particular, in this latter the calibration of the interaction constitutive relationship is described and even the influence of the slope inclination, along which the pipeline is positioned, on the interaction domain shape is accounted for (Fig. 11). The stress paths plotted in Fig. 11b are obtained by performing FEM numerical analyses (Cocchetti et al., 2009) under generalised strain controlled conditions: it is evident that two of them get the interaction domain and belong to it before reaching the ultimate conditions.

A more sophisticated way of describing the mechanical behaviour of the rigid shallow footing under monotonously increasing loading is to conceive in generalised stresses a strain hardening elasto-plastic constitutive model.

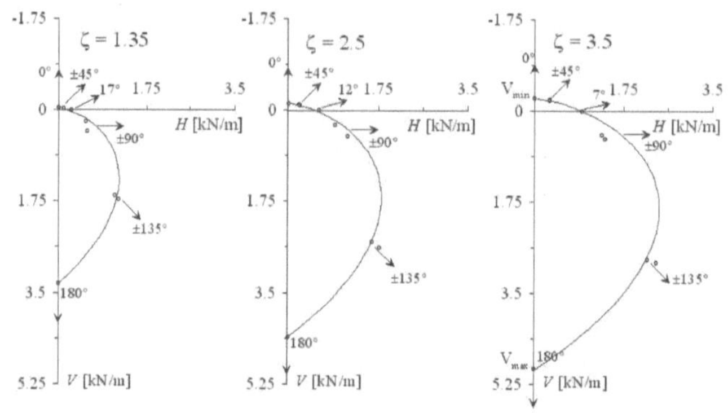


(a)

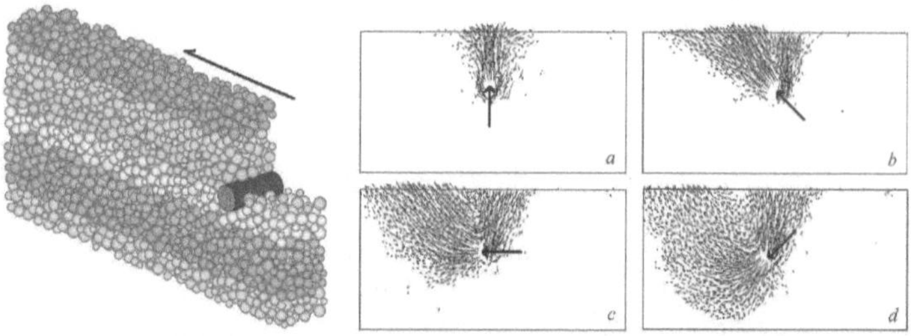

(b)

**Figure 10.** Dependence on relative depth $Z/D$ of the interaction domain related to the rigid cross section of a pipe embedded into a homogeneous loose sand stratum, evaluated by employing a DEM code, after Calvetti et al. (2004) (a); relative DEM specimen and failure mechanisms corresponding with upwards and leftwards imposed displacements (b).

Both the loading function $f$ and the plastic potential $g$, in this case, can be conveniently defined by assuming their shape to be coincident with that of $F$ and $G$: the failure locus is, therefore, a special yield locus for which the hardening parameter $V_c = V_{\max}$ (Nova and Montrasio, 1991) where $V_{\max}$ is the footing bearing capacity. The development of plastic strains causes an increment in $V_c$, according to an appropriate hardening rule definition. For instance Nova and Montrasio (1991) proposed for a strip shallow footing:

$$dV_c = \left(1 - \frac{V_c}{V_{\max}}\right) \frac{R_0}{V_{\max}} \left(dv^p + \alpha \left|du^p\right| + \gamma B \left|d\vartheta^p\right|\right) \tag{3}$$

where $B$ stands for the footing width, $R_0$ is a constitutive parameter governing the stiffness of the system that can be easily calibrated when centred

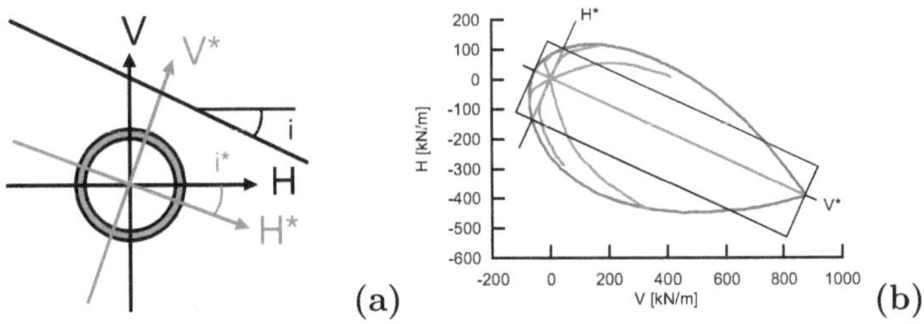


**Figure 11.** Dependence on the slope inclination $i$ of the interaction domain shape, evaluated by employing a FEM code for a rigid cross section of a pipe embedded into a homogeneous loose sand stratum, after Cocchetti et al. (2009): schematic representation of the problem (a), interaction domain and stress path in the $H-V$ plane.

vertical loads are applied, the superscript index $p$ stands for plastic, whereas $\alpha$ and $\gamma$ are constitutive parameters influencing the system response when either inclined or eccentric loads are applied. In a non-dimensional plane (by employing the non-dimensional variables introduced in Fig. 5) and by defining $\rho_c = V_c/V_{\max}$, such a hardening can be represented as in Fig. 12. As it has been clearly demonstrated by many authors, such a simple approach allows of satisfactorily reproducing the mechanical response of shallow footings under any monotonically increasing load and this statement can be supported by the comparisons with either FEM analyses or experimental test results.

Eq. (3) implies necessarily an increase in the size of the yield locus and, as a consequence, always a positive hardening. In contrast, some authors have also proposed more complex hardening rules capable of accounting for even a reduction in the yield function size and, consequently, of simulating a softening regime (Gottardi et al., 1999; Martin, 1994; Cassidy et al., 2002; Nova et al., 2008). Instead, very recently Hodder and Cassidy (2010) with regard to normally consolidated clays, whereas di Prisco and Vecchiotti (2009) with regard to loose sands, modified the aformentioned hardening rules to reproduce the mechanical interaction between rigid bodies and large deformable soils, that is to account for large displacements. In particular, di Prisco and Vecchiotti (2009) suggested for the one dimensional case the following:

$$dV_c = \left\langle 1 - \frac{V_c}{V_{\max}} \right\rangle R_0 \; dv^p + \bar{\alpha} \; dv^p \tag{4}$$

where $\bar{\alpha}$ is an additional constitutive parameter and brackets impose to consider only the positive branch of $(1 - \frac{V_c}{V_{\max}})$.

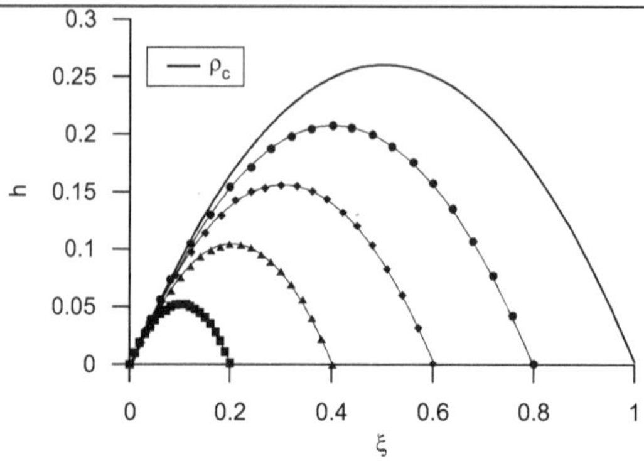


**Figure 12.** Representation of the isotropic hardening in the non-dimensional plane $h - \xi$: the yield locus evolution.

To define more precisely the loading function shape, in the 90s some authors (Gottardi et al., 1999; Butterfield and Gottardi, 2003) performed the so-called swipe-tests on rigid shallow footings, during which vertical displacements are inhibited, whilst horizontal/rotational displacement components are increased and the remaining load (rotational/horizontal) is kept constant.

An extension of this theory to visco-plasticity, according to Perzyna's approach (Perzyna, 1963) (this latter already discussed within a previous chapter of this volume), was recently suggested by di Prisco and Vecchiotti (2006) to numerically reproduce the impact of rigid boulders on granular materials. In this case, the dynamic interaction problem is approached by accounting for both the boulder mass and inertial actions within the deformable sand stratum. This is obtained through: (i) a suitable definition for the viscous nucleus, (ii) the definition, according to a radial mapping, of a sort of overstress (in Fig. 13 indicated with term distance between points $P$ and $I$ representing, respectively, the current generalised state of stress and the corresponding image point belonging to the current yield locus). As was already observed in a previous chapter, according to this theory, the point $P$ (Fig. 13) can belong to a domain external to both the yield function and the interaction domain and this implies that during the evolution of time, the load applied by the boulder to the soil can be much larger than that can be statically reached.

## 3 Shallow Foundations Under Cyclic Loading

The two previously defined approaches (the elasto-perfectly plastic and the strain hardening elasto-plastic), have been demonstrated to be capable of reproducing quite satisfactorily the mechanical response of this type of structures under monotonic tests, but, as it can be reasonably expected, they fail totally in simulating the experimental evidence when cyclic tests are taken into account. To highlight the cyclic interaction between shallow foundations and the underlying soil, in the last two decades, two different classes of experimental tests have been prevalent: (i) symmetric load controlled cyclic tests (in this case the cycles are symmetric with respect to the $V$ axis) (Pedretti, 1998; Shirato et al., 2007) and (ii) asymmetric load controlled tests (in this second case the cyclic perturbation induces a loading oscillation about an image point in the generalised stress space which does not belong to the $V$ axis; di Prisco et al. (2003c)). The first class of tests causes with time an accumulation of vertical displacements that progressively stabilise, reaching a sort of plastic adaptation regime. The second class of tests, on the other hand, induces a marked accumulation both of vertical and horizontal/rotational displacements.

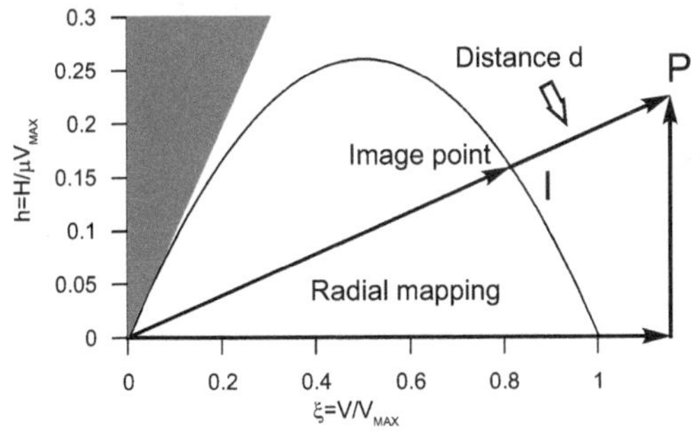


**Figure 13.** Definition of distance $d$ according to radial mapping in the non-dimensional $h - \xi$ plane.

### 3.1 Experimental Evidence

As in the previous section, to summarise the experimental test results available in the literature, not really very numerous, here some summarising statements are listed:

1. Under cyclic loading, the system dissipates energy and, in general, accumulates irreversible generalised strains;

2. When the underlying soil stratum is sufficiently rigid i.e. dense sand, during tests performed by controlling the rotation angle, by keeping constant the vertical load and recording the horizontal one, the cycles in the $M - \vartheta$ plane assume, in case of large values of tilting, a typical backbone shape (Fig. 14a). This response is mainly associated with the development of the uplift phenomenon dominating the response of the system in case of dense but not of loose sand strata (Fig. 14b).
3. In contrast, if during each cycle the ultimate load is not reached (obviously in this case the generalised stress path can be imposed under load controlled conditions) (Fig. 15), in case of both loose and dense sand strata, the mechanical response during the unloading phase of each cycle is characterised by a monotonic decrease in the rotational stiffness (Fig. 15a and Fig. 15b).

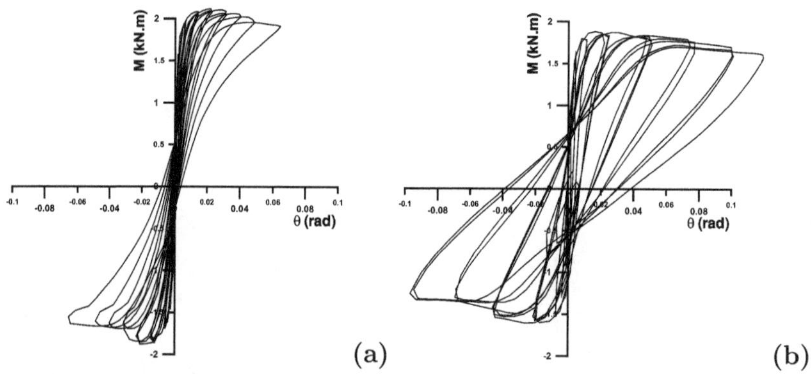


**Figure 14.** Experimental data concerning a rigid square shallow foundation cyclically tilted: (a) dense sand, (b) loose sand (PWRI, 2005).

4. The coupling among tilting/horizontal and vertical displacements is very severe when loose sand strata are tested but are even evident for dense sand strata (Fig. 15d).
5. If the experimental mechanical response of rigid footings (in case of symmetric loading paths) is interpreted by employing the well-known concepts of secant stiffness $K$ and damping factor $\eta$, (Fig. 16), a clear decay in $K$ and, at the same time, an increase in $\eta$ are evident at increasing values of either rocking or horizontal displacements. An example of these relations, obtained by processing experimental data from Pedretti (1998) and the PWRI (2005), is illustrated in Fig. 16 with reference to dense sands. Stiffness has been determined from the experimental loops, as the slope of the line joining the extreme

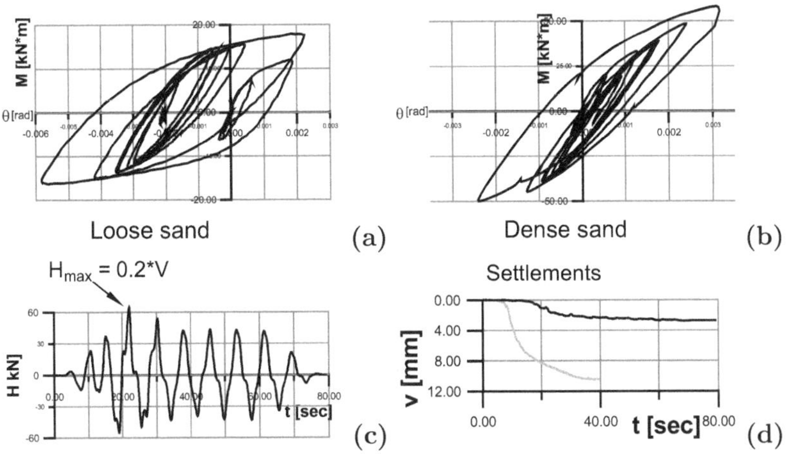


**Figure 15.** Experimental data concerning a rectangular shallow foundation under load controlled cyclic tests, after Pedretti (1998): (a) rocking moment versus rotation angle (dense sand stratum), (b) rocking moment versus rotation angle (loose sand stratum), (c) loading history versus time, (d) vertical settlements versus time for both dense (black line) and loose sand stratum (gray line).

points of the $M - \vartheta$ and $H - u$ curves, respectively. Stiffness $K$ is normalised as well with respect to the initial small strain stiffness $K_0$. In the same plots numerical simulations obtained by employing the macro-element concept are also reported but these numerical data will be commented on the following. It is worth noting that even for relatively small values of foundation rocking, for instance 1 mrad, the reduction in foundation stiffness, depending on soil relative density, ranges between about 40% and 60%. The equivalent damping ratio $\eta$ in the rocking mode, computed as the ratio between the dissipated energy $\hat{D}$ (area of the hysteresis loop) and the stored elastic energy $\Delta W$, is also plotted in the same figures. For rocking values up to 1 mrad the $\eta$ value ranges from 5% to 10%, while it significantly increases for larger rocking angles, up to 20% for dense sands and 30% for medium dense sands. When the rocking angle imposed is sufficiently large, the phenomenon of the uplift dominates the response of the system and $\eta$ stops evolving.

6. When a very large number of cycles is imposed, the accumulation rate of irreversible settlements progressively decreases: usually a sort

of stabilisation takes place, at least when no damage phenomenon is present. When the generalised stress path is symmetric, only vertical displacements accumulate with the number of cycles, but when either the initial generalised stress state or the generalised stress path is asymmetric, the ratcheting phenomenon involves not only vertical settlements but also either horizontal displacements or rocking angles. In order to exemplify this aspect, in Fig. 17 some experimental test results from di Prisco et al. (2003a) concerning small scale experimental tests on a rigid shallow footing placed on a uniform loose sand stratum are illustrated. In particular, in Fig. 17a a schematic view of the generalised stress path imposed (the tilting moment is in this case nil) is reported, whereas in Fig. 17b the loops in the $H-u$ plane are plotted. The accumulation of generalised strains is observed to be essentially a function of (i) the generalised stress path cyclically imposed, (ii) the amplitude of the cycles and (iii) the image point defined in the generalised stress space around which the loading path is imposed and in particular of its position with respect to the failure locus (full line in Fig. 18a). As is evident from Fig. 18, where all the results concern the same amplitude and only the position of cycles in the $H-V$ plane is varied, when the image point defined above approaches the limit locus (and this is particularly true for points 1 and 4 characterised by a large obliquity $H/V$), the accumulation rate increases severely.

### 3.2 Constitutive Modelling

A way of overcoming the limitations of the previously defined class of models, analogously to what is done with reference to standard constitutive relationships for the soil's REV, is to conceive either bounding surface constitutive relationships or multi-mechanism plastic approaches. For instance, the constitutive model of Nova and Montrasio (1991) cited here above was modified (di Prisco et al., 1998), by introducing within the yield locus a subloading surface (Fig. 19) and the evolution of the inner locus is conceived so that it cannot intersect the 'bounding surface'. A convenient mapping rule allows of relating any point within the yield locus to an appropriate point on it. The plastic multiplier is evaluated on the bounding surface and suitably scaled according to the distance between the current point $P_i$ and the corresponding image point belonging to the bounding surface $I_i$ (Fig. 19). When the distance reduces to zero, the scaling function converges towards unity. The inner locus is therefore exclusively employed to define the elastic domain and to determine the image point $I_i$. In this

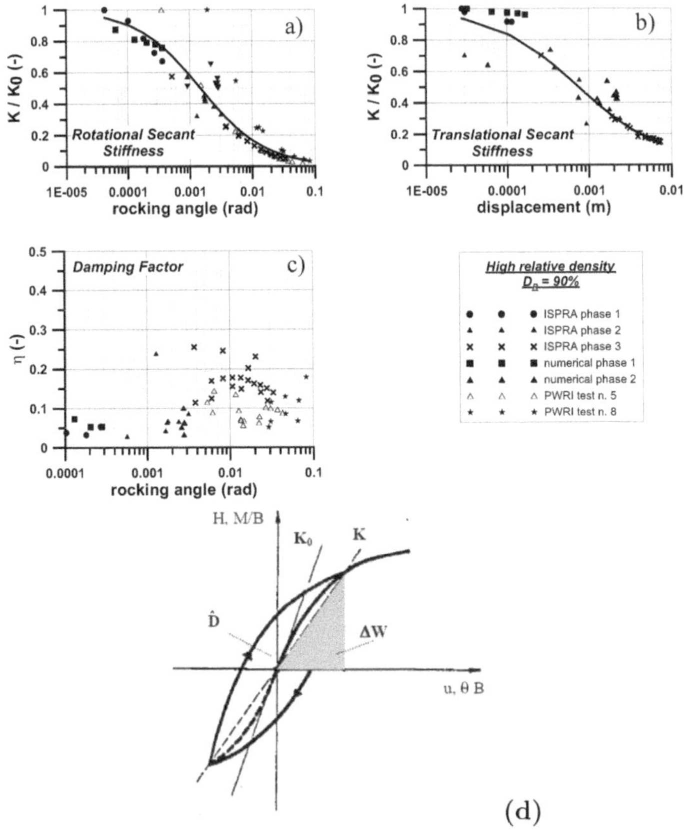


**Figure 16.** (a) Normalised rocking, (b) translational stiffness, (c) damping factor for a dense sand ($D_r = 90\%$) (after Paolucci et al. (2007) at increasing values of rocking angle and horizontal displacement, respectively) and (d) sketch for the evaluation of rocking/traslational stiffness and damping factor.

way, both the occurrence of permanent generalised strains, even when the stress point is within what is usually considered a purely elastic region, and the accumulation of plastic distortions during cyclic or transient loading can be simulated (di Prisco et al., 2003a; di Prisco et al., 2003b). The model becomes inevitably more complex and the determination of further parameters is necessary. An example of validation of the model concerning the behaviour of a plinth, 1 m wide, founded on a dense sand stratum, subject to a constant vertical load and to both cyclic horizontal load and overturning

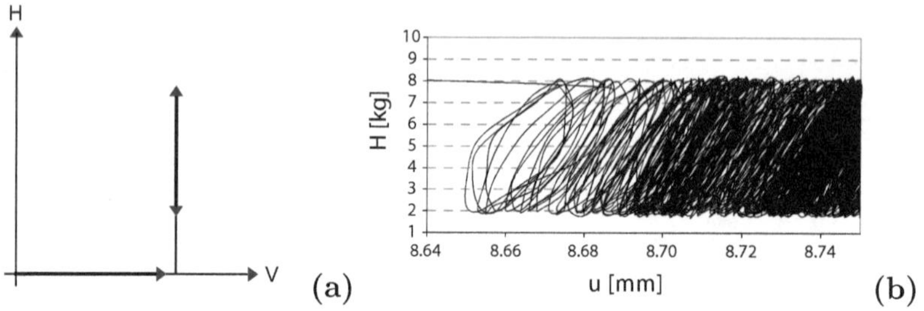


**Figure 17.** Experimental test results (after di Prisco et al. (2002)) obtained by performing $V$ constant test and by cyclically changing $H$: (a) generalised stress path, (b) horizontal load versus horizontal displacement.

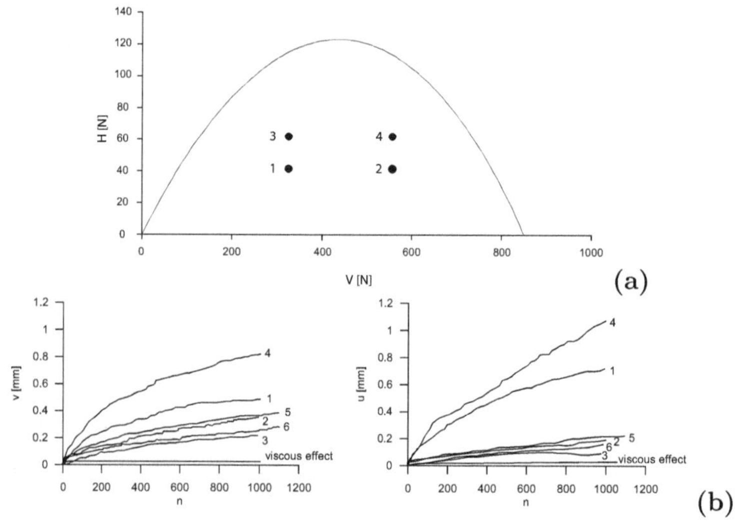


**Figure 18.** Experimental test results (after di Prisco et al. (2003c) obtained by performing $V$ constant tests and by cyclically changing $H$: (a) image point around which the cyclic generalised stress path is imposed, (b) vertical and (c) horizontal displacements versus number of cycles $n$.

moment (experimental data after Pedretti (1998)) is in Fig. 20. Loads were applied at low frequency, so that dynamic effects could be neglected. Both the overturning moment and the horizontal force are varied while their ratio is kept constant. It is evident that this constitutive approach is suitable for capturing in a satisfactory way the essential features of the experimental be-

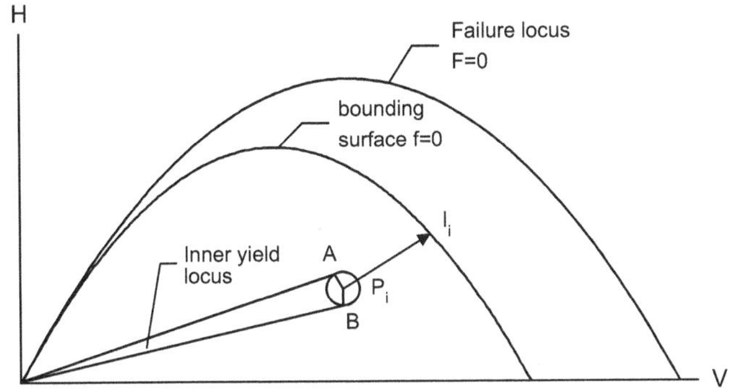


**Figure 19.** Schematic representation of both the sub-loading function and the bounding surface.

haviour of the system and this conclusion regards both dense and loose sand strata, at least in case of symmetric cycles applied on symmetric systems, that is for rigid footings placed on horizontal strata with either a homogeneous stratigraphy or a heterogeneous stratigraphy with layers stratified normally to the vertical direction. This class of models can be also employed to rapidly derive interesting heuristic information, that obviously have to be experimentally confirmed. For instance, the experimental data plotted in Fig. 16, as was previously mentioned, were obtained by imposing generalised stress paths at different initial vertical loads $V$, similar to that schematically plotted in Fig. 21a: either $H$ or $M$ are cyclically varied (in particular, in both the experimental campaigns previously cited, both variables are varied simultaneously and ratio $M/H$ is kept constant), according to progressively increasing amplitudes (Fig. 21b). The large scatter in the experimental results previously illustrated can be therefore theoretically justified throughout the use of the macroelement approach. In Fig. 22 different $K/K_0$ curves, obtained by employing the aforementioned bounding surface elasto-plastic model, are collected in relation to two values of relative densities for different values of $V$. It is evident that at decreasing values of $V/V_{\max}$, where $V_{\max}$ stands for the bearing capacity of the footing, the decay in $K/K_0$ becomes more rapid whilst the increase in damping less rapid.

Unfortunately, the previously mentioned class of models fails in reproducing three very important aspects of the mechanical response of rigid shallow foundations under repeated loading: (i) the large settlements induced by the first unloading, (ii) the reduction in stiffness during the first phase of the

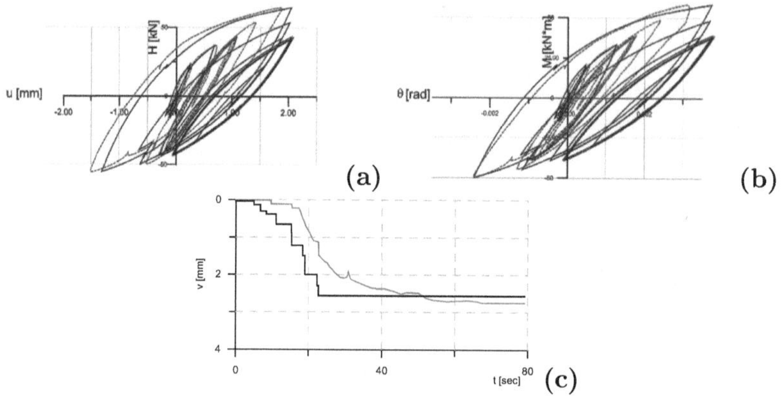


**Figure 20.** Comparison of measured (dotted lines) and calculated (full lines) displacements of a real scale foundation placed on a dense sand stratum under cyclic horizontal loading and overturning moment and constant vertical load (experimental data after Pedretti (1998)); a) horizontal load versus horizontal displacements; b) overturning moment versus rotation; c) accumulated vertical settlement versus time.

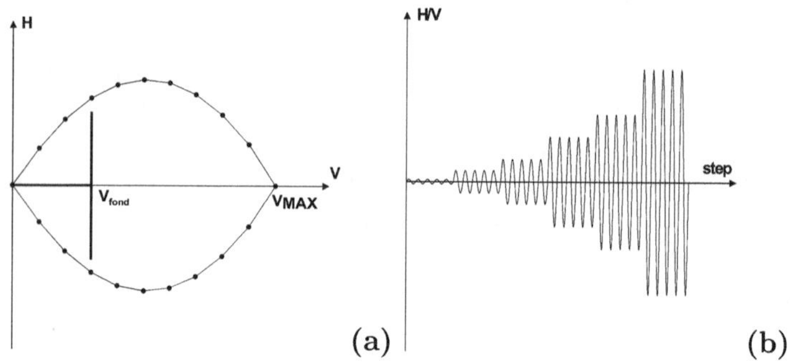


**Figure 21.** Constant vertical load path (a) and loading history with symmetric cycles of increasing amplitude (b).

unloading when the footing is previously largely tilted, (iii) the ratcheting when asymmetric loading paths are imposed.

In order to clarify the first statement, in Fig. 22 the numerical results (curves MACRO) obtained by employing the previously cited bounding surface constitutive model (di Prisco et al., 2002) are compared with those from small strain/displacement finite element numerical analyses. In particular, the

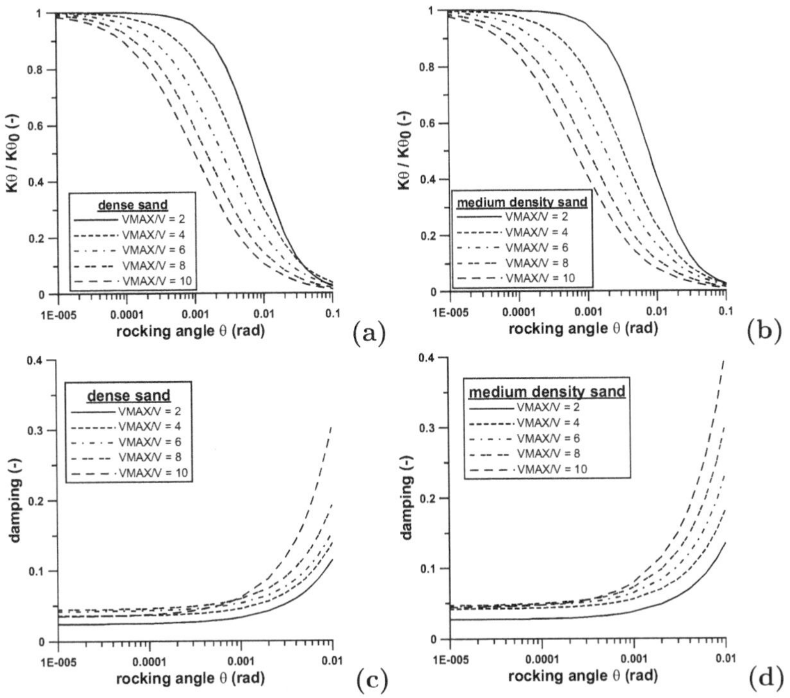


**Figure 22.** Influence of the $V/V_{\max}$ ratio on the dependence of (a, b) the secant rotational stiffness $K_\theta$ and of (c,d) the damping factor for dense and medium dense sand on the rocking angle, respectively.

curves MC are obtained by employing the FEM commercial code Tochnog (FEAT, 2004) and by assuming for the soil an elasto-perfectly plastic Mohr–Coulomb constitutive model, and the curves DP, by assigning to the soil an anisotropic strain-hardening constitutive relationship (di Prisco et al., 1993). These results are for a rigid shallow strip foundation placed on a very loose sand stratum initially subjected to a vertical load $V$ which is kept constant while the horizontal load is cyclically varied. It is evident that the numerical data obtained by employing the macro-element approach match the FEM numerical analyses results quite well during all the successive cycles but not during the first one, in particular if the DP curve is taken as target. This limit mainly derives from the isotropic hardening for the bounding surface. In fact, when the load is rightward directed, the irreversible strains take place mainly close to the right corner of the foundation, whilst at the opposite corner when the load is inclined leftward: therefore,

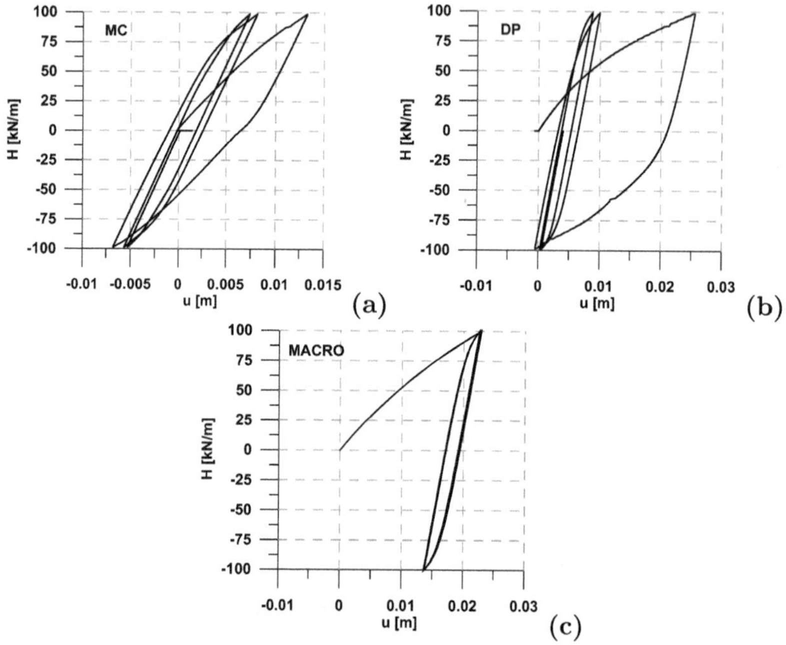


**Figure 23.** Comparison among numerical data obtained by employing the macro-element approach and FEM numerical analyses: (a), horizontal load versus horizontal footing displacements obtained by using a FEM code (MC constitutive model), (b) (DP constitutive model) and (c) di Prisco et al. (2006) macroelement constitutive relationship.

the two associated plastic mechanisms can be assumed, at first approximation, to be independent and the same can be inferred when overturning moments are applied.

To capture this aspect, as already suggested by Cremer et al. (2001), the hardening of the plastic surface should be anisotropic. These authors suggested, by following the theoretical approach conceived for cohesive strata by Salençon and Pecker (1995a,b), two different expressions for the failure locus $F$ and for the yield locus $f$ respectively:

$$F = \left( \frac{H'}{\tilde{a} \times V'^{\tilde{c}} \times (1 - V')^{\tilde{d}}} \right)^2 + \left( \frac{M'}{\tilde{b} \times V'^{\tilde{e}} \times (1 - V')^{\tilde{f}}} \right)^2 - 1 = 0 \qquad (5)$$

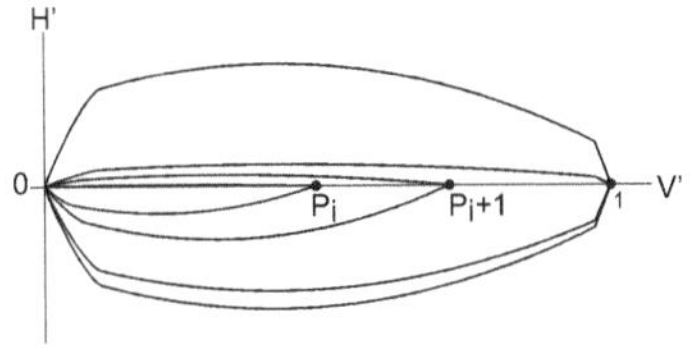


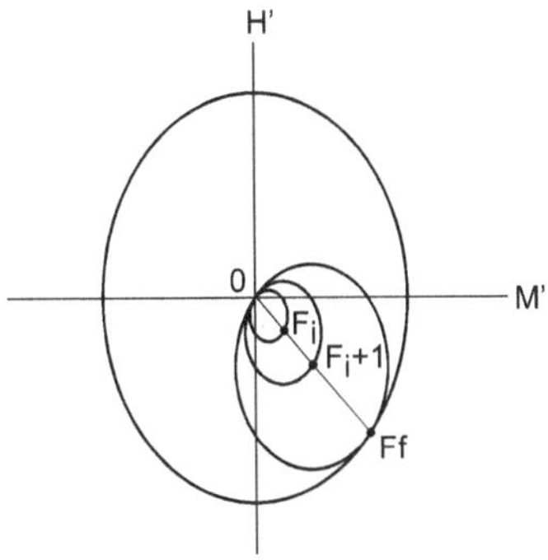


**Figure 24.** Schematic representation both of the failure locus and of the evolving yield function according to the approach of Salençon and Pecker (1995a,b), in the $H' - V'$ (a) and in the $H' - M'$ plane (b), respectively.

$$f = \left(\frac{H' - \tilde{\alpha}\Gamma_H^\gamma}{\tilde{\rho}\Gamma_H^\gamma}\right)^2 + \left(\frac{M' - \tilde{\beta}\Gamma_M^\gamma}{\tilde{\rho}\Gamma_M^\gamma}\right)^2 - 1 = 0 \tag{6}$$

where $V'$, $H'$ and $M'$ stand for normalised non-dimensional variables such as those ($\xi$, $h$ and $m$) introduced by Nova and Montrasio (1991), whereas $\tilde{a}$, $\tilde{b}$, $\tilde{c}$, $\tilde{d}$, $\tilde{e}$ and $\tilde{f}$ are constitutive parameters; $\tilde{\rho}$, $\tilde{\alpha}$ and $\tilde{\beta}$ are hardening variables, the first defining the isotropic hardening and the other two the kinematic hardening of the yield function, respectively, whereas:

$$\begin{aligned}
\Gamma^\gamma &\equiv \begin{pmatrix} 0 \\ \Gamma_H^\gamma \\ \Gamma_M^\gamma \end{pmatrix} \equiv \begin{pmatrix} 0 \\ \tilde{a}V'^{\tilde{c}}\left(\tilde{\gamma} - V'\right)^{\tilde{d}} \\ \tilde{b}V'^{\tilde{e}}\left(\tilde{\gamma} - V'\right)^{\tilde{f}} \end{pmatrix} \qquad (7) \\
\tilde{\gamma} &= \chi + (1 - \chi)\left(\tilde{\rho} + \tilde{\tau}\right) \\
\tilde{\tau} &= \sqrt{\tilde{\alpha}^2 + \tilde{\beta}^2}
\end{aligned}$$

For $\tilde{\gamma} = \tilde{\rho} = \chi = 1$, $f$ degenerates into the failure locus $F$. When the yield function isotropically hardens and a centred vertical load is applied, $\tilde{\rho} = \chi$ and $\tilde{\tau} = 0$. These quite complex analytical definitions provide a convenient description for the anisotropic hardening, as is schematised in Fig. 24, both in the $H' - V'$ and $H' - M'$ planes. An interesting application of this class of models has been recently published by Grange et al. (2007a), Grange et al. (2007b), Grange et al. (2008) and Grange et al. (2009). These authors employed a suitable evolution rule for $\tilde{\beta}$, a linear constraint between $\dot{\tilde{\rho}}$ and $\dot{\tilde{\tau}}$ ($\dot{\tilde{\rho}} = k\dot{\tilde{\tau}}$) and, to satisfactorily match the experimental results, parameter

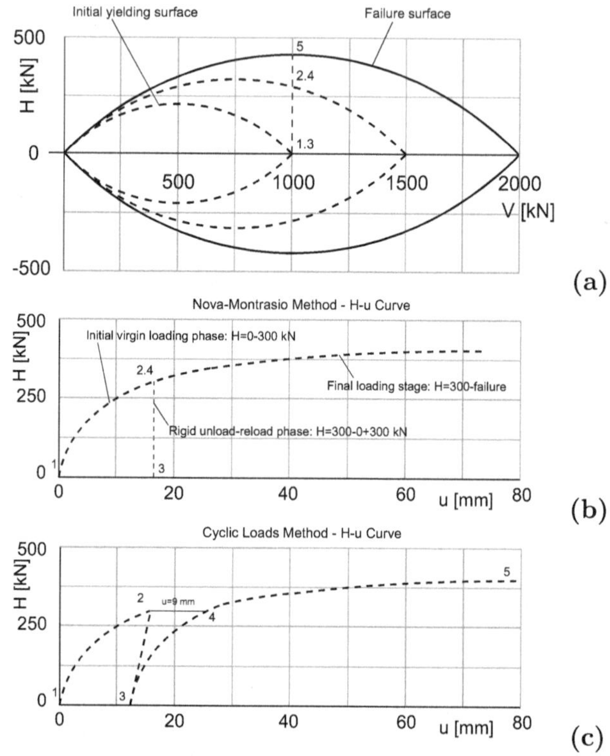


**Figure 25.** Asymmetric generalised stress path; schematic representation of the evolution of the yield function and of the failure locus for an isotropic strain hardening elasto-plastic model (a), mechanical response which could be obtained if an isotropic strain-hardening elasto-plastic model was employed (b), mechanical response is obtained by employing the bounding surface elastoplastic model (c).

$k$ was abruptly reduced. Unfortunately this choice necessarily implies the effect of previous loading histories to be lost.

The second aspect previously cited, concerning the rotational stiffness of the system during unloading after getting large values of the tilting angle, is instead more important and can be clarified by discussing once more the experimental results obtained on large scale models at the Public Works Research Institute (PWRI, 2005; Shirato et al., 2007) already plotted in Fig. 14. These experimental data were obtained by imposing to a prototype rigid structure, placed on a caisson filled with sand, a horizontal cyclically varying displacement. During the cyclic phase, the footing is therefore loaded by a constant vertical load due to the steel frame weight while both overturning moment and horizontal force cyclically vary. As the test is conducted under displacement controlled conditions, even the reduction in generalised loads can be observed. As already commented, when the soil density is sufficiently high, during the unloading a typical 's' shaped trend is observed. As was already mentioned, this mechanical response during unloading in the case of dense sand is essentially due to the uplift of the foundation: the reduction of the contact surface between the footing and the soil due to the detachment between the two materials generates a sort of damage of the system that could be described coherently for instance by introducing an elasto-plastic coupling. Some recent efforts in this direction, although not totally satisfying, are in Cremer et al. (2002), Shirato et al. (2007), Paolucci et al. (2008), Grange et al. (2008) and Grange et al. (2009). In particular, Paolucci et al. (2008) suggested a sort of pseudo-empirical damage rule, whereas Chatzigogos and Figini (2008) suggests a sort of non-linear damaging elasticity accounting for the detachment.

Finally, as far as the third issue listed above is concerned, to demonstrate that standard bounding-surface-plasticity and anisotropic strain hardening constitutive models severely overestimate the phenomenon of ratcheting when asymmetric loading paths are imposed, consider Fig. 25. In Fig. 25b and c respectively, the typical mechanical responses to the generalised asymmetric stress path of Fig. 25a, we could obtain by means of a standard isotropic hardening elastoplastic constitutive model and a bounding surface model, are illustrated. In case of elastic isotropic strain hardening (Fig. 25b), all the successive unloading/reloading do not cause the accumulation of any irreversible displacement, whereas the bounding surface approach cannot generate any loop, so that the ratcheting phenomenon is too marked. To reduce the accumulation of the irreversible generalised strains and to reproduce, from both a qualitative and quantitative point of view, the experimental evidence, already commented on wih reference to Fig. 17 and Fig. 18, the previously cited bounding surface model was further modified by the author by introducing a sort of artificial memory rotating in the $\xi - h - m$ space and distance between $I_i$ and $P_i$ of Fig. 19 was suitably redefined. Nevertheless, so far there has not been, in the literature, any convincing constitutive relationship, based on the macroelement approach, capable of quantitatively reproducing either the progressive stabilisation, or the collapse, induced by cyclic loading.

## List of Symbols

| | | |
|---|---|---|
| $^p$ | plastic component | §2.3 |
| $a$ | $= L/2$ | §2 |
| $\tilde{a}$ | constitutive parameter | §3.2 |
| $b$ | characteristic length of the plastic hinge | §2 |
| $\tilde{b}$ | constitutive parameter | §3.2 |
| $B$ | footing width | §2 |
| $\tilde{c}$ | constitutive parameter | §3.2 |
| $\tilde{d}$ | constitutive parameter | §3.2 |
| $D$ | diameter of the pipe | §2.2 |
| $\hat{D}$ | dissipated energy | §3.1 |
| $\mathbf{D}$ | incremental constitutive matrix | §2 |
| $D_r$ | relative density | §3.1 |
| $\tilde{e}$ | constitutive parameter | §3.2 |
| $f$ | yield locus | §2.3, §3.2 |
| $\tilde{f}$ | constitutive parameter | §3.2 |
| $F$ | failure locus | §2, §3.2 |
| $g$ | plastic potential in strain hardening elasto-plastic approach | §2.3 |
| $G$ | plastic potential in elasto-prefectly plastic approach | §2.3 |
| $h$ | $= H/\mu V_{\max}$ | §2.1 |
| $H$ | horizontal load | §2, §3 |
| $H'$ | normalised non-dimensional horizontal load | §3.2 |
| $i$ | slope inclination | §2.3 |
| $I_i$ | image point belonging to the bounding surface | §3.2 |
| $k$ | parameter | §3.2 |
| $K$ | secant stiffness | §3.1 |
| $K_0$ | initial small strain stiffness | §3.1 |
| $K_R$ | elastic stiffness | §2.2 |
| $K_\theta$ | secant rotational stiffness | §3.2 |
| $L$ | length of the beam | §2.3 |

| | | |
|---|---|---|
| $m$ | $= M/\psi B V_{\max}$ | §2.1 |
| $M$ | overturning moment | §2, §3 |
| $M'$ | normalised non-dimensional overturning moment | §3.2 |
| $M_U$ | ultimate moment | §2 |
| $M_0$ | moment at base of foundation when uplift of footing edge occurs | §2.2 |
| $M_1$ | moment when subgrade reaction intensity at foundation edge has reached the upper limit | §2.2 |
| $M_2$ | moment when the distribution of subgrade reaction intensity is rectangular | §2.2 |
| $n$ | number of cycles | §3.2 |
| $N$ | longitudinal action | §2 |
| $P_i$ | current state of stress point | §3.2 |
| $P_U$ | ultimate load | §2 |
| $\mathbf{q}$ | vector of generalised strain | §2 |
| $\mathbf{q}^p$ | accumulated generalised plastic strain | §2 |
| $q_y$ | upper limit of subgrade reaction intensity | §2.2 |
| $\mathbf{Q}$ | vector of generalised stress | §2 |
| $R_0$ | constitutive parameter | §2.3 |
| $T$ | torsional component | §2 |
| $u$ | horizontal displacement | §2, §3 |
| $v$ | vertical displacement | §2, §3 |
| $V$ | vertical load | §2, §3 |
| $V'$ | normalised non-dimensional vertical load | §3.2 |
| $V_c$ | hardening parameter | §2.3 |
| $V_{\max}$ | footing bearing capacity | §2, §3 |
| $Z$ | depth | §2 |
| $\alpha$ | constitutive parameter | §2.3 |
| $\tilde{\alpha}$ | hardening variable | §3.2 |
| $\bar{\alpha}$ | constitutive parameter | §2.3 |
| $\boldsymbol{\alpha}_F$ | vector describing the shape of the failure locus | §2.3 |
| $\boldsymbol{\alpha}_G$ | vector describing the plastic potential | §2.3 |

| | | |
|---|---|---|
| $\beta$ | parameter | §2.1 |
| $\tilde{\beta}$ | hardening variable | §3.2 |
| $\gamma$ | constitutive parameter | §2.3 |
| $\tilde{\gamma}$ | $= \chi + (1-\chi)(\tilde{\rho} + \tilde{\tau})$ | §3.2 |
| $\gamma_{\text{sat}}$ | saturated soil unit volume weight | §2.2 |
| $\Gamma_H^\gamma$ | $= \tilde{a} V'^{\tilde{c}} (\tilde{\gamma} - V')^{\tilde{d}}$ | §3.2 |
| $\Gamma_M^\gamma$ | $= \tilde{b} V'^{\tilde{e}} (\tilde{\gamma} - V')^{\tilde{f}}$ | §3.2 |
| $\delta$ | maximum vertical displacement of the beam | §2 |
| $\Delta W$ | stored elastic energy | §3.1 |
| $\zeta$ | $= Z/D$ relative depth | §2.2 |
| $\eta$ | damping factor | §3.1 |
| $\vartheta$ | foundation rotation | §2, §3 |
| $\vartheta_0$ | tilting angle corresponding with the initiation of the detachment | §2.2 |
| $\mu$ | parameter | §2.1, §2.3 |
| $\xi$ | $= V/V_{\max}$ | §2.1 |
| $\tilde{\rho}$ | hardening variable | §3.2 |
| $\rho_c$ | $= V_c/V_{\max}$ | §2.3 |
| $\tilde{\tau}$ | $= \sqrt{\tilde{\alpha}^2 + \tilde{\beta}^2}$ | §3.2 |
| $\chi$ | hardening variable | §3.2 |
| $\psi$ | parameter | §2.1 |
| $\boldsymbol{\psi}$ | vector describing the previous history of the system | §2 |

## Bibliography

R. Butterfield and G. Gottardi. Determination of yield curves for shallow foundations by swipe testing-dtermination des courbes de chargement limite des fondations superficielles par essais swipe. In Magnan et Droniuc, editor, *Fondations superficielles.* Presses de l'ENPC/LCPC, Paris, 2003.

R. Butterfield and G. Gottardi. A complete three dimensional failure envelope for shallow footings on sand. *Géotechnique*, 44(1):181–184, 1994.

R. Butterfield and J. Ticof. Discussion: design parameters for granular soils. In *Proceedings 7th European Conf. on Soil Mechanics, Fndn Engng.*, pages 259–262 (4), Brighton, 1979.

F. Calvetti, C. di Prisco, and R. Nova. Cyclic macro-element for soil-structure interaction: material and geometrical non-linearities. *ASCE J. Geotech. and Geoenv. Eng.*, 12:1292–1299, 2004.

M. Cassidy, B. Byrne, and G. Houlsby. Modelling the behaviour of circular footings under combined loading on loose carbonate sand. *Géotechnique*, 52(10):705–712, 2002.

C. T. Chatzigogos and R. Figini. A general macro-element for shallow foundations and applications for performance-based design. In *Proc. 19th European Young Geotechnical Engineers Conference*, Gyor, Hungary, September 2008.

G. Cocchetti, C. di Prisco, A. Galli, and R. Nova. Soilpipeline interaction along unstable slopes: a coupled three-dimensional approach. Part 1: Theoretical formulation. *Canadian Geotechnical Journal*, 46:1289–1304, 2009.

C. Cremer, A. Pecker, and L. Davenne. A model for landslide-pipe interaction analysis. *Journal of Earthquake Engineering*, 6:175–212, 2001.

C. Cremer, A. Pecker, and L. Davenne. Modelling of nonlinear dynamic behaviour of a shallow strip foundation with macro-element. *Int. J. for Num. and Anal. Meth. in Geomechanics*, 25:1257–1284, 2002.

P. A. Cundall and O. D. L. Strack. A discrete numerical model for granular assemblies. *Géotechnique*, 29:47–65, 1979.

C. di Prisco and M. Vecchiotti. A rheological model for the description of boulder impacts on granular strata. *Géotechnique*, 56(7):469–482, 2006.

C. di Prisco and M. Vecchiotti. Design charts for evaluating impact forces on dissipative granular soil cushions. *ASCE J. Geotech. and Geoenv. Eng.*, 2009. Reviewed: 18 September 2009, Manuscript Number: GTENG-484R1.

C. di Prisco, R. Nova, and J. Lanier. A mixed isotropic kinematic hardening constitutive law for sand. In D. Kolymbas, editor, *Modern approach to plasticity*, pages 83–124. Balkema, 1993.

C. di Prisco, B. Fornari, R. Nova, and S. Pedretti. A constitutive model for cyclically loaded shallow foundations. In *Proc. Of EUROMECH 385 Colloquium on Inelastic Analysis under Variable Loads: Theory and Engineering Applications*, pages 107–111, Achen, September 1998.

C. di Prisco, R. Nova, and A. Sibilia. Analysis of soil-structure interaction of towers under cyclic loading. In G. N. Pande & S. Pietruszczak, editor, *Proc. 8th European Conference on Numerical Methods in Geotechnical Engineering, NUMOG*. Balkema, Rome, Italy, 2002.

C. di Prisco, F. Montanelli, G. Caloni, and A. Savoldi. Shallow foundations on geo-reinforced sand layers: experimental results and theoretical observations. In Magnan & Droniuc, editor, *Proc. Int. Symp. On Shallow Foundations FONDSUP*, volume 1. Balkema, Paris, Novemeber 2003a.

C. di Prisco, R. Nova, F. Perotti, and A. Sibilia. Analysis of soil-foundation interaction of tower structures under cyclic loading. In M.Maugeri & R. Nova, editor, *Geotechnical analysis of seismic vulnerability of historical monuments*. Patron, Bologna, 2003b.

C. di Prisco, R. Nova, and A. Sibilia. Shallow footings under cyclic loading: experimental behaviour and constitutive modeling. In M.Maugeri & R. Nova, editor, *Geotechnical analysis of seismic vulnerability of historical monuments*. Patron, Bologna, 2003c.

C. di Prisco, R. Nova, and A. Corengia. A model for landslide-pipe interaction analysis. *Soils and Foundations*, 44:1–12, 2004.

C. di Prisco, A. Galli, and M. Vecchiotti. Comportamento meccanico di fondazioni superficiali sottoposte a carichi ciclici. In *Proc. V Convegno Nazionale dei Ricercatori di Ingegneria Geotecnica*, pages 135–147, Bari, September 2006.

FEAT. *Tochnog Professional Users Manual*. http://www.feat.nl/, 2004.

S. Gajan, B. Kutter, J. Phalen, T. C. Hutchinson, and G. R. Martin. Centrifuge modeling of load-deformation behaviour of rocking shallow foundations. *Soil Dynamics and Earthquake Engineering*, 25:773–783, 2005.

M. Georgiadis and R. Butterfield. Displacements of footings on sand under eccentric and inclined loads. *Canadian Geotechnical Journal*, 25:192–212, 1988.

G. Gottardi, G. Houlsby, and R. Butterfield. Plastic response of circular footings under general planar loading. *Géotechnique*, 49(4):453–469, 1999.

S. Grange, P. Kotronis, and J. Mazars. 3D macro-element for soil structure interaction. In *Proceedings 4th International Conference on Earthquake Geotechnical Engineering*, volume cd paper 1308, Thessaloniki, Greece, June 2007a.

S. Grange, P. Kotronis, and J. Mazars. Extensive validation of the ssi macro-element using experimental and numerical results. *European Contract LESSLOSS, Proj. No. GOCE-CT-2003-505488, Risk Mitigation for Earthquakes and Landslides, 6th Framework Prog.*, Deliverable [D8-2]: http://hal.archives–ouvertes.fr/hal–00138210, 2007b.

S. Grange, P. Kotronis, and J. Mazars. A macro-element for a circular foundation to simulate 3D soil-structure interaction. *Int. J. for Num. and Anal. Meth. in Geomechanics*, 32(10):1205–1227, 2008.

S. Grange, P. Kotronis, and J. Mazars. Numerical modelling of the seismic behaviour of a 7-story building. *Materials and structures*, 42(10):1433–1442, 2009.

M. S. Hodder and M. J. Cassidy. A plasticity model for predicting the vertical and lateral behaviour of pipelines in clay soils. *Géotechnique*, DOI: 10.1680/geot.8.P.055:Available online: 07 January 2010, 2010.

C. Martin. *Physical and numerical modelling of offshore foundations under combined loads.* PhD thesis, University of Oxford, Oxford, 1994.

R. Nova and L. Montrasio. Settlement of shallow foundations on sand. *Géotechnique*, 41(2):243–256, 1991.

R. Nova and L. Montrasio. Settlement of shallow foundations on sand: geometrical effects. *Géotechnique*, 47(1):46–60, 1997.

R. Nova, M. Parma, and R. Castellanza. Settlements of foundations on soft rocks. *Rivista Italiana di Geotecnica*, 42(2):10–22, 2008.

R. Paolucci. Simplified evaluation of earthquake induced permanent displacements of shallow foundations. *Journal of Earthquake Engineering*, 1:563–579, 1997.

R. Paolucci, C. di Prisco, M. Vecchiotti, M. Shirato, and M. Yilmaz. Seismic behyaviour of shallow foundations: large scale experiments vs. numerical modelling and implications for performance based design. In *Proc. 1st US-Italy Seismic Bridge Workshop, Eucentre*, pages 107–111, Pavia, Italy, 2007.

R. Paolucci, M. Shirato, and M. T. Yilmaz. Seismic behaviour of shallow foundations: shaking table experiments vs. numerical modelling. *Earthquake Engineering and Structural Dynamics*, 37:577–595, 2008.

Y. Le Pape and J. P. Sieffert. Application of thermodynamics to the global modelling of shallow foundations on frictional material. *Int. J. for Num. and Anal. Meth. in Geomechanics*, 25:1377–1408, 2001.

S. Pedretti. *Non-linear seismic soil-foundation interaction: analysis and modelling methods.* PhD thesis, Politecnico di Milano, Milan, Italy, 1998.

P. Perzyna. The constitutive equations for rate sensitive plastic materials. *The Quarterly of Applied Mathematics*, 20:321–332, 1963.

W. Prager. The theory of plasticity: a survey of recent achievements. In *Proc. Inst. Mech. Eng*, pages 41–57 (169), 1955.

PWRI. *Experimental study on the residual displacements of shallow foundations.* Public Work Research Institute, Technical Note, Tsukuba, Japan, 2005.

M. F. Randolph and G. T. Houlsby. The limiting pressure on a circular pile loaded laterally in cohesive soil. *Géotechnique*, 34(4):613–623, 1984.

J. Salençon and A. Pecker. Ultimate bearing capacity of shallow foundations under inclined and eccentric loads, part i: purely cohesive soil. *European J. of Mechanics, A/Solids*, 14(3):349–375, 1995a.

J. Salençon and A. Pecker. Ultimate bearing capacity of shallow foundations under inclined and eccentric loads, part ii: purely cohesive soil without tensile strenght. *European J. of Mechanics, A/Solids*, 14(3):377–396, 1995b.

M. Shirato, S. Nakatani, J. Fukui, and R. Paolucci. Large-scale model tests on shallow foundations subjected to earthquake loads. In *Proc. 2nd Japan-Greece Workshop on Seismic design, observation and retrofit of foundations*, Tokyo, April 2007.

M. Shirato, R. Paolucci, T. Kouno, S. Nakatani, J. Fukui, R. Nova, and C. di Prisco. Numerical simulation of model tests of pier-shallow foundation systems subjected to earthquake loads using an elasto-uplift-plastic macro element. *Soils and Foundations*, 48(5):693–711, 2008.

J. G. Sieffert and F. Cevaer. *Handbook of impedance functions-surface foundations.* Ouest Editions, Nantes, 1991.

X. Zeng and R. S. Steedman. Bearing capacity failure of shallow foundations in earthquakes. *Géotechnique*, 48:235–256, 1998.

# Macroelement Modelling

David Muir Wood *
* University of Bristol, United Kingdom

## 1 Macroelement Models

Numerical modelling using a full finite element or finite difference analysis may be ultimately necessary but may be a heavy-handed way of seeking insight into some aspects of a problem of geotechnical behaviour. Theoretical modelling may only be possible for rather restricted problems. Macroelement modelling may be a helpful intermediate way of introducing some realistic geotechnical nonlinearity in order, for example, to compare different constitutive possibilities or perhaps just to provide a rapid 'order-of-magnitude' estimate of response against which the results of more extensive numerical modelling—or physical modelling—can be compared. Equally, physical or numerical modelling may itself provide clues concerning mechanisms of system response which may suggest ways in which simple macroelement models might be devised. It will be seen that this has indeed been the route for the development of some of the macroelement models outlined here. We will describe three examples of analysis of soil-structure interaction using a macroelement approach. Two of these are closely related - the lateral deflection of a pile or pipeline or tunnel as the surrounding ground moves. The third is a dynamic analysis of a gravity retaining wall.

## 2 Soil-Structure Interaction

Soil-structure interaction is one of those interface topics which requires a holistic approach to the analysis or modelling. It is not possible to consider the structural and geotechnical elements of the problem separately because the system response will certainly depend on some combination of properties of both the soil and the structure. If the ground and the structure are both behaving elastically then simple configurations lead to exact analyses and results can be presented in terms of dimensionless groups which simply describe relative stiffnesses of ground and structure. For example, the response of a laterally loaded pile is described by Muir Wood (2004). The resistance of the soil as the pile moves is described using a subgrade reaction model with the relative displacement of pile and ground generating resisting stress through a series of linear Winkler springs, with no allowance for the continuous nature of the enveloping ground.

In reality, however, although the structural element may behave elastically, it is extremely unlikely that the behaviour of the soil will be elastic except for extremely small deformations. So we would like to be able to extend our analysis to include non-linearity of the soil-structure interaction while still—if at all possible—retaining the possibility of displaying the results as economically as possible, making use of appropriate dimensionless groups.

The situation is often worse than this implies because we may not actually know *anything* about the nonlinear stiffness of the soil, and yet this will be crucial in deriving dimensionless groups which encapsulate sufficient detail of the parametric description of the problem. In that case we have to make informed guesses. Our expectation is that there may well be both a stiffness element to the ground-soil interaction but also a limiting interactive load. To propose a simplification of this interaction relationship is not particularly original but what is explored here is a way in which even nonlinear interaction can be capable of analytical solution and presentation in terms of nondimensional groups, thus throwing light both on the parameters which are important, and on the character of the system response.

Two related problems will be presented: the behaviour of a flexible pile in translating ground; and the effect of a fault displacement on a buried pipeline or tunnel. The geometries considered are inevitably simplistic. (A more extensive discussion of accessible soil-structure interaction analyses is given by Muir Wood (2004).)

## 3 Pile in Displacing Ground

Situations in which both the structural and the soil materials can be described as linear elastic lend themselves to closed-form analysis. However, the behaviour of soils is at best linear elastic only to very small shear strains and the subsequent nonlinearity will certainly have an influence on the nature of the soil-structure interaction. In this section we will investigate some aspects of the structural consequences of slow movement of ground past a pile. We will model the interaction of the pile with the soil by means of a series of *nonlinear* springs—and explore the effects of this more realistic description.

The stimulus for performing this analysis was provided by a prototype problem in which a landfill through which a piled structure had been con-

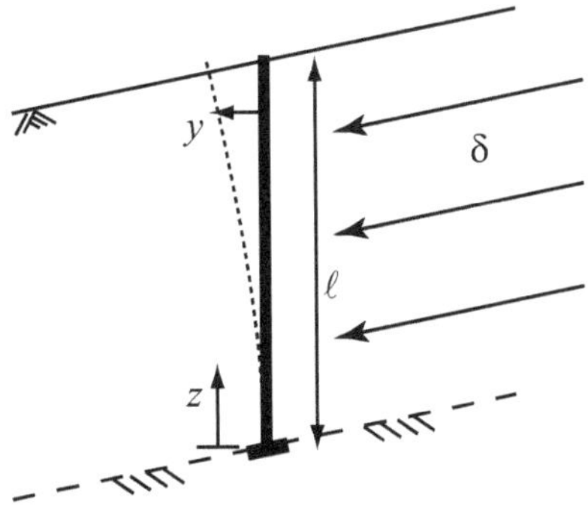


**Figure 1.** Pile loaded by translating ground.

structed was known to be sliding slowly but inexorably down the shallow slope of its underlying rock surface. There are two major unknowns: the profile of lateral displacement within the fill is not known; and the nature of the (nonlinear) fill-structure interaction is not known. It is nevertheless possible to introduce some fairly rational descriptions of these two aspects of the interaction and then perform parametric studies. To put it further in context, the structural interest was to understand the extent to which knowledge of the lateral displacement of the piles at the 'ground' surface could be used to characterise the maximum moment generated in the piles.

The analysis of active or passive lateral loading of piles using a set of subgrade reaction springs is not novel (Reese and Matlock, 1956; Poulos, 1973) but such analyses have usually assumed that the soil is elastic, or piecewise linear (though Poulos introduces the possibility of a limiting soil resistance). The problem to be analysed here is approached in a different way.

The piles are of length $\ell$, and flexural rigidity $EI$, and are fixed at their base, so that both rotation and displacement are prevented (Fig. 1). A dimensionless coordinate $\eta = z/\ell$ defines position on the pile with $\eta = 0$ at the base and $\eta = 1$ at the ground surface. The pile has no restraint at the top, $\eta = 1$ and, for the purposes of this analysis, it is assumed that the only lateral loading is provided by the moving fill.

We do not know the detail of the profile of ground movement so we give it a general profile:

$$\delta = \delta_o \eta^\alpha \tag{1}$$

where $\delta_o$ is the movement at the ground surface, at $\eta = 1$, and $\alpha$ is a parameter which characterises the profile of movement (Fig. 2). A value of $\alpha = 1$ implies linear variation of movement with depth; $\alpha > 1$ implies

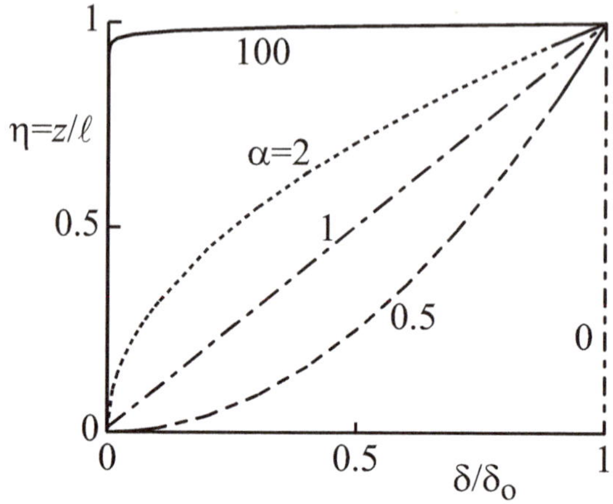


**Figure 2.** Profiles of ground displacement.

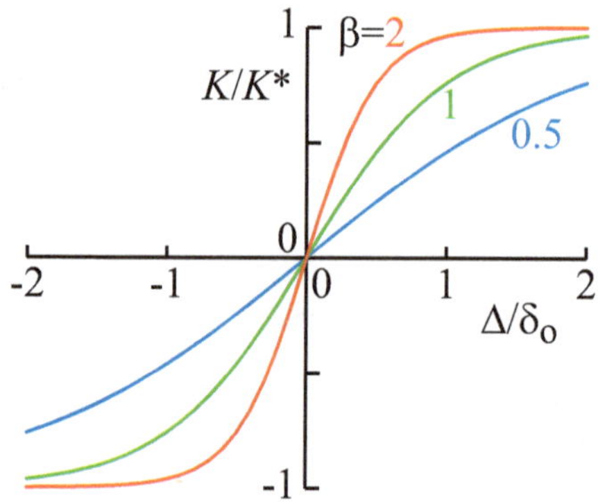


**Figure 3.** Development of pressure on pile as result of relative movement between pile and ground.

that the movement is more concentrated towards the surface (in principle $\alpha = \infty$ implies that movement is everywhere zero except for $\eta = 1$); $\alpha < 1$ implies that the movement is more concentrated towards the base of the ground ($\alpha = 0$ implies that the ground is moving as a block with $\delta = \delta_o$ at all depths). It is assumed that the presence of the individual piles does not influence the 'free-field' flow of the ground.

The detail of load generation through relative movement of ground and structure is not known—and the fact that the 'ground' in this case is actually landfill does not help. We can make some reasonable propositions. The ground-structure interaction is likely to be symmetric and dependent only on the *magnitude* of the relative movement, the difference between pile displacement $y$ and ground displacement $\delta$. As relative movement increases it is expected that the load generated will saturate to some limiting value characterised by a strength property of the ground. It is assumed that this strength varies linearly with depth. There will be some description of the stiffness of the interaction at small relative displacements but the tangential, incremental stiffness will fall as the limiting strength is approached. An appropriate expression for the lateral pressure on the pile, as a function of relative displacement $\Delta = \delta - y$ could then be (Fig. 3):

$$\frac{K}{K^*} = \tanh\left[\lambda\left(\frac{\Delta}{B}\right)\right] = \tanh\left[\beta\left(\eta^{\alpha} - \zeta\right)\right] \tag{2}$$

where $\zeta = y/\delta_o$ and $\beta = \lambda\delta_o/B$. The first part of this equation indicates that the pressure on the pile depends on the relative movement $\Delta$ of pile and soil (which can be positive or negative) normalised by a typical pile dimension $B$ related to pile diameter or pile width. The 'stiffness' of the pile:ground relationship is controlled by $\lambda$. The mobilised lateral pressure coefficient $K$ is linked through the nonlinear stiffness with a limiting asymptotic lateral pressure coefficient $K^*$ reached as the relative movement increases. Thus $\lambda$ and $K^*$ are subgrade reaction parameters.

The second part of the equation normalises the relative displacement of pile and ground with the displacement $\delta_o$ of the ground at the surface, as in (1). That then leads to the dimensionless group $\beta = \lambda\delta_o/B$ which combines information about the stiffness of the pile-ground response, $\lambda$, with a ratio of length parameters: ground movement at the free surface, $\delta_o$ and typical pile dimension, $B$.

The earth pressure coefficient $K$ is applied to the vertical stress $\gamma(\ell - z)$ at any depth down the pile in order to calculate the horizontal stress on the pile generated by the relative movement between pile and soil.

$$\frac{\sigma_h}{\gamma\ell} = K(1 - \eta) \tag{3}$$

where $\gamma$ is an appropriate unit weight for the soil.

For a material with frictional strength the limiting earth pressure coefficient $K^*$ would be linked with the passive pressure coefficient $K_p$ for the soil. In a cohesive material the undrained strength, and hence the limiting pressure, might be assumed to be constant over the length of the pile: this is the assumption made in the next section for a different structural configuration. The tanh function is symmetric and will behave identically whatever the sign of the relative movement between ground and pile. The initial stiffness of the pile-ground interaction can be presented as a coefficient of subgrade reaction (stress per unit relative displacement of pile and ground):

$$k_h = \frac{\mathrm{d}\sigma_h}{\mathrm{d}\Delta} = \lambda K^* \gamma \frac{\ell}{B}(1-\eta) \tag{4}$$

with the value $k_{ho} = \lambda K^* \gamma \ell / B$ at the base of the pile ($\eta = 0$). The relative movement required to move halfway to the limiting value (corresponding to $K/K^* = 1/2$) is $\Delta_{50} = (B \ln 3)/(2\lambda) = (\delta_o \ln 3)/(2\beta)$ or

$$\beta = \frac{\ln 3}{2} \frac{\delta_o}{\Delta_{50}} \tag{5}$$

which provides a potential link between $\beta$ and other parameters describing the problem.

We assume that the lateral stress $\sigma_h$ acts over a width $B$ related to the diameter of the pile. Deformation of the pile is then governed by the dimensionless equation

$$\frac{\mathrm{d}^4\zeta}{\mathrm{d}\eta^4} = \frac{1-\eta}{\chi} \frac{K}{K^*} = \frac{1-\eta}{\chi} \tanh\left[\beta\left(\eta^\alpha - \zeta\right)\right] \tag{6}$$

with

$$\chi = \frac{EI\delta_o}{\ell^2} \frac{1}{K^* \gamma \ell^3 B} \tag{7}$$

The problem is thus controlled by three parameters: $\alpha$ describes the profile of ground movement; $\beta$ controls the initial stiffness of interaction; and $\chi$ is the ratio of two moments, one a structural property and the other a loading characteristic. For a cantilever of length $\ell$ and flexural rigidity $EI$ subjected to a tip load which produces a tip displacement $\delta_o$, and subjected to no other loading, the root moment is $M_r = 3EI\delta_o/\ell^2$. For a cantilever of length $\ell$ subjected to lateral pressures over a width $d$ given by the limiting value of the lateral stress coefficient $K = K^*$ over the entire length, the root moment is $M_f = K^*\gamma\ell^3 B/6$. Thus $\chi = M_r/18M_f$.

The boundary conditions for the cantilever pile shown in Fig. 1 are zero deflection and slope at the base of the pile (assuming complete fixity at the base): $\zeta_{\eta=0} = [\mathrm{d}\zeta/\mathrm{d}\eta]_{\eta=0} = 0$; and zero moment and shear force at the top of the pile $[\mathrm{d}^2\zeta/\mathrm{d}\eta^2]_{\eta=1} = [\mathrm{d}^3\zeta/\mathrm{d}\eta^3]_{\eta=1} = 0$.

The deformation equation (6) is solved to give a profile of normalised displacement $\zeta$ with normalised depth $\eta$. The bending moment $M$ within the pile is given by:

$$M = \frac{EI\delta_o}{\ell^2} \frac{\mathrm{d}^2\zeta}{\mathrm{d}\eta^2} \tag{8}$$

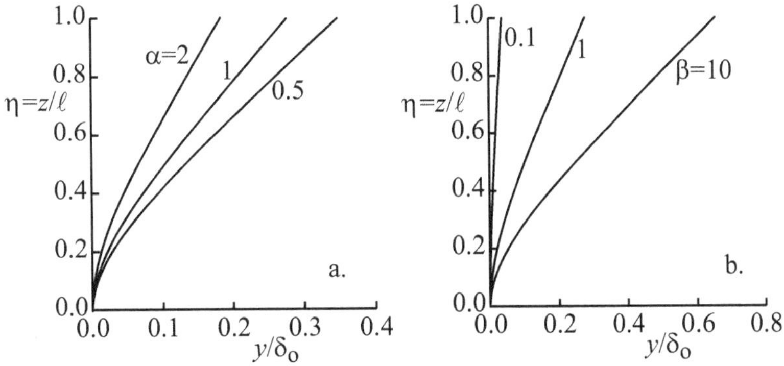


**Figure 4.** Deflected shape of pile for (a) different values of $\alpha$ ($\beta = 1$, $\chi = 0.05$); (b) different values of $\beta$ ($\alpha = 1$, $\chi = 0.05$).

and this points the way to normalisation of the bending moments in terms of relevant material and system parameters. The displacement of the tip of the pile is an output quantity, $y_{\max} = \zeta_{\eta=1}\delta_o$, which will in general be different from the ground displacement $\delta_o$. A cantilever in air whose tip is moved transversely by a distance $y_{\max}$ develops a root moment $M_{air} = 3EIy_{\max}/\ell^2$ so that the dimensionless group $\mu_1$

$$\mu_1 = \frac{M}{M_{air}} = \frac{EI\delta_o\ell^2}{3EIy_{\max}\ell^2}\frac{\mathrm{d}^2\zeta}{\mathrm{d}\eta^2} = \frac{1}{3\zeta_{\eta=1}}\frac{\mathrm{d}^2\zeta}{\mathrm{d}\eta^2} \tag{9}$$

allows us to compare the moments in a pile whose lateral movement is brought about by the translation of the ground with the maximum (root) moment in a pile in air given the same movement at the top, and thus allows us to understand the extent to which interpretation of maximum moment from the observed displacement of the piles at the ground surface is or is not unconservative. On the other hand the dimensionless group $\mu_2$:

$$\mu_2 = \frac{M}{M_f} = \frac{6EI\delta_o}{K^*\gamma\ell^5B}\frac{\mathrm{d}^2\zeta}{\mathrm{d}\eta^2} = 6\chi\frac{\mathrm{d}^2\zeta}{\mathrm{d}\eta^2} \tag{10}$$

normalises the moment with the maximum moment that can be generated when the soil is slipping past the pile and fully mobilising the resistance coefficient $K^*$ over the full length of the pile and thus indicates the progress of moment mobilisation.

The two parameters, $\beta$ and $\chi$, both involve the surface movement, $\delta_o$, of the ground: parameter $\beta$ describes the stiffness of the pile-ground interac-

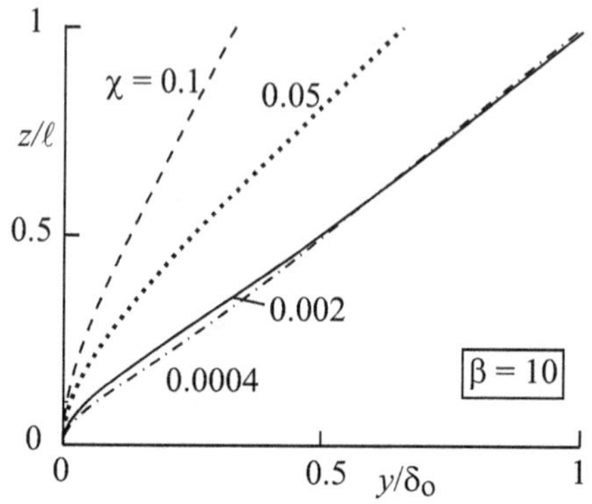


**Figure 5.** Deflected shape of pile for different values of $\chi$ ($\alpha = 1$, $\beta = 10$).

tion and $\chi$ introduces the pile flexibility. For a given pile the ratio:

$$\frac{\beta}{\chi} = \frac{\ln 3}{2} \frac{\delta_o}{\Delta_{50}} \frac{\ell^2}{EI\delta_o} \left(K^* \gamma \ell^3 B\right) = \frac{\ln 3}{2} \frac{K^* \gamma \ell^5}{EI} \frac{B}{\Delta_{50}} \tag{11}$$

forms a composite parameter which introduces pile flexibility, limiting earth pressure ratio, and the stiffness of pile-ground interaction. We can investigate the behaviour of the pile-ground system by varying $\beta$ and $\chi$ both separately and together. It is therefore helpful to have some baseline values for the parametric study.

The definition of $\beta$ introduces both the relative movement $\Delta_{50}$ required to generate half the limiting load (a key element of the ground-pile interaction), and the magnitude of the ground surface movement $\delta_o$ (which is completely independent of the pile response). Perhaps $\Delta_{50}$ might be about half the pile diameter—we could in principle perform tests to discover its value. The ground surface movement is quite unknown but as an order of magnitude one might suppose it to be of the same order as the pile diameter. Then, for pile diameter $B = 0.5$ m ($\approx 2\Delta_{50}$), $\beta = 1.1$. A reference value $\beta = 1.0$ thus seems reasonable. Stiffer pile-ground interaction (lower $\Delta_{50}$, more rapid attainment of the limiting pressure) will imply increased $\beta$. An earlier stage in the development of ground movement after installation of the piles will imply lower $\delta_o$ and hence a decreased value of $\beta$.

A reference value of $\chi$ can be estimated in the same way. We have already assumed an order of magnitude of ground movement $\delta_o = 0.5$ m. We will suppose that the (solid) piles are of diameter: $B = 0.5$ m (implying $I = \pi B^4/64 \approx 0.0031$ m$^4$); and that the Young's modulus for concrete: $E =$ 30 GPa. We also need to choose a pile length: $\ell = 10$ m. The unit weight

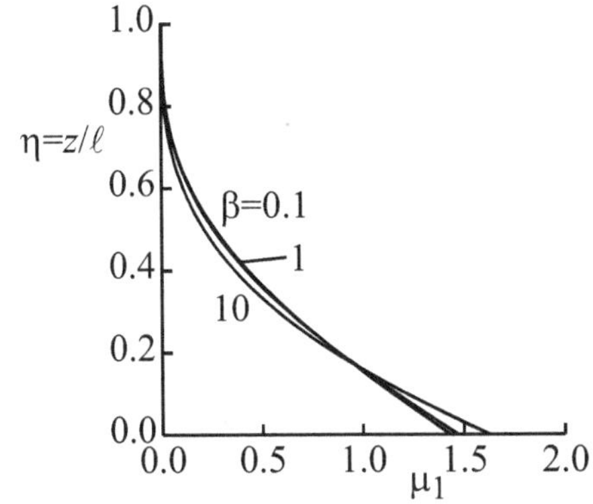


**Figure 6.** Normalised moments in pile for different values of $\beta$ ($\alpha = 1$, $\chi = 0.05$).

of the ground, which drives the development of lateral stress on the pile is: $\gamma = 10$ kN/m$^3$; and we might take the limiting lateral pressure coefficient: $K^* = 2$ (corresponding to the Rankine passive pressure ratio $K_p$ for angle of friction $\phi' \approx 20°$). Then $\chi = 0.0464$. A reference value $\chi = 0.05$ has been used for parametric studies and the effect of increasing and decreasing this has been explored over the range 0.0004 to 0.1. The value of $\chi$ is very sensitive to the details of the pile geometry: $\chi \propto B^3(\propto I/B)$ and $\ell^{-5}$ (see (7)).

The governing equation (6) is extremely nonlinear but is capable of numerical solution using standard routines. A short program has been written in MATLAB to apply the solver bvp4c to the fourth order governing differential equation. Results are shown in plots of dimensionless pile displacement $\zeta = y/\delta_o$ (Figs. 4 and 5) and of dimensionless moment $\mu_1$ (Figs. 6, 7) with dimensionless position on the pile $\eta = z/\ell$.

The effect of varying $\alpha$ is as expected (Fig. 4a): the greater the average movement of the ground (the lower the value of $\alpha$) the greater the load on the pile and the greater the pile movement and root moment. The value of $\alpha$ is not something over which an engineer has much control, but any tendency for mass movement of the ground to occur will certainly be very damaging for any structure getting in the way of the motion. A baseline value of $\alpha = 1$, corresponding to linear variation of ground translation with depth, has been used for most analyses.

The pile displacements and moments are very sensitive to the stiffness of the interaction between the pile and the ground: the higher the value of $\beta$

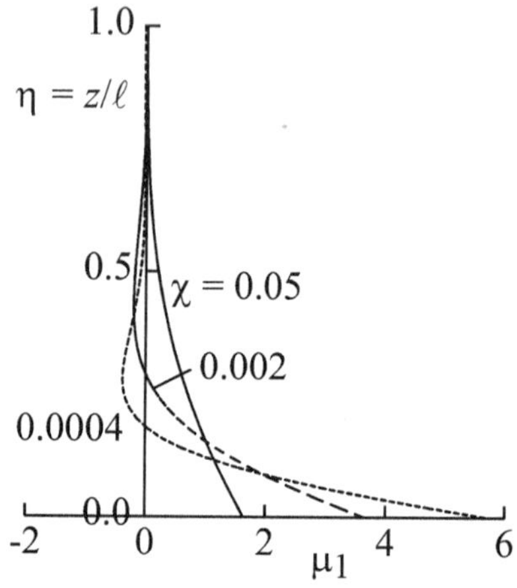


**Figure 7.** Normalised moments in pile for different values of $\chi$ ($\alpha = 1$, $\beta = 10$).

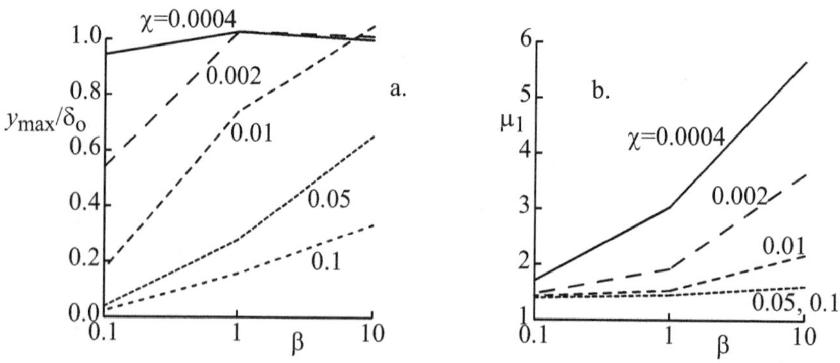


**Figure 8.** Influence of $\beta$ and $\chi$ on (a) maximum deflection and (b) maximum moment developed in pile.

the larger the displacements and moments (Figs. 4b and 6). However, there is interaction between the effects of changing $\beta$ and $\chi$, which is dependent on the pile stiffness. With a lower value of $\chi = 0.002$, the tip deflection of the pile is hardly affected by the value of $\beta$ for $\beta > 1.0$ although the greater curvature at the toe of the pile leads to much higher moments (Fig. 5).

Increase of pile stiffness through $\chi$ has the expected effect of reducing pile

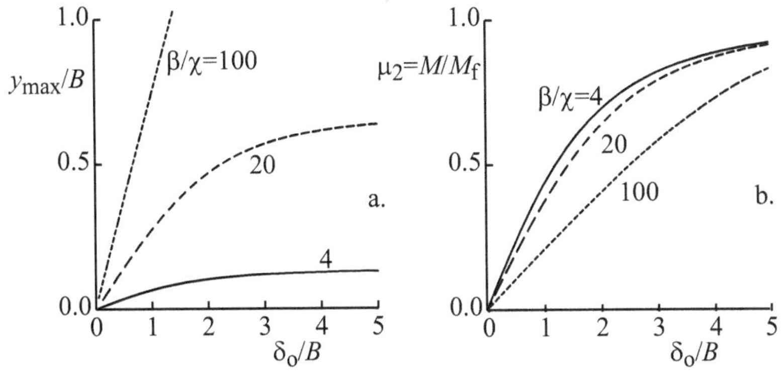


**Figure 9.** Development of (a) pile top movement and (b) pile toe moment with ground movement.

deflection (Fig. 5). However, reducing $\chi$ below 0.002 has little additional effect on the pile displacements: the pile is sufficiently flexible that the displacement at the ground surface matches the displacement of the ground itself. The profile of dimensionless moment, $\mu_1$, remains unchanged for values of pile stiffness $\chi > 0.05$.

The interaction of values of $\beta$ and $\chi$ in influencing the tip movement and root moment is shown in Fig. 8. It is clear that for a very flexible pile ($\chi = 0.0004$) the tip movement is more or less equal to the ground movement for all values of $\beta$ (Fig. 5), whereas for less flexible piles the proportion of ground movement at the tip increases with $\beta$ (or, which is equivalent, with $\delta_o$). Similarly, the scaled moment, $\mu_1$, is independent of $\beta$ for higher values of pile stiffness (the stiffer pile forces the ground to reach its limiting pressure as it flows past the pile). For the most flexible pile considered the maximum moment is nearly 6 times the free air value; for the stiffer piles it is still more than 50% higher than the value obtained for a cantilever displaced in free air. Thus one of the questions posed is clearly answered: soil-structure interaction leads to higher moments at depth (out of sight) than would be estimated from observing the ground surface displacement of the pile and treating it as an object in air. The result is of course obvious and intuitive but requires acknowledgement of the interaction between soil and structural properties. It can be concluded that observation of pile tip movement at the ground surface gives a very poor indication of the magnitude of moments in the pile.

The process of gradual mobilisation of ground-pile interaction for a given

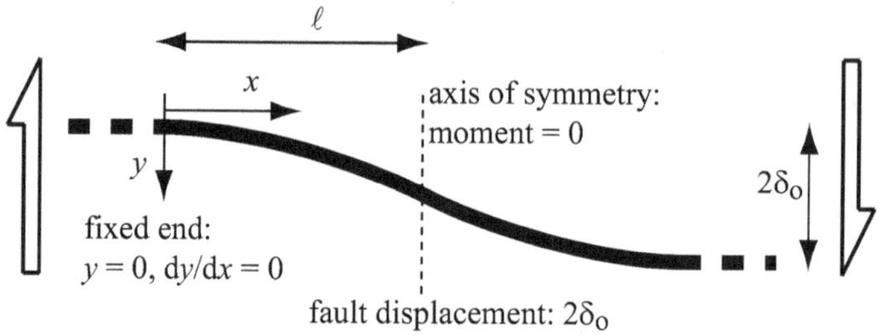


**Figure 10.** Pipeline crossing displacing fault.

pile can be followed by varying *both* $\beta$ and $\chi$ in appropriate constant proportion. and for the typical values that we have suggested here, $\beta/\chi \approx 20$. We can follow the gradual mobilisation of pile moment by looking at the variation of the normalised moment $\mu_2$ (10) with increasing normalised ground displacement $\delta_o/B \approx \beta$ (Fig. 9). As ground displacement builds up this normalised moment approaches 1: the earth pressure coefficient between the soil and the pile is close to $K^*$ over the whole length of the pile. A lower ratio $\beta/\chi$ indicates a stiffer pile: for stiffer piles the limiting moment is reached more rapidly but the displacement is of course smaller (Fig. 9).

The analysis of the displacements and moments generated in a pile by translating ground has been shown to be dependent on three dimensionless parameters giving results which can be applied generally. Defining the controlling parameters does not mean that they are all under our control. The profile of ground movement may be discoverable with appropriate instrumentation but will not be amenable to external restriction. The stiffness of the pile as a structural element is known but the detail of the interaction between the pile and ground, especially in its initial *stiffness* will be somewhat unknown and not easy to determine. The key dimensionless parameters combine structure and soil characteristics. Nevertheless, by invoking rather simple functions to describe the nature of the interaction, even these unknown quantities can be incorporated.

## 4 Pipeline Crossed by Fault

The second problem to be investigated is actually a simpler version of the previous one. We imagine a long buried flexible structure such as a pipeline or tunnel which is crossed by a fault. The fault displaces by a distance $2\delta_o$ and we want to discover the structural consequences for the pipeline. The problem is illustrated in Fig. 10. Symmetry dictates that, at the centre of the pipeline, the pipeline displacement is equal to half the overall ground displacement, and that the point of inflexion in the displacement profile along the structural element implies that the moment is zero. We might suppose that the interaction between the pipeline and the ground will be independent of position along the pipeline. This simplifies somewhat the formulation of the problem and modifies the boundary conditions. However, the complicating factor is the uncertainty concerning the length $\ell$ of the pipeline which needs to be analysed. The pile in the previous section was embedded at its lower end into a competent rock layer, and its length was thus clearly defined. The pipeline is assumed to extend to infinity in both directions. Assuming fixity at the end $x = 0$ is equivalent to proposing that the length $\ell$ is sufficiently large that there is no influence of the fault displacement at this position. For a laterally loaded pile in elastic soil, with coefficient of subgrade raction $k$, it can be shown (Muir Wood, 2004) that this might imply a length $kB\ell^4/4EI > 6^4 = 1296$, defined in terms of a dimensionless ratio of structural stiffness and stiffness of ground-structure interaction. We will expect to discover some equivalent controlling grouping for the nonlinear problem.

The pipeline is of length $\ell$, and flexural rigidity $EI$, and is fixed at $x = 0$, so that both rotation and displacement are prevented (Fig. 10). A dimensionless coordinate $\eta = x/\ell$ defines position, with $\eta = 0$ at the fixed end and $\eta = 1$ at the point of intersection with the fault. At $\eta = 1$ the moment is zero but there is a shear force in the pipeline. The only lateral loading is provided by the moving ground. The overall fault displacement is $2\delta_o$. The displacement $y$ of the pipeline is normalised with this displacement, $\zeta = y/\delta_o$ with the requirement that at $\zeta_{\eta=1} = 1$.

We assume again a tanh function (Fig. 3) to describe the load generation through relative movement of ground and structural element in terms of mobilisation of earth pressure coefficient $K$ with relative movement $\Delta$ normalised with some effective section dimension $B$ of the structural element.

$$\frac{K}{K^*} = \tanh\left[\lambda\left(\frac{\Delta}{B}\right)\right] = \tanh\left[-\beta\zeta\right] \tag{12}$$

As before $\beta = \lambda\delta_o/B = (\ln 3/2)\,(\delta_o/\Delta_{50})$ and $\lambda$ and $K^*$ are subgrade reaction parameters. The parameter $K^*$ describes the asymptotic earth pressure coefficient reached as the relative movement increases; and $K$ describes the presently mobilised earth pressure coefficient so that the pressure exerted on the structural element is:

$$\sigma_h = K\gamma D \tag{13}$$

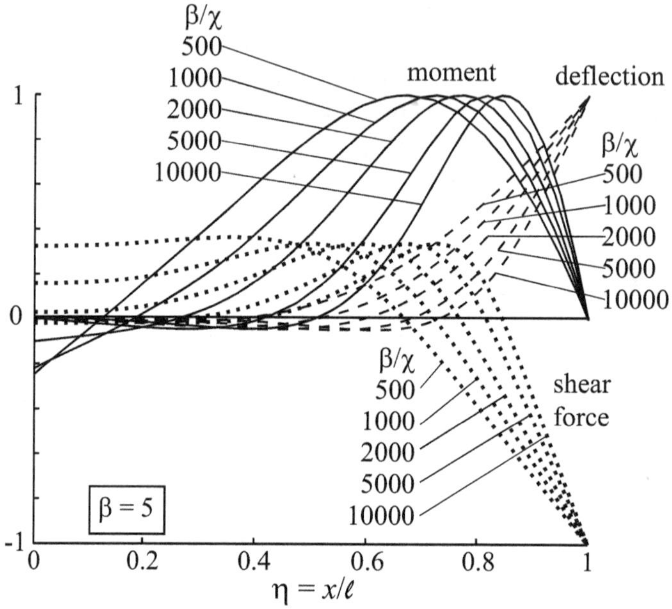


**Figure 11.** Deflection, shear force and bending moment normalised with maximum values for $\beta = 5$ or $2\delta_o/\Delta_{50} = 20/\ln 3$.

where $\gamma$ is an appropriate unit weight for the soil and $D$ is the depth of the pipeline. The limiting value $K^*$ is a passive pressure coefficient. For flow of cohesive soil with undrained strength $c_u$ around a cylindrical pile Randolph and Houlsby (1984) suggest that $\sigma_h/c_u \approx 10$. This would be a useful starting point in estimation of $K^*$ for the pipeline.

Deformation of the pipeline is governed by the dimensionless equation

$$\frac{\mathrm{d}^4\zeta}{\mathrm{d}\eta^4} = \frac{1}{\chi}\frac{K}{K^*} = \frac{1}{\chi}\tanh\left[-\beta\zeta\right] \tag{14}$$

with

$$\chi = \frac{EI\delta_o}{\ell^2}\frac{1}{K^*\gamma\ell^2 DB} \tag{15}$$

The problem is thus controlled by two parameters: $\beta$ controls the initial stiffness of ground-structure interaction; and, as before, $\chi$ is the ratio of two moments, one a structural property and the other a loading characteristic. For a cantilever of length $\ell$ and flexural rigidity $EI$ subjected to a tip displacement $\delta_o$ with no other loading, the root moment is $M_r = 3EI\delta_o/\ell^2$.

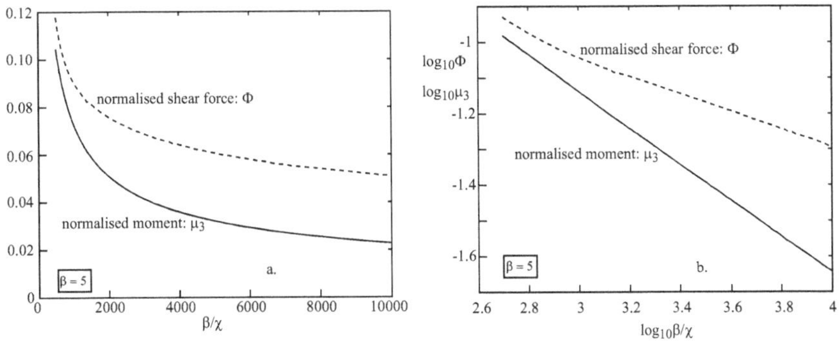


**Figure 12.** Maximum normalised shear force $\Phi$ (17) and bending moment $\mu_3$ (16) as function of $\beta/\chi$ for $\beta = 5$: plotted with (a) linear scales; (b) log:log scales.

For a cantilever of length $\ell$ subjected to lateral pressures over a width $B$ given by the limiting value of the lateral stress coefficient $K = K^*$ over the entire length, the root moment is $M_f = K^*\gamma\ell^2 DB/2$. Thus $\chi = M_r/6M_f$. (The definition of $\chi$ for the pipeline is slightly different from the definition of $\chi$ for the pile (7) because the pile is aligned vertically, with gravitationally varying stress level along its length, and the pipeline is oriented horizontally, and the loading generated between the ground and the structure is independent of position along the structure.)

The boundary conditions for the analysis (Fig. 10) are zero deflection and slope at $x = 0$: $\zeta = \mathrm{d}\zeta/\mathrm{d}\eta = 0$ for $\eta = 0$; unit normalised displacement and zero moment at $x = \ell$: $\zeta = 1$ and $\mathrm{d}^2\zeta/\mathrm{d}\eta^2 = 0$ for $\eta = 1$.

Once the deformation equation (14) has been solved to give a profile of normalised displacement $\zeta$ with normalised position $\eta$, the shear force and moment can be found by differentiation. The moment $M$ is conveniently normalised with $M_f$:

$$\mu_3 = \frac{M}{M_f} = 2\chi\frac{\mathrm{d}^2\zeta}{\mathrm{d}\eta^2} \tag{16}$$

and, similarly, the shear force $F$ is conveniently normalised with the loading $F_f$ equivalent to $M_f$: the limiting earth pressure coefficient $K^*$ mobilised along the entire length $\ell$: $F_f = K^*\gamma DB\ell$:

$$\Phi = \frac{F}{F_f} = \chi\frac{\mathrm{d}^2\zeta}{\mathrm{d}\eta^2} \tag{17}$$

Results are presented in Figs. 11-14. The general form of the variation of displacement, moment and shear force along the structural element is shown in Fig. 11. As before, the two controlling variables $\beta$ and $\chi$ both incorporate the fault displacement $\delta_o$. The ratio $\beta/\chi$ is characteristic of a specific configuration: the higher the value of $\beta/\chi$ the more *relatively* flexible is the pipeline or tunnel. The maximum values of dimensionless moment $\mu_3$ and dimensionless shear force $\Phi$ are shown in Fig. 12 as a function of $\beta/\chi$. The variation of these maximum values is shown both in linear and log:log plots: the linear relationship for the moment in Fig. 12b indicates an inverse power law relationship: $\mu_3 = (\beta/\chi)^\kappa$ where $\kappa \approx -0.5$. The shear force is not quite so simply related.

The transition from 'finite' length response to 'infinite' length response as the ratio $\beta/\chi$ increases from 500 to 10000 is shown in Fig. 11. For a fixed value of $\beta = 5$ the distributions of pipeline deflection, and dimensionless shear force $\Phi$ and dimensionless bending moment $\mu_3$ are shown as a function of position along the structure. All three quantities are normalised by their *maximum* values so that it is only the pattern that is being revealed. (The maximum values of shear force and moment are shown in Fig. 12; the maximum value of the normalised deflection comes at the point of inflection, axis of symmetry of the structure (Fig. 10) at $\eta = 1$ and is, by definition, unity. The maximum value of shear force also occurs at $\eta = 1$ where the fault crosses the pipeline.) For each group of curves the effect of the fault displacement moves further down the pipeline towards $\eta = x/\ell = 0$ as $\beta/\chi$ decreases. The peak moment moves from about $\eta = 0.8$ to $\eta = 0.6$ as $\beta/\chi$ falls from 10000 to 500. More to the point, as $\beta/\chi$ decreases, the moment and shear force at the end of the pipeline $\eta = 0$ become significantly non-zero (Fig. 13): only for values of $\beta/\chi > 5000$ are the values of moment and shear force at $\eta = x/\ell = 0$ becoming negligible. The analysis is thus only valid in this range unless there is some actual structural fixity at some known point along the pipeline.

For a laterally loaded pile in elastic ground the limit of 'infinite' pile response comes for values of the parametric group $kB\ell^4/4EI > 6^4 = 1296$. For the present analysis we have proposed that $\beta/\chi > 5000$ or

$$\frac{\beta}{\chi} = \frac{\ln 3}{2} \frac{K^* \gamma D B}{\Delta_{50}} \frac{\ell^4}{EI} > 5000 \tag{18}$$

$\Delta_{50}$ is the relative displacement required to bring the interaction pressure to half the limiting value, so that an equivalence can be drawn between the pile-soil subgrade reaction coefficient $k$ for the elastic analysis and $K^* \gamma D/2\Delta_{50}$ in the present nonlinear case. The two criteria for defining what constitutes 'infinite' length thus agree very closely ($4 \times 1296 \approx 5000$).

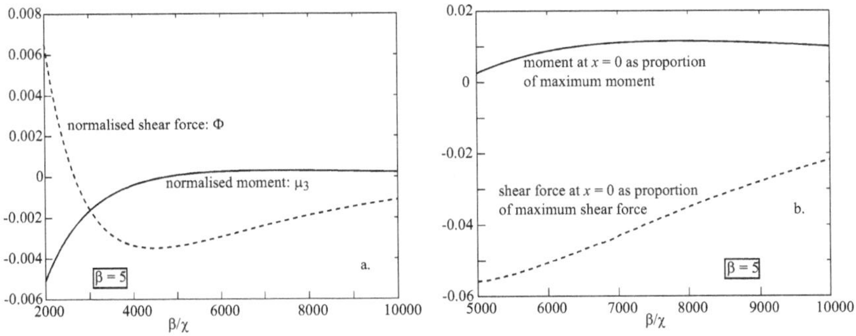


**Figure 13.** Residual shear force and moment at $\eta = x = 0$ as function of $\beta/\chi$ for $\beta = 5$: (a) normalised shear force $\Phi$ (17) and bending moment $\mu_3$ (16); (b) residual values as proportion of maximum values.

For acceptably large values of $\beta/\chi$, which means acceptably large values of the length $\ell$ for analysis, the mobilisation of shear force at the intersection of pipeline or tunnel with the displacing fault and the mobilisation of the maximum moment in the structure (Fig. 11) are shown in Fig. 14 as a function of normalised fault displacement $2\delta_o/\Delta_{50}$. For each structural resultant a group of curves is shown for values of $\beta/\chi$=20000, 15000, 10000 and 5000. As the value of $\beta/\chi$ decreases (the relative stiffness of the structural element increases) the values of the normalised moment and shear force increase for a given fault displacement. (Recall that the pipeline is analysed as a cantilever built in at $x = 0$: an equivalent analysis could be performed with the pipeline completely free at $x = 0$ analogous to a laterally loaded pile of finite length.)

## 5 Macroelement for Study of Dynamic Response of Gravity Retaining Walls

Macroelement modelling can be a helpful intermediate way of introducing some realistic geotechnical nonlinearity in order, for example, to compare different constitutive possibilities or perhaps just to provide a rapid 'order-of-magnitude' estimate of response against which the results of more extensive numerical modelling - or physical modelling - can be compared. Such speedy modelling can be particularly beneficial where the concern is to study the dynamic response of a geotechnical system and, in particular, to study the way in which that response is influenced by the nature of the dynamic

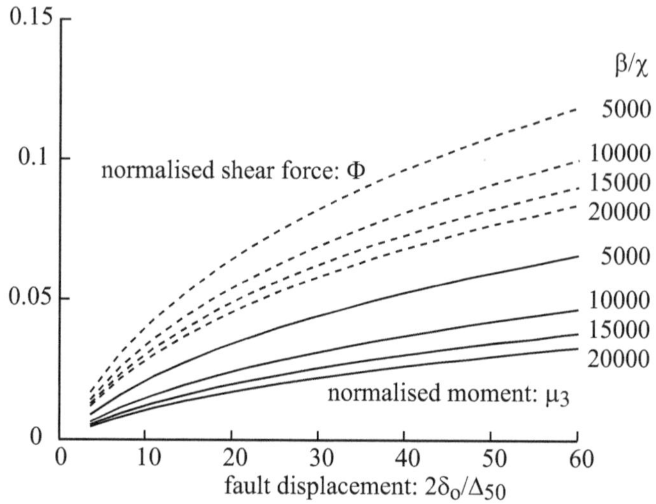


**Figure 14.** Development of shear force and moment as fault displaces: maximum normalised moment $\mu_3 = 2M/K^*\gamma\ell^2 DB$ (16); maximum normalised shear force $\Phi = F/K^*\gamma DB\ell$ (17).

input motion. This has been the motivation behind the development of the macroelement model for a gravity retaining wall that is described here.

The problem that is being tackled is illustrated in Fig. 15: we are concerned to develop a simple model for a gravity retaining wall under dynamic (seismic) loading. Separate macroelements have been developed for the interaction of the wall and its foundation and for the interaction of the wall with the retained soil. Fig. 16 defines the sign convention for the forces acting on the back of the wall, from the retained soil, and on the base of the wall, from the foundation. The positive directions of movement of the wall and the underlying ground are shown together with the positive direction of rotation of the wall. At this stage of the modelling we are concerned only with horizontal motion and rotation of the wall and will ignore vertical movements, although the macroelement footing models (see, for example, Nova and Montrasio (1991); Paolucci (1997); Gottardi et al. (1999); Martin and Houlsby (2001); Cremer et al. (2001)tell us that any sliding or rotation of the foundation will in general be associated with settlement or heave of the foundation. We will then only consider horizontal shaking of the underlying ground. We will also neglect the potential loss of contact over part of the foundation when rocking of the wall occurs. This would obviously be a

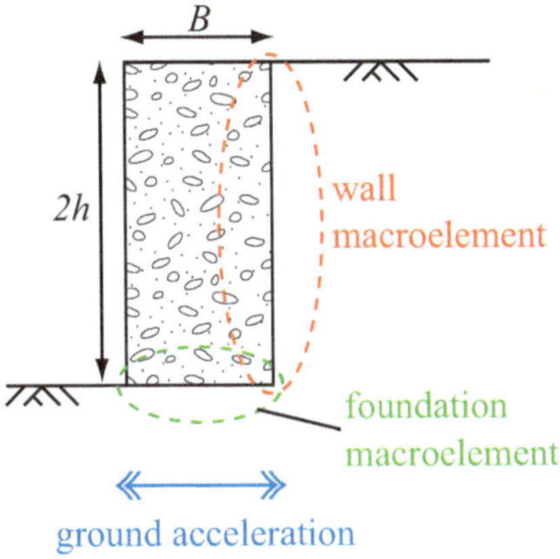


**Figure 15.** Macroelement modelling for gravity retaining wall.

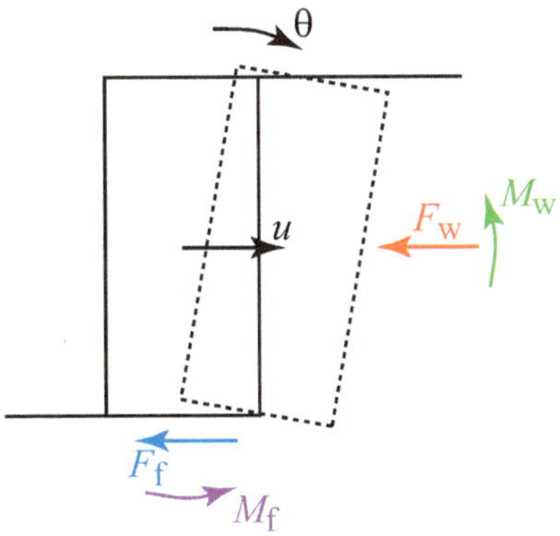


**Figure 16.** Displacement and force components.

real effect - it is included in the detailed foundation macroelement analysis of Cremer et al. (2001) - and requires the foundation model to be applied, rather subtly, to a changing area of contact. The foundation macroelement model has already been described.

Elastic-plastic macroelement footing models link the hardening of the yield surface to the settlement. In our application we simplify the problem by neglecting vertical movements, and considering motions of the underlying ground that do not involve vertical accelerations. We are concerned only

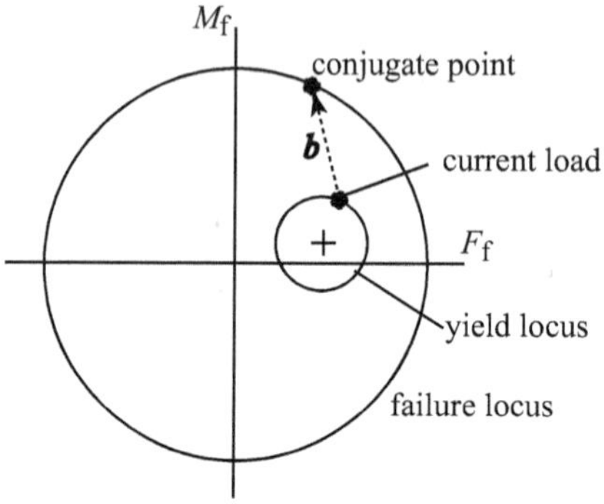


**Figure 17.** Circular failure locus and kinematic yield locus.

with horizontal foundation force $F_f$ and moment $M_f$. The vertical force $V$ on the foundation is given by the weight of the wall. The ratio of $V$ to $V_{\max}$, the bearing capacity of the foundation, controls the maximum values of horizontal force $F_f$ and moment $M_f$ that the foundation can sustain. The load components and corresponding displacement components must be properly work-conjugate. For a strip footing of width $B$ we define a force vector $\boldsymbol{F}$:

$$\boldsymbol{F} = \begin{pmatrix} F_f \\ M_f/B \end{pmatrix} \tag{19}$$

with corresponding work-conjugate displacements:

$$\boldsymbol{u} = \begin{pmatrix} u \\ B\theta \end{pmatrix} \tag{20}$$

The four key elements of an elastic-plastic constitutive model are: the elastic properties; a yield function; a flow rule or plastic potential; and a plastic hardening rule. We will assume that there is a circular interaction relationship between the limiting values of force and moment (Fig. 17). As noted, hardening of the overall yield surface has been deliberately excluded: this is equivalent to assuming that the vertical load generated by the wall is about half the bearing capacity of the foundation. The yield and failure locus for the resulting perfectly plastic material is then a circle. We further assume that the plastic potential is coincident with this yield/failure function for combinations of horizontal load $F_f$ and moment $M_f$.

The elastic-perfectly plastic approach suggests that for all load combinations inside the yield/failure surface the response is elastic whereas once a load combination reaches the yield/failure surface the response is perfectly plastic. While such a model might provide a crude representation of the behaviour of the foundation when monotonically loaded it will be inadequate for dynamic problems where there will be plastic dissipation of energy even for load combinations which do not bring the foundation to perfectly plastic failure. Consequently we introduce a 'bubble' yield surface (Al-Tabbaa and Muir Wood, 1989) which can move around with the current loads, within the failure surface which itself retains the perfect plasticity character and thus remains of constant size. We assume for convenience that the yield surface and failure surface have the same shape (Fig. 17) with the ratio of sizes being a constituive parameter $r$. The current centre of the 'bubble' is $\tilde{\boldsymbol{F}}$.

Within the yield surface ('bubble') the response is elastic. We can adopt the expressions proposed by Gazetas (1991) for the elements of the stiffness matrix $\boldsymbol{K}^e$ linking changes in force $\delta\boldsymbol{F}$ with increments of displacement $\delta\boldsymbol{u}$:

$$\boldsymbol{F} = \boldsymbol{K}^e\boldsymbol{u} = \begin{pmatrix} 2G/(2-\nu) & \sim 0 \\ \sim 0 & \pi GB^2/2(1-\nu) \end{pmatrix} \boldsymbol{u} \tag{21}$$

where $G$ and $\nu$ are shear modulus and Poisson's ratio for the foundation soil. For shallow foundations the off-diagonal terms in the stiffness matrix can probably be neglected.

When the load combination reaches the boundary of the 'bubble' irrecoverable displacements $\delta\boldsymbol{u}^p$ occur. We assume that they occur in the direction of the outward normal to the 'bubble' at the load point: the yield surface acts as a plastic potential. The 'bubble' may have to move in order to accommodate the new load state: the change in load state relative to the change in the centre of the 'bubble' must be tangential to the yield surface. We can define a conjugate point $\boldsymbol{F}_c$ on the failure surface having the same position relative to the centre of this surface (the origin of the load plane, Fig. 17) as the position of the current load state relative to the centre of the 'bubble'. The unit vector $\hat{\boldsymbol{b}}$ is in the direction linking the current load state and the conjugate point (Fig. 17). It is convenient, but not essential, to adopt a translation rule for the 'bubble' that implies that the 'bubble' can never overlap or intersect the failure surface. This implies that the shift in the centre of the 'bubble' must be in the direction of $\hat{\boldsymbol{b}}$. This shift then accommodates any change in load which is in the direction of $\hat{\boldsymbol{n}}$.

The magnitude of the plastic displacement is controlled by the magnitude of the translation of the centre of the 'bubble'. This translation is caused by the component of the load increment that is orthogonal to the 'bubble' at the current load and introduces a plastic hardening modulus $H$ is the plastic hardening modulus which can in general be chosen to be some

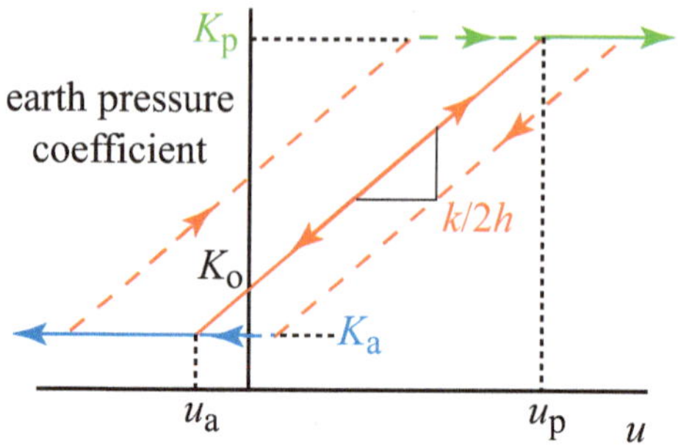


**Figure 18.** Trilinear relationship between earth pressure coefficient and relative displacement of wall and soil (dashed lines indicate possibility of updating $u_a$ and $u_p$ when direction of relative displacement changes).

function of the length of the vector $\boldsymbol{b}$ (Fig. 17). We are able to generate a stiffness relationship linking changes in load components with changes in displacement:

$$\delta \boldsymbol{F} = \boldsymbol{K}^{ep} \delta \boldsymbol{u} \tag{22}$$

This can be operated numerically in the usual way with an elastic prediction to test whether the displacement increment involves yielding of the foundation, and a subsequent plastic correction (or use of combined elastic-plastic stiffness relationship) to calculate an acceptable change in load.

We have thus defined three possible regimes of foundation response depending on whether the behaviour is elastic (for load changes inside the 'bubble'); kinematic hardening plastic (for load changes on the 'bubble'); or perfectly plastic (for load changes on the failure surface). In this third case, the 'bubble' makes contact with the failure surface, and the only possible plastic load changes are tangential to both the 'bubble' and the failure surface at the point of contact. The 'bubble' must translate parallel to this tangent so that the two surfaces remain in contact.

## 5.1 Wall Macroelement

The interaction of the wall with the retained soil is treated separately from the foundation of the wall. It is assumed that at every level the soil-wall interaction is represented by a trilinear Winkler spring relationship between earth pressure coefficient and relative displacement (normalised by wall height) (Fig. 18). The local response is assumed linearly elastic between perfectly plastic active and passive limits. This relationship can be characterised by these limiting pressures and by the elastic stiffness separating them. In addition it is necessary to specify a reference displacement, which could be the actual value of the displacement required to reach the full passive earth pressure or full active earth pressure, $u_p$ or $u_a$ as shown in Fig. 18, or could be an indication of where the initial earth pressure coefficient $K_o$ lies between the limiting values. Allowing this reference displacement to vary can provide a modest amount of kinematic hardening in the wall macroelement model.

The relationship between force or moment and displacement and rotation of the wall is then deduced by integration of the local response. There are nine possible types of response depending on the values of displacement and rotation relative to the reference displacement. These response regimes are shown in Figs. 19 and 20. In each case the moment is calculated at the mid-height of the wall.

When all the wall-soil interaction is elastic (Fig. 20a), $E$ in Fig. 19, the resulting force and moment are:

$$\frac{F_w}{\frac{2}{3}\gamma h^2} = 3K_o + k(3u - h\theta) \tag{23}$$

$$\frac{M_w}{\frac{2}{3}\gamma h^3} = -[K_o + k(u - h\theta)] \tag{24}$$

For this response regime it is thus simple to generate a stiffness relationship:

$$\begin{pmatrix} \delta F_w \\ \delta M_w/h \end{pmatrix} = \frac{2}{3}k\gamma h^2 \begin{pmatrix} 3 & -1 \\ -1 & 1 \end{pmatrix} \begin{pmatrix} \delta u \\ h\delta\theta \end{pmatrix} \tag{25}$$

When all the wall-soil interaction is perfectly plastic with full mobilisation of active earth pressures (Fig. 20b), $A$ in Fig. 19, the resulting force and moment are:

$$\frac{F_w}{\frac{2}{3}\gamma h^2} = 3K_a \tag{26}$$

$$\frac{M_w}{\frac{2}{3}\gamma h^3} = -K_a \tag{27}$$

When all the wall-soil interaction is perfectly plastic with full mobilisation of passive earth pressures (Fig. 20c), $P$ in Fig. 19, the resulting force and moment are:

$$\frac{F_w}{\frac{2}{3}\gamma h^2} = 3K_p \tag{28}$$

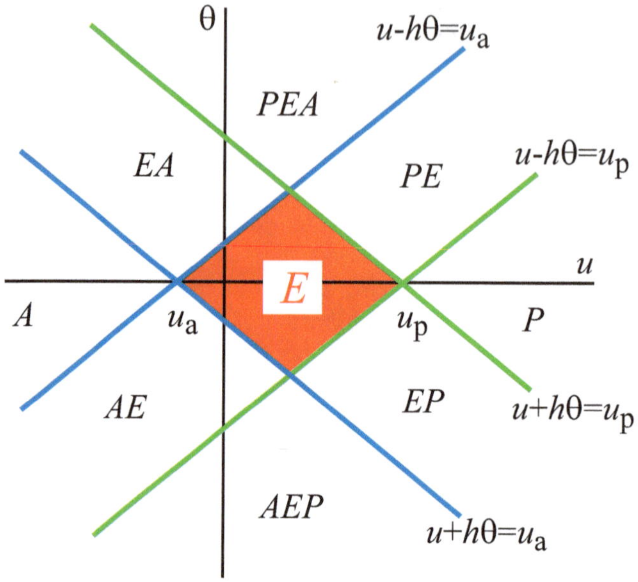


**Figure 19.** Response regimes defined by combinations of wall displacement and rotation.

$$\frac{M_w}{\frac{2}{3}\gamma h^3} = -K_p \tag{29}$$

The six other regimes have combinations of active and/or passive and elastic conditions over the height of the wall. For regime $PE$ (Fig. 19) the wall-soil interaction is fully passive towards the top but elastic towards the base (Fig. 20d). The resulting force and moment are most conveniently described using intermediate parameters:

$$\lambda = 1 + \frac{u - u_p}{h\theta} \tag{30}$$

$$K_1 = K_o + k(u - h\theta) \tag{31}$$

which is the earth pressure coefficient at the bottom of the wall.

$$\frac{F_w}{\frac{2}{3}\gamma h^2} = K_p(1 + \lambda + \lambda^2) + K_1(2 - \lambda - \lambda^2) \tag{32}$$

$$\frac{M_w}{\frac{2}{3}\gamma h^3} = -K_p\lambda^3 + K_1(\lambda^3 - 1) \tag{33}$$

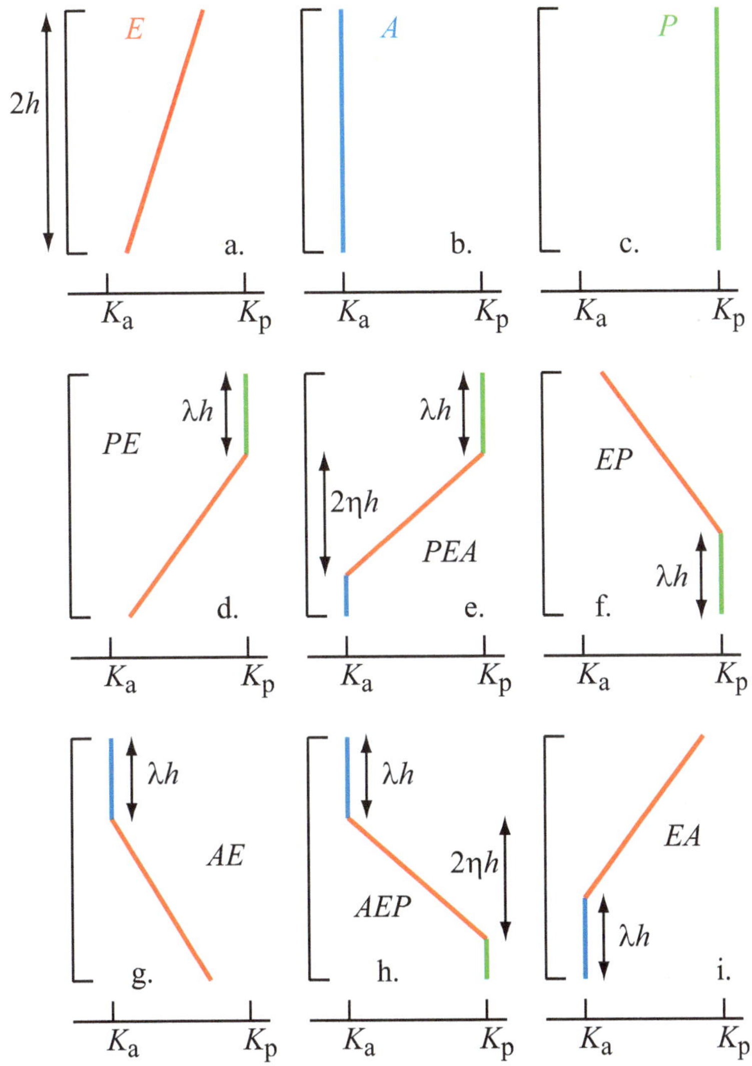


**Figure 20.** Regimes of wall response: (a) fully elastic $E$; (b) fully active $A$; (c) fully passive $P$; (d) $PE$; (e) $PEA$; (f) $EP$; (g) $AE$; (h) $AEP$; (i) $EA$.

For regime $AE$ (Fig. 19) the wall-soil interaction is fully active towards the top but elastic towards the base (Fig. 20g). The resulting force and moment are described using intermediate parameters:

$$\lambda = 1 + \frac{u - u_a}{h\theta} \tag{34}$$

$$K_1 = K_o + k(u - h\theta) \tag{35}$$

which is the earth pressure coefficient at the bottom of the wall.

$$\frac{F_w}{\frac{2}{3}\gamma h^2} = K_a(1 + \lambda + \lambda^2) + K_1(2 - \lambda - \lambda^2) \tag{36}$$

$$\frac{M_w}{\frac{2}{3}\gamma h^3} = -K_a\lambda^3 + K_1(\lambda^3 - 1) \tag{37}$$

For regime $EP$ (Fig. 19) the wall-soil interaction is fully passive towards the bottom but elastic towards the top (Fig. 20f). The resulting force and moment are described using intermediate parameters:

$$\lambda = 1 + \frac{u_p - u}{h\theta} \tag{38}$$

$$K_1 = K_o + k(u + h\theta) \tag{39}$$

which is the earth pressure coefficient at the top of the wall.

$$\frac{F_w}{\frac{2}{3}\gamma h^2} = K_p(2 + 2\lambda - \lambda^2) + K_1(1 - \lambda)^2 \tag{40}$$

$$\frac{M_w}{\frac{2}{3}\gamma h^3} = -K_p(1 + \lambda - 2\lambda^2 + \lambda^3) + K_1\lambda(1 - \lambda)^2 \tag{41}$$

For regime $EA$ (Fig. 19) the wall-soil interaction is fully active towards the bottom but elastic towards the top (Fig. 20i). The resulting force and moment are described using intermediate parameters:

$$\lambda = 1 + \frac{u_a - u}{h\theta} \tag{42}$$

$$K_1 = K_o + k(u + h\theta) \tag{43}$$

which is the earth pressure coefficient at the top of the wall.

$$\frac{F_w}{\frac{2}{3}\gamma h^2} = K_a(2 + 2\lambda - \lambda^2) + K_1(1 - \lambda)^2 \tag{44}$$

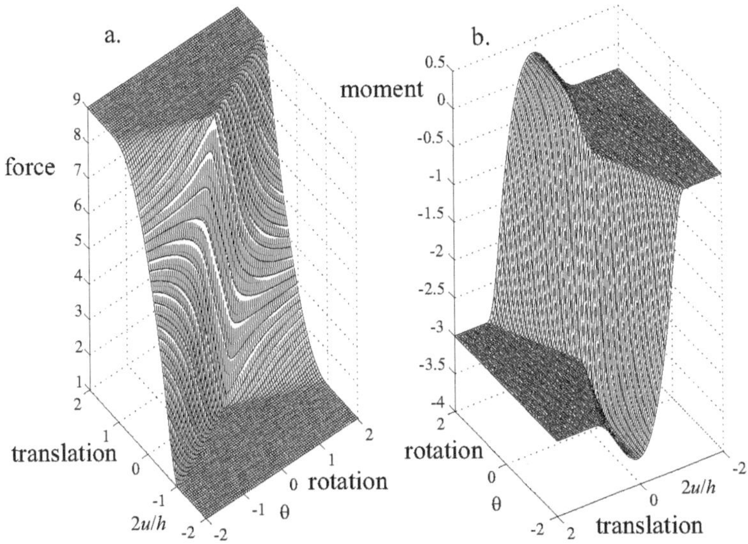


**Figure 21.** Dependence of (a) wall force and (b) moment on relative displacement and rotation of wall.

$$\frac{M_w}{\frac{2}{3}\gamma h^3} = -K_p(1 + \lambda - 2\lambda^2 + \lambda^3) + K_1\lambda(1 - \lambda)^2 \tag{45}$$

For regime $PEA$ (Fig. 19) the wall-soil interaction is fully passive at the top and fully active at the base and elastic in between (Fig. 20e). Again intermediate parameters are introduced:

$$\lambda = 1 + \frac{u - u_p}{h\theta} \tag{46}$$

$$\eta = \frac{u_p - u_a}{2h\theta} \tag{47}$$

$$\frac{F_w}{\frac{2}{3}\gamma h^2} = \alpha_1 K_p + \alpha_2 K_a \tag{48}$$

where

$$\alpha_1 = 3\lambda^2 + 3\lambda\eta + \eta^2 \tag{49}$$

$$\alpha_2 = 3 - \alpha_1 \tag{50}$$

$$\frac{M_w}{\frac{2}{3}\gamma h^3} = \alpha_3 K_p + \alpha_4 K_a \tag{51}$$

where

$$\alpha_3 = \alpha_1 - 4\lambda^3 - 6\lambda^2\eta - 4\lambda\eta^2 - \eta^3 \tag{52}$$

$$\alpha_4 = -1 - \alpha_3 \tag{53}$$

For regime $AEP$ (Fig. 19) the wall-soil interaction is fully active at the top and fully passive at the base and elastic in between (Fig. 20h). Again intermediate parameters are introduced:

$$\lambda = 1 + \frac{u - u_a}{h\theta} \tag{54}$$

$$\eta = \frac{u_p - u_a}{2h\theta} \tag{55}$$

$$\frac{F_w}{\frac{2}{3}\gamma h^2} = \alpha_1 K_a + \alpha_2 K_p \tag{56}$$

where

$$\alpha_1 = 3\lambda^2 - 3\lambda\eta + \eta^2 \tag{57}$$

$$\alpha_2 = 3 - \alpha_1 \tag{58}$$

$$\frac{M_w}{\frac{2}{3}\gamma h^3} = \alpha_3 K_a + \alpha_4 K_p \tag{59}$$

where

$$\alpha_3 = \alpha_1 - 4\lambda^3 + 6\lambda^2\eta - 4\lambda\eta^2 + \eta^3 \tag{60}$$

$$\alpha_4 = -1 - \alpha_3 \tag{61}$$

The force exerted by the soil on the wall is explicitly determined by the actual values of wall displacement and rotation at any time. The complete surfaces linking force and moment with displacement and rotation are shown in Figs. 21a, b. While the nine pairs of expressions for force and moment could be differentiated in order to convert them to stiffness relationships it is numerically more practical to use them directly rather than incrementally.

Some element of kinematic hardening is introduced by updating the reference displacement (at all levels) whenever the combination of displacement and rotation takes the wall macroelement into the perfectly plastic zones $A$ or $P$. The effect of this updating is that the macroelement is effectively never within either of these zones, but only ever on the limiting boundary. Although Fig. 18 introduces six quantities, each of which may conveniently be introduced in certain of the governing equations, there are only in fact four independent quantities: if the stiffness $k$ and the 'initial' earth pressure coefficient $K_o$ are specified then the values of the displacements required to reach the passive or active limits are controlled by the values of those limits.

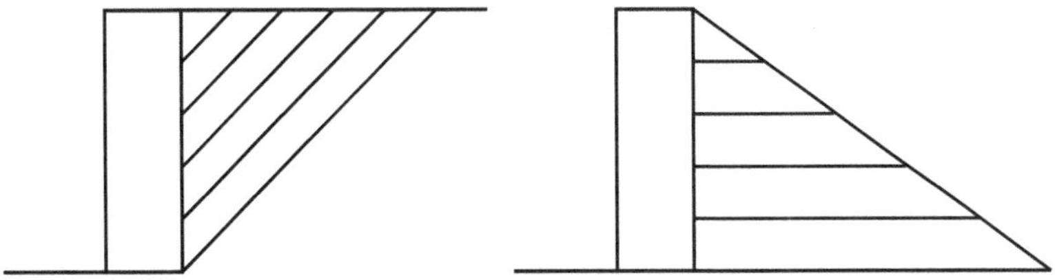


**Figure 22.** Allowance for inertia of retained soil.

### 5.2 Inertial Effects

The inertia of the wall is included in the normal way. We assume that the wall is of rectangular section with height $2h$ and width $B$. It is made of material with density $\rho_w$. Its mass is then $2\rho_w hB$ per unit length and its moment of inertia about its centre of mass is $2\rho_w Bh^3/3$.

There will be inertial effects associated with the movement of the soil in the foundation and behind the wall. Numerical studies by Paolucci (1997) and Cremer et al. (2001) show that the inertia of the foundation soil can be neglected. Soil velocities decay very rapidly away from the footing itself. This is then a convenient simplification.

The inertia of the soil behind the wall cannot be neglected - accelerations of this soil are of the same order as the accelerations of the wall itself. Calculation procedures such as that used by Nadim (1980) propose that the soil movements occur as a result of sliding on inclined planes behind the wall (Fig. 22a). It might be supposed that the inclination of such sliding planes would change with the direction of motion of the wall (active movement away from the soil encouraging steeper failure planes than passive movement towards the soil) (Zarrabi-Kashani, 1978). In actual dense granular materials, the initiation of failure will tend to lead to dilation and reduction of strength so that any failure zones which have developed may attract future deformation. (Such a hypothesis lies behind the adapted Newmark 'sliding block' procedure suggested by Koseki et al. (1998)). For wall movements which do not actually mobilise the full strength of the soil the nature of the deformation mechanism is less well defined. Here, as a pragmatic first order approximation, it will be assumed that the inertial contribution of the soil is constant and can be represented by 'attaching' horizontal layers of soil to the back of the wall (Fig. 22b). Following Nadim we assume that slding occurs on an inclined plane from the back of the wall up to the free surface. We will neglect vertical inertia and simply assume that the length of the attached soil layer at each level on the wall is $z\sqrt{2}$, where $z$ is the distance below the top of the wall. The mass of the wall is then increased by $2\sqrt{2}\rho_s h^2$. The moment of inertia of the wall about its centre of mass is increased by $4\sqrt{2}\rho h^4/9$. Realistically we are not concerned with large rotations of the wall so the conversion of sliding horizontal layers of soil to a triangular mass of attached soil does not really introduce any further significant approximations.

### 5.3 Solution Procedure

The wall is assumed to be initially at rest, with zero velocity and acceleration, so that horizontal and moment equilibrium can be imposed. This controls the initial horizontal force and moment imposed on the foundation which must balance those imposed by the backfill soil. Then at subsequent times any translational or rotational acceleration of the wall arises because of imbalance of the forces and moments acting on the wall from the foundation and from the retained soil generated as a result of relative movement of the wall and the underlying ground.

Within each calculation cycle the motion of the wall is calculated as follows. The velocity and acceleration (translational and rotational) of the wall at the end of the previous time step are integrated to give the new horizontal displacement and rotation of the wall. The horizontal displacement of the underlying ground is known. The relative displacement (translation and rotation) of the wall can then be used to calculate the new horizontal force and moment on both the foundation (using the incremental relationships) and the wall (using the actual values of relative displacement). The equation of motion is then written in terms of these initial values together with the incremental relative displacement in order to calculate the new displacements during the time step. The absolute acceleration of the wall is $\ddot{\boldsymbol{x}} = \ddot{\boldsymbol{u}} + \ddot{\boldsymbol{x}}_g$ where $\boldsymbol{x}_g = (x, 0)^T$ is the displacement of the underlying ground which provides the disturbing input motion for the wall.

Then we can write:

$$\boldsymbol{M}\ddot{\boldsymbol{u}} + \boldsymbol{\lambda}\dot{\boldsymbol{u}} + \boldsymbol{K}_f(\boldsymbol{u} - \boldsymbol{u}_o) + \boldsymbol{F}_o = -\boldsymbol{M}\ddot{\boldsymbol{x}}_g \tag{62}$$

where $\boldsymbol{F}_o$ represents the sum of the various force components acting on the wall at the start of the time step and $\boldsymbol{u}_o$ is the relative displacement of the wall at the start of the time step. The stiffness term $\boldsymbol{K}_f$ includes, at present, only the incremental elastic-plastic stiffness of the foundation because the formulation of the incremental stiffness for the interaction with the retained soil is more cumbersome.

If the force and moment on the foundation lie within the elastic 'bubble' and the wall displacement regime is in the elastic $E$ region, then the motion

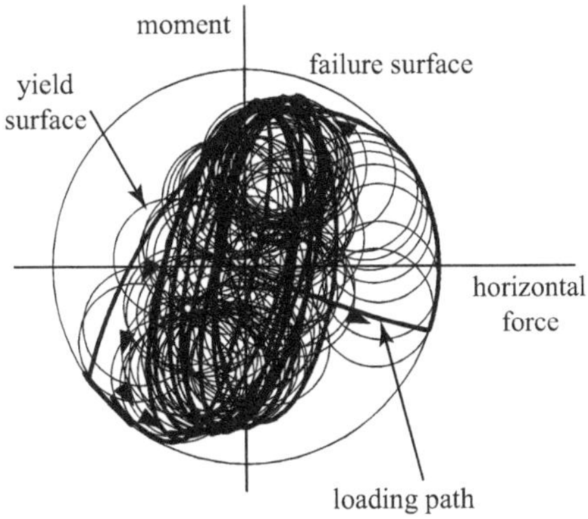


**Figure 23.** Response of foundation macroelement: movement of 'bubble' yield surface within failure surface.

is undamped. Practically, then, it is necessary to introduce some damping $\boldsymbol{\lambda}$ in order to remove energy from the system. The damping matrix $\boldsymbol{\lambda}$ is:

$$\boldsymbol{\lambda} = \begin{pmatrix} C_{xx} & 0 \\ 0 & C_{\theta\theta} \end{pmatrix} \tag{63}$$

and the damping coeficients that have been used have been adapted from those proposed by Gazetas (1991) for surface foundations.

The solution of (62) is treated in a semi-implicit manner using a modified Newton-Raphson technique adapted from Crisfield (2001).

## 5.4 Results of Analysis

Three examples of the operation of the macroelement model are shown here. First the response of the foundation macroelement on its own (without retained soil) is presented to illustrate the combination of kinematic hardening and perfect plasticity in the response. The wall is given width $B = 3$ m and height $2h = 5$ m. The angle of friction of the foundation soil (which is used to ensure that the design of the wall is plausible) is 25°. The size of the 'bubble' yield surface is taken as $r = 0.3$. A harmonic horizontal input motion is assumed with amplitude $0.5g$ and frequency 2 Hz. The movement of the 'bubble' within the failure surface is shown in Fig. 23.

In order to add in the wall macroelement a value is required for the 'spring' stiffness $k$ in Fig. 18. Typically, it is suggested in the literature that the wall displacements required to reach fully active conditions and fully passive conditions are 0.1% and 1% of wall height for a wall retaining dense sand. (Of course the actual values would be influenced by any initial compaction stress locked into the soil, but these values give an order of magnitude.) Then, with passive and active pressure coefficients calculated from angle of friction as usual (we are ignoring wall friction) we can deduce that $k \approx (K_p - K_a)/0.009 \approx 111(K_p - K_a)$. Calculations have been performed with $\phi = 40°$ and with $k = 487$.

The double macroelement model is then also subjected to harmonic input of acceleration amplitude $0.2g$ and frequency 2 Hz. It is instructive to look at the path in the regime map of Fig. 24 (compare Fig. 19). The path lies almost entirely in the $AE$ region, with a brief excursion along the $A$ boundary. Whenever the path reaches the $A$ boundary the reference displacements $u_a$ and $u_p$ are updated and become permanent displacements of the wall. This is illustrated in Fig. 24. There is a small permanent active, outward displacement in the first cycle; thereafter the wall develops no further permanent displacement.

Earthquake motion tends to be of quite short duration with a very few cycles of high acceleration and many cycles of much lower acceleration with a typical duration of $10 - 20$ s depending on the location and nature of the earthquake. Once the motion becomes large enough to move into the regimes of nonlinear response of the wall interacting with the foundation and the retained soil, then the detail of the movement of the wall depends on the exact time within the earthquake that particular large pulses of ground acceleration occur.

This can be illustrated in two ways. In Fig. 25 the same wall model with the same initial conditions is analysed using an actual time history of earthquake acceleration imposed in four different ways: with the time reversed and with the sign of the accelerations reversed. Reversing the sign of the acceleration can have the effect of changing the direction of the most damaging pulse of the input from active to passive loading.

We can also explore this aspect of the system nonlinearity by showing the dependence of the translation and rotation of the wall on the amplitude of the earthquake (as indicated by the maximum acceleration) relative to the critical steady unidirectional acceleration required to generate active failure of the wall for the same time history applied with different scales.

The acceleration histories for four earthquakes are shown in Fig. 26. The system being analysed is extremely nonlinear. The input motion that is generated by an earthquake is extremely irregular. Put together, the overall system response is somewhat chaotic. There is no linear variation of displacement or rotation of the wall with this relative magnitude of the input

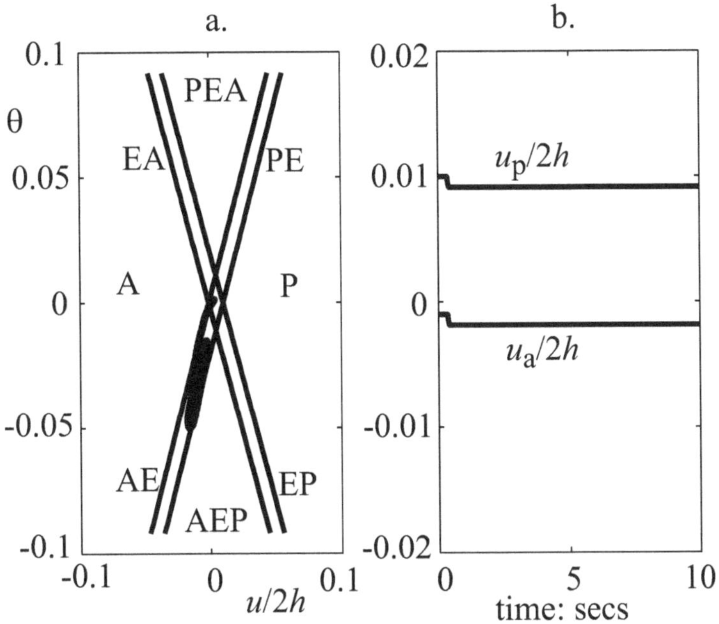


**Figure 24.** Response of retaining wall to harmonic input: (a) displacement path in response regime diagram; (b) development of permanent displacement.

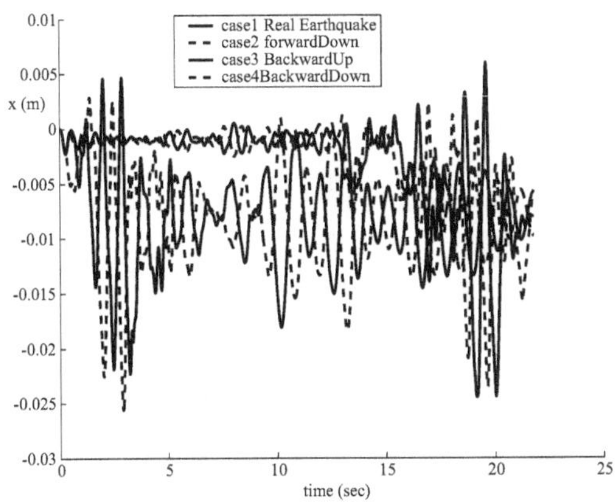


**Figure 25.** Influence of direction and sign of earthquake motion on wall displacement (after Kalasin, 2004).

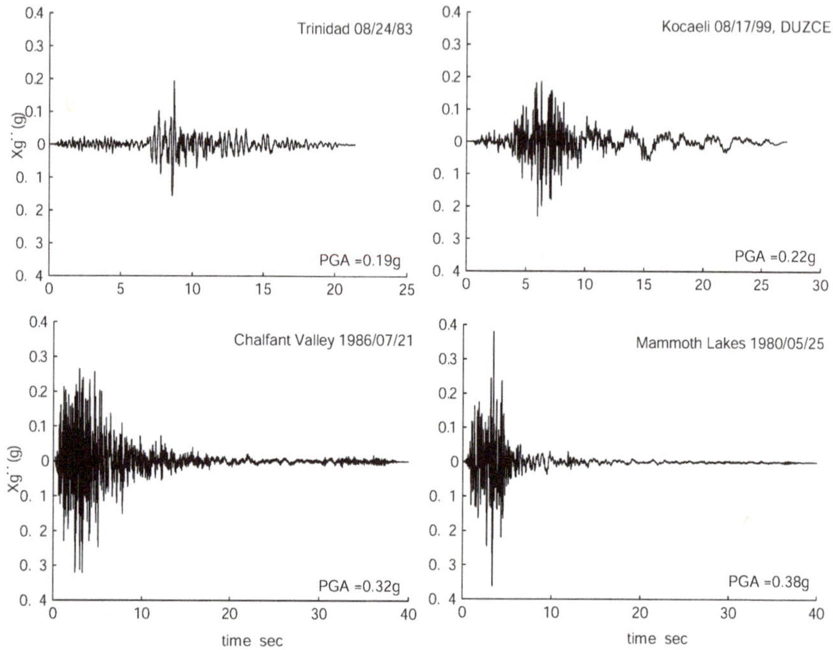


**Figure 26.** Acceleration time histories.

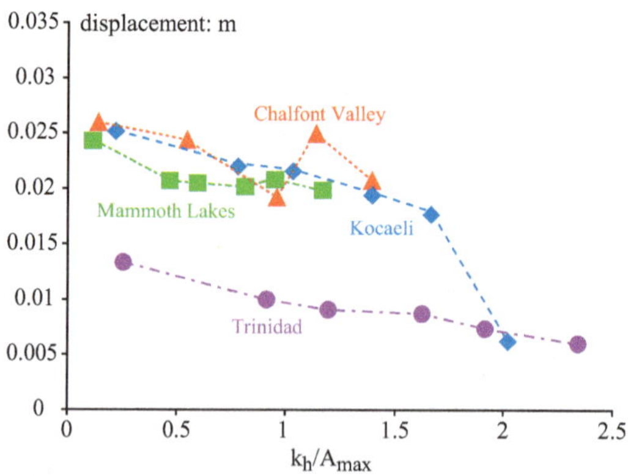


**Figure 27.** Influence of relative earthquake acceleration on wall displacement (after Kalasin, 2004).

motion (Figs. 27, 25). In fact, for some time histories of input motion there is not even a monotonic variation of movement with relative acceleration input magnitude.

The response of this simple macroelement system shows rather clearly that, so far as *performance* of the system is concerned - by which, in the present context, we mean primarily the permanent displacement developed during the seismic event - the occurrence of high acceleration pulses of short duration is not necessarily devastating. The macroelement model also allows us to explore parametric variations of both the geotechnical features of the system and the input motion itself in order to understand their interaction.

### 5.5 Wall Model: Discussion

A macroelement model has been developed which combines two separate macroelements to describe the interaction of a gravity retaining wall with its foundation and with the retained soil. The foundation macroelement introduces the key features observed in nonmonotonic response of foundations under combinations of vertical, horizontal and moment loading and accommodates kinematic plasticity in order to include the dissipation of energy that will accompany reversal of loading direction. This macroelement is a subset of more sophisticated foundation macroelements described elsewhere.

The macroelement for the retained soil adopts a traditional 'Winkler spring' approach to the development of earth pressure, treating each level of the wall as independent of every other level, and allowing the local earth pressure coefficient to vary between fully active and fully passive limits with an elastic intermediate response. The behaviour of the whole wall is then the integration of these Winkler springs. Nine regimes of response can be identified. A modest degree of kinematic hardening has been introduced into this wall model.

The response of the combined macroelement model for the wall with its foundation and with the retained soil has been studied for various dynamic excitations of the underlying ground. When realistic earthquake excitations are used the nonlinearity of the overall system response becomes clear. The benefit of such a simple model for parametric study of effects of geotechnical and dynamic variables has been demonstrated.

## 6 Conclusions

The nature of the problems analysed is not expecially complex and the solution procedures are rather straightforward. However, the several examples illustrate how one might introduce plausible nonlinear models where the data of actual behaviour are unavailable and nevertheless present results in terms of clearly defined dimensionless groups—the use of which evidently increases the applicability. Parametric study then becomes efficient.

Cremer et al. (2001) write in the context of development of a macroelement to describe the behaviour of a footing under combined vertical, horizontal and moment loading: *'It is well known that an alternative model of the foundation behaviour obtained by the finite element method, with suitable nonlinear constitutive laws and special contact elements, requires a high degree of modelling competence and is time consuming. The macroelement provides a practical and efficient tool, which may replace efficiently, in a first approach, a costly finite element soil model, and which ensures the accurate integration of the effect of soil-structure interaction.'*

Similar sentiments may be expressed here. The nonlinear Winkler spring approach adopted for interaction of pile or pipeline with the ground blatantly disregards the continuity of the soil surrounding the pile or other structural element but introduces sufficient allusion to actual soil behaviour to allow one to understand and interpret aspects of the overall system response.

One of the important possibilities of analyses of the sort described here is the pedagogic potential to educate engineers in the reality of soil-structure interaction. We all know that it is important to consider the soil and structural elements as a *system* and that it is usually not possible to produce sensible results unless we do so. The use of dimensionless parameters for analysis—whether of linear or nonlinear problems—provides a basis for stating whether the effects of interaction are in fact important and, more especially, emphasises that the response is controlled by the *relative* stiffnesses of the ground and structure. (A salutary case history of development of stresses in a flexible integral bridge abutment is outlined by Muir Wood and Nash (2000)).

If practising engineers are not as aware as they should be of the importance of soil-structure interaction then we educators must take some of the blame. We atomise the teaching from the moment that the students arrive, with separate courses on structural mechanics, fluid mechanics, soil mechanics and (in my experience) retain and deepen these divisions as the degree programmes proceed with only limited time being allocated to itegrative activities which show the consequences of taking a broader system approach. Simple analytical methods such as those presented here can help the campaign to improve this awareness.

## List of Symbols

| | | |
|---|---|---|
| $\boldsymbol{b}$ | vector from current load to conjugate point | §5 |
| $\hat{\boldsymbol{b}}$ | unit vector in the direction of $\boldsymbol{b}$ | §5 |
| $B$ | width of the wall | §5 |
| $c_u$ | undrained soil strength | §4 |
| $C_{xx}$ | damping coefficient | §5 |
| $C_{\theta\theta}$ | damping coefficient | §5 |
| $D$ | depth of pipeline | §4 |
| $E$ | Young's modulus | §3, §4 |
| $\tilde{f}$ | yield surface | §5 |
| $F$ | shear force | §4 |
| $\tilde{\boldsymbol{F}}$ | current centre of the 'bubble' yield surface | §5 |
| $\boldsymbol{F}_c$ | conjugate point on the failure surface | §5 |
| $F_f$ | horizontal foundation force | §4, §5 |
| $\boldsymbol{F}_o$ | force components acting on the wall at start of time step | §5 |
| $F_w$ | force of backfill on wall | §5 |
| $G$ | shear modulus | §5 |
| $h$ | height of wall | §5 |
| $I$ | second moment of area of structural section | §3, §4 |
| $k$ | pile-soil or wall-soil subgrade reaction coefficient | §5 |
| $k_h$ | initial stiffness of the pile-ground interaction | §3 |
| $K$ | mobilised lateral pressure coefficient | §3, §4 |
| $K_a$ | active earth pressure coefficient | §5 |
| $\boldsymbol{K}_f$ | stiffness matrix | §5 |
| $K_o$ | initial earth pressure coefficient | §5 |
| $K_p$ | passive earth pressure coefficient | §3, §5 |
| $K^*$ | asymptotic earth pressure coefficient | §3, §4 |
| $K_1$ | earth pressure coefficient at the bottom of the wall | §5 |
| $\boldsymbol{K}^e$ | elastic stiffness matrix | §5 |
| $\boldsymbol{K}^{ep}$ | elastic-plastic stiffness matrix | §5 |
| $\ell$ | length of pile | §3, §4 |

| | | |
|---|---|---|
| $\boldsymbol{M}$ | mass matrix | §5 |
| $M_f$ | moment at root of cantilever pile under full soil loading | §3, §4 |
| $M_f$ | foundation moment | §5 |
| $M_{\max}$ | maximum moment | §3, §4 |
| $M_r$ | moment at root of pile | §3, §4 |
| $M_w$ | moment applied by backfill on wall | §5 |
| $\hat{\boldsymbol{n}}$ | unit vector normal to the yield surface | §5 |
| $r$ | relative size of yield surface and failure surface | §5 |
| $u$ | displacement | §5 |
| $\boldsymbol{u}_o$ | relative displacement of wall at start of time step | §5 |
| $\delta\boldsymbol{u}^p$ | plastic displacement increment | §5 |
| $u_a$ | displacement required to reach full active earth pressure | §5 |
| $u_p$ | displacement required to reach full passive earth pressure | §5 |
| $\ddot{\boldsymbol{x}}$ | absolute acceleration of the wall | §5 |
| $\boldsymbol{x}_g$ | displacement of the underlying ground | §5 |
| $y$ | displacement | §3, §4 |
| $y_{\max}$ | maximum displacement | §3 |
| $z$ | distance below the top of the wall | §5 |
| $\alpha$ | parameter characterising ground movement profile | §3 |
| $\alpha_1$ | intermediate parameter | §5 |
| $\alpha_2$ | intermediate parameter | §5 |
| $\alpha_3$ | intermediate parameter | §5 |
| $\alpha_4$ | intermediate parameter | §5 |
| $\beta$ | dimensionless parameter: pile-ground stiffness | §3, §4 |
| $\gamma$ | unit weight of the soil | §3, §4 |
| $\delta$ | ground movement | §3, §4 |
| $\delta_o$ | movement at the ground surface | §3, §4 |
| $\Delta$ | relative displacement of structure and ground | §3, §4 |
| $\Delta_{50}$ | relative movement required to move halfway to the limiting value | §3, §4 |
| $\zeta$ | dimensionless displacement | §3, §4 |
| $\eta$ | dimensionless coordinate | §3, §4 |

| | | |
|---|---|---|
| $\eta$ | intermediate parameter | §5 |
| $\theta$ | wall rotation | §5 |
| $\lambda$ | intermediate parameter | §5 |
| $\boldsymbol{\lambda}$ | damping matrix | §5 |
| $\mu_1$ | dimensionless bending moment | §3, §4 |
| $\mu_2$ | dimensionless bending moment | §3, §4 |
| $\mu_3$ | dimensionless bending moment | §4 |
| $\nu$ | Poisson's ratio | §5 |
| $\rho_w$ | density of the material of the wall | §5 |
| $\sigma_h$ | horizontal stress | §3, §4 |
| $\phi'$ | angle of friction | §3 |
| $\Phi$ | dimensionless shear force | §4 |
| $\chi$ | dimensionless structural stiffness | §3, §4 |

## Bibliography


A. Al-Tabbaa and D. Muir Wood. An experimentally based 'bubble' model for clay. In S. Pietruszczak and G. N. Pande, editors, *Numerical Models in Geomechanics NUMOG III*, pages 91–99. Elsevier Applied Science, London, 1989.

C. Cremer, A. Pecker, and L. Davenne. Cyclic macro-element for soil-structure interaction: material and geometrical nonlinearities. *International Journal for Numerical and Analythical Methods in Geomechanics*, 25(13):1257–1284, 2001.

M. A. Crisfield. *Non-linear finite element analysis of solids and structures. 1: Essentials*. John Wiley & Sons, 2001.

G. Gazetas. Foundation vibrations. In H.-Y. Fang, editor, *Foundation engineering handbook*, pages 553–593. Van Nostrand Reinhold, New York, 1991.

G. Gottardi, G. T. Houlsby, and R. Butterfield. Plastic response of circular footings on sand under general planar loading. *Géotechnique*, 49(4):453–469, 1999.

T. Kalasin. *Dynamic macroelement model for gravity retaining walls*. PhD thesis, University of Bristol, Bristol, 2004.

J. Koseki, F. Tatsuoka, Y. Munaf, M. Tateyama, and K. Kojima. Modified procedure to evaluate active earth pressure at high seismic loads. *Soils and Foundations*, Special issue No. 2 on Geotechnical aspects of the January 17 1995 Hyogoken-Nambu Earthquake:209–216, 1998.

C. M. Martin and G. T. Houlsby. Combined loading of spudcan foundations on clay: numerical modelling. *Géotechnique*, 51(8):687–699, 2001.

D. Muir Wood. Experimental inspiration for kinematic hardening models of soil. *Journal of Engineering Mechanics*, 130(6):656–664, 2004.

D. Muir Wood and D. F. T. Nash. Earth pressures on an integral bridge abutment: a numerical case study. *Soils and Foundations*, 40(6):23–28, 2000.

F. Nadim. Tilting and sliding of gravity retaining walls during earthquakes. Master's thesis, Department of Civil Engineering, MIT, Boston, 1980.

R. Nova and L. Montrasio. Settlements of shallow foundations on sand. *Géotechnique*, 41(2):243–256, 1991.

R. Paolucci. Simplified evaluation of earthquake-induced permanent displacements of shallow foundations. *Journal of Earthquake Engineering*, 1(3):563–579, 1997.

H. G. Poulos. Analysis of piles in soil undergoing lateral movement. *ASCE Journal of the Soil Mechanics and Foundations Division*, 99(SM5):391–406, 1973.

M. F. Randolph and G. T. Houlsby. The limiting pressure on a circular pile loaded laterally in cohesive soil. *Géotechnique*, 34(4):613–623, 1984.

L. C. Reese and H. Matlock. Non-dimensional solutions for laterally loaded piles with soil modulus assumed proportional to depth. In *Proc. 8th Texas Conf. on Soil Mechanics and Foundations*, page 41, 1956.

K. Zarrabi-Kashani. Sliding of gravity retaining wall during earthquakes considering vertical acceleration and changing inclination of failure surface. Master's thesis, Department of Civil Engineering, MIT, Boston, 1978.

# Offshore Design Approaches and Model Tests for Sub-Failure Cyclic Loading of Foundations

Mark F. Randolph [*]

[*] Centre for Offshore Foundation Systems, University of Western Australia, Australia

## 1 Introduction

Cyclic loading is an important aspect of offshore design because the environmental loading during extreme storm conditions generally dominates compared with the permanent loading. In most cases the focus for assessing the effect of cyclic loading is on quantifying the reduction in shear strength (and hence foundation capacity), although assessment of cumulative deformations under cyclic loading may also be critical for some designs. The chapter describes how a systematically planned programme of laboratory tests may be used to construct failure envelopes, or in a more general sense contours of cumulative shear strain as a function of normalised shear stress levels and number of cycles.

Strain rates associated with cyclic loading are much higher than are typically used for monotonic laboratory tests, and this will be evident in the contour plots presented later. The increase in resistance due to higher strain rate in part compensates for the damage due to cyclic shearing. In most offshore design, rather than implement a sophisticated model for the soil that simulates the effects of cyclic loading into a numerical (e.g. FE) analysis, the soil strength is adjusted globally by applying a reduction factor based on the severity of the cyclic loading.

There are many idealisations that have to be made to achieve this, one of which is the representation of a storm loading sequence, with varying magnitude and period of the cyclic loading, by a uniform harmonic cyclic shear stress (Fig. 1). Depending on the failure mode associated with the foundation, different stress paths will be followed in different parts of the failure mechanism. In principle this may be addressed by conducting appropriate element tests (triaxial compression, simple shear, triaxial extension) to evaluate damage due to the cyclic loading differentially through the various mechanism (Fig. 2), and then adjusting soil strengths in a numerical analysis of the foundation (Andersen and Lauritzsen, 1988; Andersen, 1991). In practice, anisotropy (both in monotonic response and cyclic damage) is often idealised by focusing on the element response obtained from simple shear tests, which is generally representative of the average response.

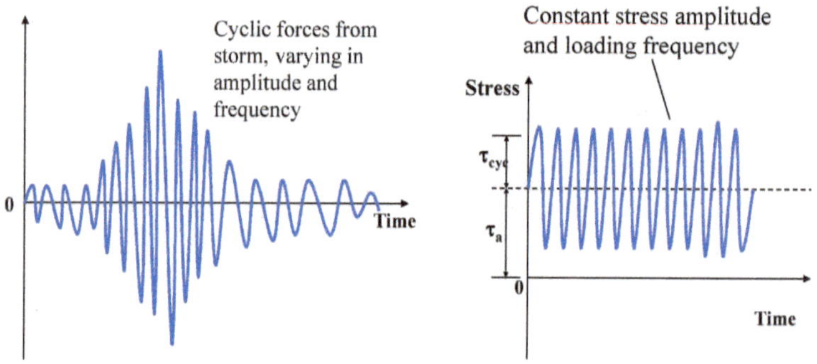


**Figure 1.** Contrast between field and laboratory cyclic shearing.

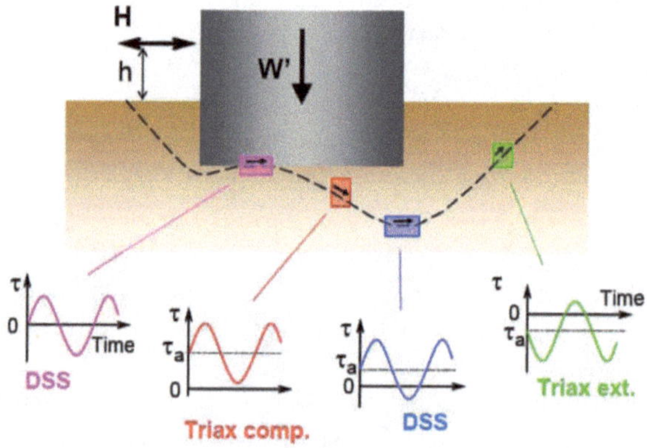


**Figure 2.** Map of different soil tests and design applications (Andersen et al., 2008), where $H$ is the horizontal force, $h$ describes its application point, $W'$ is the buoyant weight of an ideal offshore foundation, and $\tau$ represents the local induced shear stress.

Shallow caisson foundations, such as used for the Draupner E and Sleipner platforms in the North Sea, and which are viewed as potential foundations for tripod-mounted offshore wind turbines in moderate water depths, are subjected to high cyclic vertical loads because of the relatively large lever arm through which horizontal loading is applied (Fig. 3). This can result in biased 2-way cyclic loading of the caisson foundations, where the extreme load is negative (i.e. uplift). Model tests have indicated that the stress reversal triggers significant excess pore pressures and strength loss but, provided the average load is still well positive, the net effect is to induce cumulative settlement rather than uplift (Bye et al., 1995; Byrne and Houlsby, 2002; Kelly et al., 2006).

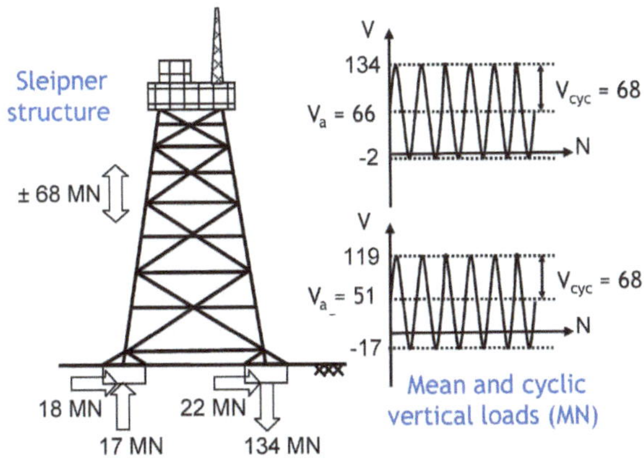


**Figure 3.** Example loading regimes for a 4-legged jacket structure, where symbol $V$ stands for the vertical load, $V_a$ for the average value for $V$, $V_{\mathrm{cyc}}$ for the cyclic amplitude, and $N$ the number of cycles.

## 2 Cyclic Shear Strength

The soil supporting a given foundation will be capable of redistributing stresses away from zones where cyclic-induced strains are accumulating. This provides some justification for simplifying the complete six-degree of freedom design loading into an equivalent one-dimensional cyclic stress regime applied to the soil. The nature of the cyclic stress regime may be classified under four different types, ranging from symmetric 2-way shear stresses (comparatively rare, although the most commonly applied in laboratory testing programmes) to asymmetric 2-way, ideal 1-way and biased

1-way (Fig. 4). The last category, where the shear stresses do not reverse in sign, is the least damaging type of cyclic loading, generally leading to mild accumulation of strains, but limited excess pore pressure development.

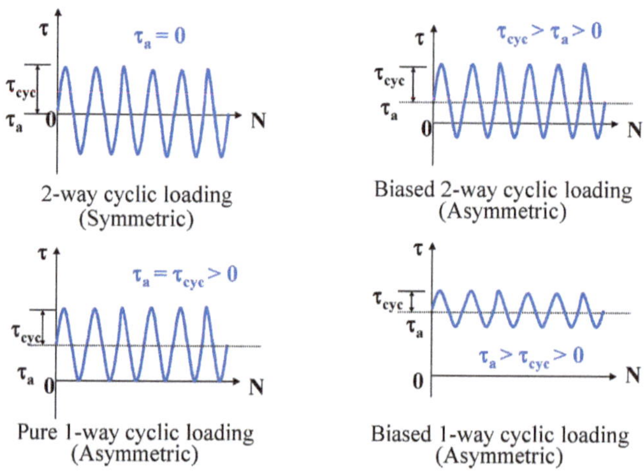


**Figure 4.** Classification of cyclic loading regimes.

Fig. 5 to Fig. 7 show examples of a monotonic test and two cyclic tests at different stress levels carried out under undrained conditions on a silty seabed sand. The monotonic test shows the very high undrained strength that can be mobilised due to dilation induced negative pore pressures. The phase transformation point is critical, and in many ways represents a better measure of monotonic strength against which to normalise the cyclic behaviour, than any notional 'maximum' shear stress mobilised at a particular strain level. Here, the phase transformation occurs at a shear stress of about 20 kPa ($\tau/\sigma'_{vc} = 0.27$, where $\sigma'_{vc}$ is the vertical consolidation effective stress) (Fig. 5), and 2-way cyclic shearing at a stress level of $\pm 15$ kPa ($\tau_{cyc}/\sigma'_{vc} = 0.2$) leads to failure in about 25 cycles (Fig. 6). Liquefaction occurs during the mid-part of each cycle as failure is approached, accompanied by a rapid increase in cyclic shear strain. At the lower cyclic shear stress level of $\tau_{cyc}/\sigma'_{vc} = 0.15$, similar behaviour occurs eventually, but only after more than 500 cycles (Fig. 7).

The precise point of 'liquefaction' failure cannot be identified explicitly, since although the excess pore pressure reaches the vertical stress during the mid-point of the later cycles, dilation occurs as the sample is sheared under increasing shear stress. From a design point of view, a more critical measure is the level of shear strain.

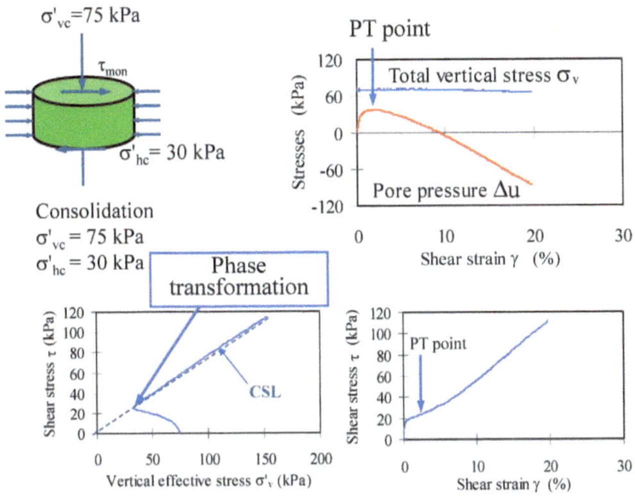


**Figure 5.** Monotonic simple shear test on carbonate silt. In the first picture above, $\sigma'_{\text{hc}}$ is the horizontal consolidation effective stress.

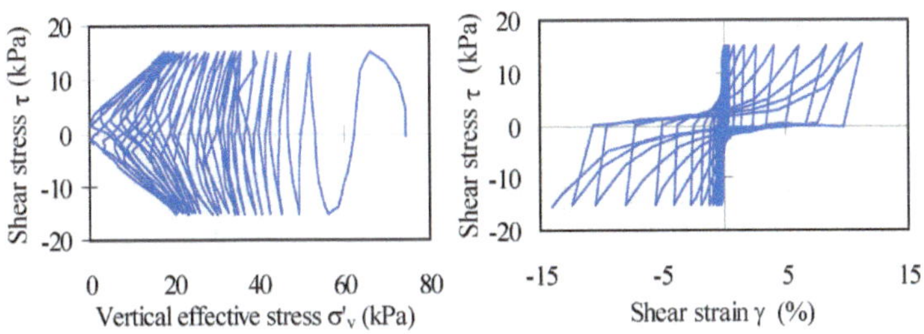


**Figure 6.** Cyclic simple shear test on carbonate silt with $\tau_{\text{cyc}}/\sigma'_{\text{vc}} = 0.2$.

A series of cyclic tests at different shear stress levels allows construction of (accumulated) strain and excess pore pressure contours as a function of number of cycles. The manner in which such diagrams are constructed is

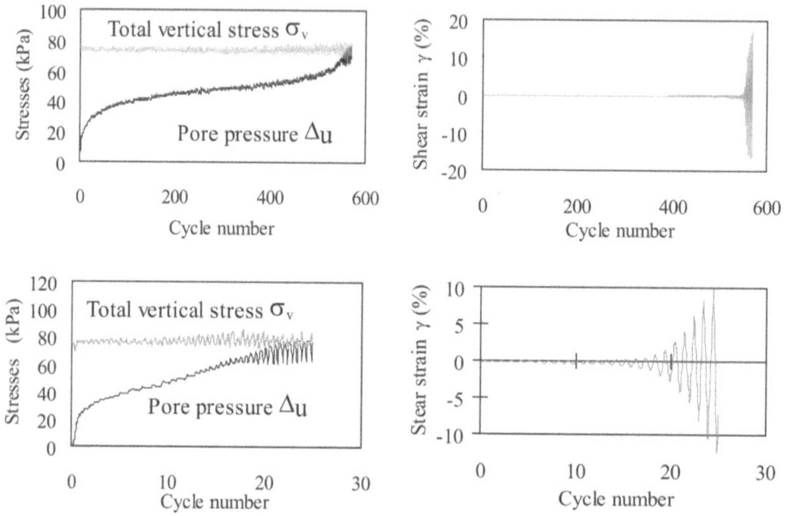


**Figure 7.** Comparison of cyclic responses for $\tau_{cyc}/\sigma'_{vc} = 0.15$ (top) and 0.2 (bottom).

illustrated in Fig. 8, with the information extracted from 1 monotonic and 4 cyclic tests. The cyclic stress levels are normalised by the shear strength measured in the monotonic test (possibly at a specified shear strain level, rather than a true peak). Then, in each test the number of cycles required to develop a particular strain level is plotted, and finally contours for each accumulated shear strain level are drawn through the data.

Cyclic loading tests are typically carried out at a frequency of 0.05 to 0.1 Hz, representing typical frequencies of wave loading in the offshore environment. Because of the much higher strain rates during a cyclic loading test (potentially exceeding 1%/s) compared with a monotonic test (typically 5%/hour for simple shear tests), it is possible for the cyclic stress ratio, $\tau_{cyc}/s_u$, to exceed unity, as illustrated by data on Drammen clay shown in Fig. 9 (Andersen, 1991).

The corresponding procedure for excess pore pressure development is shown in Fig. 10, this time with the cyclic shear stress level expressed as $\tau_{cyc}/\sigma'_{vc}$. Liquefaction corresponds to an excess pore pressure ratio of $\Delta u/\sigma'_{vc} \sim 1$.

A typical design storm loading will comprise a series of waves (and hence

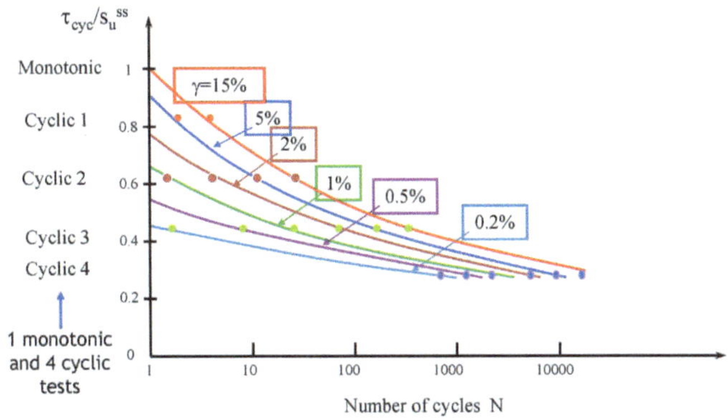


**Figure 8.** Construction of shear strain contour diagrams from cyclic tests. Here $\tau_{\text{cyc}}/s_u^{SS}$ stands for the cyclic shear stress imposed, relative to the monotonic strength in simple shear.

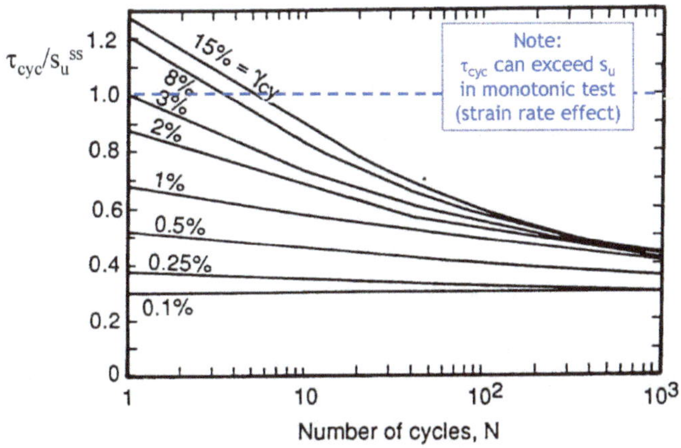


**Figure 9.** Example cyclic simple shear data on Drammen clay, showing effect of higher strain rates in cyclic tests (after Andersen, 1991.)

loads) of increasing magnitude, but decreasing number, with the peak design loading usually occurring only once. Rather than model the complete loading sequence, offshore design often makes use of an 'equivalent' number of cycles (of the peak design loading) to represent the cumulative damage

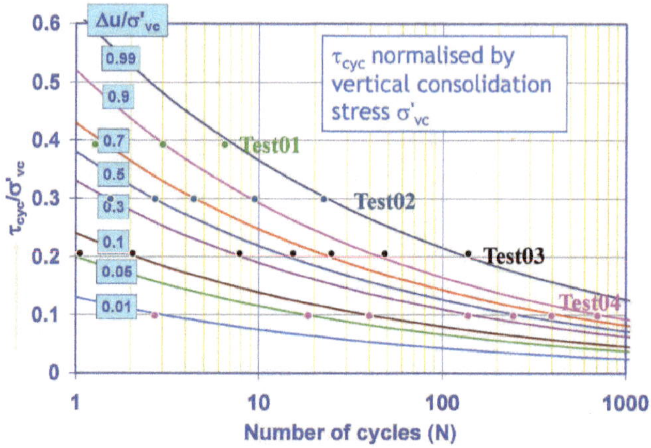


**Figure 10.** Construction of excess pore pressure contour diagrams from cyclic tests.

that occurs under the full sequence. Either the shear strain or excess pore pressure contour diagrams may form the basis for establishing the equivalent number of cycles, and the procedure is illustrated in Fig. 11 (Andersen et al., 1992). Starting with the smallest (most frequent) loading level, the excess pore pressure ratio that would develop under that cyclic shear stress is estimated by plotting the data point at the corresponding values of $\tau_{\mathrm{cyc}}/\sigma'_{\mathrm{vc}}$ and the number of cycles. The notional contour for that magnitude of excess pore pressure ratio is then traced back (parallel to the closest actual contour) to reach the next higher cyclic shear stress level in the storm sequence. That point represents an equivalent number of cycles at this cyclic shear stress level to give the same degree of damage as the larger number of cycles at the lower cyclic shear stress level. The process is now repeated, but with the end point from each stage plotted by adding the number of cycles at the new cyclic shear stress level to the (deduced) equivalent number of cycles obtained so far (from the previous loading levels). The process finishes with the peak design load level, and the final point represents the equivalent number of cycles for that design load level, and that particular storm sequence. Typically the equivalent number of cycles will be in the range 10 to 20 cycles (although the procedure illustrated in Fig. 11 leads to $N_{\mathrm{eq}} = 60$).

The equivalent number of cycles may be used to establish a 'cyclic' stress

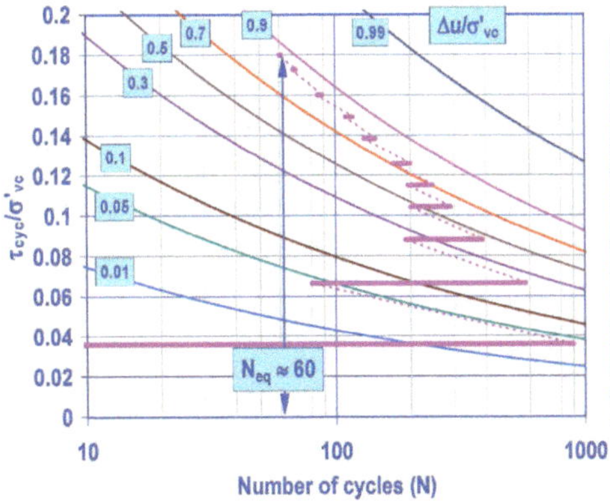


| Design Storm | | |
|---|---|---|
| No. | $H/H_{max}$ (%) | $\tau/\sigma'_v$ |
| 1 | 100 | 0.180 |
| 2 | 96 | 0.173 |
| 4 | 89 | 0.160 |
| 8 | 83 | 0.149 |
| 15 | 77 | 0.139 |
| 30 | 70 | 0.126 |
| 50 | 64 | 0.115 |
| 90 | 58 | 0.104 |
| 200 | 49 | 0.088 |
| 500 | 37 | 0.067 |
| 900 | 20 | 0.036 |

**Figure 11.** Application of Miner's rule to evaluate equivalent number of cycles.

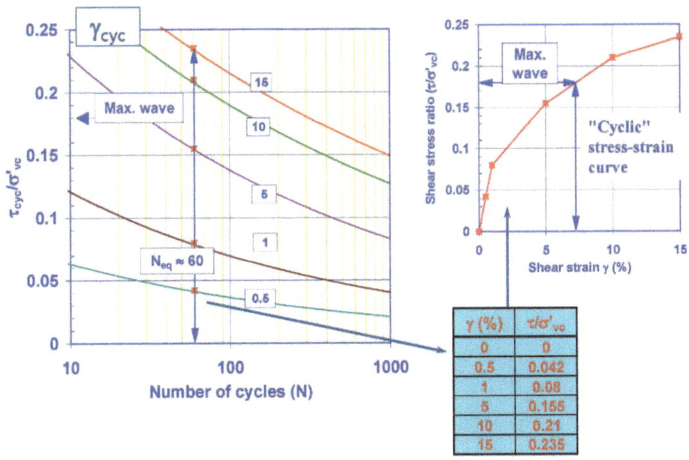


| γ (%) | $\tau/\sigma'_{vc}$ |
|---|---|
| 0 | 0 |
| 0.5 | 0.042 |
| 1 | 0.08 |
| 5 | 0.155 |
| 10 | 0.21 |
| 15 | 0.235 |

**Figure 12.** Derivation of an equivalent 'cyclic' stress-strain curve.

strain curve, which represents the envelope of cyclic (or accumulated) shear strain developed over that number of cycles for different cyclic shear stress levels Fig. 12. The cyclic stress strain curve may be used in a numerical (e.g. FE) analysis to evaluate the cyclic response of the foundation, in place of a non-linear monotonic response.

The discussion so far has been based around symmetric 2-way cyclic loading, although the same principles apply equally well to more general, asymmetric cyclic loading. Just as two shear stress magnitudes are required to specify asymmetric loading (the average shear stress, $\tau_a$, and the cyclic shear stress, $\tau_{\rm cyc}$), so two magnitudes of shear strain are also relevant: the accumulated permanent strain, $\gamma_p$ (measured at mid-cycle - sometimes referred to as average strain, $\gamma_a$) and the cyclic shear strain, $\gamma_{\rm cyc}$. This is illustrated in Fig. 13 (Andersen et al., 2008).

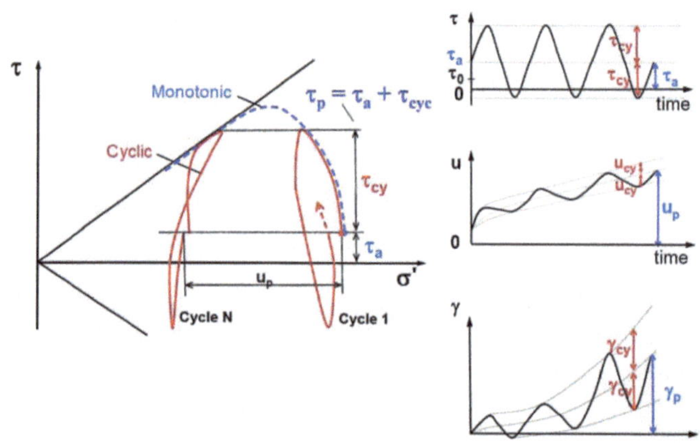


**Figure 13.** Accumulation of permanent and cyclic shear strains and excess pore pressure (after Andersen et al., 2008).

Failure under cyclic loading may be defined according to a specified maximum level of shear strain. Combinations of average and cyclic shear stress that lead to failure under different numbers of cycles are shown for triaxial conditions in Fig. 14, where failure in this case has been defined as a shear strain level of 15% (Andersen et al., 2008). Regions (at either edge) limited by the average accumulated shear strain are distinguished from those (in the centre) that are limited by cyclic shear strain. Equivalent failure contours for cyclic simple shear tests are shown in Fig. 15 (Andersen et al., 2008).

An alternative way of representing the data is to replace the cyclic shear stress axis by the peak (or failure) shear stress, $\tau_f = \tau_a + \tau_{\rm cyc}$. This shows more clearly the extent to which rate effects allow shear stresses exceeding the monotonic shear strength to be sustained for a small number of cycles (see Fig. 16). Note that for cyclic strain rates that may be 3 or 4 orders of

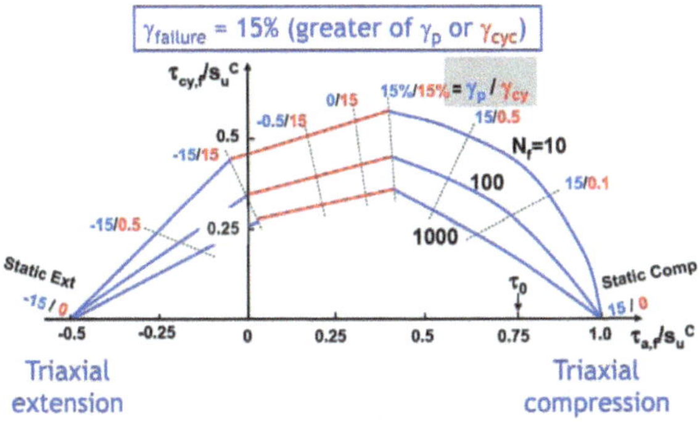


**Figure 14.** Failure envelopes from cyclic triaxial tests on Drammen clay (after Andersen et al., 2008).

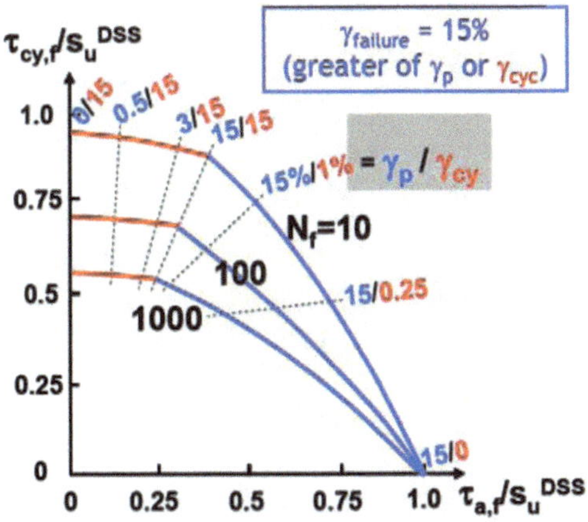


**Figure 15.** Failure envelopes from cyclic simple shear tests on Drammen clay (after Andersen et al., 2008), where $s_u^{DSS}$ is the monotonic shear strength in simple shear.

magnitude greater than typical shear strain rates for monotonic tests, and rate dependency of 10 to 15% per log cycle, the strength at cyclic strain rates may be 30 to 60% greater than the monotonic strength.

Depending on the application, much lower shear strain levels may be

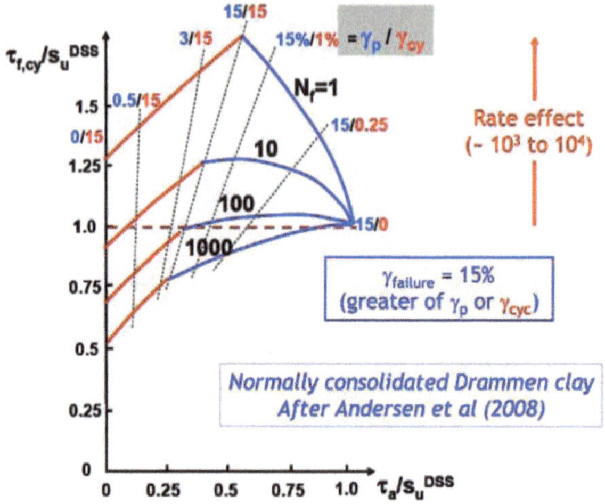


**Figure 16.** Alternative form of failure envelopes plotting maximum shear stress against the average shear stress.

permitted in a design, particularly where sensitive equipment such as an offshore wind turbine has to be supported. In such cases it is more useful to consider contours of accumulated and cyclic shear strain under different combinations of average and cyclic shear stress and a specified (equivalent) number of cycles. An example is shown in Fig. 17 for simple shear tests on Drammen clay, for the case of 10 cycles (Andersen, 1991).

A very large programme of laboratory tests is required to develop a complete picture of the cyclic response of a given soil, and few offshore projects justify sufficient tests to achieve this given that the testing programme must also address variations due to depth (different strata) and also a region that may span an area of several square kilometres. As is evident from the previous slides, the most extensive database for cyclic response is probably that developed in the 1980s for Drammen clay (Andersen, 1991). A typical approach is to undertake sufficient tests to allow comparisons with the Drammen clay database, and provide a basis for generic adjustment of the failure contours.

The range of failure contours (defined as a shear strain level of 15%) for normally consolidated clays of different plasticity index is shown in Fig. 18 for simple shear tests respectively, for $N_{eq} = 10$ and taking failure as when

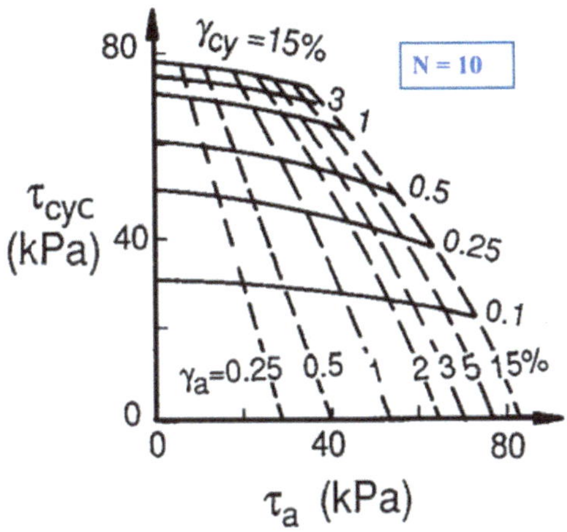


**Figure 17.** Contours of cumulative ($\gamma_a$) and cyclic ($\gamma_{cyc}$) shear strain from simple shear tests on Drammen clay (Andersen, 1991).

either the cumulative (average) or cyclic shear strain reaches 15% (Andersen et al., 2008). Under simple shear conditions, there is remarkable consistency in the cyclic shear strength envelopes of all clays, with the only exception being the extreme example of Storebaelt, with a plasticity index of 7 to 12%. The effect of plasticity index was more marked for triaxial conditions (Andersen et al., 2008). In general though, for both testing conditions, high plasticity clays tend to show greater strain rate dependency, and hence show greater cyclic shear strength, while the reverse is true for low plasticity clays.

## 3 Application to Foundation Response

As noted earlier, the actual loading applied to a foundation is multi-dimensional, and even in one vertical plane will consist of vertical, horizontal and moment components. Since the average and cyclic proportions for each load component will rarely be the same, a procedure is needed to choose an appropriate combination that represents the design loading conditions adequately. For shallow foundations, this may be achieved using the yield envelope as a measure of the extent to which the strength of the foundation soils is mobilised.

Fig. 19 shows a slice through the three-dimensional yield envelope in

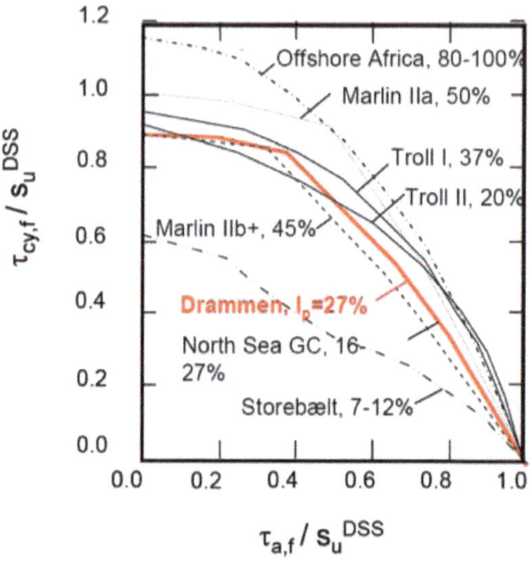


**Figure 18.** Comparison of failure envelopes for clays of different plasticity index (Andersen et al., 2008).

$V - M - H$ space for a shallow skirted foundation. The yield envelope is expressed in normalised terms, as $H/As_{u0}$ plotted against $V/As_{u0}$, where $A$ and $s_{u0}$ are the foundation plan area and shear strength at skirt tip level respectively, and represents a two-dimensional slice through the three-dimensional envelope for the appropriate moment, $M$. Three loading conditions are plotted, showing the permanent load, $V$ (assuming $H$ and $M$ are both zero), the average load for the design storm conditions and the corresponding peak loads.

The material factor, $\gamma_{m,p}$, against failure under the peak storm load conditions may be expressed as the ratio of the radial distance to the peak load, divided by the radial distance to the yield envelope, passing through the peak load. The point at which failure would occur if the peak loads were factored by $1/\gamma_{m,p}$ defines the mode of failure in terms of the incremental plastic movements, $\delta_v$ and $\delta_h$, with $\delta_v/\delta_h = -dH/dV$ (the gradient of the yield envelope) from the associated flow rule.

At peak load, $V_p$, $H_p$, the upper bound theorem leads to

$$V_p\delta_v + H_p\delta_h = \frac{\tilde{W}}{\gamma_{m,p}} \tag{1}$$

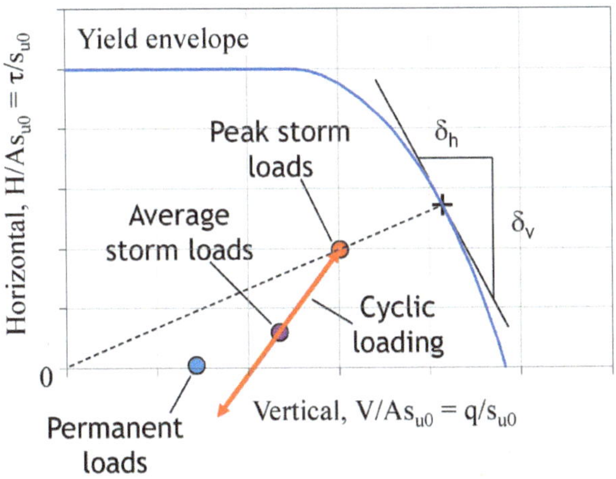


**Figure 19.** Yield envelope approach to foundation design.

where $\tilde{W}$ is the internal plastic work (volume integration of shear strains and $s_u$). It is more convenient to rewrite this as

$$V_p + H_p \left( \frac{\delta_h}{\delta_v} \right)_p = \frac{\tilde{W}/\delta_v}{\gamma_{m,p}} = \frac{\tilde{W}'}{\gamma_{m,p}} \tag{2}$$

where $\tilde{W}'$ is the work per unit vertical displacement increment, for the given failure mechanism (corresponding to peak conditions).

The material factor for average storm load conditions, but relative to the failure mechanism relevant for the peak loads (so same ratio, $\delta_h/\delta_v$ as above), may be estimated from the work equation

$$V_a + H_a \left( \frac{\delta_h}{\delta_v} \right)_p = \frac{\tilde{W}'}{\gamma_{m,a}} \tag{3}$$

Since $\tilde{W}'$ is unaltered, we may write

$$\frac{\gamma_{m,a}}{\gamma_{m,p}} = \frac{V_p + H_p \left( \frac{\delta_h}{\delta_v} \right)_p}{V_a + H_a \left( \frac{\delta_h}{\delta_v} \right)_p} \tag{4}$$

and the cyclic load reversal is therefore expressed as

$$\frac{\dfrac{1}{\gamma_{m,p}}-\dfrac{1}{\gamma_{m,a}}}{\dfrac{1}{\gamma_{m,a}}}=\frac{\gamma_{m,a}}{\gamma_{m,p}}-1=\frac{V_p+H_p\left(\dfrac{\delta_h}{\delta_v}\right)_p}{V_a+H_a\left(\dfrac{\delta_h}{\delta_v}\right)_p}-1=\frac{V_p-V_a+(H_p-H_a)\left(\dfrac{\delta_h}{\delta_v}\right)_p}{V_a+H_a\left(\dfrac{\delta_h}{\delta_v}\right)_p} \tag{5}$$

This equation gives the relative magnitudes of the cyclic and average shear stresses to be applied in laboratory testing, or in interpolating from contour maps of damage.

As noted earlier, laboratory testing for offshore projects tends to focus on simple shear testing, partly to maximise the use of very expensive soil samples, and partly because simple shear response is representative of the average stress-strain response. However, some foundation types or loading conditions lend themselves to triaxial testing, particularly when the focus is on settlement or low-level cumulative shear strains, rather than on stability. Fig. 20 shows the general arrangement from a recent project, where a caisson foundation for a wind farm development was embedded through clay to tip into medium dense sand. The load level was relatively low in comparison to the caisson capacity, but cumulative settlements under vertical cyclic loading was a potential design problem. (Note that horizontal loading of the caisson was assumed to be taken out entirely within the clay layer.)

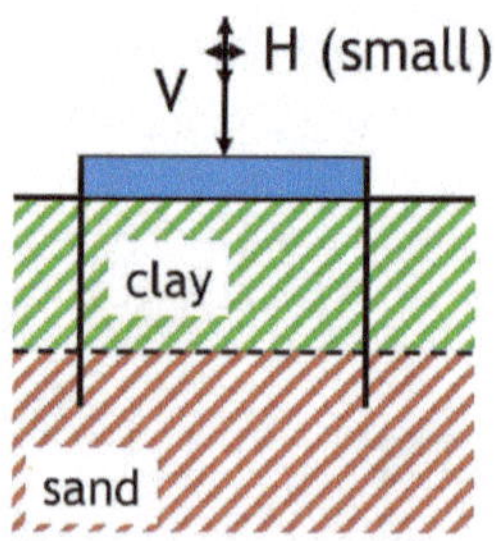


**Figure 20.** Schematic of caisson design.

A triaxial sample of the sand was consolidated under stresses representing the average (long time scale) design storm conditions, and then the cyclic component was applied in an undrained manner. The response is shown in Fig. 21, showing a cumulative axial strain of 0.2%. A centrifuge model test

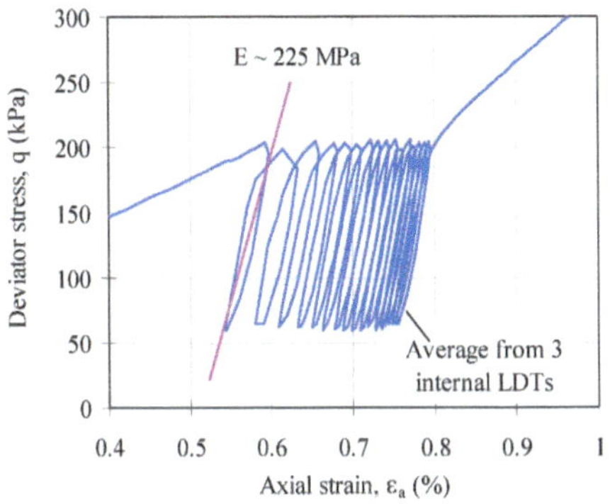


**Figure 21.** Cumulative axial strain from triaxial test on sand.

of the foundation gave rise to 16 mm of settlement under the corresponding cyclic loading regime, which was 0.23% of the foundation diameter. The comparison is somewhat approximate (and perhaps fortuitous), but suggests a general consistency of the results at the element level and for the model foundation.

## 4 Role of Model Testing in Design

Physical modelling can play a major role in the design process, either through validation of numerical approaches or directly, for example, by indicating safe levels of cyclic loading that may be applied without adverse effects. Even where field conditions are not reproduced accurately, numerical (or other) design techniques should be able to match the results of model tests accurately, particularly since the soil stratigraphy will usually be simpler than in the field, and more accurately characterised. The interplay of different input into the design process is illustrated in Fig. 22.

Physical modellers are mostly sceptical of the ability to simulate soil response through numerical analysis, particularly when it comes to cyclic loading. In both approaches there are limitations in modelling the soil response, either due to shortcomings or simplifications in the constitutive models adopted for numerical analysis, or because of the use of reconstituted soil in model tests. Issues of time scale, with high strain rates and yet relatively slow cyclic loading in comparison to consolidation times, can lead to errors in physical modelling.

Numerical analysis is often restricted to 2-dimensional analysis, although

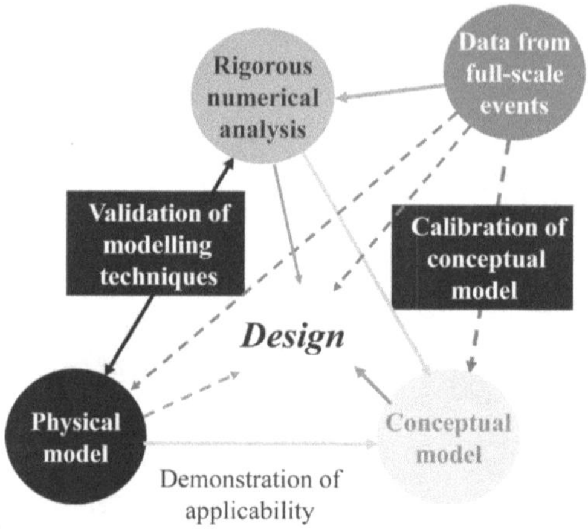


**Figure 22.** Interplay of different input into the design process.

this is becoming less of an issue with the increasing power of computers. Physical modelling captures the relevant 3-dimensional geometry, but may have size limitations (particularly with centrifuge modelling). In both approaches, detailed visualisation is possible, with unlimited access to kinematics, strain and stress paths throughout the domain for numerical modelling, but more limited internal measurements in model tests. Image processing and PIV techniques in model testing (White et al., 2003) has allowed a major improvement in quantifying mechanisms.

Centrifuge modelling, as opposed to small scale prototype experiments conducted in the field or laboratory, allows true scale modelling of stresses and shear strength relative to each other and to the stiffness of structural members within the soil. It is also a cost effective means of conducting limited parametric studies of the response of a foundation system to cyclic loading of different magnitudes.

At the simplest level, model tests allow the load-displacement response of a foundation or anchoring system to be quantified for a given strength profile in the soil. It is essential that consistent and accurate methods are used to obtain the strength profile, for example using different types of penetrometer, and that any changes in shear strength or sand density through the course of tests is monitored.

Detailed observations using PIV enable accurate measurements of deformations at the soil surface of down planes of symmetry, revealing the spatial decay of deformations and internal features such as shear bands. It is also possible to model complex geometries or construction techniques such as pile or anchor installation, dynamic compaction, suction installed caissons; complex soil response such as crushing or strain softening and the effects of cyclic loading, creep or large deformations may be reproduced faithfully by using the appropriate soil for the field conditions.

The second half of this chapter shows examples of model tests undertaken to explore the effects of cyclic loading, consider anchoring systems first, where the focus is on assessing any reduction in capacity of the system due to cyclic loading. Because anchoring systems will tend to pull out of the soil as deformations accumulate, capacity is a critical issue. Foundation systems loaded in compression will then be considered, where the focus is more on cumulative deformations that may lead to a serviceability limit state.

# 5 Anchoring Systems

## 5.1 Suction Anchors

Examples of model testing for two different anchoring systems, suction anchors and dynamically embedded anchors, are considered. The first example is drawn from the relatively early period of experience with suction anchors - large caissons that are installed using under pressure, pumping water out from within the caisson cavity. In moderate water depths, catenary moorings may be used without leading to an excessive footprint of a floating system, and these provide primarily horizontal loading of the suction anchor, with loading angles generally around 25° or less.

The example is from an actual project, the Laminaria field in the Timor Sea operated by Woodside Energy (then Woodside Offshore Petroleum) (see Fig. 23). Catenary moorings were used to secure the FPSO (floating production, storage and offloading) facility in a water depth of 380 m, with suction anchors used to anchor the moorings. Originally, the anchors were proposed with removable lids, used during installation but with the anchors then left open topped; this led to a concern as to the extent to which the anchors might undergo vertical movement during cyclic loading. Eventually, partly as a result of the model testing, the 5.4 m diameter anchors were designed with permanent lids and a maximum embedment depth of 12.7 m (Fig. 24).

The padeye to which the catenary mooring is attached is located at of two thirds of the embedment depth, on the basis that this is the centre of resistance for soil profiles where the strength increases approximately proportionally with depth. This will give maximum lateral capacity, with

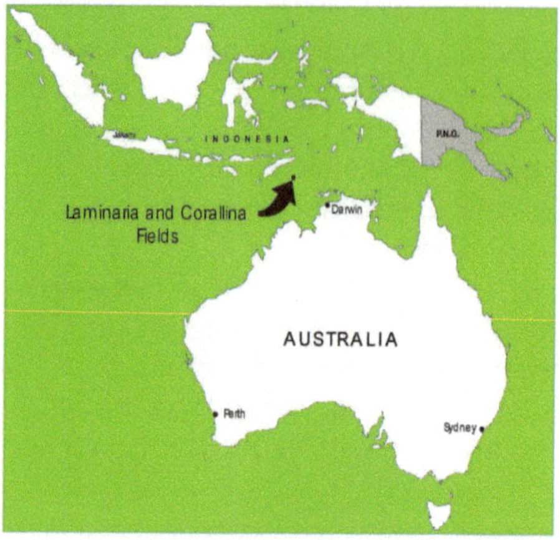


**Figure 23.** Location of Laminaria field in the Timor Sea.

**Figure 24.** Laminaria suction anchors in transit.

pure translation of the caisson rather than forward or backward rotation (Fig. 25). (Note, the padeye position should be chosen so that the line of action of the mooring chain crosses the central axis of the caisson at the depth of the centre of resistance.)

The model tests were conducted on the beam centrifuge at UWA, as described by Randolph et al. (1998). The general layout of the tests, with two tests being conducted in each 'strongbox' containing the soil, is shown in Fig. 26, and the model suction caisson, with a scaling ratio of 1:120, is shown in Fig. 27. Note that the suction caisson has a perforated cap and was jacked into the seabed (in flight, at 120 g, where g stands for the gravity acceleration) rather than being installed by suction. Extensive testing subsequently has shown that the method of installation (suction or jacking) has negligible effect on the axial capacity (Jeanjean et al., 2006; Chen and Randolph, 2007a).

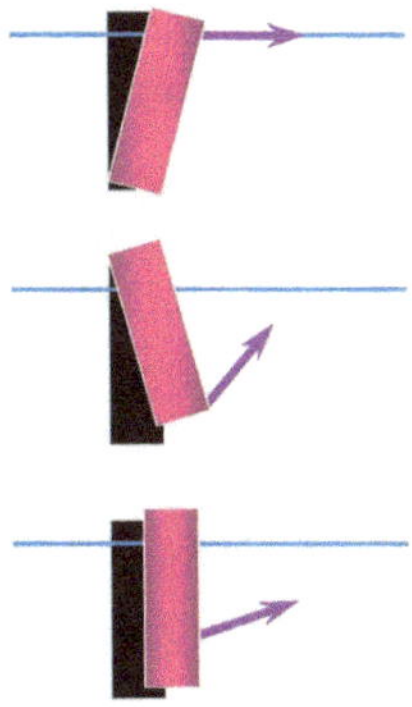

**Figure 25.** Effect of padeye depth on suction anchor response.

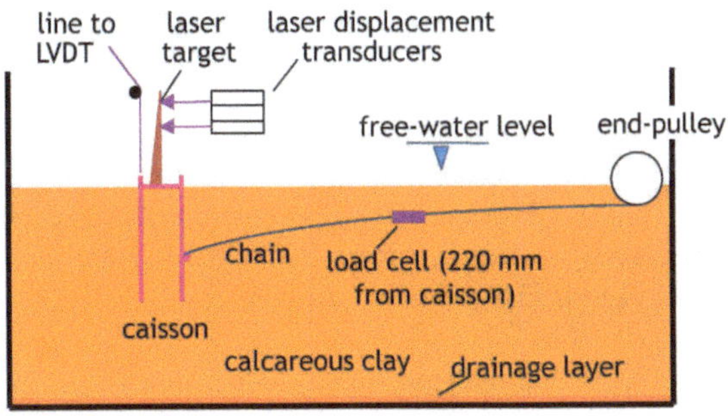


**Figure 26.** General arrangement for centrifuge model tests on suction anchors.

The seabed sediments at Laminaria consist of a carbonate mud, and bulk samples were recovered from the seabed for the model tests. Two soil samples were prepared, one a normally consolidated sample (consolidated

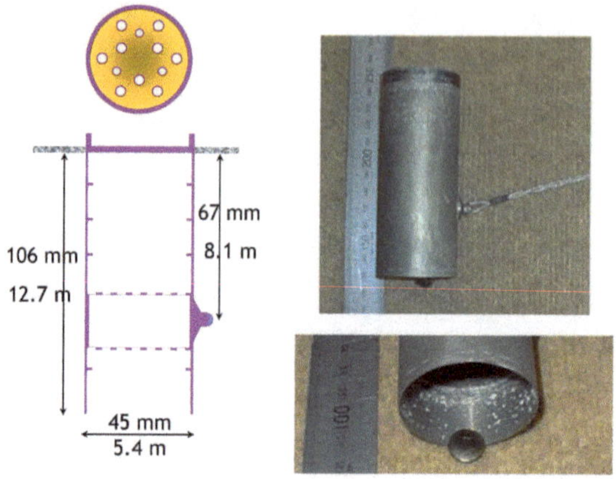


**Figure 27.** Schematic and photograph of model suction anchors.

in flight on the centrifuge), and the other light overconsolidated by pre-consolidation in a press on the laboratory floor in order to match the field profile better (Fig. 28). The strength profiles were measured using a T-bar penetrometer (Stewart and Randolph, 1994).

Monotonic and cyclic tests were carried out in each of the two soil samples. Fig. 29 illustrates the failure zones that developed in the normally consolidated and lightly over consolidated samples. In the former, the soil behind the caisson remained attached to the caisson initially, but then after a movement of about 30 or 40% of the caisson diameter the suction was lost and a gap appeared (accompanied by a reduction in the lateral resistance, see Fig. 30). The photograph shows evidence of significant forward movement of the soil behind the caisson prior to the gap developing. For the over consolidated sample, a gap appeared right from the start of movement of the caisson, leaving a sharp back-face.

The monotonic response of the caisson in the normally consolidated sample is shown in Fig. 30. The peak capacity corresponds well with a 2-sided upper bound solution based on Murff and Hamilton (1993), while after the gap appears the capacity reduces to the 1-sided bound solution (Randolph et al., 1998). The rotation response indicates forward rotation initially, and then a reduction in rotation once the gap appears.

The corresponding cyclic response is shown in Fig. 31. Pure 1-way cyclic loading was applied, initially for 100 cycles between zero and the design load of 4.6 MN. The maximum load level was then increased to the factored de-

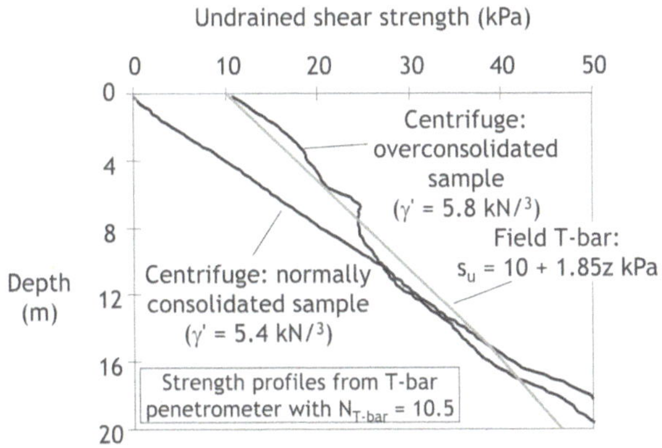


**Figure 28.** Comparison of shear strength profiles from field and centrifuge model tests. Here, $\gamma'$ is the effective unit volume weight.

**Figure 29.** Post-failure photographs of model suction caisson in normally consolidated (left) and lightly overconsolidated (right) samples.

sign of 6 MN and a further 200 load cycles applied. On further increase of the load level to 7.2 MN, the chain parted after 16 cycles. The monotonic loading response appears to provide a bounding envelope to the cyclic response.

The cumulative horizontal translation of the caisson with the three packages of cyclic loading is shown in Fig. 32. The movements are relatively small, and appear to be stabilising at the intermediate load level, even though this is close to the 1-sided failure load of the caisson.

Corresponding monotonic and cyclic responses for the caisson in the lightly over consolidated sample are shown in Fig. 33, with the cyclic load-

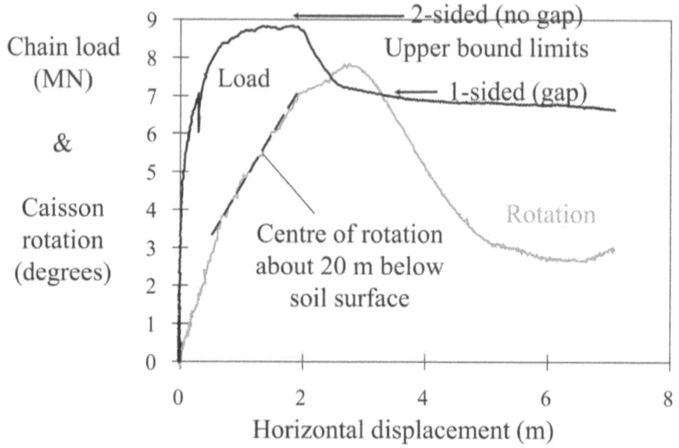


**Figure 30.** Monotonic response of suction caisson in normally consolidated sample.

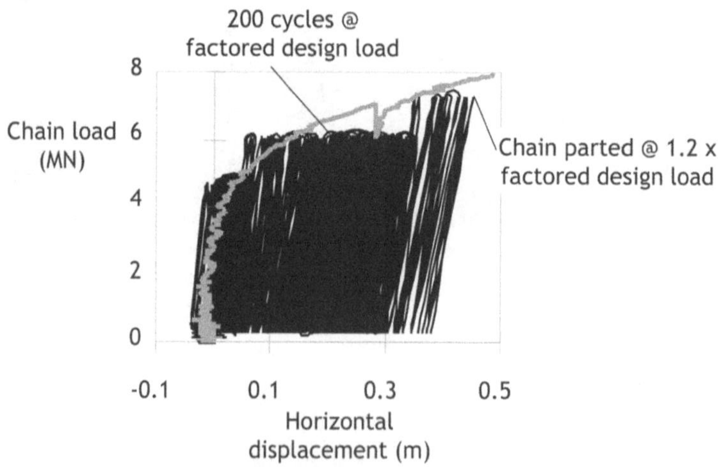


**Figure 31.** Cyclic response of suction caisson in normally consolidated sample.

ing comprising 100 cycles at each of 5.5 MN, 7.1 MN and 9.2 MN. The anchor started to fail after 10 cycles at the highest load level, which corresponds to the peak resistance measured in the monotonic test. The overall

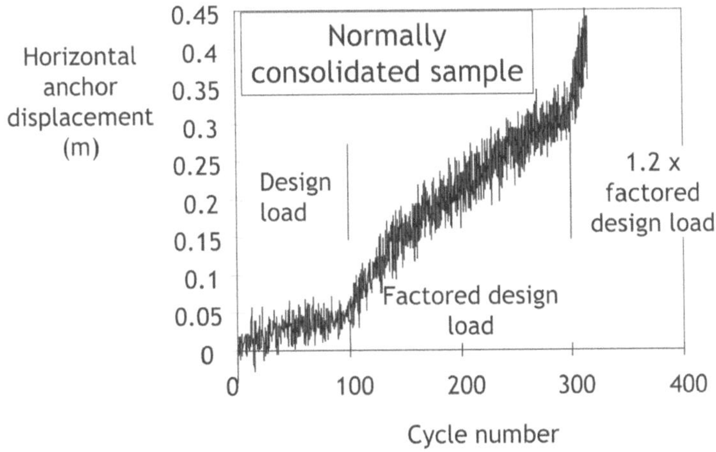


**Figure 32.** Cumulative horizontal displacement of suction caisson under cyclic loading.

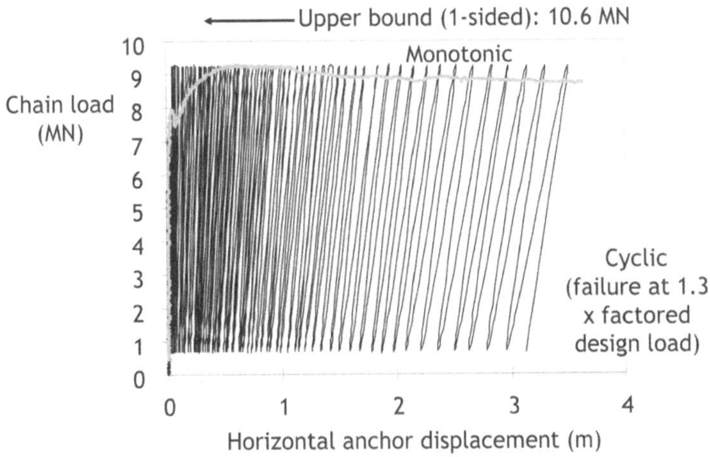


**Figure 33.** Overall response of suction caisson in lightly overconsolidated sample.

cyclic response, including the detail shown in Fig. 34, shows that the monotonic curve again closely matches the monotonic response. This is also true of the rotational response, indicating that the same failure mechanism under monotonic loading pertains for the incremental motion under cyclic loading (Fig. 35).

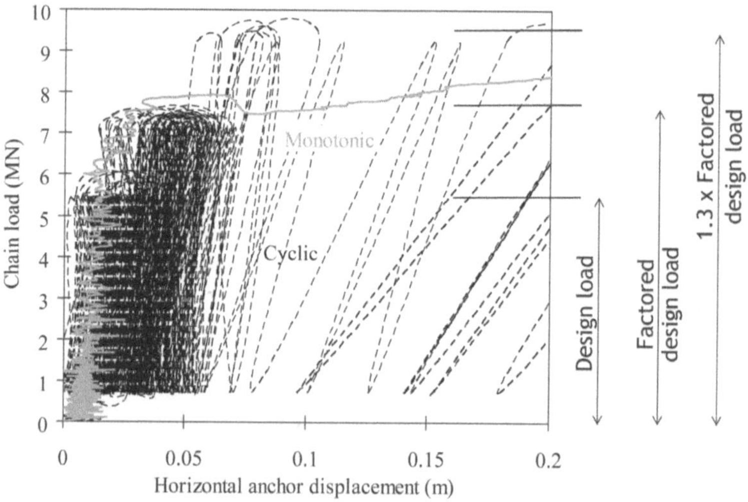


**Figure 34.** Detailed cyclic response of suction caisson in lightly overconsolidated sample.

Of particular interest was the relative magnitudes of pre-failure vertical and horizontal movement. This is shown in Fig. 36 for the tests in the over consolidated sample, with upward vertical movement of about 50% of the horizontal movement initially. Interestingly, the test on the over consolidated sample showed much smaller vertical movements, even though other aspects (such as the rotation) were well matched. The high ratio of vertical to horizontal movement from the monotonic test gave rise to concern for the open topped caisson design.

In summary, the cyclic loading tests showed a robust response with little evidence of damage to the soil as a result of cyclic loading. The caisson withstood high levels of cyclic loading without significant movements until the factored design load was exceeded. The measured capacities were reasonably consistent with those predicted using the upper bound approach, even though the rotation of the caisson in over consolidated soil showed evidence of a different mechanism with the centre of rotation well below

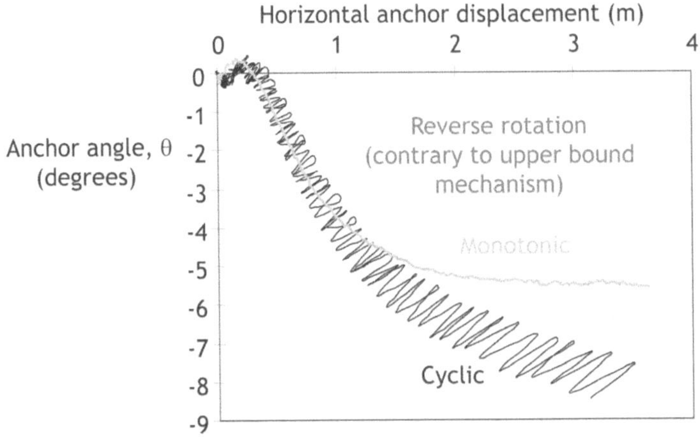


**Figure 35.** Rotational response of suction caisson in lightly overconsolidated sample.

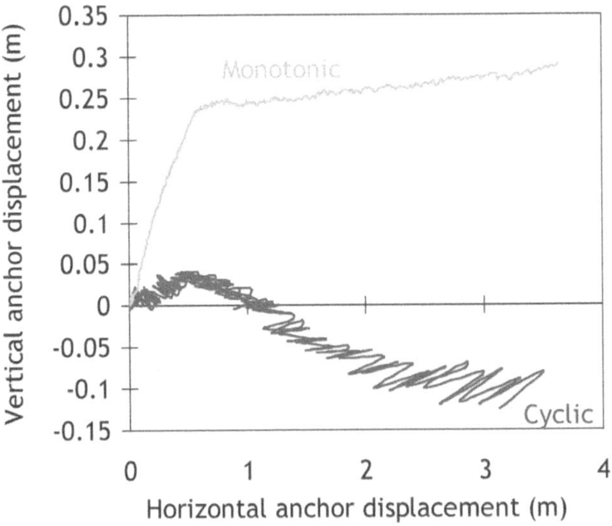


**Figure 36.** Relative displacements of suction caisson in lightly overconsolidated sample.

the centre of gravity (Fig. 35). The overall conclusion from the study was to revise the design and adopt a closed top for the caisson rather than a removable lid.

The effect of pure vertical cyclic loading was investigated by Chen and Randolph (2007b), with 1-way cyclic loading applied to a centrifuge model caisson with embedment ratio of 4. They found that, in three slightly different lightly overconsolidated clay samples, cyclic load levels of 72 to 86% were required to trigger significant incremental upward movements and failure (see Fig. 37). This is a slightly higher range than reported previously by Clukey et al. (1995), where failure occurred at load levels as low as 61 to 89%, although some tests showed no failure at load levels of 67 to 69%. An additional feature of the latter tests was the inclusion of a small (and varying) loading angle of up to 6° from the vertical.

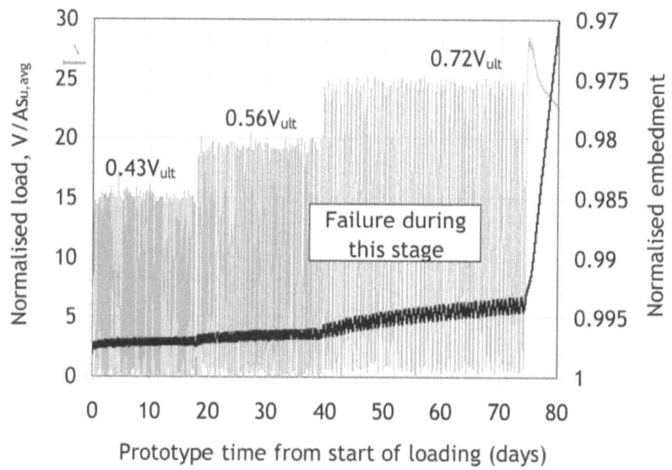


**Figure 37.** One-way vertical cyclic loading response of suction caisson in kaolin (Chen and Randolph, 2007b).

## 5.2 Dynamically Embedded Anchors

There has been a rapid growth over the last decade in the use of dynamically embedded torpedo-shaped anchors. These have been spearheaded by Petrobras for anchoring mobile offshore drilling units off the coast of Brazil (Medeiros, 2001), although an alternative design for use in the North Sea has been developed by Lieng et al. (1999) (see Fig. 38, top left and right respectively). Both those anchors have the mooring chain attached at the top of the anchor on the centreline. More recently a dynamically embedded anchor has been developed by Delmar for use in the Gulf of Mexico, with the mooring chain attached to a revolvable shank part way down the shaft of the anchor (Zimmerman et al., 2009) (see Fig. 38, bottom left).

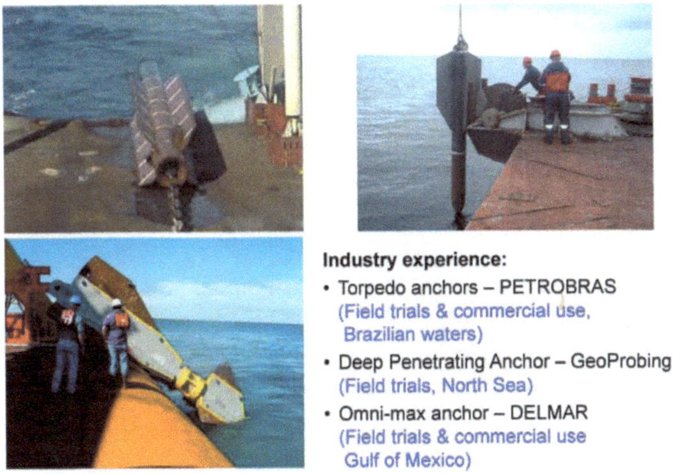


**Figure 38.** Commercial use of dynamically embedded anchors.

Anchors such as these are released from 50 to 100 m above the seabed with sufficient length of chain above them to the release point to allow full penetration. The anchors reach a terminal velocity of 25 to 30 m/s at the seabed, and penetrate with the tip embedded 2 to 3 anchor lengths. Although loading is usually applied at around 40° from the horizontal, vertical pullout through the entry wound is a critical design issue. Cyclic loading tests with pure vertical loading were undertaken by Richardson (2008) on the beam centrifuge at UWA, with anchors scaled by 1:200 (see Fig. 39).

A typical test result is shown in Fig. 40, with (inevitably) 1-way cyclic loading applied at a maximum load level of 80% of the monotonic capacity. Failure was not triggered at this level in four separate tests; post-cyclic monotonic tests gave capacities in the range 98 to 109% of the monotonic capacity without cyclic loading. Note that the large displacement to mobilise the anchor load is primarily associated with stretching of the anchor chain, rather than movement of the anchor itself.

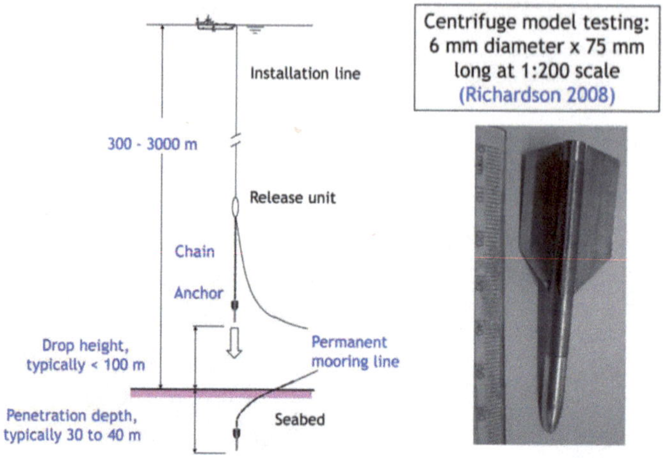


**Figure 39.** Schematic of dynamic anchor embedment and 1:200 model anchor.

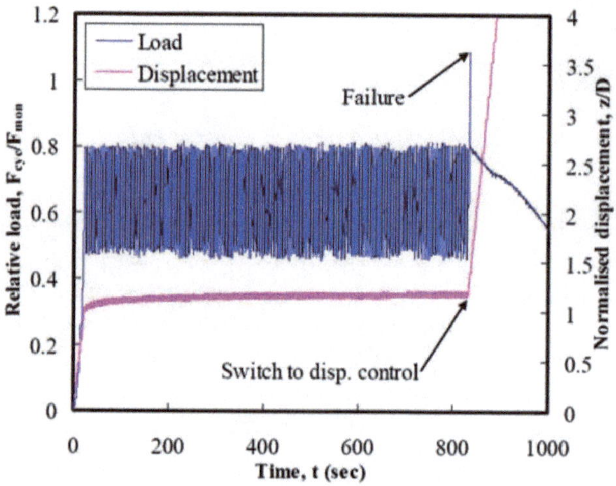


**Figure 40.** Typical cyclic loading response of torpedo anchor (Richardson, 2008).

## 6 Skirted Foundation in Calcareous Silt

Attention is now turned to shallow foundations loaded in compression, where accumulation of deformations under cyclic loading is a potential serviceability design issue, even though the capacity is not compromised. Such caissons, or skirted foundations, are used to support subsea equipment such as manifolds, but are also being considered as foundations for tripod supported offshore wind farms.

An example is taken from the doctoral research of Watson (1999), with details of the cyclic model testing reported by Watson and Randolph (2006). Centrifuge model tests at a scaling ratio of 1:100 were undertaken on a caisson foundation, 75 mm in diameter with skirts of penetrating by half the diameter into normally consolidated calcareous silt (Fig. 41). Vertical and horizontal loading were applied through a rigid loading arm, and the tests discussed here were conducted under a fixed vertical load, $V$, of 40% of the monotonic capacity. Symmetric 2-way horizontal cyclic loading was applied at different load levels, with normalised loads ranging from $\Delta H/As_{u0}$ of $\pm 0.2$ to 2, where $A$ is the plan area of the foundation and $s_{u0}$ is the shear strength at the level of the skirt tips. A typical monotonic response is shown in Fig. 41, showing significant non-linearity in response for $\Delta H/As_{u0} > 1$.

It was found that vertical settlement accumulated during the horizontal cyclic loading, with an almost linear increase with the number of cycles, but with the rate of increase strongly affected by the cyclic loading level (Fig. 42).

The tests were interpreted in a manner similar to that for laboratory element tests, with 'fatigue' contours developed showing different magnitudes of cumulative vertical settlement as a function of cyclic horizontal load level and the number of cycles (see Fig. 43). A different form of contour plot was used to express the amplitude of horizontal movement, since this tended to reach a plateau beyond about 40 cycles; hence it was easier to construct contours showing the cyclic movement $h_{\mathrm{cyc}}/D$ as a function of cyclic horizontal load level for different numbers of loading cycles ($h$ is the total horizontal displacement, $h_{\mathrm{cyc}}$ is the horizontal displacement induced by the cyclic loading, and $D$ is the foundation diameter).

Although the main focus was on compression loading, tests were also conducted with ambient tensile loading applied to the foundation. Perhaps not surprisingly, for a given level of cyclic loading, cyclic horizontal and cumulative vertical displacements were significantly larger for tensile loading compared with compression Fig. 44. The horizontal movements tended to accelerate for load levels of $\Delta H/As_{u0} \geq 1.4$ if the static vertical loading was tensile. Further studies of shallow skirted foundations loaded in tension have been reported recently by Acosta-Martinez and Gourvenec (2008).

Using contour diagrams of the form shown in Fig. 43, empirical methods

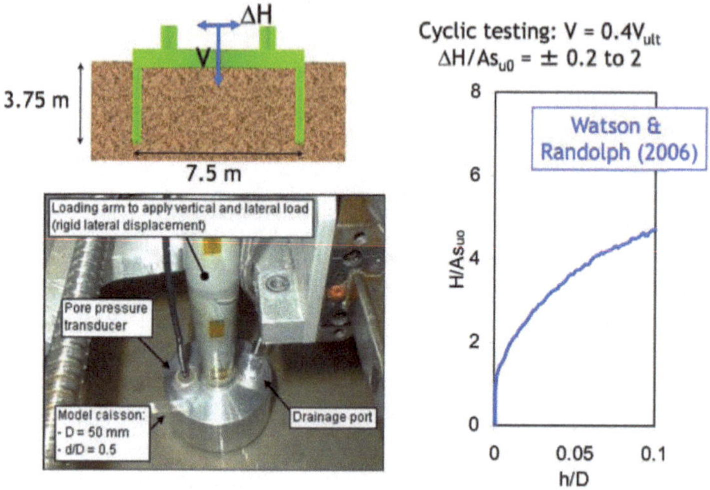


**Figure 41.** Model caisson and response in normally consolidated calcareous silt.

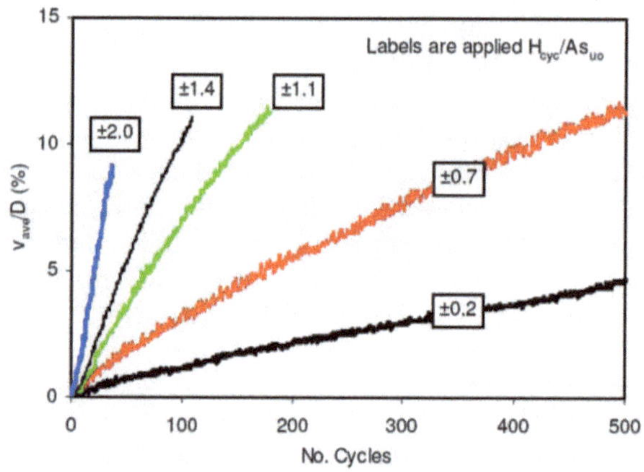


**Figure 42.** Cumulative settlement $v_{ave}$ of caisson under cyclic horizontal loading (Watson and Randolph, 2006).

were developed by Watson (1999) for predicting foundation response under general loading conditions, such as experienced during storm loading.

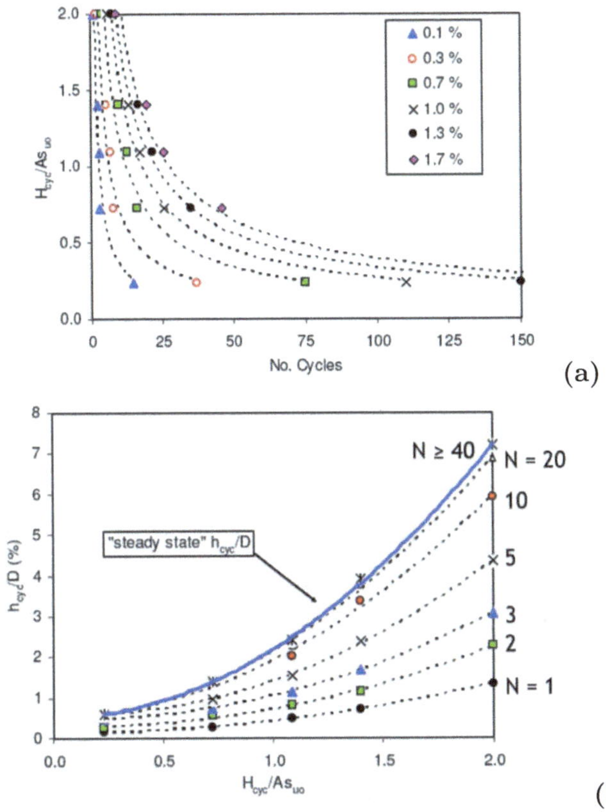


**Figure 43.** Design charts for cyclic displacements of caisson (Watson and Randolph, 2006): (a) Contours of average vertical displacement, $v_{\mathrm{ave}}/D$; (b) Horizontal cyclic displacement, $h_{\mathrm{cyc}}/D$.

An example is shown in Fig. 45, showing the cyclic loading history, and the comparison between observed and predicted vertical settlement, again maintaining the vertical loading constant at $V = 0.4V_{\mathrm{ult}}$, where $V_{\mathrm{ult}}$ is the failure load under pure vertical load.

Full details of the methodology for predicting the cyclic response are given in Watson (1999). Essentially, the approach is based on Miner's Rule, where for each level of cyclic loading the accumulated settlement from previous loading is attributed to an equivalent number of cycles at the current

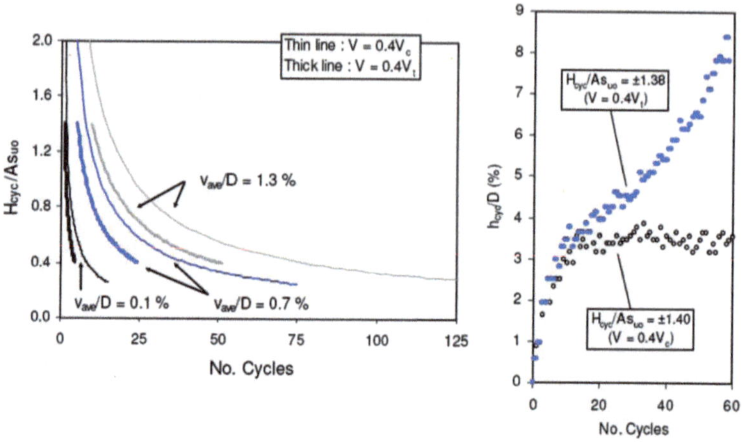


**Figure 44.** Comparison of caisson response under horizontal cyclic loading, for tensile and compressive vertical loads (Watson and Randolph, 2006).

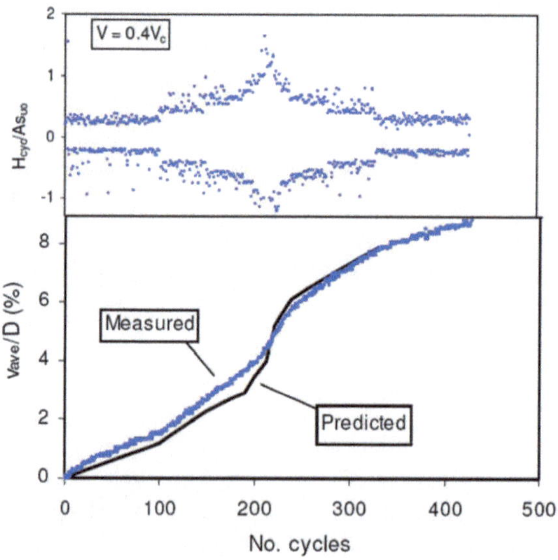


**Figure 45.** Estimation of cumulative vertical displacement of caisson under full storm horizontal cyclic loading (Watson and Randolph, 2006).

load level. Additional settlement from the (real) additional cycles at the current load level may then be calculated. The approach needed to be modified to take account of work-hardening due to increasing embedment (resulting in increased vertical and horizontal capacities) in order to reach the reasonable agreement between predicted and measured accumulation of vertical settlement (Fig. 45).

In this parametric study of the effects of cyclic loading on cumulative deformations, the style of loading is similar to that for programmes of laboratory testing. A monotonic test to failure ($H_{\mathrm{ult}}$) is useful to set the scale for cyclic loading, and then separate cyclic model tests are carried out at different magnitudes of $\Delta H/H_{\mathrm{ult}}$. Fatigue contours are then constructed to capture the effect of cyclic loading level and numbers of cycles on the vertical and horizontal deformations. Finally, a Miner's rule approach is used to integrate the cumulative effect of a cyclic loading spectrum with varying magnitude of loading.

## 7 Summary

For practical offshore design, the effects of cyclic loading are generally based on estimating cumulative 'permanent' and cyclic shear strains, and cumulative excess pore pressure, from cyclic triaxial and simple shear tests. Simple shear tests are favoured over triaxial tests because of the small sample size, greater flexibility in the applied shear stress regimes and because they represent average stress paths well.

Rather than follow a full storm sequence, damage from cyclic loading is assessed using an equivalent number (often less than 20) of cycles at a shear stress level reflecting the peak storm loading. The equivalent number of cycles is derived from contour plots of cumulative damage, and may then be used to develop a 'cyclic' stress strain response.

A limited number of site specific cyclic loading tests may be sufficient to justify adoption of generic curves for cyclic response derived from other soils. The procedure is still relatively robust, since similar cyclic strength envelopes have been observed for different soils with a wide range of plasticity indices.

The relative magnitude of average and cyclic shear stresses may be assessed for a given application by considering the relative material factors with respect to the multi-dimensional yield envelope for the foundation. This allows a consistent choice, reflecting the average shear stress mobilisation under different loading conditions.

Model tests have a useful role to play in design in terms of identifying potential failure mechanisms or quantifying effects of cyclic loading that are difficult to capture through numerical analysis. Centrifuge model tests allow correct scaling of strength and stress levels in the soil. However, the most difficult aspect of cyclic loading to capture is the balance between accumulating damage (for example excess pore pressures and strain softening) and healing due to consolidation, since at model scale the time scale of cyclic loading is relatively long in comparison to consolidation times.

The two main focuses in the latter part of this chapter were on (potential) loss in tensile capacity of anchors due to cyclic loading, and cyclic induced deformations for shallow foundations. Both suction anchors and dynamically embedded anchors are relatively robust in respect of cyclic loading, with safe threshold levels under purely vertical loading that exceed 70% of the monotonic capacity. Cyclic loading of dynamically embedded anchors showed no tendency for incremental pullout at cyclic load levels of 80% of the monotonic capacity, and generally showed higher post-cyclic failure loads than when tested monotonically without cycling.

Cyclic induced deformation of shallow foundations were explored and quantified in a similar manner to laboratory element tests. Horizontal cyclic loading much greater than the base sliding capacity ($\Delta H/As_{u0} > 1$) tended to show an acceleration in both horizontal movement and cumulative vertical settlement. Fatigue contour diagrams of cyclic-induced deformation as a function of cyclic loading level and numbers of cycles provided a reasonable basis to assess the effect of a wave loading spectrum with varying magnitude of cyclic load using a Miner's rule approach.

## 8 Acknowledgements

This chapter draws from the activities of the Centre for Offshore Foundation Systems (COFS), established under the Australian Research Council's Research Centres Program and currently supported as a Centre of Excellence by the State of Western Australia and through grants FF0561473 and DP0665958 from the Australian Research Council. The author would like to acknowledge the significant contribution from colleagues at COFS and elsewhere in the world in the material presented.

## List of Symbols

| | | |
|---|---|---|
| $a$ | average load conditions | §1, §2, §3 |
| cyc | cyclic load conditions | §1, §2, §6 |
| $p$ | peak (or failure) load conditions | §3 |
| $A$ | foundation plan area | §3, §6 |
| $D$ | foundation diameter | §6 |
| $E$ | Young's modulus | §3 |
| g | gravity acceleration | §5.1 |
| $h$ | application point of the horizontal load | §1 |
| $h$ | horizontal displacement | §6 |
| $H$ | horizontal load | §1, §3, §6 |
| $H_{\text{ult}}$ | failure load under horizontal load | §6 |
| $M$ | overturning moment | §3 |
| $N$ | number of cycles | §1 |
| $N_{\text{eq}}$ | equivalent number of cycles | §2 |
| $q$ | deviatoric stress | §3 |
| $s_u$ | shear strength | §2 |
| $s_{u0}$ | shear strength at skirt tip level | §3, §6 |
| $s_u^{SS}$ | monotonic strength in simple shear | §2 |
| $s_u^{DSS}$ | monotonic shear strength in simple shear | §2 |
| $u$ | pore pressure | §2 |
| $\text{v}_{\text{ave}}$ | cumulative settlement of caisson | §6 |
| $V$ | vertical load | §1, §3, §6 |
| $V_{\text{ult}}$ | failure load under pure vertical load | §6 |
| $\tilde{W}$ | internal plastic work | §3 |
| $\tilde{W}'$ | work per unit vertical displacement increment | §3 |
| $W'$ | buoyant weight of an ideal offshore foundation | §1 |
| $\gamma$ | shear strain | §2 |
| $\gamma'$ | effective unit volume weight | §5.1 |
| $\gamma_p$ | accumulated permanent shear strain | §2 |
| $\gamma_{m,a}$ | material factor for average load conditions | §3 |

| | | |
|---|---|---|
| $\gamma_{m,p}$ | material factor for peak load conditions | §3 |
| $\delta_h$ | incremental horizontal plastic movement | §3 |
| $\delta_v$ | incremental vertical plastic movement | §3 |
| $\varepsilon_a$ | axial strain | §3 |
| $\sigma'_{\mathrm{hc}}$ | horizontal consolidation effective stress | §2 |
| $\sigma'_{\mathrm{vc}}$ | vertical consolidation effective stress | §2 |
| $\sigma'_{\mathrm{v}}$ | vertical effective stress | §2 |
| $\tau$ | shear stress | §1 |
| $\tau_f$ | peak (or failure) shear stress | §2 |

## Bibliography

H. E. Acosta-Martinez and S.M. Gourvenec. Response of skirted foundations for buoyant facilities subjected to cyclic uplift loading. In *Proc. 18th Int. Offshore and Polar Eng. Conf., ISOPE 08*, volume 2, pages 705–712, Vancouver, Canada, 2008.

K. H. Andersen. Foundation design of offshore gravity structures. In *Cyclic Loading of Soils*, London, U.K., 1991. Eds O'Reilly & Brown, Blackie.

K. H. Andersen and R. Lauritzsen. Bearing capacity for foundation with cyclic loads. *ASCE J. Geotech. Eng. Div.*, 114 (5):540–555, 1988.

K. H. Andersen, R. Dyvik, Y. Kikuchi, and E. Skomedal. Clay behaviour under irregular cyclic loading. In *Proc. Int. Conf., Behaviour of Offshore Structures*, pages 937–950 (2), London, U.K., 1992.

K. H. Andersen, T. Lunne, T. J. Kvalstad, and C. F. Forsberg. Deep water geotechnical engineering. In *Proc. 24th Nat. Conf. of Mexican Soc. of Soil Mechanics*, pages 1–57, Aguascalientes, Mexico, 2008.

A. Bye, C. T. Erbrichand B. Rognlier, and T. I. Tjelta. Geotechnical design of bucket foundations. In *Proc. Offshore Technology Conf.*, volume OTC 7793, Houston, Texas, 1995.

B. W. Byrne and G. T. Houlsby. Experimental investigations of the response of suction caissons to transient vertical loading. *ASCE J. Geotech. and Geoenv. Eng.*, 128(11):926–939, 2002.

W. Chen and M. F. Randolph. Radial stress changes and axial capacity for suction caissons in soft clay. *Géotechnique*, 57(6):499–511, 2007a.

W. Chen and M. F. Randolph. Uplift capacity of suction caissons under sustained and cyclic loading in soft clay. *ASCE J. Geotech. and Geoenv. Eng.*, 133(11):1352–1363, 2007b.

E. C. Clukey, M. J. Morrison, J. Garnier, and J. F. Corté. The response of suction caissons in normally consolidated clays to cyclic TLP loading conditions. In *Proc. Offshore Technology Conf.*, volume OTC 7796, Houston, Texas, 1995.

P. Jeanjean, D. Znidarcic, R. Phillips, H. Y. Koand S. Pfister, and K. Schroeder. Centrifuge testing on suction anchors: double-wall, stiff clays, and layered soil profile. In *Proc. Offshore Technology Conf.*, volume OTC 18007, Houston, Texas, 2006.

R. B. Kelly, G. T. Houlsby, and B. W. Byme. Transient vertical loading of model suction caissons in a pressure chamber. *Géotechnique*, 56(10): 665–675, 2006.

J. T. Lieng, F. Hove, and T. I. Tjelta. Deep penetrating anchor: Sub-seabed deepwater anchor concept for floaters and other installations. In *Proc. 9th Int. Offshore and Polar Eng. Conf.*, pages 613–619, Brest, 1999.

C. J. Medeiros. Torpedo anchor for deep wate. In *Proc. Deep Offshore Technology Conf.*, Rio de Janeiro, Brazil, 2001.

J. D. Murff and J. M. Hamilton. P-ultimate for undrained analysis of laterally loaded piles. *ASCE J. Geotech. Eng. Div.*, 119(1):91–107, 1993.

M. F. Randolph, M. P. O'Neill, D. P. Stewart, and C. Erbrich. Performance of suction anchors in fine-grained calcareous soils. In *Proc. Offshore Technology Conf.*, volume OTC 8831, Houston, Texas, 1998.

M. D. Richardson. *Dynamically installed anchors for floating offshore structures.* PhD thesis, University of Western Australia, Australia, 2008.

D. P. Stewart and M. F. Randolph. T-bar penetration testing in soft clay. *ASCE J. Geotech. Eng. Div.*, 120(12):2230–2235, 1994.

P. G. Watson. *Performance of skirted foundations for offshore structures.* PhD thesis, University of Western Australia, Australia, 1999.

P. G. Watson and M. F. Randolph. A centrifuge study into cyclic loading of caisson foundations. In *Proc. Int. Conf. on Physical Modelling in Geotechnics*, volume 1, pages 693–699, Hong Kong, China, 2006.

D. J. White, W. A. Take, and M. D. Bolton. Soil deformation measurement using particle image velocimetry (PIV) and photogrammetry. *Géotechnique*, 53(7):619–631, 2003.

E. H. Zimmerman, M. W. Smith, and J. T. Shelton. Efficient gravity installed anchor for deepwater mooring. In *Proc. Offshore Technology Conf.*, volume OTC 20117, Houston, Texas, 2009.

# Cyclic Interface Shearing in Sand and Cemented Soils and Application to Axial Response of Piles

Mark F. Randolph *

* Centre for Offshore Foundation Systems, University of Western Australia, Australia

## 1 Introduction

Estimating the shaft capacity of piles driven into sand is an area of considerable uncertainty, because of the complex processes involved and the sensitivity of the normal effective stress acting on the pile shaft to minor volume changes within the sand. The starting point for calculating values of shaft friction $\tau_s$ for piles in non-cohesive soil is the expression

$$\tau_s = \sigma'_n \tan\delta = K\sigma'_{\mathrm{v0}} \tan\delta = \beta\sigma'_{\mathrm{v0}} \tag{1}$$

where $\sigma'_n$ is the normal effective stress acting round the pile shaft after installation, $K$ is the stress ratio, $\sigma'_{\mathrm{v0}}$ is the in situ effective vertical stress and $\delta$ is the angle of friction between pile and soil. The latter quantity may be measured in interface shear tests for the particular pile material. Kishida and Uesugi (1987) reported a detailed study of the effects of surface roughness, and showed how the interface friction angle may be related to the friction angle of the soil in terms of a normalized roughness coefficient, defined as the maximum roughness of the pile surface (over a gauge length of $d_{50}$ for the soil) normalized by the value of $d_{50}$. For typical pile surfaces (oxidized mild steel or concrete), the normalized roughness coefficient will exceed 0.05, and the coefficient of friction at the interface will lie in the range 0.75 to 1 times that for the soil itself. An alternative assumption, where interface shear data are not available, is to assume that the interface friction angle $\delta$ may be approximated as $\phi_{\mathrm{cv}}$, the critical state angle of friction. This may be justified on the basis that no dilation is to be expected between the sand and the wall of the pile.

For driven piles in sand and other soils of high permeability, it has long been realized that the magnitude of shaft friction at a given depth can reduce as the pile is driven further, with the net effect that the average friction along the pile shaft can reach a limit and even reduce as the pile embedment increases (Vesic, 1977). That is the basis for the design method in the API (2007) design guidelines for offshore structures Fig. 1, although the logic behind the approach has never been justified.

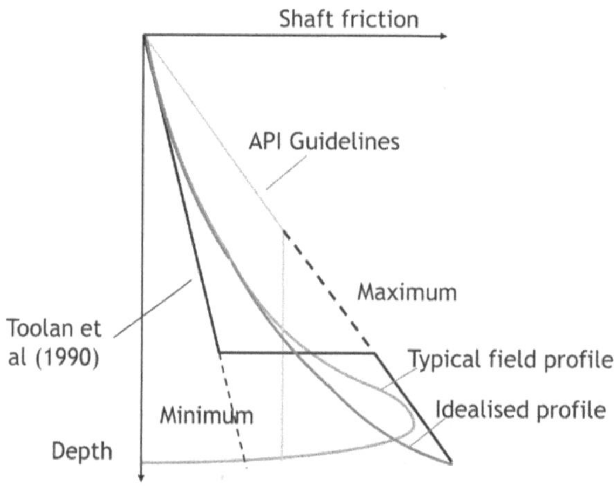


**Figure 1.** Design approaches for estimating shaft friction for piles in sand.

Over the last 20 years, however, more rational design approaches have been developed, based on improved understanding of the main mechanisms. Actual measurements of the variation of friction along the piles were obtained through carefully instrumented pile tests undertaken by the research group at Imperial College (Lehane et al., 1993). The phenomenon of 'friction degradation' is illustrated in Fig. 2 with profiles of shaft friction measured in the three instrument clusters at different distances ($h$) from the tip of a 6 m long, 0.1 m diameter, pile as it is jacked into the ground. For comparison, the cone profile is plotted on the same scale, but with $q_c$ factored down by 100 (where $q_c$ stands for the cone resistance). The shaft friction measured at $h/D = 4$ (where $D$ is the pile diameter), in particular, follows the shape of the $q_c$ profile closely, allowing for differences in cone and pile diameter. Comparison of the profiles from the instrument clusters at $h/D = 4$ and $h/D = 25$ shows that the friction measured at the latter position is generally less than 50 % of that measured close to the pile tip.

The above data lent support to an alternative design approach for driven piles proposed by Toolan et al. (1990), where a minimum (lower bound) friction ratio, $\beta$, was adopted over most of the pile length, apart from a region

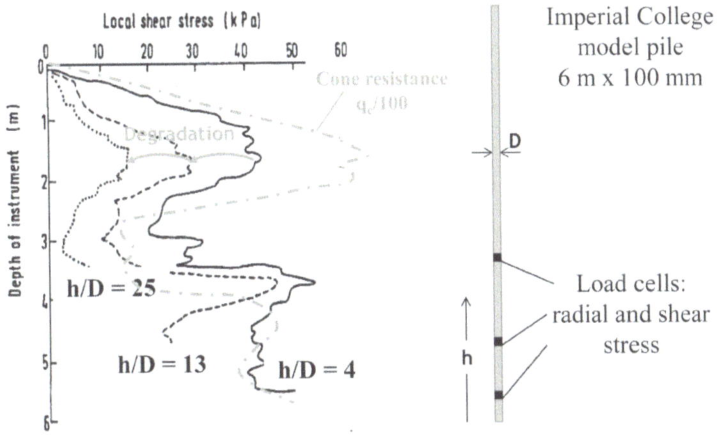


**Figure 2.** Experimental evidence of friction degradation (Lehane et al., 1993).

near the pile tip, where an upper limit was postulated (Fig. 1).

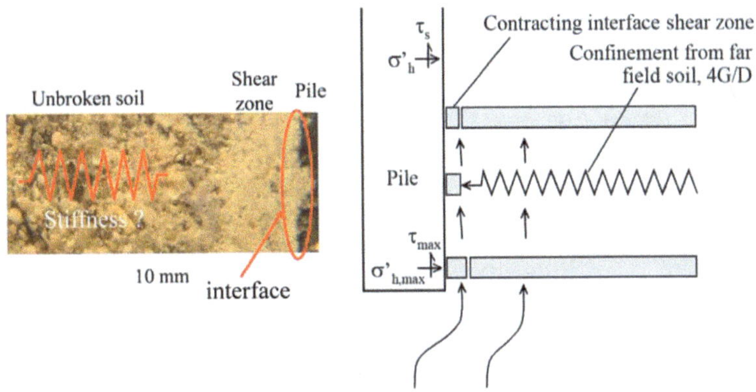


**Figure 3.** Visual evidence (left) and schematic model (right) of mechanism for friction degradation for piles in sand (after White and Bolton, 2004).

The physical basis for friction degradation is the gradual densification of soil adjacent to the pile shaft under the cyclic shearing action of installation. This process is enhanced by the presence of crushed particles from the passage of the pile tip (Fig. 3), which gradually migrate through the matrix of uncrushed material (White and Bolton, 2004). Quantification of the path taken by individual soil particles shows evidence of 'relaxation' towards the pile shaft in the vicinity of the pile tip (White, 2002).

The far-field soil acts as a spring, with stiffness equal to $4G/D$ (where $G$ is the soil shear modulus), so that any densification close to the pile results in reduced radial effective stress (Fig. 3). The operative value of $G$ will be high, since the soil is heavily over-consolidated having moved through the zone of high stress close to the pile tip during installation and is being unloaded. Simple calculations show that reduction in the width of the shear zone by as little as 1 grain diameter could account for near total loss of the normal effective stress acting on the pile shaft (White and Bolton, 2004).

## 2 Constant Normal Stiffness Direct Shear Tests

Investigation of the friction degradation mechanism may be undertaken by means of direct shear tests conducted under conditions of constant normal stiffness (CNS) (Johnston et al., 1987; Shahrour and Rezaie, 1999). External measurements of the shear and normal stresses may be enhanced by observations through a glass-sided shear box (Fig. 4).

Results of an extensive series of tests of this nature have been reported by DeJong et al. (2003, 2006). Tests were performed using a direct shear device modified to enable interface shearing and image capture for PIV analysis (Fig. 5). The interface shear box was 60 mm × 100 mm with a typical specimen height of 20 mm. A 12 mm thick Perspex window extended along the length of one side of the shear box. The top platen was fixed to the loading piston to prevent particle movements in the upper portion of the specimen due to platen rotation. The frame of the upper part of the shear box was constrained with rollers to prevent vertical movement while allowing unrestrained horizontal movement. Global displacement and stress measurements were acquired at every 0.01 mm of horizontal displacement.

A Nikon 995 digital camera mounted on a bracket at a lens to image plane distance of about 10 cm and equipped with a tether control was used to acquire digital images through the Perspex window. 1600 × 1200 pixel images with a typical field of view of 35.0 mm by 26.25 mm captured the central 1/3 of the 100 mm specimen in addition to the top platen and interface plate. Images were obtained throughout the monotonic test (t059m) and during cycles 1-5, 21-25, and 41-45 of the displacement controlled cyclic tests at 0.1 mm interface displacement intervals.

Two sands were used during testing (Fig. 6). IMDEX 16-30 sand is a subrounded silica sand with a $d_{50}$ of 0.72 mm. LeGendre 16-30 sand is a well-graded angular carbonate sand from the Northwest Shelf, Australia, with a $d_{50}$ of 0.83 mm. All samples were prepared by tamping to a medium relative density of 60-65%. The CNS boundary conditions were an initial normal stress of $\sigma_{n0} = 100$ kPa, and normal stiffness of $k = 250$ kPa/mm.

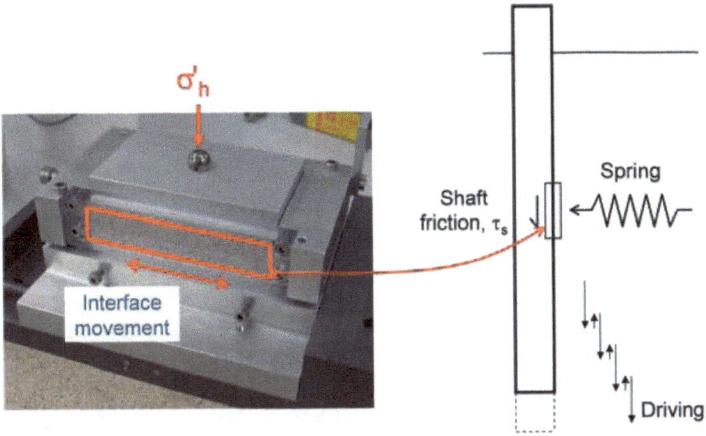


**Figure 4.** Glass-fronted constant normal stiffness shear box and axial pile response.

Fig. 7 highlights typical CNS cyclic interface shear behaviour at relatively small cyclic displacements for the silica sand. An initial increase in shear stress to 150 kPa (corresponding to a friction coefficient of 1.2) during displacement to +1 mm mirrored the monotonic behaviour. Following displacement reversal at ±1 mm, a rapid decrease to 30-40% of the peak shear stress was observed within the first 0.1 mm of displacement, followed by a non-linear transition zone over about 0.3 mm of displacement, which led into a region in which the shear stress increased at a constant rate. The shear stress then increased more rapidly as the displacement reversal limit (±1 mm) was approached. As the number of cycles increased, the initial decrease in shear stress occurred more rapidly, the transition zone narrowed, and the shear stress increased more rapidly near the displacement reversal limit.

In the global displacement plot, a characteristic 'saddle' behaviour was observed following initial dilation of 0.06 mm. Between displacement reversals, contraction occurred within approximately the first 35% (0.7 mm)

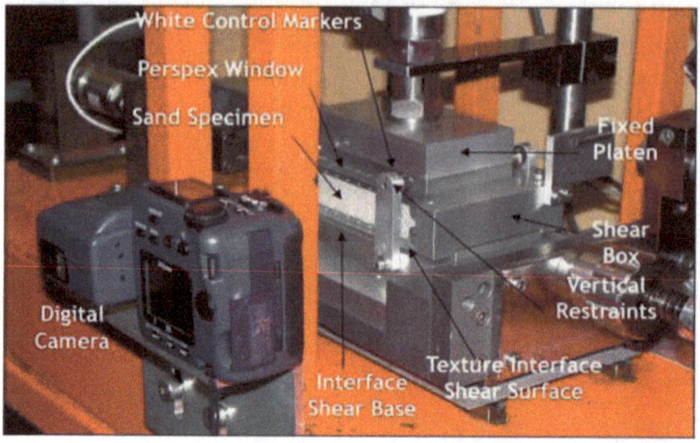


**Figure 5.** Apparatus for conducting constant normal stiffness shear box tests including digital imaging of the soil response.

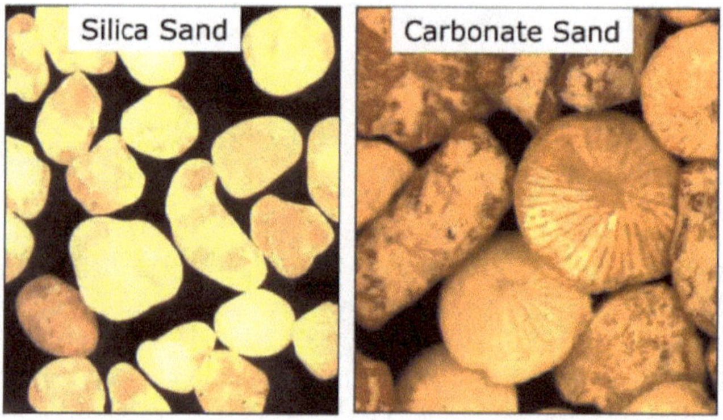


**Figure 6.** Appearance of silica and carbonate sand grains.

followed by dilation over the remaining 65% (1.3 mm) (see schematic inset on Fig. 7). The rate of contraction was higher than the dilation rate, resulting in cumulative contraction of 0.28 mm over 45 cycles, with a corresponding reduction in normal stress from 100 kPa to 36 kPa.

The characteristic displacement behaviour within three regions of the specimen for each cycle interval is shown in Fig. 8, where the horizontal coordinate is relative to the initial position and the vertical coordinate is

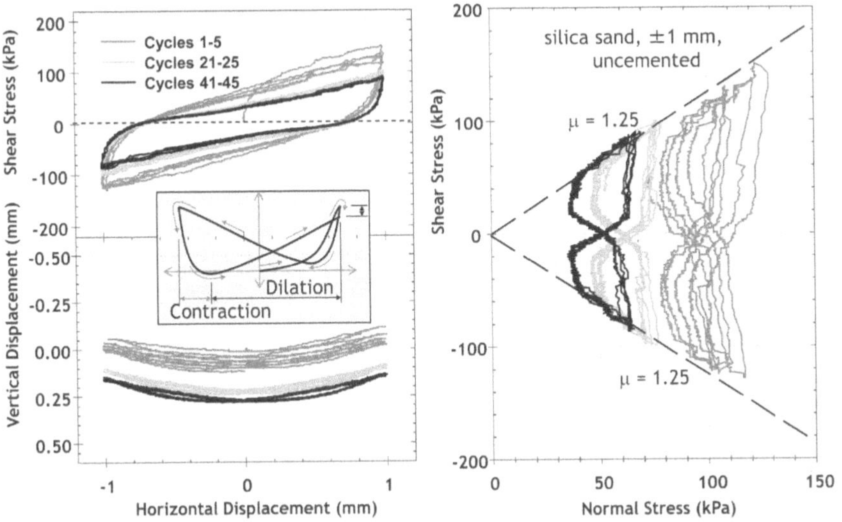


**Figure 7.** Global response of silica sand for small cyclic displacements.

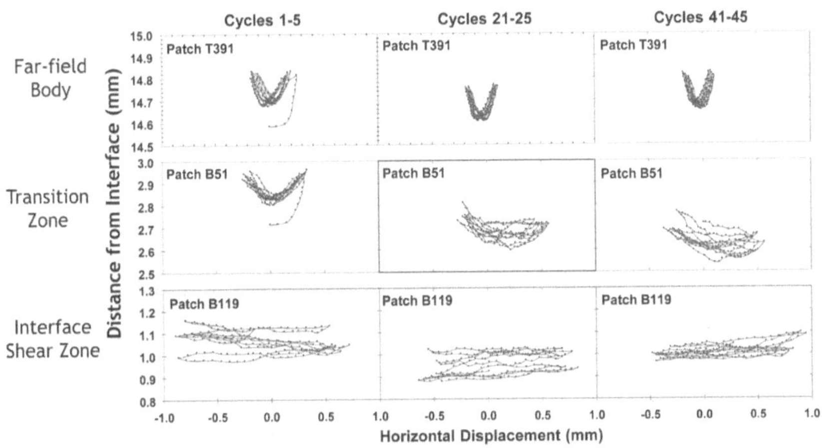


**Figure 8.** Characteristic displacement paths in different zones (silica sand).

relative to the interface. Patches T391 (14.7 mm), B51 (2.7 mm), and B119 (1.0 mm) correspond to the upper part of the sample, outside the shear zone, on the border of the shear zone, and within the shear zone respectively. The dilation associated with initial shearing to +1.0 mm followed by the 'saddle' contractive-dilative behaviour in subsequent cycles is clearly evident in T391. In contrast, patch B119 is characterized by horizontal random movement with minor vertical displacements due to proximity to the rough interface. In cycles 1-5 the displacement behaviour of patch B51, which lies initially 2.7 mm from the interface shear zone, was similar to T391, but with larger horizontal displacements (0.6 mm). In the later cycles behaviour was more random, containing increased horizontal displacements (0.8 mm), and a displacement pattern more similar to patch B119. This transition is indicative of the widening of the shear zone to include patch B51, and matches with the observed increase in shear band thickness.

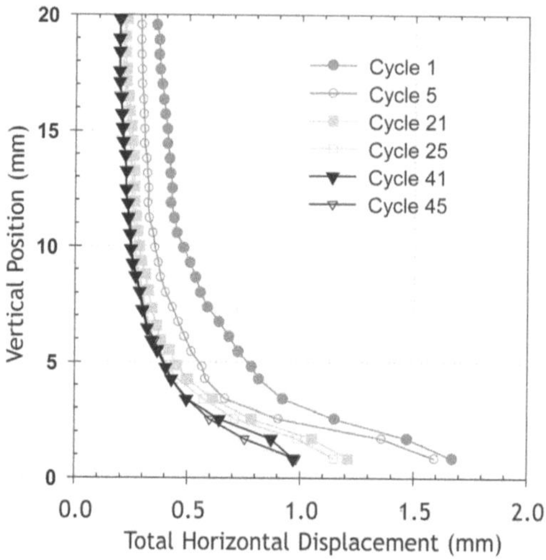


**Figure 9.** Profile of total horizontal displacements (silica sand).

Summary results from the PIV analysis were generated by averaging displacements within each patch row. The average horizontal displacements between the limit reversals (+1.0 mm to -1.0 mm) in cycles 1, 5, 21, 25, 41, and 45 are plotted in Fig. 9. The shear strain within the lower portion of the shear zone (0.0 mm to 3.57 mm) is consistent between all cycles and averages 0.38 (0.34 to 0.44) for the cycles plotted. The percent slip averaged 8.9% throughout cycling. The volumetric strain was relatively constant, averaging 0.014. The shear zone thickness increased from 3.6 mm in cycle 1 to 5.0 mm and 5.6 mm in cycles 21 and 41, respectively. Above the shear zone, the soil moved horizontally, reflecting compression of the sample against the end walls of the box. The amplitude of this horizontal movement decreased from 0.40 mm to 0.23 mm between cycles 1 and 45 as the peak shear stress, and hence the horizontal compression of the sample reduced.

Corresponding results for cyclic shearing of the carbonate sand are shown in Fig. 10 and Fig. 11. Initial shearing to +1.0 mm resulted in a maximum shear stress of 140 kPa, with the specimen initially contracting by 0.02 mm before dilating to 0.04 mm. The maximum shear stress achieved at $\pm 1.0$ mm decreased to 71 kPa after 45 cycles. However, the maximum stress ratio observed at $\pm 1.0$ mm averaged 1.2 for all cycles due to continuous decrease in the average normal stress per cycle. A total contraction of 0.35 mm resulted in a decrease in the normal stress to 20 kPa, at zero shear stress - notably lower than observed with the silica sand. The minimal initial dilation is indicative of increased particle rearrangement due to crushing as increased angularity would increase particle frustration and inhibit rearrangement.

PIV analysis showed that the shear zone is significantly larger than in silica sand but less sharply defined (Fig. 11). The average shear strains for cycles 1-5, 21-25, and 41-45 within 4 mm of the interface averaged 0.33, 0.22, and 0.17, respectively. The percent slip averaged 5.4%, 32.7%, and 47.0% for cycles 1-5, 21-25, and 41-45, respectively. Increased slip is attributed to increased particle rearrangement enabled by particle crushing adjacent to the fully rough interface. It is noted that the silica sand glued to the interface surface is significantly harder than the carbonate sand and will not wear. Unlike previous observations, the upper bound of the shear zone decreased, from about 10 mm to 6 mm between cycles 1 and 21 beyond which it remained relatively constant. The depth of the shear zone, and the horizontal movement due to compression of the ends of the shear box, are both lower than in the silica sand tests.

Comparison of the results for the silica and carbonate sands emphasises the importance of the soil mineralogy on the global interface response and the micro-mechanical behaviour. Particle crushing was observed close to the interface for the carbonate sand, coupled with increased slip at the interface with cycling (Fig. 12). This resulted in accelerated degradation compared with silica sand.

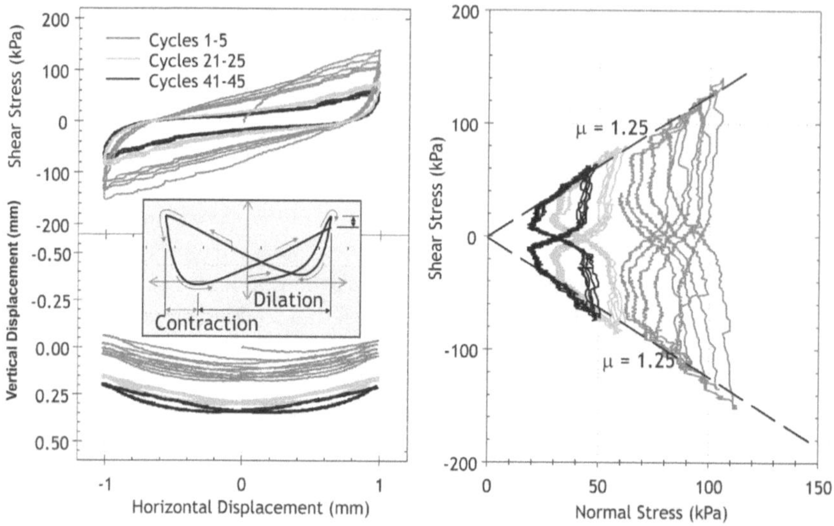


**Figure 10.** Global response of carbonate sand for small cyclic displacements.

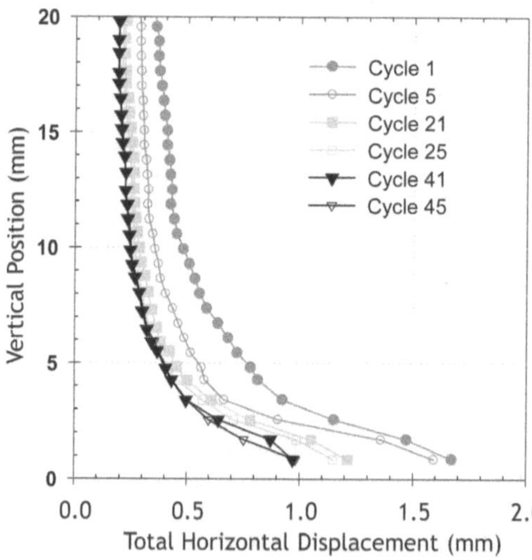


**Figure 11.** Profile of total horizontal displacements (carbonate sand).

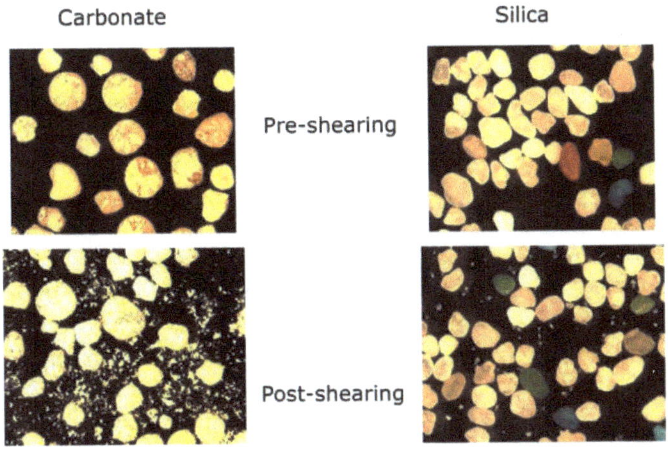


**Figure 12.** Degree of grain crushing due to shearing.

## 2.1 Prediction of Cyclic Response Under CNS Conditions

The progressive reduction in normal and shear stress during interface cycling under CNS conditions arises from the cumulative contraction of a well defined shear band adjacent to the pile-soil interface. GeoPIV analysis quantitatively demonstrates that the specimen deformation is concentrated within a thin interface shear band, typically several mean particle diameters ($d_{50}$) in thickness. The progressive cycling of the interface leads to a net contraction of the interface shear band, and hence a reduction in interface normal stress. It is this reduction in normal stress that governs friction degradation.

The system can therefore be modelled as an interface, of thickness $h_s$, that is dilating and contracting within each cycle leading to a cumulative net contraction by distance $\Delta h$ (DeJong et al., 2006). After a specimen contraction of $\Delta h$ from an initial specimen void ratio of $e_0$, the void ratio of the interface shear band, $e_{sb}$, is:

$$e_{sb} = \left(1 - \frac{\Delta h}{h_s}\right)(1 + e_0) - 1 \tag{2}$$

Permissible values of $e_{sb}$ lie between the maximum and minimum void ratios of the specimen. An additional limit is the no-tension condition for drained (or dry) sand, so $\Delta h$ cannot exceed $\sigma_{n0}/k$ (where $k$ is the CNS stiffness).

DeJong et al. (2006) provide the following discussion in respect of the void ratio limits, $e_{\min}$ and $e_{\max}$:

> Conventional practice is to define the $e_{\min}$ and $e_{\max}$ limits following the ASTM (2004) procedures, with $e_{\min}$ obtained by vibrating a specimen under a known weight corresponding to a normal stress of ~14 kPa, while $e_{\max}$ is obtained by a sand deposition into a known volume using a funnel or an open cylinder about 0.6 times the diameter of the mould. Since the sand falls onto a slope at the angle of repose, the resulting state can be likened to a critical state at a very low stress (equal to about 1 kPa at mid-depth of specimen). Correlations assembled by Cubrinowski and Ishihara (2000) suggest that the critical state voids ratio as effective stress tends to zero is slightly lower than $e_{\max}$ and corresponds to a relative density of about 10% in clean silica sands (e.g. Lee and Seed, 1967; Verdugo and Ishihara, 1996). Others have related the CSL to $e_{\max}$ at a nominal stress of typically 1 kPa (e.g. Klotz and Coop, 2002). Void ratios slightly higher than $e_{\max}$ can be achieved through other deposition conditions and can occur locally in laboratory specimens following shearing (e.g. Frost and Jang, 2002). Void ratios lower than $e_{\min}$ may be achieved at high stress levels and are associated with particle breakage. Breakage increases the coefficient of uniformity, which in turn reduces $e_{\min}$. For CNS interface testing, an appropriate $e_{\min}$ value may be defined as the void ratio at which further cyclic loading at a given stress level will induce no additional volume change. The ASTM $e_{\min}$ test provides a suitable value of $e_{\min}$ for the stress level of ~14 kPa; the variation of $e_{\min}$ with stress level is discussed below. In CNS tests, during monotonic shearing to failure, the deforming soil close to the interface will shear towards a critical state.

A simple model was proposed by DeJong et al. (2006) whereby the specimen contraction over a single cycle was defined as a fixed fraction, $1/N_{\text{char}}$, of the potential change in density during contraction to $e_{\min}$.

$$\frac{\Delta e_{sb}}{\Delta N} = e_{sb(N+1)} - e_{sb(N)} = -\frac{1}{N_{\text{char}}}\left(e_{sb(N)} - e_{\min}\right) \tag{3}$$

where $\Delta e_{sb}/\Delta N$ is the change in shear band void ratio per cycle, $e_{sb(N+1)}$ is the shear band void ratio at cycle $N+1$, and $e_{sb(N)}$ is the shear band void ratio at cycle $N$. Integrating over $N$ cycles yields:

$$e_{sb(N+1)} = e_{\min} + \left(e_0 - e_{\min}\right) e^{-N/N_{\text{char}}} \tag{4}$$

where $N_{\text{char}}$ is a characteristic number of cycles. Combining Eq. (2) with the CNS condition ($\Delta h = \Delta\sigma'_n/k$) enables the evolution in $\Delta h$ with $N$ to be found.

A reasonable fit to the specimen contraction data from the cyclic CNS tests is found by using the shear band thickness ($h_s$) reached at the end of each test and a characteristic number of cycles ($N_{\text{char}}$) equal to 15 (Fig. 13). The expression follows the degradation trend reasonably well aside from differences due to dilation in the first cycle which are not explicitly accounted for in the analysis. Only test t075c (carbonate sand) diverges from the model in the later cycles. This divergence arises because the $e_{\text{min}}$ value (0.639) for calcareous sand compared to the initial state ($e_0 = 0.758$) requires a specimen contraction of 0.41 mm, which exceeds the no-tension limit of $\sigma'_{n0}/k$ by 0.01 mm. Therefore, Eq. (4) does not fully capture behaviour in the later cycles of test t075c.

The above contraction model can be explored further to predict the lower limit of the interface normal stress at the end of a cyclic test. The CNS spring stiffness, $k$, links specimen contraction with loss of interface normal stress, $\sigma'_n$, defining the test path in specimen height$-\sigma_n$ space. Noting that the specimen contraction is limited to the shear band, the test paths can be recast as straight lines in $e_{sb} - \sigma'_n$ space with gradient $(1 + e_0)/h_s k$ (see Fig. 14) by combining the CNS condition with the idealized interface system (Eq. 2) to give:

$$e_{sb} = e_0 - \left(\frac{\sigma'_{n0} - \sigma'_n}{kh_s}\right)(1 + e_0) \tag{5}$$

Eq. (5) reveals the link between shear band thickness, confining stiffness and the degradation of interface normal stress with change in voids ratio. For a given change in $e$, any increase in shear band thickness ($h_s$) or CNS stiffness ($k$) leads to a greater loss of interface normal stress, since the test path gradient is flatter in $e_{sb} - \sigma'_n$ space. A corresponding upper bound value of $\sigma'_n$ can be defined on the test path at the intersection with the CSL, which represents a critical state condition, which is achieved following sufficient monotonic shearing.

The degradation of $\sigma'_n$ with cycling can be predicted by combining Eq. (4) and (5), leading to:

$$\sigma'_n = \sigma'_{n0} - kh_s \frac{(e_0 - e_{\text{min}})\left(1 - e^{-N/N_{\text{char}}}\right)}{(1 + e_0)} \tag{6}$$

It should be noted that the void ratio $e_{sb}$ referred to in Eq. (4) is the minimum value within a cycle, so Eq. (6) relates to the minimum normal stress within a cycle, not the value at failure.

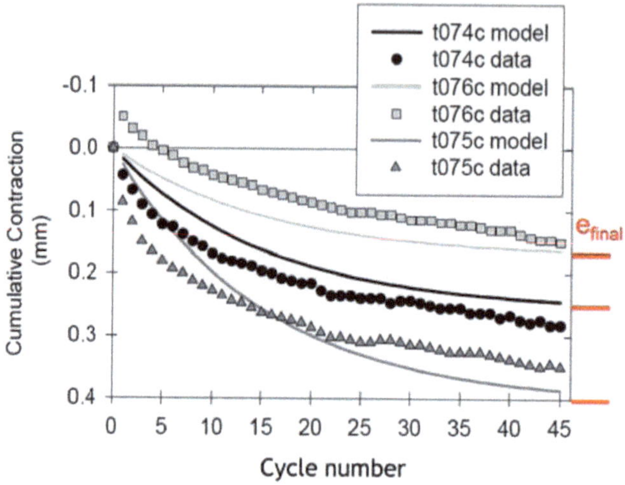


**Figure 13.** Comparison of measured and predicted contraction during cyclic CNS shearing.

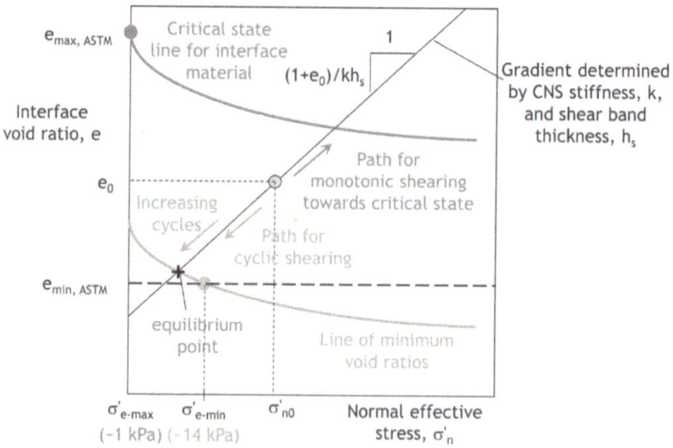


**Figure 14.** Predictive model to estimate minimum normal stress (after DeJong et al., 2006.

The minimum value of $\sigma_n'$ that can be achieved by cycling is the intersection of the test path with the line $e_{\min}$. It is likely that the variation in

$e_{min}$ with normal stress $\sigma'_n$ will reflect the conventional (monotonic) critical state line (CSL). It is found that the CSL is straight in $e - \ln \sigma'$ space at high stresses, and flattens at low stresses, tending to a value close to the $e_{max}$ value. For more breakable sands, the transition in the CSL occurs at lower stresses, reflecting the earlier onset of particle breakage.

Two forms of the $e_{min}$ line in $e - \ln \sigma'_n$ space may be considered: a stress-independent value; and an alternative $e_{min}$ line running parallel to the CSL and passing through $e_{min}$ at a stress of 14 kPa (Fig. 14).

Particle breakage is an additional consideration, since that will tend to reduce the minimum density of the sand. The manner in which this will affect the resulting stress paths has been explored by White and Bolton (2004) for the particular conditions relevant for their tests, as indicated in Fig. 15. After crushing, the material in the shear zone is denser than the maximum density of the original material, and the range of possible densities has moved towards lower specific volumes.

## 3 Design Approaches for Shaft Capacity of Driven Piles

A general framework for shaft resistance of driven piles in sand has been discussed by White (2005), building on the concept of friction degradation due arising from cyclic shearing during pile installation. Examples include values of shaft friction deduced from back-analysis of pile driving performance for the North Rankin piles (Dolwin et al., 1988; Randolph, 1988) as indicated in Fig. 16. White and Lehane (2004) presented a series of centrifuge model tests where the piles were installed either by intermittent jacking, or by a pseudo dynamic installation method with cyclic reversals and advances of penetration. The loss of normal stress measured at the pile-soil interface correlated well with the number of jacking or displacement reversal cycles.

The influence of installation cycles on friction degradation is illustrated in Fig. 17 (White, 2005) from field data on driven (high number of cycles, $N$) and jacked (low number of cycles) piles. From a practical perspective, the number of blows required to install the pile is not known a priori, and hence it is necessary to link degradation in friction to the normalised distance $(h/D)$ measured from the pile tip, rather than the number of shearing cycles.

Different approaches to quantify the variation of the normal effective stress or shaft friction have been proposed by Randolph et al. (1994) and by Jardine and Chow (1996), although Randolph (2003a) has shown that both approaches yield similar profiles of shaft friction. The former approach expresses the stress ratio, $K$ (or friction ratio, $\beta$) as an exponential function

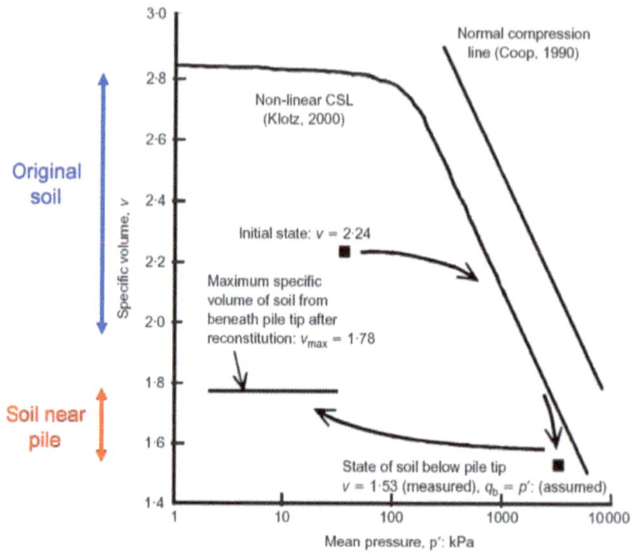


**Figure 15.** Effect of particle crushing on the specific volume range.

of $h/D$, ranging between a maximum value, $K_{\max}$, at the pile tip, and a minimum value, $K_{\min}$, according to

$$K = K_{\max} - (K_{\max} - K_{\min})\, e^{-\mu h/D} \tag{7}$$

where the rate of degradation is controlled by the parameter, $\mu$, which may be taken as $\sim 0.05$. $K_{\max}$ may be estimated as a proportion of the normalised cone resistance, typically 2% of $q_c/\sigma'_{v0}$ for closed-ended piles reducing to 1% for open-ended piles, and $K_{\min}$ lies in the range 0.2 to 0.4, giving a minimum friction ratio, $\tau_s/\sigma'_v$, of 0.1 to 0.25 (Toolan et al., 1990).

Where data from a cone friction sleeve are available, the friction ratio may be used to refine estimates of $K_{\max}$ for closed-ended piles, provided an appropriate value for the interface friction angle, $\delta$, between cone and soil is adopted. However, it should also be noted that data from friction sleeves are very sensitive to wear of the sleeve, and may therefore underestimate $K_{\max}$.

Jardine and Chow (1996) adopted a power law function of $h/D$, but also allowed for dilation during subsequent shearing to failure, giving an express

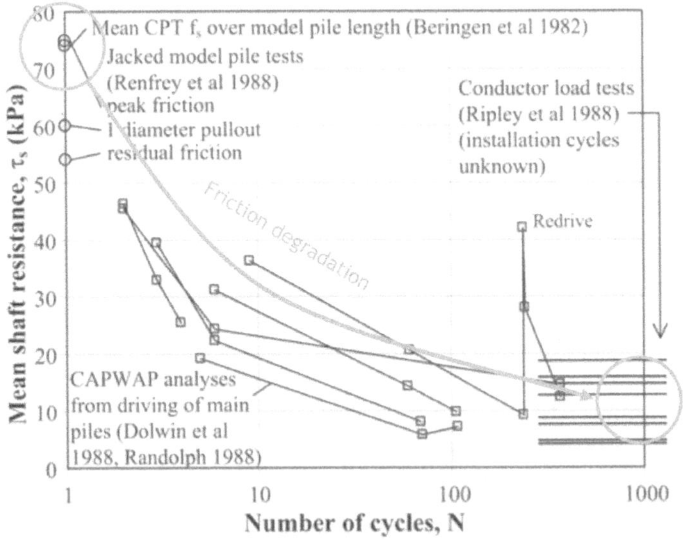


**Figure 16.** Mean shaft friction deduced for piles in carbonate sands at North Rankin, North-West Shelf of Australia (after White, 2005).

for shaft friction of

$$\tau_s = \left(\frac{q_c}{45}\left(\frac{\sigma'_{\text{v0}}}{p_a}\right)^{0.13}\left(\frac{D}{h}\right)^{0.38} + \Delta\sigma'_{\text{rd}}\right)\tan\delta_{\text{cv}} \qquad (8)$$

where $p_a$ stands for atmospheric pressure (100 kPa) and $\Delta\sigma'_{\text{rd}}$ is the radial stress change due to dilation.

For open ended piles, they suggested replacing the external diameter, $D$, with the equivalent diameter, $D_{\text{equiv}}$, of a solid pile with the same volume of steel (per unit length).

Two differences between the above two relationships should be noted. The first expression allows for a minimum value of the stress ratio, $K$, which would be approached at large values of $h/D$, or large numbers of installation cycles. This is consistent with the ideas discussed earlier in relation to CNS shear tests. The second point is that it is necessary to stipulate a lower limit of $h/D$ in Eq. (8) in order to avoid infinite shear resistance. The proposed limit was $h/D = 4$, which corresponded to the

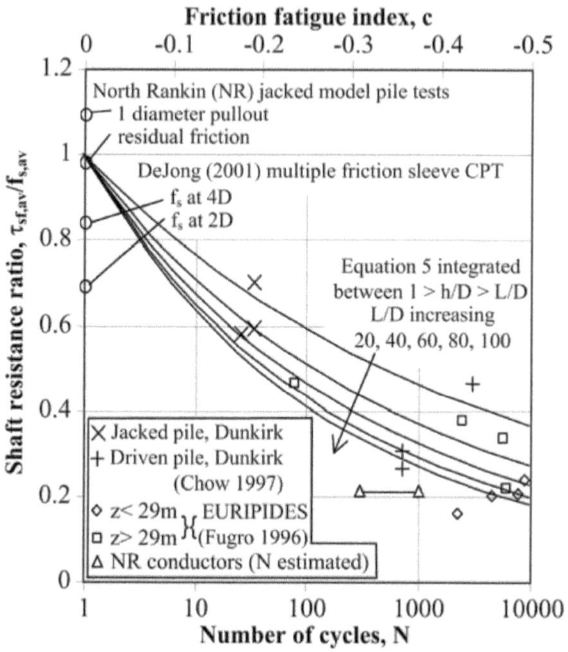


**Figure 17.** Influence of number of stress reversal cycles on shaft friction of displacement piles (White, 2005).

lower instrumentation cluster in the Imperial College instrumented pile.

A number of alternative methods have now been incorporated in a revised commentary to the American Petroleum Institute (API, 2007) guidelines for the design of fixed steel structures. The approach for shaft friction, $\tau_s$, at any depth, $z$, may be expressed as

$$\tau_s = u\bar{q}_c \left(\frac{\sigma'_{\text{v0}}}{p_a}\right)^a A_r^b \left[\max\left(\frac{L-z}{D}, v\right)\right]^{-c} (\tan\delta_{\text{cv}})^d \left[\min\left(\frac{L-z}{D}\frac{1}{v}, 1\right)\right]^{\text{e}} \tag{9}$$

where $\sigma'_{\text{v0}}$ is the vertical effective stress, normalised by atmospheric pressure, $\delta_{\text{cv}}$ is the steady state interface friction angle between pile and soil, $\bar{q}_c$ is an appropriately average value of cone resistance, $L - z$ is distance ($h$) from the tip of the pile, $A_r$ is the area ratio for the pile, and the parameters $a$

to e, $u$ and $v$ are given in Tab. 1 for the three cone-based methods in the commentary (Fugro-05: Kolk et al., 2005; Simplified ICP-05: Jardine et al., 2005; Offshore UWA-05: Lehane et al., 2005a).

A comparison of these three methods with the traditional approach for estimating the capacity of driven piles in sand showed significant improvements in accuracy, with much lower coefficients of variation of the ratio of predicted to measured capacity, for the current database of pile load tests in the public domain (Lehane et al., 2005b).

**Table 1.** Parameter values for shaft friction estimation in Eq. (9).

| Method | | Parameters | | | | | | |
|---|---|---|---|---|---|---|---|---|
| | | $a$ | $b$ | $c$ | $d$ | e | $u$ | $v$ |
| Fugro-05 | compression | 0.05 | 0.45 | 0.90 | 0 | 1 | 0.043 | $2\sqrt{A_r}$ |
| | tension | 0.15 | 0.42 | 0.85 | 0 | 0 | 0.025 | $2\sqrt{A_r}$ |
| Simplified ICP-05 | compression | 0.1 | 0.2 | 0.4 | 1 | 0 | 0.023 | $4\sqrt{A_r}$ |
| | tension | 0.1 | 0.2 | 0.4 | 1 | 0 | 0.016 | $4\sqrt{A_r}$ |
| Offshore UWA-05 | compression | 0 | 0.3 | 0.5 | 1 | 0 | 0.030 | 2 |
| | tension | 0 | 0.3 | 0.5 | 1 | 0 | 0.022 | 2 |

Various database studies have been conducted by the research groups involved in the development of the methods listed in Tab. 1. Database exercises are a necessary stage in the calibration of axial pile capacity design methods, due to the empiricism of the recommended equations. The methods listed in Tab. 1 all give broadly similar performance when compared to the field load test measurements, which is to be expected since all were calibrated against similar databases. There is a trend for the more recently-developed methods to give better agreement with the current database, partly because of having been calibrated against the more recent versions. Typically, the database studies reveal a mean ratio of predicted to measured capacity of $1 \pm 0.1$ with a standard deviation of ~0.2-0.3.

The methods all adopt broadly the same form of expression for shaft and base resistance, with some differences in the mechanisms accounted for. However, the empirical fitting parameters differ, as shown in Tab. 1. As a result, the methods give significantly different capacity predictions for field scale piles - varying by up to 50%, with a scatter that depends on the mean value and shape of the cone resistance profile and the pile geometry. A designer can then be faced with a design load which exceeds the capacity calculated according to one of the new design methods, but which is acceptable according to another, even after reducing this capacity by a factor of safety based on the database statistics.

This conflict presents difficulties for designers, highlighting the need for caution when applying these new design methods. The API Commentary notes that "*more experience is required with all these new methods before any single one can be recommended for routine design*". A conservative approach - perhaps overly so - would be to adopt the lesser of the capacities predicted by each method.

## 4 Axial Shaft Capacity of Grouted Piles in Carbonate Sand

It has been recognised since the early 1980s that extremely low values of shaft friction can result for piles driven into calcareous sands, a classic example being the piles for the North Rankin A platform on the North-West Shelf of Australia (Poulos et al., 1988). The mechanism responsible may be attributed to an extreme form of shaft friction degradation, as a result of the fragile grains and high compressibility of carbonate soils.

A typical profile of cone resistance for sediments between 50 and 120 m depth on the North-West Shelf is shown in Fig. 18. A lower bound cone resistance in carbonate silts, with $q_c \sim 0.04z$ MPa (with $z$ the depth in m), is broken up by stronger material comprising cemented calcarenite, with cone resistance of 10 to 50 MPa. Laboratory element tests (triaxial, simple shear) tend to reflect $N_{kt}$ (the cone factor) values of about 13 but with considerable scatter. Direct shear tests carried out under conditions of constant normal stiffness show lower strengths, which is consistent with the tendency for the material to collapse (or crush) on shearing at stress levels relevant for pile design. Back-analysis of results from (driven) conductor tests, or the driving records of driven piles, suggests shaft friction values of less than 10 kPa, even at depths of 100 m (see earlier, Fig. 16).

In most deposits of carbonate material, more consistently cemented strata are encountered at depth. On the North-West Shelf, typical values of cone resistance range from 10 to 50 MPa, and design profiles based on taking the moving average cone resistance are mostly in the range 20 to 35 MPa (see Fig. 19). Well cemented material of this nature is ideal for foundation piles, provided good load transfer can be achieved from the pile to the cemented sediments.

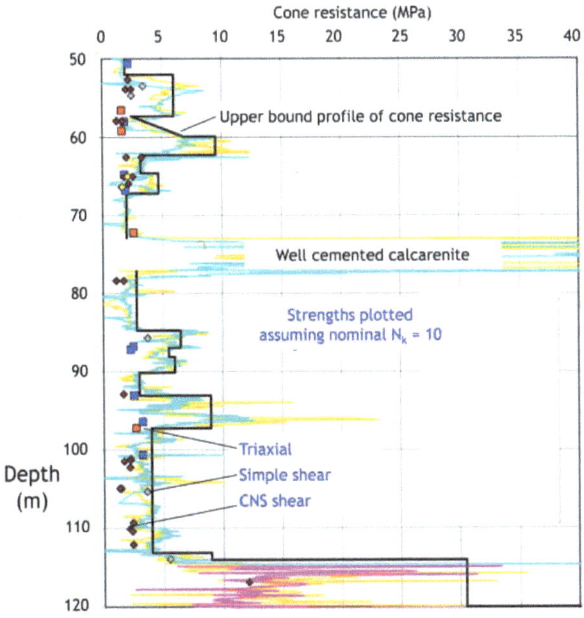


**Figure 18.** Cone resistance profile in variably cemented carbonate sands on North-West Shelf of Australia.

## 4.1 Grouted Pile Construction

An alternative construction technique to driven piles is the equivalent of an onshore 'cast in situ' pile, referred to offshore as a 'drilled and grouted' pile, and comprises a steel tubular insert pile grouted into a pre-drilled hole. The existence of uncemented sediments at shallow depth generally precludes construction of grouted piles from the seabed surface. Instead, a primary pile is first driven deep enough to prevent collapse of any uncemented strata. Then, a rotary drill is advanced through the primary pile and extended to sufficient depth to accommodate the grouted insert pile. This comprises a steel tubular, which is lowered to the bottom of the borehole and then the annulus between the tubular and the drilled hole is grouted, as is the connection with the primary pile (see Fig. 20). For ease of construction, it is simpler if the complete pile (inside and outside the tubular) is grouted, but this can cause considerable thermal shrinkage, sufficient to fail the grout-sediment connection. One approach to mitigate this is to fill the inside of the tubular with precast cement blocks, thus limiting the total volume of grout.

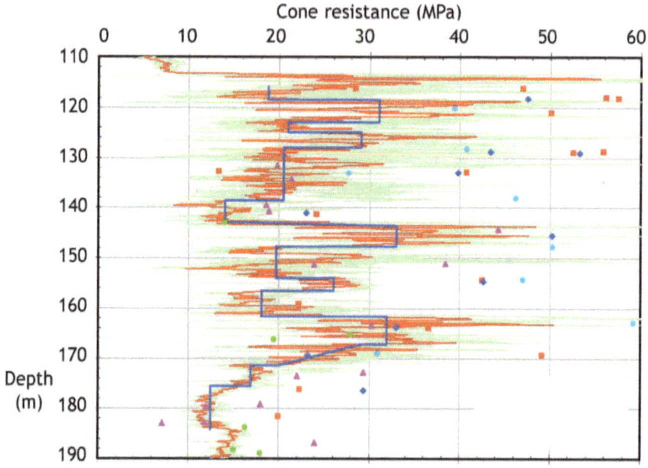


**Figure 19.** Cone resistance profile in deeper, more cemented, carbonate material on North-West Shelf of Australia.

Typical dimensions are illustrated on Fig. 21. The primary pile will generally be 2.5 to 2.8 m in diameter, and may extend as deep as 120 m. The drilled borehole will need to leave sufficient clearance between the drill bit and the inside of the primary pile, to give an external diameter of 2 to 2.3 m. The tubular should be sized to allow a grout annulus of around 150 mm. Note that the length for this type of pile may be as much as 180 m. Weld beads are used on the outside of the tubular insert and inside of the primary pile at its lower end, in order to ensure adequate load transfer to the grout.

A modified form of construction, the 'grouted driven pile', where a pipe pile is first driven to full penetration and then grout is forced out through pre-installed valves (Rickman and Barthelemy, 1988) has never been pursued in spite of apparent economic advantages in reduced construction costs; the main obstacle has been a suitable means of assuring the quality of the grouting process.

Drilled and grouted construction achieves pile installation without the destructive processes of pile driving. Any temporary loss in effective stress in the ground is assumed to be (largely) recovered by means of the head of

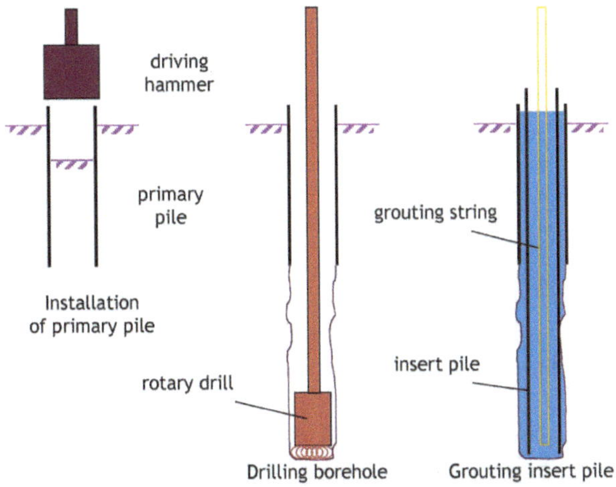


**Figure 20.** Construction of offshore grouted insert pile.

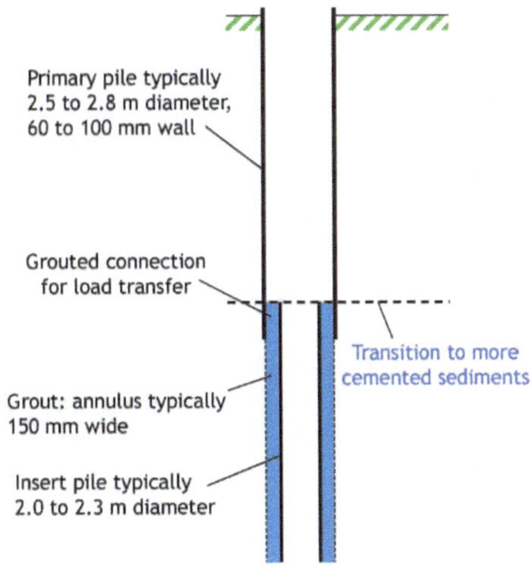


**Figure 21.** Typical dimensions of a drilled and grouted insert pile.

grout during the grouting operation. In addition, inevitable roughness of the drilled hole will encourage initial increases in the horizontal stresses as the pile is brought to failure.

For onshore design of rock-socketed piles, it is customary to correlate the shaft friction with the unconfined compression strength, $q_u$, of the rock, or equivalent shear strength, $s_u = q_u/2$ (Kulhawy and Phoon, 1993; Randolph et al., 1998). The correlation parameter has been shown to depend on the roughness of the socket, but also reduces with increasing pile diameter (Seidel and Haberfield, 1995).

In offshore design, there has been a tendency to correlate peak shaft friction with the cone resistance. Correlations have been developed through a combination of field data and laboratory tests. The two most common laboratory tests are (a) rod shear tests (essentially a model pile test, Jewell and Randolph, 1988), and (b) constant normal stiffness (CNS) direct shear tests (Johnston et al., 1987). The principal of the latter test is that it simulates the changes in normal stress that accompany volumetric expansion or collapse of the material in the shear zone surrounding the pile.

Typical correlations from model pile tests showing a decreasing trend with increasing cone resistance, as (Fig. 22):

$$\frac{\tau_p}{q_c} \approx 0.02 + 0.2e^{-0.04q_c/p_a} \tag{10}$$

but with the asymptotic ratio of 0.02 varying in the range 0.02 to 0.04 (Abbs, 1992; Joer and Randolph, 1994). The asymptotic range applies once the cone resistance exceeds about 10 MPa.

Correlations of peak shaft friction such as in Eq. (10) have been developed from a variety of laboratory and field load tests. Since field tests are mostly not feasible with the budget and time constraints of a project, it is necessary to rely on laboratory testing to establish drilled and grouted design parameters, not just for the peak shaft friction, but also for values of residual shaft friction and the shape of the transition from peak to residual. The constant normal stiffness (CNS) shear test is the obvious choice, given its simplicity and the limited amount of material required, although in earlier work rod shear tests were also used extensively. On average the CNS tended to give lower values of peak shear stress, and thus provide a more conservative approach for design.

### 4.2 Response to Cyclic Loading

Under monotonic displacement, piles grouted into cemented carbonate soils exhibit a very brittle reduction in shear resistance, and this appears to degrade gradually over a considerable displacement (Fig. 23).

Under cyclic displacements, a very low shear resistance is measured, in some cases as low as 1% of the peak shear stress. The low shear resistance

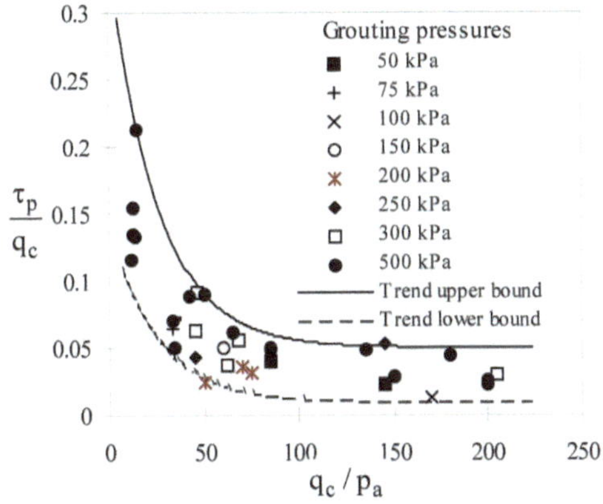


**Figure 22.** Correlation of shaft friction with cone resistance for cemented sediments (Joer and Randolph, 1994).

extends through the major portion of the displacement range over which failure has been reached. This has often been referred to as a 'gap' zone (analogous to the physical gap that might appear around a pile loaded laterally). On emerging from this 'gap' zone, the shear resistance builds back up towards the monotonic backbone curve, but with a lower failure shear stress, reflecting the additional plastic shearing within the cyclic gap zone.

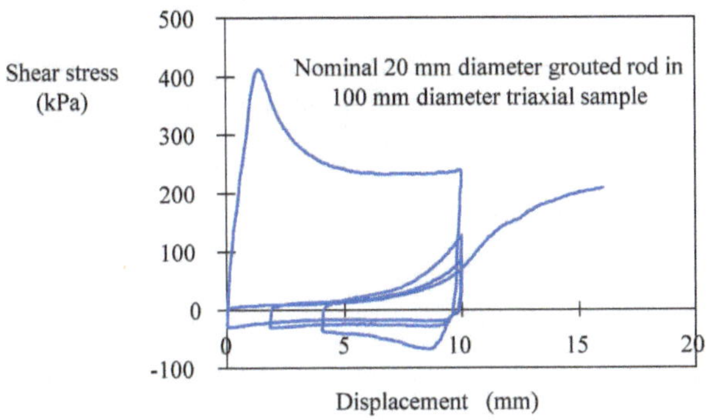


**Figure 23.** Typical response from rod shear test in calcarenite.

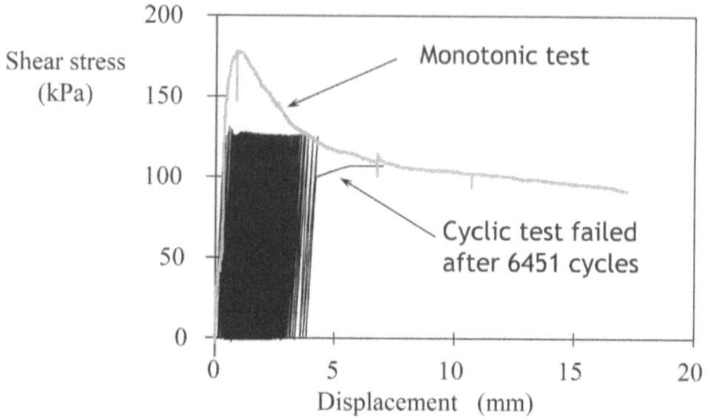


**Figure 24.** Rod shear test response to one-way cyclic loading.

Pre-failure cyclic loading can also lead to accumulation of displacement (assuming a bias, rather than purely symmetric 2-way cycles) and ultimately failure (Fig. 24). Relatively high shear stress levels, as a proportion of the monotonic resistance, are required for such failure. However, such high levels can occur in the upper portion of a grouted pile, since elastic compression of the pile shaft will lead to initial concentration of load transfer near the pile head.

The brittleness of cemented carbonate sediments such as calcarenite and calcisiltite, and the vulnerability to damage during cyclic shearing, leads to the potential for progressive failure of axially loaded piles. Under monotonic loading, failure will occur near the pile head and propagate downwards, so that the maximum pile load that may be mobilised is much less than the ideal capacity of a rigid pile. The challenge is to quantify the strain softening behaviour from appropriately devised monotonic and cyclic laboratory shear tests.

### 4.3 Field Grouted Pile Tests

Two key programmes of grouted pile tests were conducted at a site in South Australia near Overland Corner, to provide field-scale data from which to evaluate design parameters and calibrate analysis tools (Randolph et al., 1996). The first programme, in 1987 to 1988, was undertaken to underpin the design of grouted insert piles for the Goodwyn A platform on the North-West Shelf of Australia. The programme was based around tests on a number of short pile segments, ranging in diameter from 0.4 to 2 m, but also included one 15 m long pile (0.4 m in diameter) in order to explore the potential for progressive failure under cyclic loading (see Fig. 25). A second programme of tests in 1988 addressed grouted driven piles, 0.4 and 0.95 m in diameter and 10 m long; these tests proved the concept but without the construction technique ever having been taken up in practice.

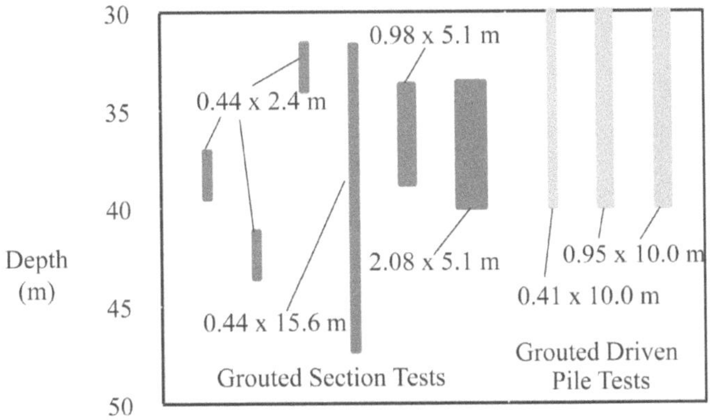


**Figure 25.** Program of field grouted pile tests at Overland Corner, SA.

**Table 2.** Summary of geotechnical parameters for Overland Corner, SA.

| Parameter | Range of values | Comments |
|---|---|---|
| Cone resistance, $q_c$ (MPa) | 10 - 18 | Average of 14 MPa |
| Grouted anchors, $\tau_p$ (kPa) | 526 - 614 | Residual: 25 - 40% of peak |
| Hydraulic fracture, $p_f$ (kPa) | 1600 - 2300 | Overburden: 500 - 800 kPa |
| Pressuremeter modulus, $G_{ur}$ (MPa) | > 350 | Limited by compliance |
| Pressuremeter strength, $s_u$ (kPa) | 600 - 700 | |
| CNS shear box tests, $\tau_p$ (kPa) | 160 - 320 | Residual: 50 - 100 kPa |
| Rod shear tests, $\tau_p$ (kPa) | 300 - 600 | Residual: 100 - 300 kPa |

Tab. 2 provides a summary of key geotechnical parameters for the Overland Corner limestone. The average cone resistance was around 14 MPa, and the average peak shear stress ranged from 240 kPa (CNS) to 450 kPa (rod shear) and 570 kPa (small diameter grouted anchor rods in the field). Pressuremeter tests provided data on the unload-reload modulus from which to assess appropriate stiffness values for CNS testing.

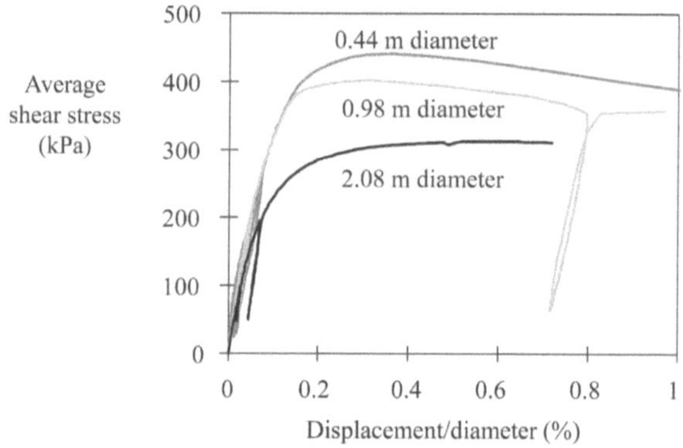


**Figure 26.** Monotonic loading response of grouted sections.

The response of the grouted sections under monotonic loading showed a marked effect of diameter (see Fig. 26). The peak shaft friction ranged from 440 kPa for the smallest diameter, down to 310 kPa for the largest diameter. This variation is consistent with the concept of constant normal stiffness of the response at the pile-soil interface (and as simulated in CNS tests). The more ductile response of the largest pile segment appeared to be carried through into the post-peak region, with less obvious degradation in resistance. However, that evidence should be treated with caution, as the largest diameter segment was also the shortest (in terms of length to diameter ratio), and so increasing end resistance (even though a soft base was incorporated) may have contributed to the sustained resistance.

Limiting values of shaft friction from the various grouted pile tests ranged from 310 kPa up to over 600 kPa (the smaller diameter grouted driven pile) (Fig. 27). As a ratio of the cone resistance, the peak shaft friction values ranged from around 2.5% to over 4%.

## 5 Non-Linear Load Transfer Analysis - RATZ

Numerical analysis is required in order to evaluate the pile response under typical operating conditions, which must consider extreme events, such as 1 in 1000 or 1 in 10,000 return period cyclonic storms, but also the accumulation of damage during a 30 year operating life. For design purposes, a simplified analysis approach is generally adopted, with the pile modelled as an elastic column (using a finite difference or finite element approach, with the pile discretised into elements) and interaction with the soil captured by

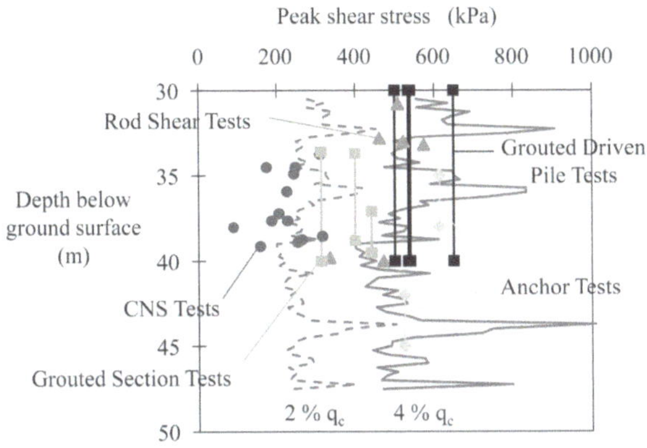


**Figure 27.** Strengths and peak shaft friction measured at Overland Corner, SA.

discrete 'load transfer' springs attached to each pile element.

In clay soils, quite simple load transfer curves are adopted, with a non-linear pre-peak curve that closes matches an inverted parabola, and mild strain softening by up to 30% (e.g. API, 2007). It is relatively uncommon to carry out detailed analysis of the cyclic response for driven piles in clay, with the possible exception of piles anchoring tension leg platforms, where the consequence of failure is severe.

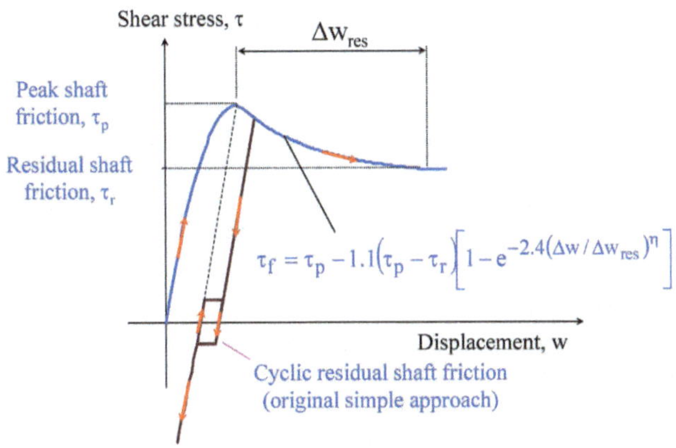


**Figure 28.** Form of axial load transfer curve in RATZ.

For grouted piles in carbonate soils, however, cycle-by-cycle analysis of piles subjected to storm loading has become routine, and software such as RATZ (Randolph, 2003b) have been specifically calibrated against the field tests at Overland Corner. The general form of load transfer response implemented in RATZ is illustrated in Fig. 28. The monotonic response comprises an initial elastic response up to $\xi\tau_p$, followed by an inverted parabola (with apex at the peak) to failure. The post-peak response to the residual shaft friction, $\tau_r$, is captured by a two parameter exponential decay curve:

$$\tau_f = \tau_p - 1.1\,(\tau_p - \tau_r)\left[1 - e^{2.4(\Delta w/\Delta w_{\text{res}})^\eta}\right] \tag{11}$$

The two parameters dictate the relative abruptness of the initial decay ($\eta$), and the additional displacement to reach residual conditions ($\Delta w_{\text{res}}$). Values of $\eta$ less than unity lead to an abrupt change of gradient, with initial infinitely steep decay, at the peak.

Modelling of the low residual cyclic resistance within the so-called gap zone is achieved by nominating a low shaft friction, $\tau_{cr}$, within the plastic region (defined by extrapolating from the displacement at peak friction as indicated) and an abrupt recovery to the current failure shaft friction on emerging from the gap zone.

Pre-failure cyclic loading is modelled using a form of kinematic hardening, similar to a bubble model, where the stress at which yield would occur on reversal of loading tracks behind the minimum (i.e. current) shear stress according to Fig. 29:

$$\tau_y = \tau_{\text{min}} + 0.5\,(1+\xi)\,(\tau_p - \tau_{\text{min}}) \tag{12}$$

This formulation leads to well-defined shakedown limits for 1-way loading (where $\tau_{\text{min}}$ is zero) of elastic $\tau_{\text{elastic}} \leq 0.5\,(1+\xi)\,\tau_p$ and 2-way loading (where $\tau_{\text{min}}$ is the negative of the elastic limit) of elastic $\tau_{\text{elastic}} \leq (1+\xi)\,\tau_p/\,(3-\xi)$.

The relationship in Eq. (12) and the resulting variation of the safe cyclic range is similar to that proposed for metals by Goodman, and is shown in Fig. 30 in the form of a conventional cyclic stability diagram, where the yield criterion becomes a straight line (shown for $\xi = 0.333$). The minimum 1-way and 2-way cyclic amplitudes for which purely elastic response would result are obtained with $\xi = 0$, giving $\tau_{\text{cyc}}/\tau_p = \pm 0.25$ and $\pm 0.33$ respectively.

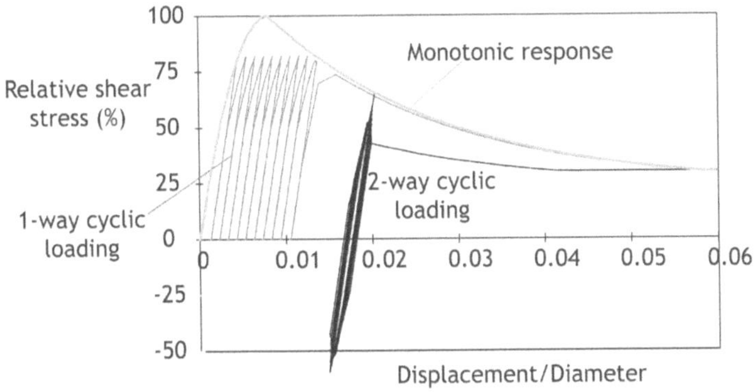


**Figure 29.** RATZ load transfer response under 1-way and 2-way cyclic loading.

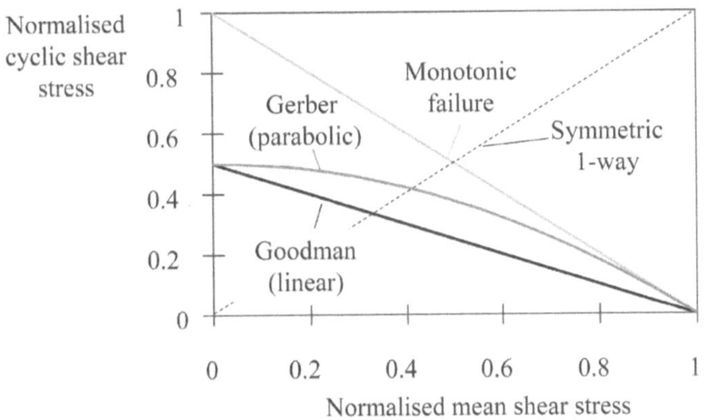


**Figure 30.** Example cyclic stability diagrams simulated using RATZ.

Two other cyclic algorithms are provided in RATZ, one more conservative where the minimum yield stress is reduced to $\xi\tau_p$, and one less conservative that corresponds to the Gerber criterion used in metal fatigue studies, which is equivalent to the parabolic curve shown in Fig. 30.

Calibration of the load transfer algorithm in RATZ has been undertaken in respect of element (CNS and rod shear) tests and also in comparison with the field tests at Overland Corner. Fig. 31 shows a comparison of

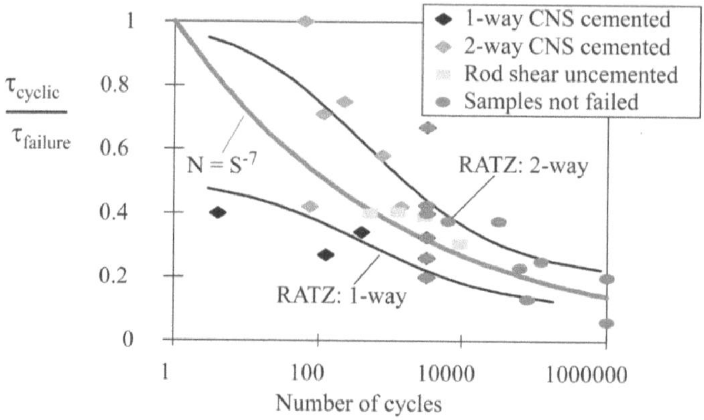


**Figure 31.** Fatigue curves predicted by RATZ load transfer algorithm.

the 'fatigue life' predicted using the pre-failure cyclic algorithm described above with results from 1-way and 2-way cyclic element tests. Reasonable consistency may be observed, in spite of some scatter in the experimental data. The trend for 2-way cyclic loading may be expressed approximately as

$$N_{\text{failure}} \approx \left(\frac{\tau_{\text{cyc}}}{\tau_p}\right)^{-7} \tag{13}$$

A comparison of the predicted response of a grouted segment test subjected to large displacement cycles is shown in Fig. 32. The RATZ simulation captures the essence of the measured response, but in a somewhat crude form, with abrupt changes in resistance at the extremes of the cyclic range, and a constant cyclic residual resistance. Improvements in the algorithm are discussed later.

A similar comparison for the full (15 m long) pile tested at Overland Corner is shown in Fig. 33, focusing on failure of the pile under increasing levels of cyclic loading. The simulation matches the measured response well, particularly in respect of the onset of failure.

The concentration of load transfer in the upper part of (compressible) piles at low load levels, and the gradual mobilisation of shaft resistance further down the pile as the load is increased and failure starts to propagate down the pile, are illustrated in Fig. 34 for the first three stages of cyclic loading of the 15 m long pile referred to above. The field test was instru-

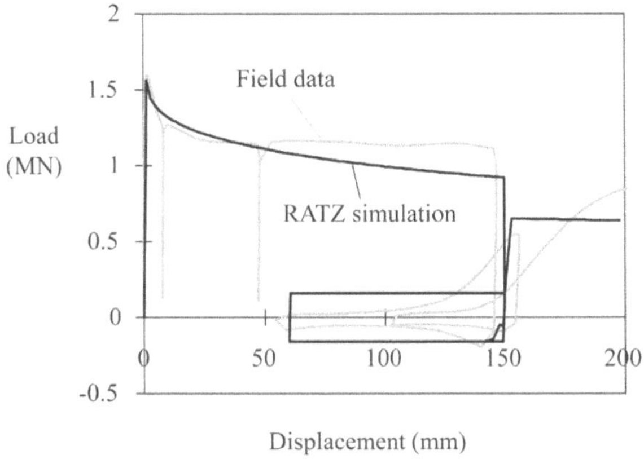


**Figure 32.** Modelling response of 0.44 m diameter grouted section test.

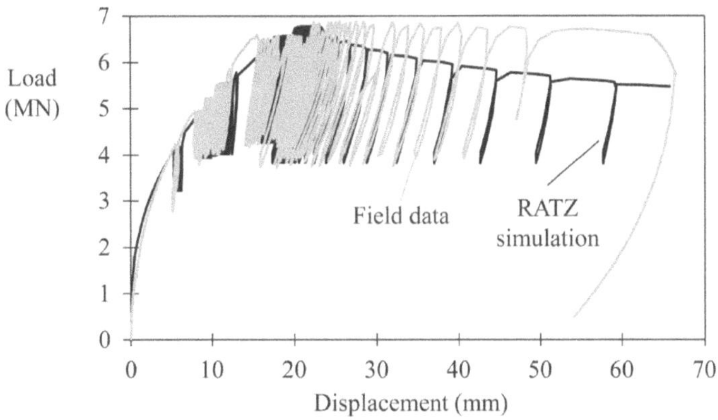


**Figure 33.** Simulation of cyclic response of 15 m long grouted section.

mented to reveal the profile of load in the pile, and the analysis matches the measured data well.

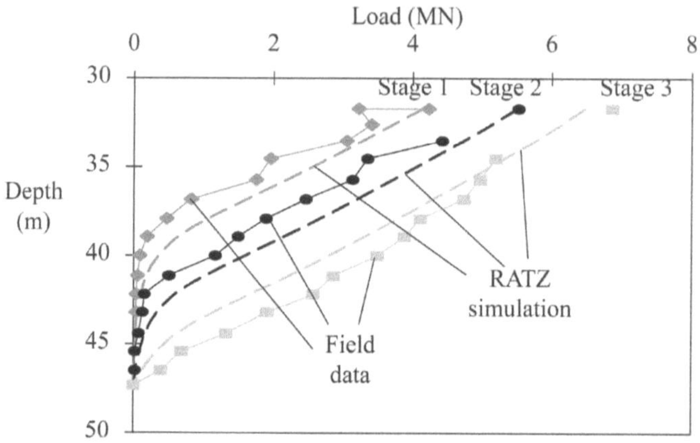


**Figure 34.** Load distributions at different stages of cyclic loading for 15 m long grouted section at Overland Corner, SA.

## 5.1 Improved Cyclic Algorithms - Cyclops

Continuing expansion of the offshore oil and gas industry in Australia, and increasingly arduous design requirements for drilled and grouted insert piles, has prompted further refinement of the cyclic load transfer algorithms, resulting in the software Cyclops (Advanced Geomechanics, 2007). Few projects can justify the high cost (in the region of US$10 million) of site specific pile load tests, so projects must rely primarily on laboratory (principally CNS) test data and correlations with the cone resistance profile.

Monotonic and cyclic CNS tests, with particular focus on fixed displacement cyclic tests at increasing displacement amplitudes, allow correlations to be developed for the peak and residual shaft friction in relation to the cone resistance, $q_c$. Fitting of the cyclic tests allows adjustment of the internal cyclic load transfer parameters, but a significant difficulty remains in respect of scaling of displacements between laboratory and field. Experience from the field tests conducted at Overland Corner still underpins choice of appropriate post-peak displacement to reach residual conditions.

Another aspect of modern design that has required careful consideration is how to capture the effect of the many millions of load cycles that a pile would be subjected to during its operational life. The approach that has been taken includes:

- modelling extreme events (such as a 1 in 10,000 year cyclonic storm) in detail;

- establishing a threshold level of cyclic loading that causes no further degradation in shaft resistance;

- modelling 30-year life environmental loading for levels exceeding that threshold.

Analysis software must be capable of simulating many thousands of load cycles (for example, up to 50,000 cycles) and retain numerical stability. Hand in hand with the approach of undertaking cycle-by-cycle analysis is the need to refine the cyclic algorithm, particularly in respect of the response in the plastic 'gap' zone, and the transition region as the soil element is sheared beyond that zone.

An overview of the load transfer algorithm in Cyclops is shown in Fig. 35. The algorithm within the gap zone simulates a non-linear S-shaped response, with the maximum shear stress mobilised at the extremes of the zone expressed in terms of a bias parameter, $b$, representing the proportion of the current failure shear stress (typically $< 0.5$). The value of the bias parameter may be taken as constant, or may decrease within the first few cycles, as is typically seen in CNS test data. Beyond the gap zone, a transitional region allows the shear resistance to increase back to the current failure level, allowing for degradation that has occurred due to plastic displacement within the cyclic gap zone.

A feature of the monotonic degradation curve expressed in Eq. (11) is that, for values of the shape parameter, $\eta$, less than unity, there is an abrupt change of gradient at peak Fig. 36. This can give rise to numerical instability, which is ameliorated by introducing a smooth transition to the required degradation curve. The displacement over which this occurs can be adjusted, but is usually set so that it the main degradation curve is rejoined within about 5% of the displacement to peak.

An example response from the refined cyclic algorithm within the gap zone, and the transition beyond it, is shown in Fig. 37. The proportion of the current failure shear stress that is mobilised at the extremes of the gap zone gradually decreases with increasing number of cycles, and a smooth response within and beyond the gap zone is ensured.

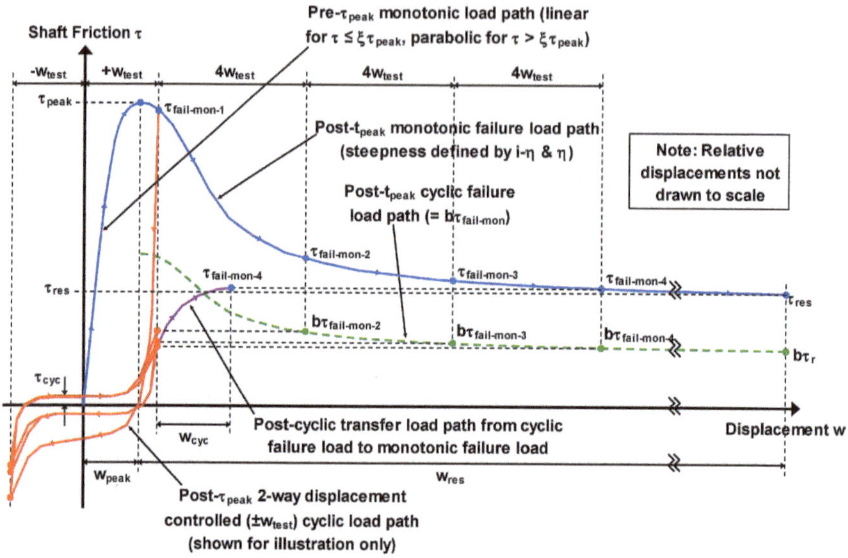


**Figure 35.** Summary of load transfer algorithm features in Cyclops.

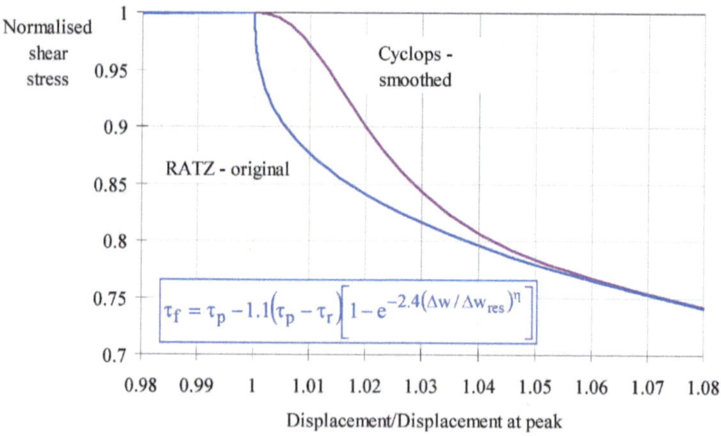


**Figure 36.** Smoothing of immediate post-peak response in Cyclops.

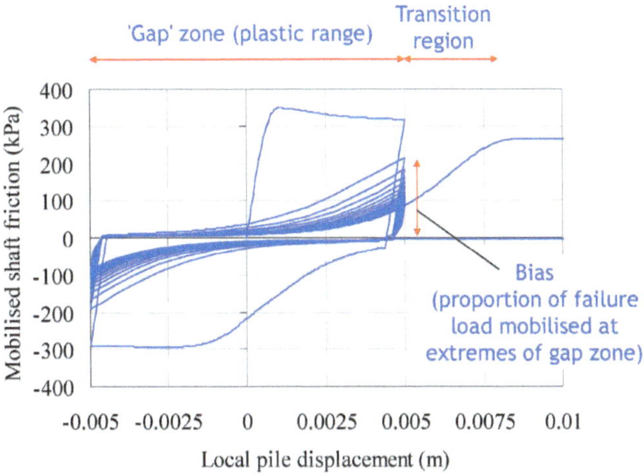


**Figure 37.** Simulation of typical CNS shearing response.

### 5.2 Calibration Against Laboratory and Field Data

Fig. 38 shows the results of a CNS test where a series of post-failure cyclic displacements of ±5 mm have been applied. The data have then been fitted by simulating the element test using the pile analysis software, Cyclops. The left hand plot shows the actual shear stress-displacement response, while the right hand plot shows the match to the gradual reduction in the shear resistance mobilised at the extreme of each cycle. Adopting a varying bias provides improved simulation of the test data. After the 25 cycles, the sample was subjected to further monotonic shearing up to the limits of the apparatus. The load transfer fit includes modelling of the transition back to the monotonic shearing curve.

Simulation of the complete response from one of the grouted segment tests at Overland Corner is shown in Fig. 39. Three simulation curves are shown for different combinations of the displacement to residual, $\Delta w_{\text{res}}$, and the shape parameter, $\eta$. All three response curves provide a reasonable match, and the approach has been to adopt as low a value of $\Delta w_{\text{res}}$ (and the corresponding $\eta$) that preserves a good match, in this case taking $\Delta w_{\text{res}}$ as 0.5 m in order to match the initial post-peak response.

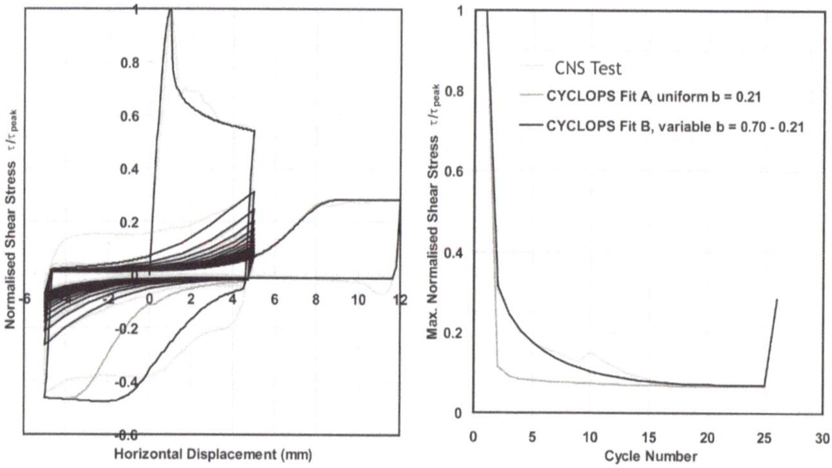


**Figure 38.** Matching of CNS test data with Cyclops.

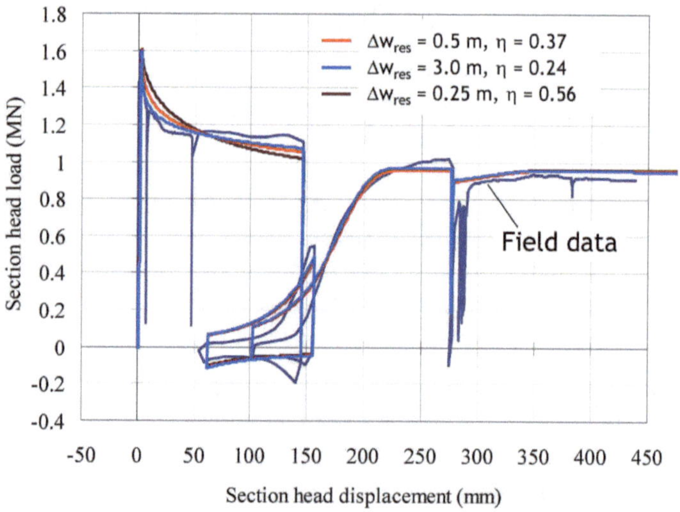


**Figure 39.** Simulation of complete grouted section test from Overland Corner, SA.

## 6 Example Design Analysis for Grouted Insert Pile

A typical design storm may comprise several thousand load cycles, with a peak load that is perhaps 3 times the ambient (calm sea) load on the pile. The majority of load cycles will be less than 70% of the peak, with only 3 or 4 cycles exceeding that level (Fig. 40). In order to establish a suitable threshold for the 30-year lifetime loading, the extreme design storm is applied first, after which constant amplitude load cycles (symmetric 2-way about the ambient load) are applied at increasing levels until further degradation occurs. That level represents the cyclic threshold, which in the present example corresponds to about 4% of the extreme design load, or 11% of the maximum cyclic amplitude in the extreme event. The cyclic threshold reduces the number of load cycles that need to be considered from over 150 million to a more manageable 13,500.

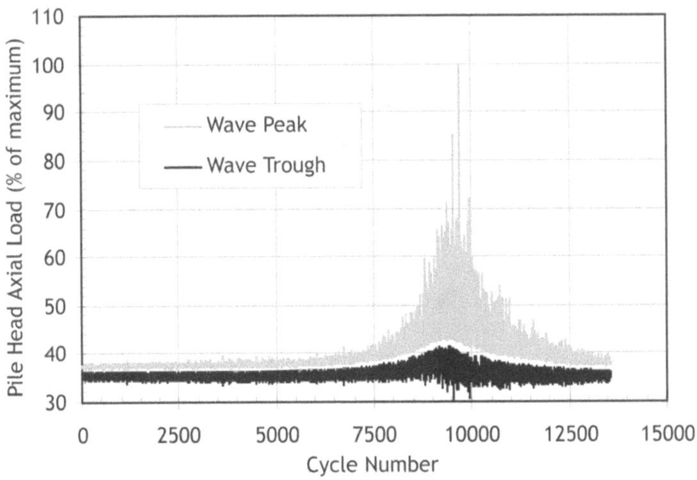


**Figure 40.** Typical 10,000 year storm loading spectrum.

An example of the resulting response of a drilled and grouted pile under monotonic and cyclic loading is shown in Fig. 41, taken from an actual design study. The load, $P$, has been normalised by the ideal 'rigid pile' capacity, $Q_r$, while the displacement is normalised by the diameter, as $w/D$. It may be seen that the peak monotonic capacity is only 64% of the rigid pile capacity because of progressive failure. Following cyclic loading to simulate the design storm, plus a full 30-year 'lifetime cycle' of more moderate environmental loading, the pile capacity is further reduced, to just over 50%

of the ideal capacity. This compares with the peak design load, which is just under 40%, indicating a reserve strength ratio of about 1.2.

Fig. 42 shows the pattern of degradation of shaft friction at the end of the cyclic loading stages. The upper part of the pile, down to just over 50% of the grouted length, has been reduced to residual shaft friction, while the bottom 36% has suffered no degradation. The transition zone over which partial degradation has occurred is therefore around 14% of the grouted length of the pile.

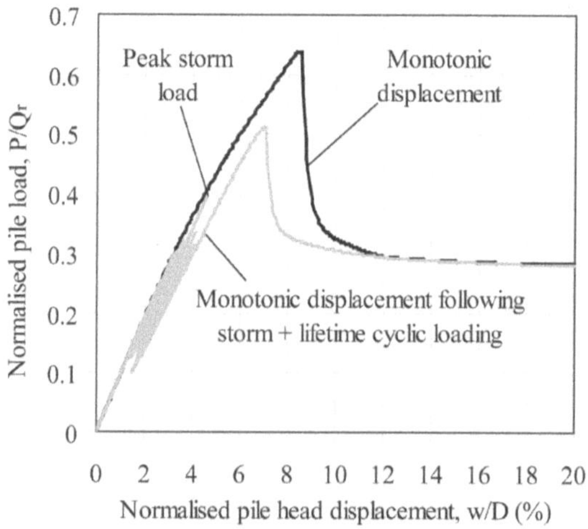


**Figure 41.** Example analysis showing effects of cyclic loading on pile capacity.

From a design perspective, it is important to establish the robustness of such calculations. This may be achieved from a numerical viewpoint by varying the control parameters in the numerical algorithm, which follows an explicit integration approach. From a geotechnical viewpoint, parameters have to be chosen with a reasonable degree of conservatism, and the sensitivity of the results to variations in the parameters assessed. The various parameters are usually viewed against a background of the cone resistance profile for the site in question, not because of an expectation of any precise correlation for each parameter with the cone resistance, but in order to gauge whether extremes of the parameters deduced for each layer reflect extremes in the cone resistance, or whether other patterns emerge.

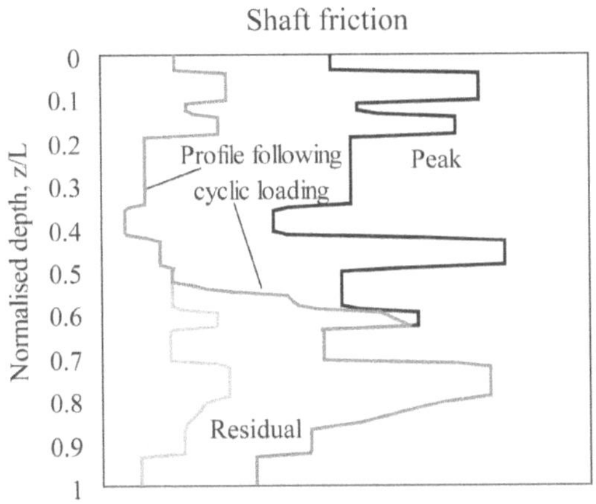


**Figure 42.** Profile of shaft friction following cyclic loading.

## 7 Summary

For practical offshore design, the effects of cyclic loading are generally based on estimating cumulative 'permanent' and cyclic shear strains, and cumulative excess pore pressure, from cyclic triaxial and simple shear tests. Simple shear tests are favoured over triaxial tests because of the small sample size, greater flexibility in the applied shear stress regimes and because they represent average stress paths well.

The shaft friction mobilised along piles driven into sand degrades with distance from the pile tip or, more accurately, the number of shearing cycles imposed on the particular zone of soil during pile installation. Degradation may be expressed by expressing the shaft friction either as a proportion of the vertical effective stress or, in order to allow for variations in sand density, as a proportion of the local cone resistance.

Quantitative studies of the mechanism of friction degradation have been undertaken through cyclic direct shear tests conducted under constant normal stiffness (CNS) conditions. These results show cyclic dilation and contraction within each cycle, but a net accumulation of contraction within a shear band for 5 to 8 particles wide. Some grain crushing occurs during this process, but probably much less than would occur in the real (pile) situation, where significant crushing occurs under the high tip pressures.

A prediction model based on an assumed path in void ratio - normal effective stress space provides a reasonable fit to the CNS data, and suggests that there should be a stable minimum shaft friction after many shearing cycles (or large normalised distance from the pile shaft).

Modern design methods for pile shaft friction are based on the cone resistance profile, and incorporate friction degradation as a power law function of the normalised distance measured upwards from the pile tip. These do not include an explicit minimum value of shaft friction, and this needs to be addressed in future work.

Grouted piles provide significantly higher shaft capacity in carbonate sediments compared with driven piles, with peak shaft friction in the range 2 to 4% of the cone resistance, compared with potentially less than 10 kPa for driven piles. An interesting phenomenon during post-peak cyclic displacement is the extremely low cyclic residual friction within the current plastic (or 'gap') zone. The resistance increases again once shearing occurs beyond that zone. Very similar responses of this nature are observed in laboratory tests (rod shear and CNS tests) and also in field tests on grouted pile segments.

The state-of-the-art for grouted pile design involves cycle-by-cycle analysis of extreme storm events and the life-time environmental loading on the piles. This is achieved through non-linear load transfer analysis with the load transfer algorithms validated and calibrated against tests at field scale. Refinement of the algorithms over the last 20 years has allowed close simulation of field tests and laboratory CNS test data.

Site specific design parameters may be obtained by matching the response from CNS tests, correlating peak and residual shaft friction with the cone resistance profile, and adjusting internal variables in the load transfer algorithm to match the cyclic data.

Uncertainty still exists in scaling displacements (particularly the post-peak rate of strain-softening) from laboratory to field. Adopting displacements to residual directly from measurements at laboratory scale (i.e. 20 mm or less) would be excessively conservative. The field data suggest that a lower limit of about 0.5 m is more reasonable, while still on the conservative side.

## 8 Acknowledgements

This chapter draws from the activities of the Centre for Offshore Foundation Systems (COFS), established under the Australian Research Council's Research Centres Program and currently supported as a Centre of Excellence by the State of Western Australia and through grants FF0561473 and DP0665958 from the Australian Research Council. The author would like to acknowledge the significant contribution from colleagues at COFS and elsewhere in the world in the material presented.

## List of Symbols

| | | |
|---|---|---|
| $_0$ | initial state | §1, §2 |
| $_{\min}$ | minimum value | §2, §3 |
| $_{\max}$ | maximum value | §2, §3 |
| $a$ | parameter | §3 |
| $A_r$ | area ratio for the pile | §3 |
| $b$ | parameter | §3 |
| $c$ | parameter | §3 |
| $d$ | parameter | §3 |
| $d_{50}$ | mean particle diameter | §1, §2 |
| $D$ | diameter of the pile | §1, §3, §6 |
| $D_{\text{equiv}}$ | equivalent diameter | §3 |
| e | parameter | §3 |
| $e$ | void ratio | §2.1 |
| $e_{sb}$ | void ratio of the interface shear band | §2.1 |
| $e_{sb(N)}$ | shear band void ratio at cycle $N$ | §2.1 |
| $f$ | parameter | §3 |
| $G$ | shear modulus of the soil | §1 |
| $h$ | distance from the pile tip | §1, §3 |
| $h_s$ | shear band thickness | §2.1 |
| $\Delta h$ | cumulative interface contraction | §2.1 |
| $k$ | CNS spring stiffness | §2 |
| $K$ | stress ratio | §1, §3 |
| $L$ | $= z + h$ | §3 |
| $N$ | number of cycles | §2.1, §3 |
| $N_{\text{char}}$ | characteristic number of cycles | §2.1 |
| $N_{kt}$ | cone factor | §2.1, §4 |
| $p'$ | mean pressure | §2.1 |
| $p_a$ | atmospheric pressure | §3, §4.1 |
| $P$ | load | §6 |
| $q_c$ | cone resistance | §1, §3, §4, §5 |

| | | |
|---|---|---|
| $\bar{q}_c$ | average cone resistance | §3 |
| $q_u$ | unconfined compression strength | §4.1 |
| $Q_r$ | ideal 'rigid pile' capacity | §6 |
| $s_u$ | shear strength | §4.1 |
| $u$ | parameter | §3 |
| $v$ | parameter | §3 |
| $\mathrm{v}$ | specific volume | §2.1 |
| $w$ | displacement | §5, §6 |
| $\Delta w_{\mathrm{res}}$ | displacement to reach residual conditions | §5 |
| $z$ | depth | §3 |
| $\beta$ | $= K \tan \delta$ friction ratio | §1, §2, §3 |
| $\delta$ | interface friction angle | §1, §3 |
| $\delta_{\mathrm{cv}}$ | steady state interface friction angle | §3 |
| $\eta$ | shape parameter | §5 |
| $\mu$ | parameter controlling the rate of degradation of $K$ | §3 |
| $\xi$ | initial elastic response factor | §5 |
| $\sigma_n$ | normal stress | §2 |
| $\sigma'_n$ | normal effective stress | §1, §2 |
| $\sigma'_{\mathrm{v}}$ | effective vertical stress | §3 |
| $\sigma'_{\mathrm{v0}}$ | in situ effective vertical stress | §1, §3 |
| $\Delta\sigma'_{\mathrm{rd}}$ | radial stress change due to dilation | §3 |
| $\tau_{cr}$ | low shaft friction within the plastic region | §5 |
| $\tau_{\mathrm{cyc}}$ | cyclic shaft friction | §5 |
| $\tau_{\mathrm{elastic}}$ | elastic shaft friction | §5 |
| $\tau_f$ | post-peak residual shaft friction | §5 |
| $\tau_p$ | peak shaft friction | §4.1 |
| $\tau_r$ | residual shaft friction | §5 |
| $\tau_s$ | shaft friction | §1, §3 |
| $\phi_{\mathrm{cv}}$ | critical state angle of friction | §1 |

## Bibliography

A. F. Abbs. Design of grouted offshore piles in calcareous soils. In *Proc. 6th Australia-New Zealand Conf. on Geomech.*, pages 128–132, Christchurch, 1992.

Advanced Geomechanics. *Cyclops - software for cyclic axial loading of piles.* Perth, Australia, 2007.

API. *Recommended Practice for Planning, Designing and Constructing Fixed Offshore Platforms-Working Stress Design.* American Petroleum Inst, Washington, RP2A-WSD 22nd edition, 2007.

ASTM. *Annual book of ASTM standards.* ASTM, West Conshohocken, PA, 2004.

M. Cubrinowski and K. Ishihara. Flow potential of sandy soils with different grain compositions. *Soils and Foundations*, 40(4):103–119, 2000.

J. T. DeJong, M. F. Randolph, and D. J. White. Interface load transfer degradation during cyclic loading: a microscopic investigation. *Soils and Foundations*, 43(4):81–93, 2003.

J. T. DeJong, M. F. Randolph, and D. J. White. Microscale observation and modelling of soil-structure interface behaviour using PIV. *Soils and Foundations*, 46(1):15–28, 2006.

J. Dolwin, M. S. Khorshid, and P. van Goudoever. Evaluation of driven pile capacity - methods and results. In *Proc. Int. Conf. on Calc. Sediments*, volume 2, pages 409–428, 1988.

J. D. Frost and D. J. Jang. Evolution of sand microstructure during shear. *ASCE J. Geotech. and Geoenv. Eng.*, 126(2):116–130, 2002.

Fugro. Axial pile capacity design method for offshore driven piles in sand. Technical Report No. P1003 to API, Issue 3, Fugro Engineers BV, August 2004.

R. Jardine, F. Chow, R. Overy, and J. Standing. *ICP Design Methods for Driven Piles in Sands and Clays.* Thomas Telford Publishing, London, U.K., 2005.

R. J. Jardine and F. C. Chow. New design methods for offshore piles. In *Marine Tech. Directorate*, volume MTD 96/103, London, 1996.

R. J. Jewell and M. F. Randolph. Cyclic rod shear tests in calcareous sediments. In *Engineering for Calcareous Sediments*, volume 1, pages 215–222, Perth, Australia, 1988.

H. A. Joer and M. F. Randolph. Experimental modelling of the shaft capacity of grouted driven piles in calcareous soil. In *FHWA Conf. on Design and Construction of Deep Foundations*, volume 2, pages 873–887, Florida, 1994.

I. W. Johnston, T. S. K. Lam, and A. F. Williams. Constant normal stiffness direct shear testing for socketed pile design in weak rock. *Géotechnique*, 37(1):83–89, 1987.

H. Kishida and M. Uesugi. Tests of the interface between sand and steel in the simple shear apparatus. *Géotechnique*, 37(1):45–52, 1987.

E. U. Klotz and M. R. Coop. On the identification of critical state lines for sands. *ASTM Geotechnical Testing Journal*, 24(3):1–14, 2002.

H. J. Kolk, A. E. Baaijens, and M. Senders. Design criteria for pipe piles in silica sands. In *Proc. Int. Symp. on Frontiers in Offshore Geotechnics (ISFOG)*, pages 711–716, Perth, Australia, 2005. Taylor and Francis, London.

F. H. Kulhawy and K. K. Phoon. Drilled shaft side resistance in clay soil to rock. In *ASCE Design and Performance of Deep Foundations*, pages 172–183, New York, 1993. ASCE Geotechnical Special Publication 38.

K. L. Lee and H. B. Seed. Drained strength characteristics of sands. *ASCE Journal of Soil Mechanics and Foundation Engineering*, 93(SM6):117–141, 1967.

B. M. Lehane, R. J. Jardine, A. J. Bond, and R. Frank. Mechanisms of shaft friction in sand from instrumented pile tests. *ASCE J. of Geotech. Engng.*, 119(1):19–35, 1993.

B. M. Lehane, J. A. Schneider, and X. Xu. The UWA-05 method for prediction of axial capacity of driven piles in sand. In *Proc. Int. Symp. on Frontiers in Offshore Geotechnics (ISFOG)*, pages 683–689, Perth, Australia, 2005a. Taylor and Francis, London.

B. M. Lehane, J. A. Schneider, and X. Xu. A review of design methods for offshore driven piles in siliceous sand. Technical Report Report N. GEO: 05358, University of Western Australia, Geomechanics Group, 2005b.

H. G. Poulos, M. F. Randolph, and R. M. Semple. Evaluation of pile friction from conductor tests. In *Engineering for Calcareous Sediments*, volume 2, pages 599–605, Perth, Australia, 1988.

M. F. Randolph. The axial capacity of deep foundations in calcareous soils. In *Proc. Int. Conf. on Calc. Sediments*, volume 2, pages 837–858, 1988.

M. F. Randolph. 43rd Rankine Lecture: Science and empiricism in pile foundation design. *Géotechnique*, 53(10):847–875, 2003a.

M. F. Randolph. *RATZ - Load transfer analysis of axially loaded piles, User Manual, Version 4.2.* Centre for Offshore Foundation Systems, University of Western Australia, 2003b.

M. F. Randolph, J. Dolwin, and R. D. Beck. Design of driven piles in sand. *Géotechnique*, 44(3):427–448, 1994.

M. F. Randolph, H. A. Joer, M. S. Khorshid, and A. M. Hyden. Field and laboratory data from pile load tests in calcareous soil. In *Proc. 28th Offshore Technology Conf.*, volume OTC 7992, Houston, Texas, 1996.

M. F. Randolph, H. A. Joer, and D. W. Airey. Foundation design in cemented sands. In *Proc. 2nd Int. Sem. on Hard Soils, Soft Rocks*, volume 3, pages 1373–1387, Naples, Italy, 1998.

J. P. Rickman and H. C. Barthelemy. Offshore construction of grouted driven pile foundations. In *Engineering for Calcareous Sediments*, volume 1, pages 313–319, Perth, Australia, 1988.

J. Seidel and C. M. Haberfield. The axial capacity of pile sockets in rocks and hard soils. *Ground Engineering*, 28(2):33–38, 1995.

I. Shahrour and F. Rezaie. Experimental study of the behaviour of calcareous sand: structure interface. In *Proc. Engineering for Calcareous Sediments*, pages 69–77, 1999.

F. E. Toolan, M. L. Lings, and U. A. Mirza. An appraisal of API RP2A recommendations for determining skin friction of piles in sand. In *Proc. 22th Offshore Technology Conf.*, volume OTC 6422, Houston, Texas, 1990.

R. Verdugo and K. Ishihara. The steady state of sandy soils. *Soils and Foundations*, 36(2):81–91, 1996.

A. S. Vesic. Design of pile foundations. Technical Report Synthesis of Highway Practice N. 42, National Co-operative Highway Research Program, Transportation Research Board, National Research Council, Washington DC, 1977.

D. J. White. *An investigation into the behaviour of pressed-in piles.* PhD thesis, University of Cambridge, 2002.

D. J. White. A general framework for shaft resistance on displacement piles in sand. In *Proc. Int. Symp. on Frontiers in Offshore Geotechnics (ISFOG)*, pages 697–703, Perth, Australia, 2005. Taylor and Francis, London.

D. J. White and M. D. Bolton. Displacement and strain paths during pile installation in sand. *Géotechnique*, 54(6):375–398, 2004.

D. J. White and B. M. Lehane. Friction fatigue on displacement piles in sand. *Géotechnique*, 54(10):654–658, 2004.

# Evaluation of the Remoulded Shear Strength of Offshore Clays and Application to Pipeline-Soil and Riser-Soil Interaction

Mark F. Randolph *

* Centre for Offshore Foundation Systems, University of Western Australia, Australia

## 1 Introduction

Quantifying the sensitivity of fine grained sediments is an important aspect of site characterisation. The sensitivity gives an indication of the likely effects of pre-peak cyclic loading, while the remoulded undrained shear strength is directly relevant for various offshore applications (see examples in Fig. 1). It is customary to assume that the shaft friction acting during the installation of piles, suction caissons or anchors may be taken as the remoulded undrained shear strength (Semple and Gemeinhardt, 1981). Equally the cyclic bearing resistance of laterally loaded piles or pipelines and catenary riser systems may be expressed in terms of the remoulded shear strength of the seabed sediments at shallow depth. A further example is for the operative strength of material in debris resulting from a submarine slide, which may also be estimated directly from the remoulded shear strength.

The implicit assumption is that the water content of the sediments are not changed during the particular design application (which may not be true in some cases), so that the remoulding merely destroys any bonding and other natural structure of the material.

Natural soils show rate-dependent shear strength and also a reduction in strength as they are sheared and remoulded. Different design applications involve varying degrees of these factors, and it is useful to consider a schematic 'map' of soil tests and design applications (Fig. 2) in respect of strain rates and degree of soil remoulding involved in the application (Randolph et al., 2007).

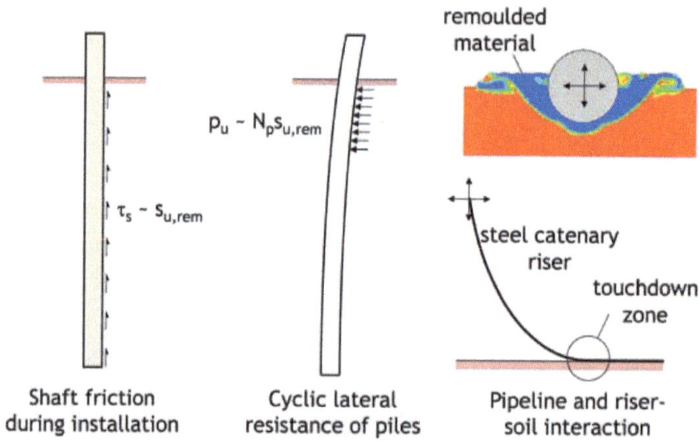


**Figure 1.** Example applications of remoulded shear strength.

The remoulded undrained shear strength ($s_{u,\text{rem}}$) may be estimated from indirect measurements such as by means of field penetrometer tests, by direct measurement such as in a field vane test or laboratory element test on remoulded material, or by correlation with the intact shear strength ($s_u$) by means of a sensitivity ($S_t$). However, because of strain rates that may vary by several orders of magnitude between different forms of soil test and different applications, the use of a (fixed) sensitivity can be very misleading.

## 2 Measurement of Remoulded Shear Strength

The most common methods of estimating the remoulded shear strength of fine-grained sediments (i.e. those where the water content will typically remain constant during the remoulding process) include:

1. unconsolidated undrained (UU) triaxial tests on remoulded material;
2. field or laboratory vane shear tests, carried out to sufficiently large rotations of the vane;
3. fall cone tests on remoulded material; and
4. sleeve friction ($f_s$) from cone penetration tests.

More recently, full-flow (T-bar and ball) penetrometers have become to be used widely in the offshore industry, particularly at the shallow depths relevant for pipeline design and for shallow anchoring systems, where the penetration resistance for remoulded conditions is measured following typi-

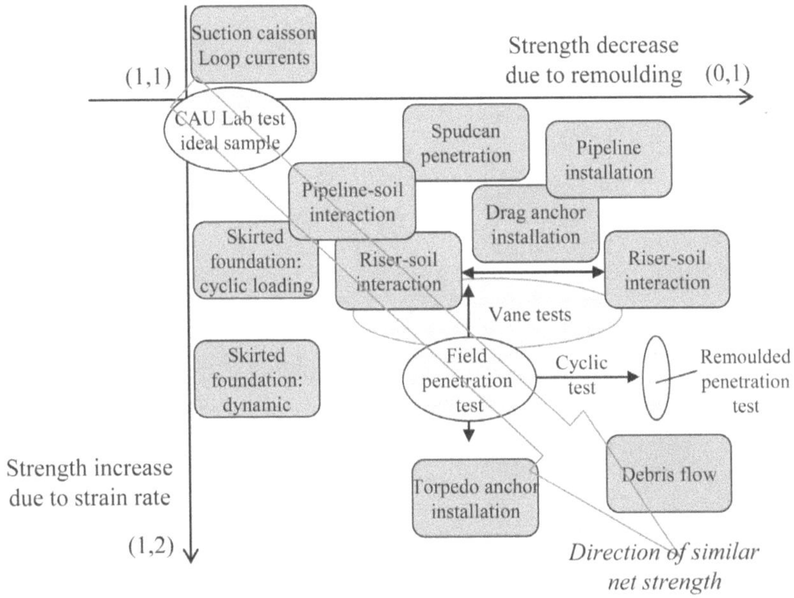


**Figure 2.** Map of different soil tests and design applications.

cally 10 cycles of penetrating and extracting the probe over a short interval. These tests are discussed in detail later in this chapter.

The different forms of test listed above are in approximate order of (typical) strain rates for the tests, which range from $\sim 10^{-4}$ s$^{-1}$ (UU), $10^{-1}$ s$^{-1}$ (vane), 10-30 s$^{-1}$ (fall cone) and 40 s$^{-1}$ (cone sleeve friction). These span 5 or 6 orders of magnitude, and thus may be expected to yield rather different values of remoulded shear strength. Typical rates of strain rate dependency are around 10 to 15% per log cycle (see Fig. 3), with a tendency for increased rate dependency at high strain rates (Biscontin and Pestana, 2001; Andersen et al., 2008).

At face value, the cone sleeve friction is an attractive method to evaluate the remoulded shear strength in order to estimate the shaft friction for installation of piles and other cylindrical objects. However, friction sleeve data from cone tests are notoriously inconsistent and current recommendations are not to rely on such data (Lunne and Andersen, 2007). The very high strain rates associated with frictional resistance on the cone sleeve, with the cone penetrating at 0.5 diameters per second, are perhaps not appreciated. Even where the cone sleeve data are reliable, the measurements might be applicable to the dynamic conditions of pile installation (transient velocities of the order of 1 diameter per second), but much less so for suction caisson installation where typical installation rates would be three orders of magnitude lower (e.g. 1 to 3 diameters per hour).

The field vane shear test is used widely in some offshore regions, in particular in the Gulf of Mexico where its use far exceeds cone penetration testing (Quiros and Young, 1988). In typical offshore practice, the vane is pushed to the required depth at a rate of 20 mm/s and then left for no more than 5 minutes before being rotated at 0.1 or 0.2°/s. The remoulded strength is measured using a similar procedure after the vane is rotated for up to one revolution at a rotation rate of 0.6°/s. This remoulding rotation, which is limited by time constraints because of the high costs of offshore vessel hire, is significantly less than the ten turns recommended for onshore practice and hence will overestimate the remoulded strength and underestimate the sensitivity.

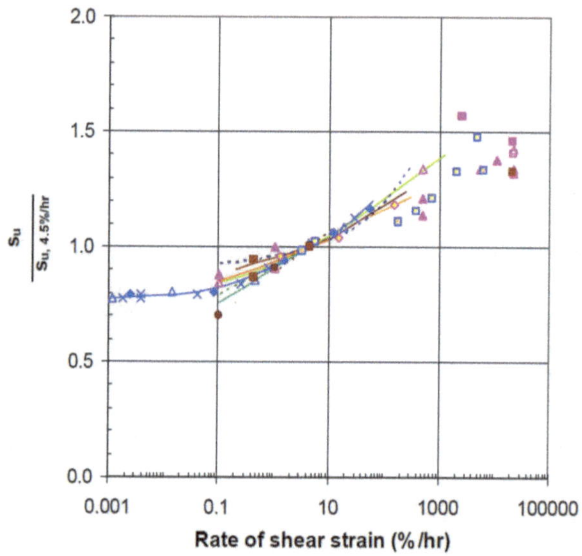


**Figure 3.** Laboratory data showing effect of strain rate on shear strength (Andersen et al., 2008).

Laboratory UU and fall cone tests carried out on remoulded material are often used to evaluate the remoulded shear strength. The two tests cover the extremes of strain rate, and hence should not be expected to provide the same value of shear strength, except that the fall cone test result has often been correlated against UU (and other low strain rate) test data (Hansbo, 1957; Karlsson, 1961; Wood, 1985). Often though, the fall cone is used to provide measurements of the sensitivity, by comparing deduced strengths for intact and remoulded specimens. This eliminates the uncertainty over the calibration constant for the fall cone shear strength, although does not lead to a specific value for the remoulded shear strength.

## 3 Full-Flow Penetrometer Testing

Over the last decade, increasing offshore use has been made of penetrometers with an enlarged tip, in particular cylindrical T-bar (Stewart and Randolph, 1994; Randolph et al., 1998) and ball (Peuchen et al., 2005; Kelleher and Randolph, 2005) penetrometers (see Fig. 4). Projected areas of the tips are generally about 10 times that of the shaft behind the tip, so T-bar of 40 mm diameter by 250 mm long, and ball of ~80 mm diameter attached to a stub shaft of 25 mm diameter (Peuchen et al., 2005), or alternatively 113 mm diameter attached directly to the cone shaft (35.7 mm diameter) (Chung and Randolph, 2004).

A key aspect of full-flow penetrometers, apart from their simple geometry, which allows direct calculation of resistance factors, is the ability to carry out cyclic tests comprising (typically) 10 cycles of penetration and extraction over a short depth range of 0.5 to 1 m. This allows direct measurement of penetration resistance under remoulded conditions, and hence the remoulded shear strength. The extraction resistance is monitored in addition to the penetration resistance, as the ratio of $q_{\mathrm{out}}/q_{\mathrm{in}}$ provides a first measure of the consistency of the data and of the sensitivity of the sediments (Fig. 5).

Cyclic testing can also reveal errors in the testing, for example due to load cell drift caused by temperature changes between the deck of the site investigation vessel and the seabed. The cyclic penetration and extraction resistances should become essentially symmetric about zero, allowing adjustment of the load cell zero (which can be significant in some cases, see example data in Fig. 6.

The advent of full-flow penetrometers has allowed a more thoughtful approach to in situ testing. As will be clear later, interpretation of the penetration resistance (for both intact and remoulded conditions) requires separation of the effects of strain rate dependency of shear strength from those of partial softening (remoulding). These effects can be quantified by planning the test appropriately, including varying the penetration rate in order to quantify strain rate dependency (and even consolidation charac-

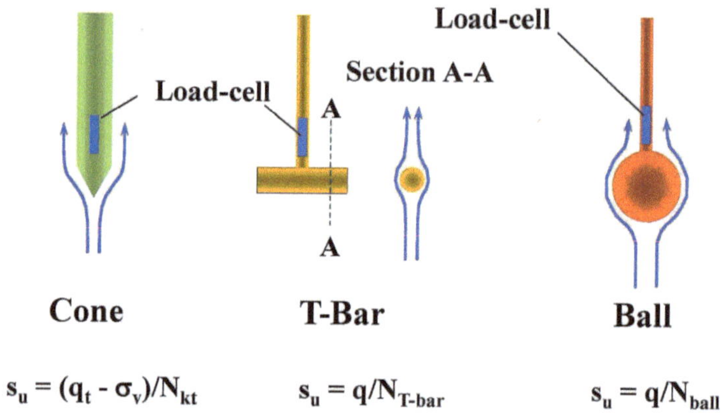


**Figure 4.** Schematic of cone, T-bar and ball penetrometers. Note: $s_u$ is the shear strength; $q_t$ and $q$ the total and net bearing resistance; $\sigma_\text{v}$ the total overburden stress and $N_k$ the various bearing factors.

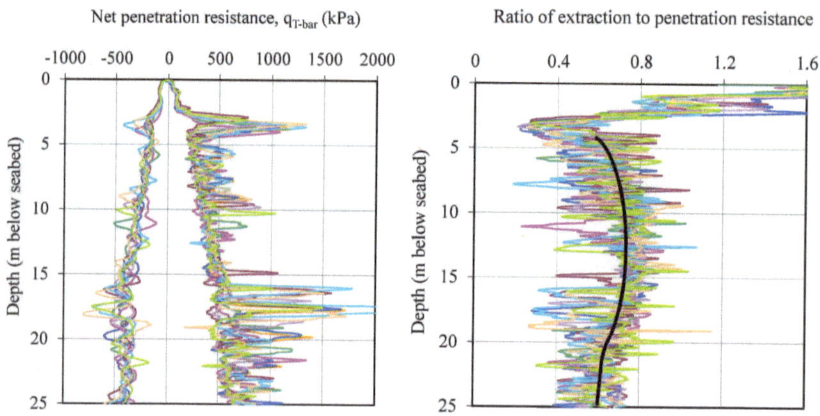


**Figure 5.** Example offshore data from T-bar penetrometer tests.

teristics if the rate can be reduced sufficiently for the given soil type), and carrying out cyclic tests in order to evaluate - and ultimately eliminate - the effects of softening with each passage of the probe. Potential ways in which field penetrometer tests can be designed in order to optimise measured data are indicated on Fig. 7.

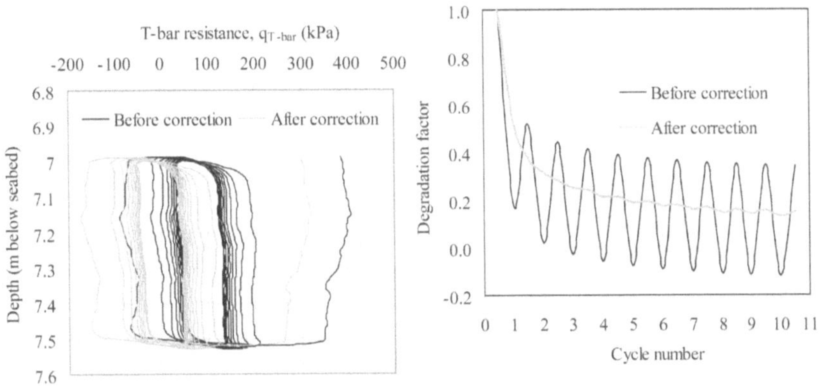


**Figure 6.** Data from cyclic T-bar penetrometer tests.

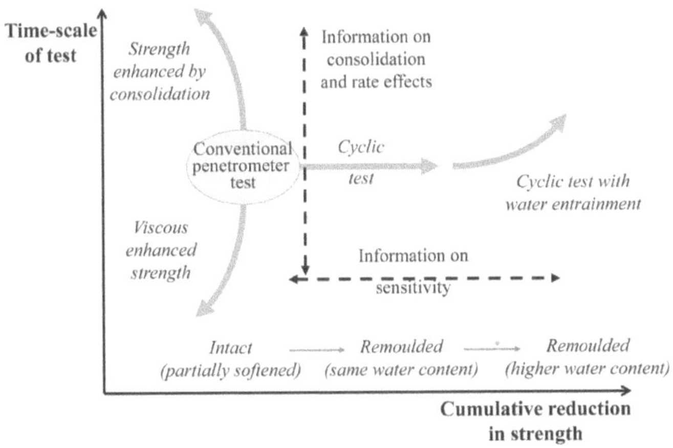


**Figure 7.** Design of field tests to optimise application of measured data.

Field data from penetrometer tests carried out in lightly overconsolidated estuarine clay at Burswood, Western Australia show remarkably similar penetration and extraction profiles for a variety of different penetrometer shapes (Fig. 8). The clay has sensitivity measured from vane shear tests mostly around 4, but with some layers showing sensitivity as high as 10. The ratio of extraction resistance to penetration resistance is reasonably uniform at around 0.6, indicating that the soil is not fully remoulded locally during a single pass of the penetrometer.

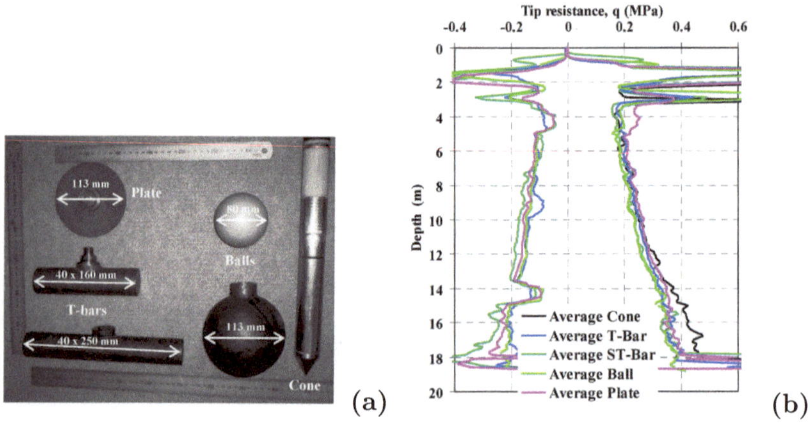


**Figure 8.** Different penetrometer shapes (a) and resulting penetration and extraction profiles (b) (Chung and Randolph, 2004).

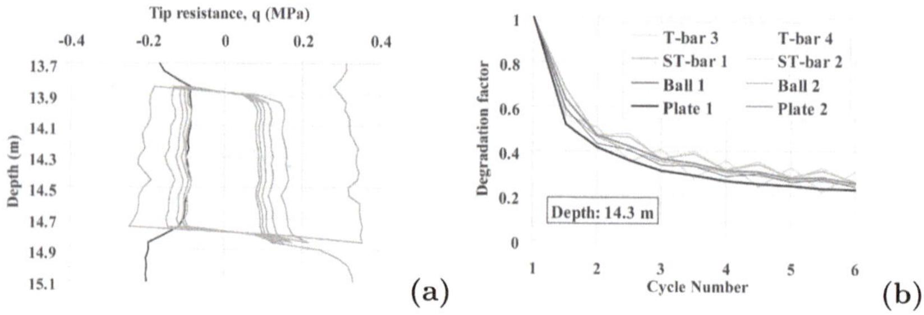


**Figure 9.** Cyclic penetrometer data: (a) T-bar test at 14 m depth; and (b) degradation curves (b) (Chung and Randolph, 2004).

Cyclic full-flow penetrometer tests show that most degradation occurs over the first 5 to 10 cycles of penetration and extraction, and the recommendation for cyclic tests is to standardise on 10 cycles. Data from different penetrometers show very similar trends (Fig. 9), and the final degradation factor (ratio of post-cyclic penetration resistance to the initial penetration resistance) is less than the inverse of the soil sensitivity.

The discrepancy between what might be called the 'penetration resistance sensitivity' and the soil sensitivity, $S_t$, appears to become more pronounced as the soil sensitivity increases (Fig. 10).

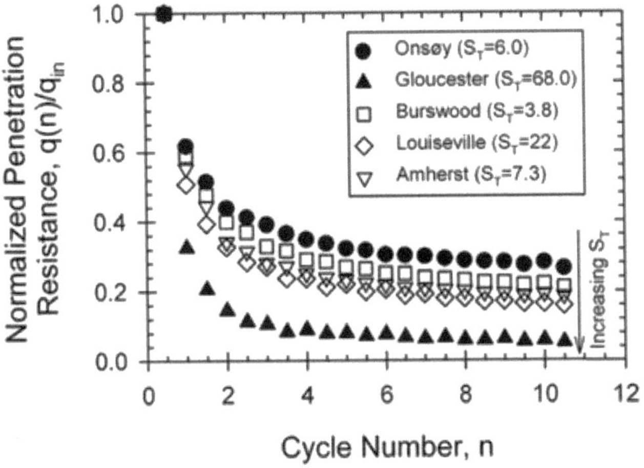


**Figure 10.** Cyclic degradation curves in high sensitivity clays (Yafrate et al., 2009).

Field cyclic penetrometer tests provide a measurement of the penetration resistance in remoulded soil, which may be used directly in certain design situations. In other applications, however, it is more convenient to evaluate a remoulded shear strength, and that requires use of an appropriate resistance factor, $N_{\text{rem}}$, so that the remoulded shear strength is obtained from

$$s_{u,\text{rem}} = \frac{q_{\text{rem}}}{N_{\text{rem}}} \sim \frac{q_{n=10}}{N_{\text{rem}}} \tag{1}$$

Field data for the resistance factor, $N_{\text{rem}}$, calculated with respect to field vane tests following remoulding of the soil, indicate that the value of $N_{\text{rem}}$ increases with the sensitivity of the soil according to

$$N_{\text{rem}} \approx N_k + \frac{\Delta N_k}{(1 + S_t/f)^{-3}} \tag{2}$$

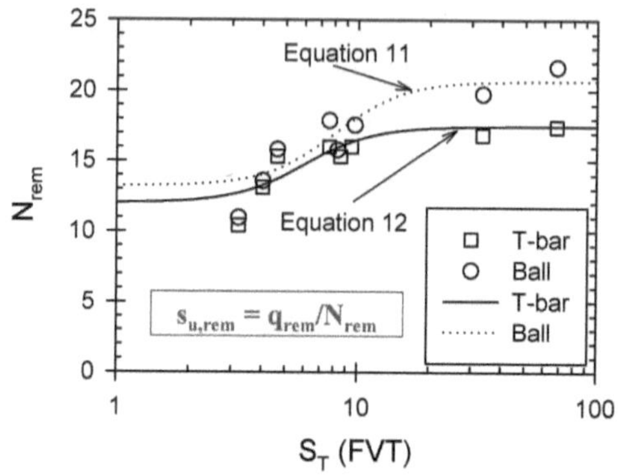


**Figure 11.** Variation of resistance factor for remoulded conditions (Equations and data from Yafrate et al., 2009).

where $N_k$ is the value for intact conditions for soils of low sensitivity (and hence also applies for remoulded conditions) typically 12 to 13, $\Delta N_k$ is the maximum increase for highly sensitive soils (range of 5.5 to 7.5 for the two curves marked as Equation 11 and Equation 12 in Fig. 11, from Yafrate et al., 2009), and the factor $f$, which determines the mid-point of the S-shaped fits, is around 6 to 8. The reasons for this transition in $N_{\text{rem}}$ have been explored further by means of numerical analysis of full-flow penetrometers, as discussed below.

## 4 Numerical Modelling of Full-Flow Penetrometers

The relatively simple geometry of T-bar and ball penetrometer, by comparison with the cone, has allowed much more sophisticated analysis of the response, with the aim of avoiding reliance on purely empirical correlations to obtain resistance factors, $N_k$. Building on plasticity solutions, which give very tight bounds on the resistance factors (Randolph et al., 2000; Martin and Randolph, 2006), secondary features such as the strain rate dependence of shear strength and strength degradation due to remoulding have been incorporated in the solutions (Einav and Randolph, 2005).

Extensive finite element analyses have been undertaken using a large deformation approach that allows penetrations of several diameters to be simulated. A comprehensive description of the methodology and modelling details has been given by Zhou and Randolph (2007, 2009a,b), where the robustness of the method was also demonstrated.

The LDFE (large deformation finite element) analysis is implemented using the RITSS approach (remeshing and interpolation with small strain: Hu and Randolph, 1998), which falls in the category of 'Arbitrary Lagrangian-Eulerian' (ALE) methods. The whole process consists of a series of small-strain analysis increments, followed by remeshing, and then interpolation of the field quantities (stresses and material properties) on the Gauss points from the old mesh to the new mesh (Fig. 12). The sequence of small-strain increments, remeshing and interpolation can be repeated until the required displacement has been reached. In remeshing step a partial remeshing technique (Zhou and Randolph, 2009b) was adopted to facilitate interpolation operations, where only part of the domain (region near the T-bar or ball) was re-discretised, while for the remainder the nodal coordinates were merely updated according to the displacements. This approach limits numerical diffusion of the solution.

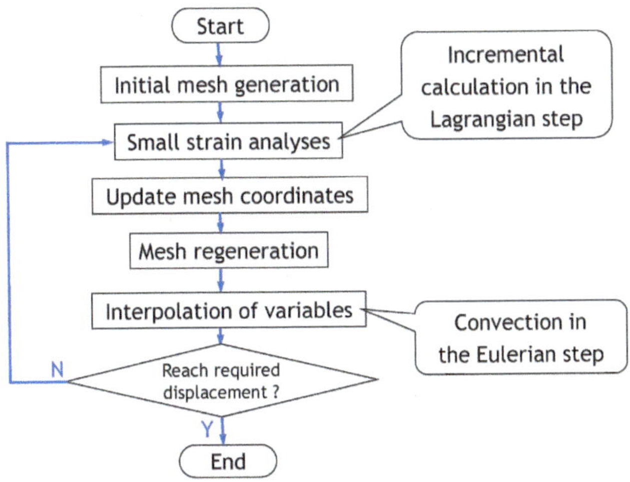


**Figure 12.** Flow chart for large deformation finite element analysis (Hu and Randolph, 1998).

### 4.1 Strain Rate-Dependent and Strain-Softening Model

When simulating cyclic penetration and extraction of the shafted ball penetrometer in strain rate-dependent and strain-softening soil, a simple modified elastic-perfectly plastic Tresca model was adopted, following Einav and

Randolph (2005). The undrained shear strength at individual Gauss points is modified according to the current rate of maximum shear strain and accumulated absolute plastic shear strain, given by

$$s_u = \left[1 + \mu \log\left(\frac{\max\left(\left|\dot{\gamma}_{\max}\right|, \dot{\gamma}_{\text{ref}}\right)}{\dot{\gamma}_{\text{ref}}}\right)\right]\left[\delta_{\text{rem}} + \left(1 - \delta_{\text{rem}}\right) e^{-3\xi/\xi_{95}}\right] s_{u0} \quad (3)$$

where the first bracketed term represents the enhancement due to high strain rates relative to a reference value (here taken as 1%/hour, or $3 \times 10^{-6}$ s$^{-1}$), following a logarithmic law with a rate parameter, $\mu$. The maximum shear strain rate, $\dot{\gamma}_{\max}$, is defined as:

$$\dot{\gamma}_{\max} = \frac{\Delta\varepsilon_1 - \Delta\varepsilon_3}{n\Delta t} \quad (4)$$

where $\Delta\varepsilon_1$ and $\Delta\varepsilon_3$ are the cumulative major and minor principal strains, respectively, over the $n$ increments (between remeshing steps). The notional time period, $\Delta t$, for each increment may be written as

$$\Delta t = \frac{\delta/d}{v_p/d_{\text{field}}} \quad (5)$$

where $\delta$ is the specified displacement for each increment, $d$ and $d_{\text{field}}$ are respectively the penetrometer diameters in the FE computations and field tests (80 mm for the ball), and $v_p$ is the penetration rate in field (20 mm/s).

The second term models the strength degradation from the intact condition to a fully remoulded state according to an exponential function of accumulated plastic shear strain, $\xi$. The quantity $\xi_{95}$ denotes the value of $\xi$ for the soil to undergo 95% remoulding, thus reflecting the relative ductility of the soil. Typical values of $\xi_{95}$ have been estimated in the range of 10 to 50 (i.e. 1000% to 5000% shear strain) by comparison with cyclic penetration and extraction data (Randolph, 2004; Einav and Randolph, 2005). $\delta_{\text{rem}}$ is the strength ratio of soil between fully remoulded and intact state (inverse of sensitivity, $S_t$). $s_{u0}$ is the original shear strength at the reference shear strain rate prior to any softening.

The strength degradation model adopted is somewhat simplistic. For a structured or cemented soil, part of the intact strength may be very brittle, requiring only small plastic shear strains to destroy it, while much larger strains may still be required to remould the soil fully. This type of strength reduction would need a two-stage degradation model such as:

$$s_u = \left[\delta_{\text{rem}} + \left(1 - \delta_{\text{str}} - \delta_{\text{rem}}\right) e^{-3\xi/\xi_{95}} + \delta_{\text{str}} e^{-3\xi/\xi_{95,\text{str}}}\right] s_{u0} \quad (6)$$

where the subscript 'str' relates to the part of the strength that is due to structure or cementation. The response of this type of model, for different values of $\xi_{str}$, and assuming that $\xi_{95} = 100\xi_{95,str} = 15$, is shown in Fig. 13. This type of strength model would result in a relatively greater proportion of strength reduction during initial penetration than during subsequent extraction and penetration cycles.

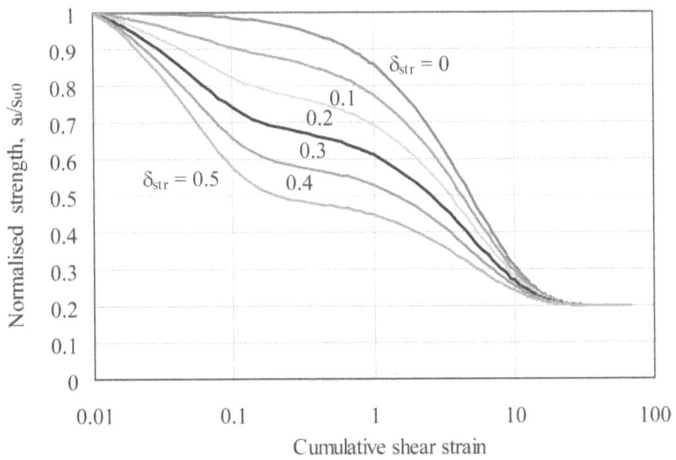


**Figure 13.** Degradation using a two-stage model.

## 4.2 Numerical Modelling of Cyclic Penetrometer Tests

LDFE techniques are required for modelling penetrometer tests in order to follow correctly the flow of soil past the penetrometer, allowing for gradual softening of the material due to the cumulative shear strains. The effects of high strain rate, which for typical field rates are around 15 to 20% per second (4 or 5 orders of magnitude greater than in a slow laboratory test), and gradual softening compensate for each other. When the two effects are combined, the resulting penetration resistance is relatively close to the plasticity solution for ideal material response (Fig. 14).

Strength degradation leads to periodic generation of shear bands, softening within the shear band, and then discarding of the shear band as a new one is generated. This leads to an oscillation in the penetration resistance (Zhou and Randolph, 2007) and a heterogeneous pattern of degradation in the wake of the penetrometer (see Fig. 15). The improved stability of the degradation patterns (with reduced numerical diffusion) obtained using par-

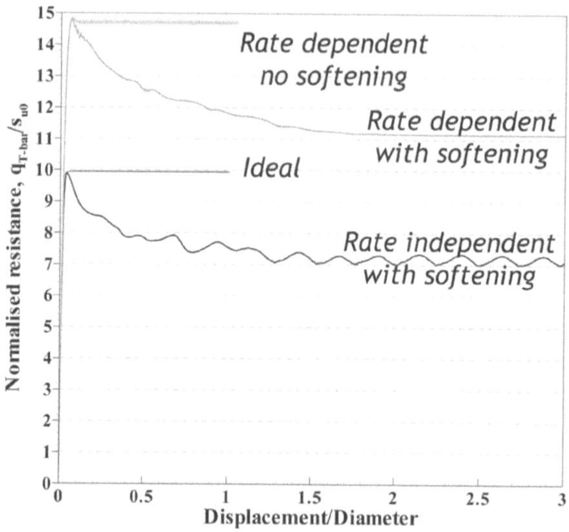


**Figure 14.** Numerical analysis of T-bar penetration.

tial remeshing of the zone adjacent to the T-bar, rather than full remeshing, is evident from this figure.

As the soil flows past the penetrometer, it undergoes an average shear strain, $\xi_k$, with values typically 3.3 (330%) for the ball to 3.7 for the T-bar (Zhou and Randolph, 2009a). Although it has been customary to plot the degradation of resistance during a cyclic penetrometer test starting from a cycle number of 0.5 for initial penetration, and 1 during first extraction, it is more logical to label the initial cycle as 0.25, and the first extraction as 0.75 (Fig. 16). In this way, an ideal 'no softening' (but rate dependent) value of $N_k$ may be estimated approximately as

$$N_{k,\text{no softening}} \approx \sqrt{\frac{N_k}{N_{k,\text{ext}}}} N_k \tag{7}$$

A parametric study performed by means of LDFE analysis (Zhou and Randolph, 2009b) indicates that this method of estimating $N_{k,\text{no softening}}$ is accurate to within about 10%, with the accuracy best for ductile soil response (high values of $\xi_{95}$) and worst for brittle response (small $\xi_{95}$). Cyclic penetration and extraction tests using full-flow penetrometers eventually

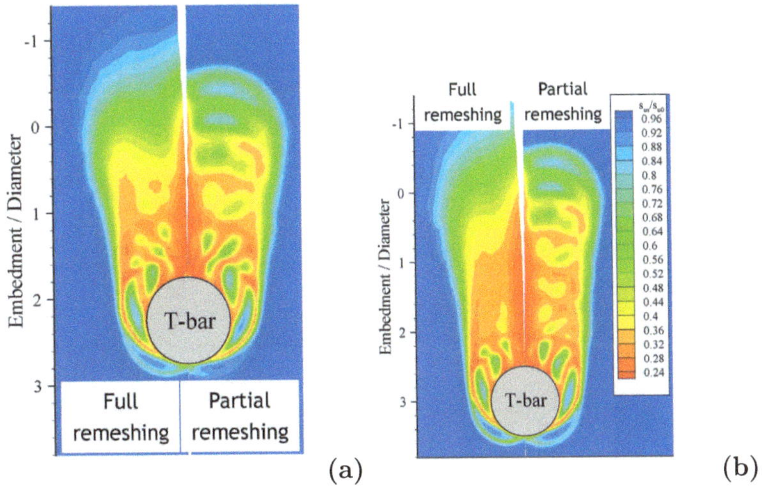


**Figure 15.** Contours of soil softening during T-bar penetration after (a) 2.5 diameters and (b) 3 diameters penetration.

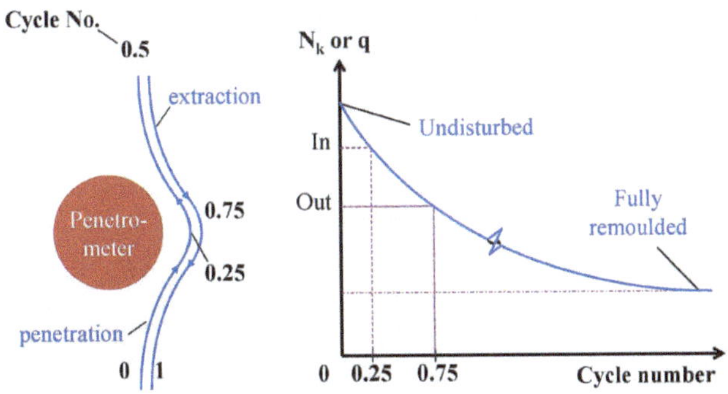


**Figure 16.** Strength degradation during penetration and extraction cycles of a full-flow penetrometer.

lead to fully remoulded conditions in the cylindrical column (or rectangular slot) of soil through which the cyclic test is conducted, and a reduced penetration resistance (see Fig. 17a).

Degradation curves obtained from the LDFE analyses (Fig. 17b) were

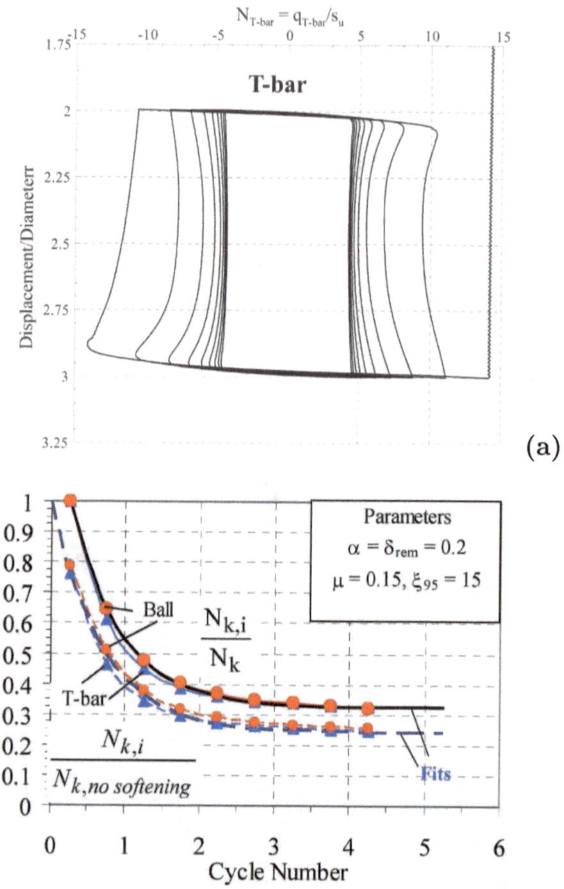


(b)

**Figure 17.** Numerical analysis of full-flow penetrometer: (a) penetration and extraction resistance, (b) degradation response (Zhou and Randolph, 2009b).

fitted using a similar exponential softening curve as in Eq. (6), based either on the 'ideal' (no softening) value of $N_k$, or the partially softened value obtained during initial penetration.

Numerical prediction of degradation curves have been compared with those from field tests (Yafrate and DeJong, 2005) (Fig. 18). The sensitivities obtained from fall cone tests for the three sites are noted in the figure. It may be seen that the agreement is reasonably good for Onsoy clay with

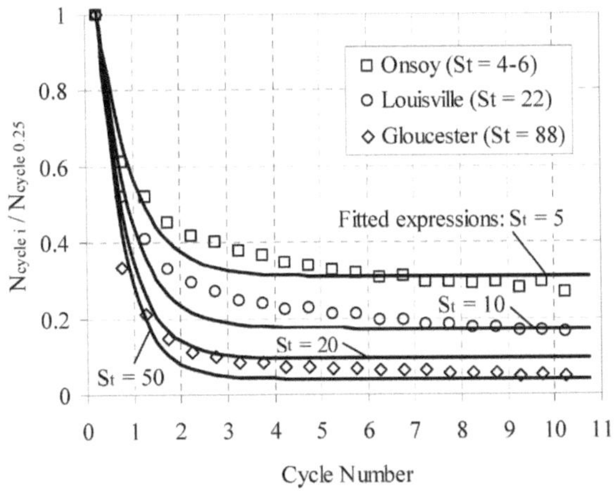


**Figure 18.** Comparison of fitted relationship for $N_{k,i}/N_k$ with field data from Yafrate and DeJong (2005) (Zhou and Randolph, 2009b).

quoted $S_t$ of 4-6. However, for the highly sensitive clays at Louisville and Gloucester, the field data show a better match with curves derived from lower sensitivities of 10 for Louisville (instead of the fall cone value of 22) and 50 for Gloucester (instead of 88). For all three sites, the shape of the degradation response is more gradual than from the numerical analysis, reflecting limitations in the degradation function adopted. The natural material appears to possess a brittle component of strength that is lost during initial penetration, after which further strength loss occurs quite gradually.

Once the soil has been remoulded, to a shear strength of $s_{u,\text{rem}}$, it can be assumed that little further softening occurs, and so the resistance factor will be different from during initial penetration. There are actually three main considerations with regard to post-cyclic resistance:

- further softening within each (half) cycle is essentially negligible, so that the remoulded resistance factor, $N_{k,\text{rem}} = q_{k,\text{rem}}/s_{u,\text{rem}}$, will increase relative to the equivalent $N_k$ during initial penetration;

- the interface friction between the penetrometer and the soil is likely to be a higher ratio (perhaps 70 to 90%) of the remoulded shear strength compared with during initial penetration, again leading to an increase in $N_k$;

- the width of the remoulded region is generally less than the extent of the ideal penetration mechanism, so that some shearing of soil that is not fully remoulded continues to occur at the edge of the failure mechanism.

In fact, theoretical considerations, supported by numerical analysis of cyclic penetrometer tests, suggest that, as the soil softens during the first 1 or 2 cycles, the width of the failure mechanism reduces in order to take advantage of the softened soil close to the penetrometer. Gradually though, the width of the softened zone expands and the mechanism width grows accordingly. This effect can lead to continuing slow degradation in resistance over many cycles. Fig. 19 shows contours of shear strength (softened value normalised by the unsoftened magnitude) and incremental displacements for a T-bar during the middle of the $5^{th}$ penetration for a cyclic range of 3 diameters. A zone of fully softened soil extends to a radius of 1.96 times the T-bar radius, but the flow mechanism extends further, to a radius of 2.28 times that of the T-bar.

Numerical analyses for soil with sensitivity of up to 100 shows that the width of the mechanism contracts very significantly initially, taking advantage of the much weaker remoulded soil close to the penetrometer. With subsequent cycles, the zone of remoulded material gradually expands, but the penetration resistance factor, relative to the remoulded shear strength remains very high.

From a practical point of view, where the remoulded penetration resistance is measured after a limited number of cycles (usually 10), the resistance factor for remoulded conditions, $N_{\rm rem}$, will be significantly higher than for intact conditions, and will increase with increasing sensitivity of the soil. By contrast, the resistance factor for intact conditions will decrease with increasing soil sensitivity. Depending on whether the penetration resistance is compared with the remoulded shear strength measured in a slow test, such as a laboratory UU test, or a fast test such as a vane shear or fall cone test, numerical predictions of resistance factors for the ball, for soil of moderate sensitivity, are around 22 (slow element test) or 15 (fast element test). These values are reasonably consistent with test data reported by Low et al. (2010) shown in Fig. 20.

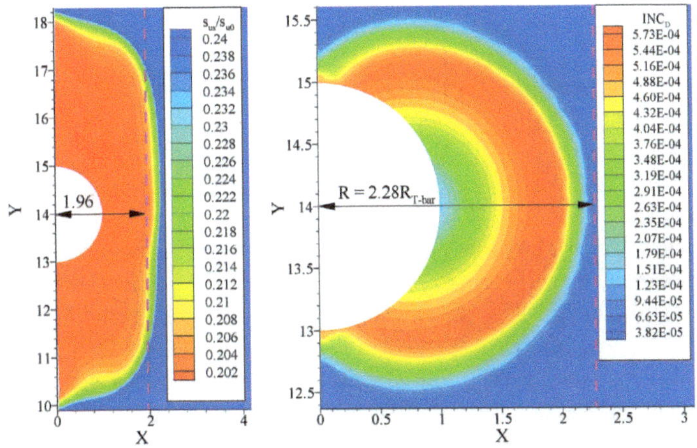


**Figure 19.** Final extent of remoulded zones around a T-bar after cycling (Zhou and Randolph, 2009b).

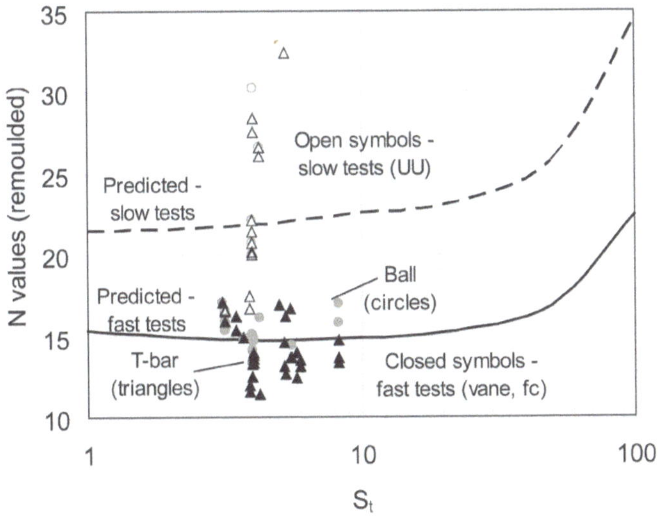


**Figure 20.** Variation of remoulded resistance factor with sensitivity (data from Low et al., 2010).

# 5 Application to Pipeline-Soil and Riser-Soil Interaction

As offshore hydrocarbon developments extend into deeper waters located further from shore, the pipelines and risers that link the production facility to the wells, the offloading point, and in some cases provide an export route to shore, represent an increasingly important part of the development infrastructure.

In shallow water the development facility is usually fixed to the seabed, and the production and export pipelines are usually connected to the facility by vertical 'risers'. The pipelines are often buried to provide protection from over-trawling and prevent thermal buckling. In deep water, trenching and burial is usually uneconomical and there is no need for trawl protection. Instead, pipelines are usually laid on the seabed without overlying protection. However, without the restraint provided by burial, alternative solutions to control the thermal expansion of the pipeline are required.

To connect a floating facility to seabed pipelines, a cost-effective solution is to use a steel catenary riser (SCR), which is essentially a continuation of the pipeline hung from the facility. The manufacture and installation of an SCR is considerably less costly than an equivalent flexible riser, which has a composite structure. A typical field layout is shown in Fig. 21, with flowlines linking the various wells, ultimately connecting to a group of SCRs suspended from the floating production vessel. Flowlines within the field, and also any export pipelines, are laid from specialist vessels, with the steel pipeline forming a catenary from the lay ramp down to the seabed. Thus, during the lay operation, the pipeline configuration is essentially similar to that for a steel catenary riser. Depending on the sea-state, the pipeline will be subjected to cyclic motions at the point where it contacts the seabed, leading to dynamic embedment.

An SCR contacts the seabed at a fixed (average) position, and so any cyclic motion caused by waves acting on the supporting facility, or current-induced vortex induced vibration, will give rise to softening of the seabed sediments and increasing embedment of the riser in what is referred to as the 'touchdown zone'. Trenches of several diameters deep have been observed to form within a few months of operation, as shown in Fig. 22, where the SCR is indicated by the dashed lines (Bridge and Howells, 2007).

A common solution to the thermal expansion of on-bottom pipelines is controlled lateral buckling, which may involve ~10 diameters of lateral movement across the seabed, in order to relieve compressive axial stresses. Sleepers placed on the seabed prior to laying the pipeline are one means

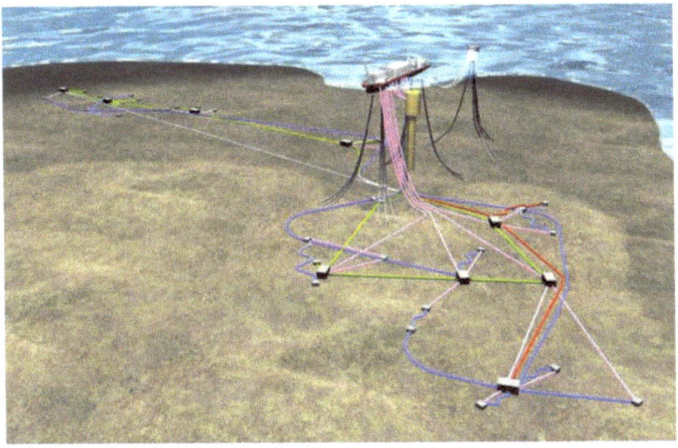

**Figure 21.** Typical field layout in deep water.

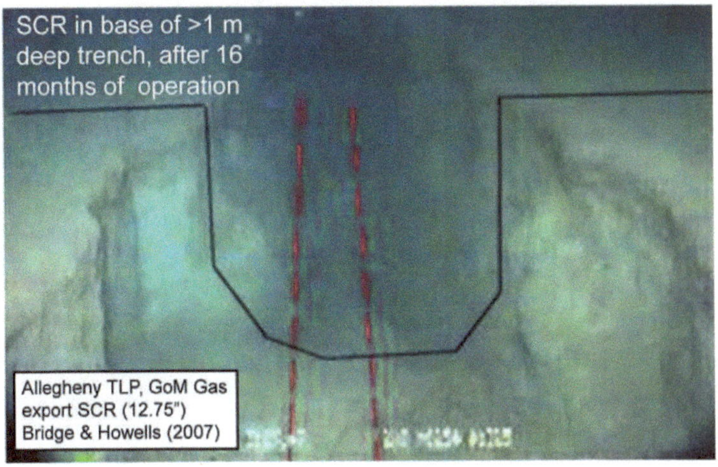


**Figure 22.** Deep trench formation in the touchdown zone of an SCR.

of encouraging lateral buckles to form at specific locations (see Fig. 23, modified from Jayson et al., 2008).

A variety of techniques may be used to investigate and quantify pipeline response under monotonic and cyclic loading. These include physical modelling of pipe elements, on the laboratory floor, in a geotechnical centrifuge of even directly on the seabed (Hill and Jacob, 2008). Post-lay surveys of pipelines also offer valuable feedback on the embedment. Ultimately, the aim is to develop analytical approaches, either through simple conceptual models that might start from plasticity solutions of pipeline response under combined vertical and horizontal loading (Randolph and White, 2008a; Merifield et al., 2008a,b) or through finite element analysis (Wang et al., 2010).

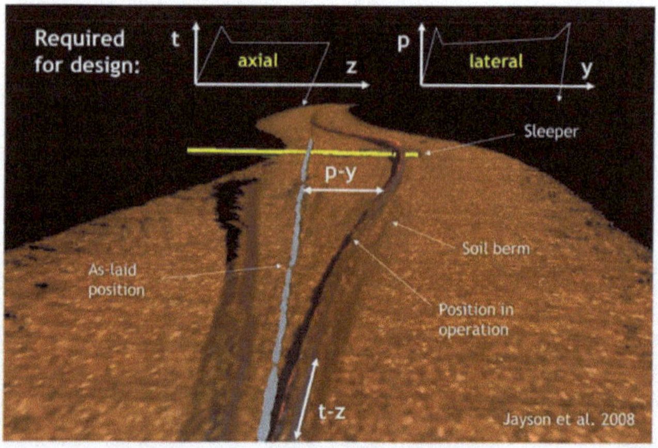


**Figure 23.** Thermally induced lateral buckling of a deep water pipeline.

## 6 Static Pipeline Penetration

It is convenient to express the vertical penetration resistance of a pipe in terms of the net vertical load, $V$, the pipeline diameter, $D$, and the shear strength at the pipeline invert, $s_{u,\text{invert}}$ as (Randolph and White, 2008b)

$$\frac{V}{Ds_{u,\text{invert}}} = a\left(\frac{w}{D}\right)^b + f_b\frac{A_s}{D^2}\frac{\gamma' D}{s_{u,\text{invert}}} \tag{8}$$

where $w$ is the nominal embedment of the invert below the seabed, $A_s$ is the nominal cross-sectional area below the level of the seabed (thus contributing resistance due to buoyancy), and $\gamma'$ is the effective unit weight of the soil.

This may also be expressed in terms of more familiar bearing capacity factors as (Merifield et al., 2009)

$$\frac{V}{Ds_{u,\text{invert}}} = N_c + N_b \frac{\gamma' w}{s_{u,\text{invert}}} \quad \text{where} \quad N_c = a\left(\frac{w}{D}\right)^b \quad \text{and} \quad N_b = f_b \frac{A_s}{Dw} \tag{9}$$

Values of the coefficients $a$ and $b$ that define the $N_c$ term have been provided by Aubeny et al. (2005) for different pipe-soil roughness and soil strength distributions. Typically a combination of $a = 6$ and $b = 0.25$ is sufficiently accurate.

The buoyancy multiplier, $f_b$, accounts for heave adjacent to the pipeline, as illustrated by results from large deformation finite element analyses, comparing results from two different software packages, for soil with a linearly varying shear strength of $s_u = s_{\text{um}} + \rho z$ (see Fig. 24) where $s_{\text{um}}$ is the mudlines strength intercept and $\rho$ is the gradient with depth, $z$. The maximum height of heave, relative to the original seabed, is just over 50% of the (nominal) pipeline penetration.

Although the additional soil above the mudline has little effect on the penetration resistance term, $N_c$, the buoyancy effect is increased significantly by the heave, and should not be neglected in the calculation of total vertical resistance. A buoyancy factor of $f_b = 1.5$ is needed to match the LDFE results (corresponding to a value of $N_b \sim 1$ at $w/D = 0.5$), with the buoyancy resistance then representing up to 16% (at an embedment of $0.5D$) of the total resistance.

LDFE analysis has proved a useful tool to derive penetration resistance curves, including allowing for the effect of heave. However, comparisons with data from centrifuge model tests show that additional adjustments must be made to account for strain rate effects and partial softening of the soil as the pipe penetrates (Zhou et al., 2008; Wang et al., 2010).

## 7 Pipeline Embedment Due to Cyclic Displacements

Additional embedment of the pipeline, over and above the static penetration, will occur during the lay operation. There are two effects: the first is that the local contact force, $V$, will be greater than the (submerged) pipe weight because of bending effects in the touchdown zone (Randolph and White, 2008b); the second is due to cyclic displacement of the pipeline from wave-induced motion of the supporting vessel. The cyclic displacement of the pipeline will lead to local remoulding of the seabed sediments and, if the lateral displacements are sufficient, a ploughing action.

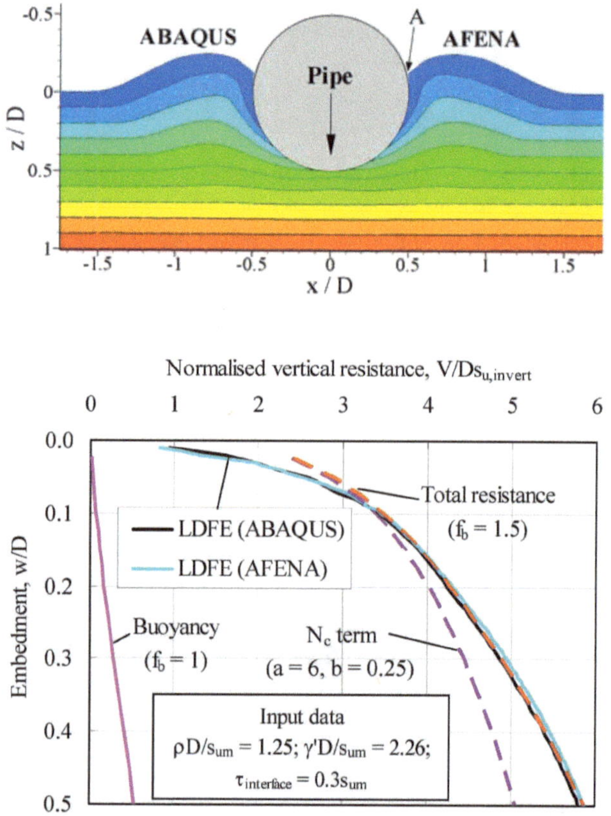


**Figure 24.** Comparison of two large deformation finite element analyses of pipeline penetration (Randolph et al., 2008). Here $\tau_{\text{interface}}$ is the shear stress mobilized along the lateral surface of the pipe.

The situation during pipe lay is similar to that for a steel catenary riser (SCR), with the main difference being that the touchdown zone for the pipeline gradually moves, as more pipeline is laid. Typical lay rates are 10 to 40 m per hour so that, provided there is no hold up due to equipment breakages or bad weather, a given segment of pipeline might see only a few hundred cycles.

The design of an SCR depends critically on the fatigue stresses generated in the region where the SCR touches down on the seabed due to vessel movement and hydrodynamic loading of the catenary. Different levels of sophistication may be applied to quantify the response in the vertical plane. Most commercial software is limited to a linear elastic seabed response. Non-linear models for the vertical riser-soil response have been proposed, incorporating the uplift resistance arising from tensile 'suction' between the riser and the seabed (Bridge et al., 2004). As the soil is softened under the action of repeated cycles of displacement, buoyancy effects can become more significant, leading to a banana shaped response.

Model tests, either on the laboratory floor or in a geotechnical centrifuge, have been undertaken in order to investigate the effects of cyclic penetration and extraction. Two examples are shown in Fig. 25, for cyclic vertical motion at shallow embedment where the effect of buoyancy is very apparent (Hodder et al., 2008), and Fig. 26 with incremental penetration by several diameters (Aubeny et al., 2008).

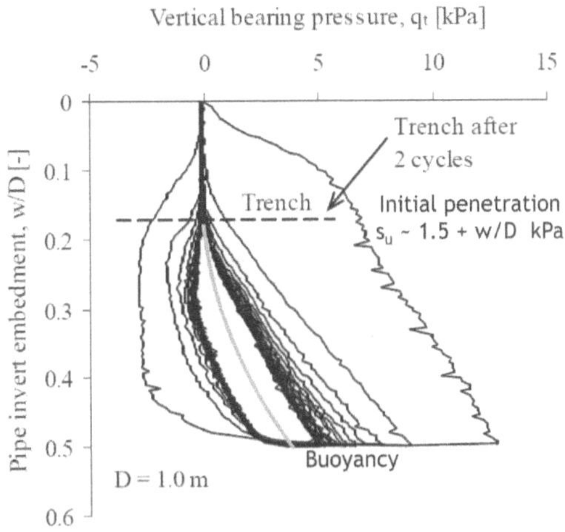


**Figure 25.** Response of pipeline during vertical cycles (Hodder et al., 2008).

### 7.1 Non-Linear Seabed Model for Riser Analysis

A phenomenological model has been developed by Randolph and Quiggin (2009) to model the non-linear interaction of SCRs in the touchdown zone. The model adopts limiting backbone resistance curves in penetration ($P_u(z)$) and uplift ($P_{u,\text{suc}}(z) = f_{\text{suc}}P_u(z)$) and incorporates a hyperbolic response that models high uplift-penetration stiffness at small displacements, with a stiffness that is $K_{\max}$ times the ultimate penetration resistance, $P_u/D$ (equivalent average pressure on the riser of diameter $D$) at that penetration. The high stiffness for small reversals in movement is an important feature for assessing riser fatigue (Bridge et al., 2004).

The main features of the model are illustrated in Fig. 27. The basic hyperbolic model may be expressed as

$$P(z) = P_0 + \frac{\zeta - \zeta_0}{A(z) + \zeta - \zeta_0}\left[P_u(z) - P_0\right]$$

$$\text{where} \quad A = \frac{P_u(z) - P_0}{P_u(z)}; \; \zeta = K_{\max}\frac{z}{D}, \tag{10}$$

where $P_0$ and $\zeta_0$ are the values of $P$ and $\zeta$ at the last reversal point.

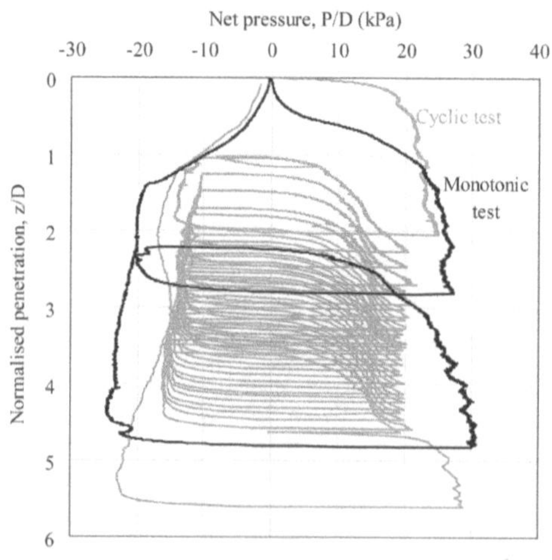


**Figure 26.** Cyclic response of deeply embedded pipeline (Aubeny et al., 2008).

This gives the intended initial stiffness of $K_{\max}P_u/D$ and ultimate asymptotic resistance of $P_u$. Although softening of the soil is not modelled explicitly, the form of the model gives rise to incremental penetration as the riser undergoes cycles of uplift and repenetration. Release of 'suction' resistance during uplift is modelled using an exponential decay term, controlled by

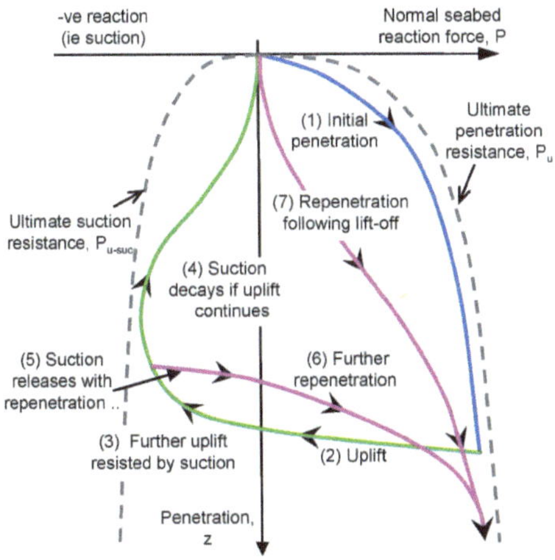


**Figure 27.** Non-linear model for riser-soil interaction (Randolph and Quiggin, 2009).

the parameter $\lambda_{suc}$, which limits the uplift resistance in accordance with the displacement beyond the point at which the resistance changes to negative (suction); the repenetration curve reflects the form of the uplift curve, and also introduces a slight delay (controlled by the parameter $\lambda_{rep}$) in rejoining the backbone curve of ultimate penetration resistance, hence giving rise to incremental penetration with each displacement cycle.

The model has been shown to simulate data from laboratory element tests reported by Aubeny et al. (2008) with reasonable accuracy. It has been implemented in the riser analysis software, Orcaflex (Orcina, 2008), and a typical response for an element of riser within the touchdown zone is illustrated in Fig. 28 for horizontal vessel motions of $\pm 3$ m.

Example fatigue studies have been presented by Randolph and Quiggin (2009), showing the fatigue life for 0.23 m diameter SCR in water depth of 1600 m, for a typical Gulf of Mexico life-time wave spectrum. For a linear seabed with stiffness, $k$ (in units of modulus), the fatigue life reduces from 800 years for a stiffness of 200 kPa, to just over 350 years for a stiffness of 3500 kPa (see Fig. 29). For the non-linear soil model, the fatigue life is a function of how much suction can be sustained during uplift, and also of the seabed strength profile. As the strength intercept at the mudline increases from 0 to 3.5 kPa, the fatigue life shows a similar trend to that for the linear seabed model. The results are consistent, since $K_{max}$ has been taken as 200, so that the effective small displacement stiffness for a given mudline strength intercept, $s_{\mathrm{um}}$, is

$$k' = K_{\max}\frac{P_u}{D} \approx 5K_{\max}s_{\mathrm{um}} = 1000s_{\mathrm{um}} \tag{11}$$

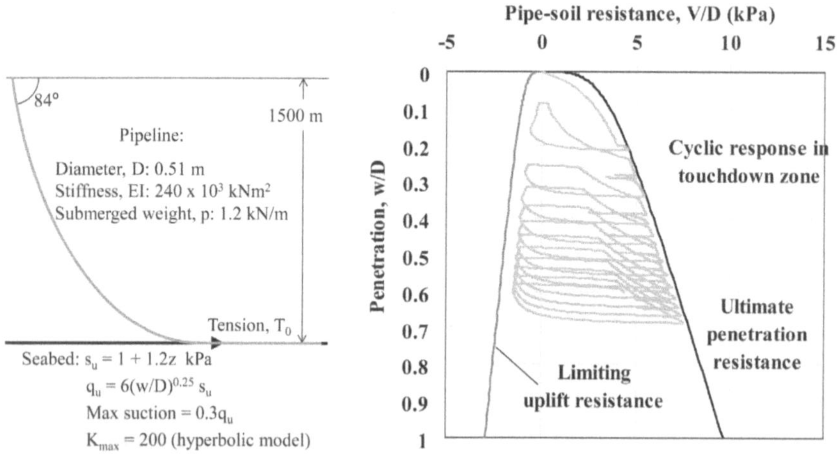


**Figure 28.** Example response of riser in the touchdown zone.

As may be seen, the proportionality constant of 1000 matches the relative scales in the plot. In reality, the cyclic motion of the riser in the touchdown zone will remould the soil, and also encourage water entrainment. It is therefore more reasonable to assume a zero strength intercept at the mudline for fatigue assessment. That leads to much increased fatigue life, in excess of 800 years for a shear strength gradient of 3 kPa/m, assuming suction resistance limited to 20% of the ultimate penetration resistance.

## 7.2 Centrifuge Model Tests for Assessing Pipeline Embedment

Model tests on pipeline segments have been used extensively for design purposes to evaluate the axial and lateral resistance, and also to quantify the tendency to embed under the action of vertical and horizontal cyclic motion. A good illustration of how such tests may be used to develop predictive models that allow for gradual softening of the soil is provided by Cheuk and White (2010). They report tests undertaken in two contrasting soils, kaolin clay and a high plasticity clay from offshore West Africa.

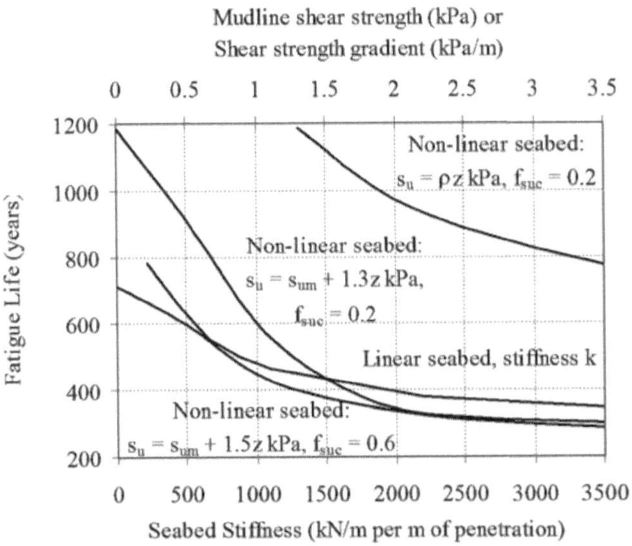


**Figure 29.** Effect of riser-soil interaction model on SCR fatigue life.

Cyclic T-bar tests, without breaking the surface of the sample, showed sensitivities in respect of the penetration resistance of around 2 (Fig. 30). However, it is likely that where the soil is remoulded with free access to water (as for a pipeline at the seabed surface) the actual loss in strength may be significantly higher. The T-bar tests also give an indication of the number of cycles required to remould the soil.

A typical response under displacement limited horizontal cycles, increasing from $\Delta u/D \sim \pm 0.05$ to 0.2, with constant vertical load applied to then pipe segment, is shown in Fig. 31. The static embedment is around $0.1D$, at which point $V/Ds_u$ is just over 3. As the embedment increases under the cyclic motion, $V/Ds_u$ decreases, while $\Delta H/Ds_u$ (where $H$ stands for the horizontal force per unit length) increases initially as the pipe becomes more deeply embedded, but then decreases. A summary of the vertical load and penetration paths for four tests, two in each soil type and at various vertical loads, is given in Fig. 32. It is clear that the additional penetration under cyclic motion is much greater than might be estimated purely on the basis of the sensitivities measured in the T-bar tests.

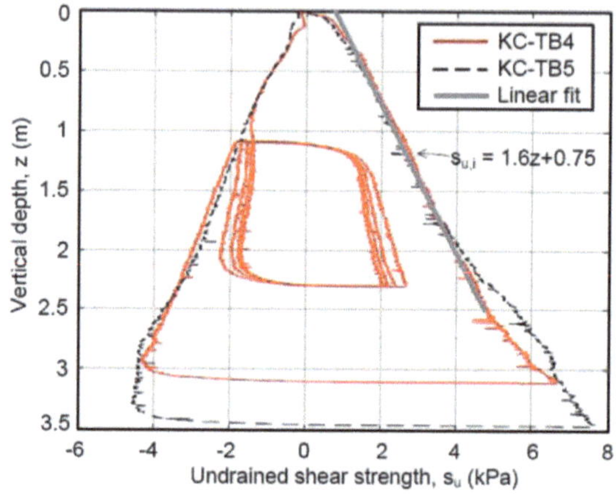


**Figure 30.** Cyclic T-bar tests in kaolin (Cheuk and White, 2010).

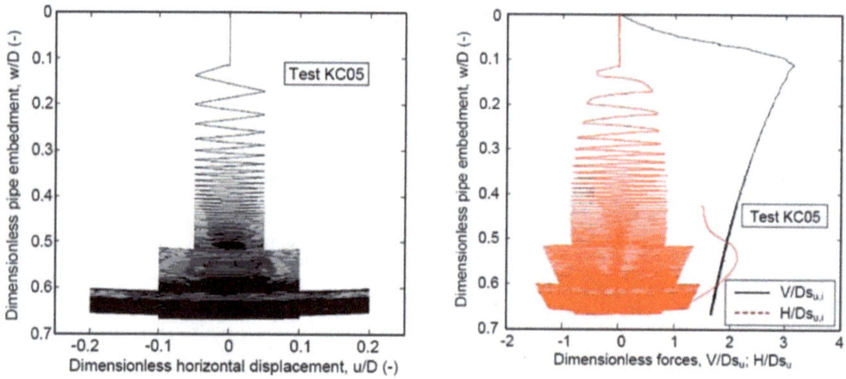


**Figure 31.** Response of pipeline segment to horizontal cycles (Cheuk and White, 2010).

### 7.3 Conceptual Model for Predicting Pipeline Embedment

In order to predict pipeline embedment during horizontal cyclic motions, it is necessary to combine a number of concepts including:

- yield functions in vertical-horizontal load space for varying embedment;

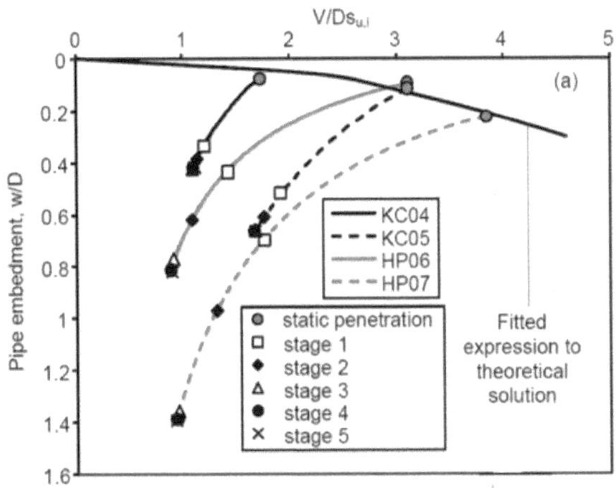


**Figure 32.** Pipeline embedment paths under cyclic motion (Cheuk and White, 2010).

- flow rules corresponding to the yield functions;
- softening of the soil due to remoulding resulting from cumulative shear strain.

An approach involving these concepts is described by Cheuk and White (2010), who proposed a softening model similar to that adopted for analysis of full-flow penetrometers tests, in the form:

$$s_{us}/s_u = \delta_{rem} + (1 - \delta_{rem})\, e^{-3\zeta^* \sum u/D / \sum (u/d)_{95}} \tag{12}$$

The normalised (cumulative) horizontal displacement, $u/D$, is taken as a surrogate for cumulative shear strain, and an adjustment factor, $\zeta^*$, is introduced to allow for the gradual embedment of the pipe into fresh (unsoftened) soil. Thus, at any stage during embedment an attempt is made to estimate an average degree of softening in the soil surrounding the pipe.

Yield functions for unbonded and bonded pipes are taken from Randolph and White (2008a) and Merifield et al. (2008a,b), examples of which are shown in Fig. 33 for bonded and unbonded (immediate breakaway conditions at the trailing edge of the pipe) cases in uniform strength soil or for a normalised strength gradient of $kD/s_{um} = 5$. While it is not clear whether the pipe should be regarded as bonded or unbonded, the assumption is that, during cyclic horizontal displacements, the load point will lie near the parallel point of the yield envelope (where $H/Ds_u$ is a maximum).

Flow rules arising from the yield envelopes for bonded and unbonded pipes follow similar gradients of $dw/du$ against $V/V_{\max}$ over much of the vertical load range (Fig. 34). The critical point is where $dw/du = 0$ (the parallel point), which occurs for $V/V_{\max} \sim 0.4$ for the unbonded case. This suggests that, even without softening, the pipe would penetrate to a depth where the uniaxial ($H = 0$) penetration resistance was some 2.5 times the submerged pipe weight. With full softening, the final embedment would be to a depth where the original (unsoftened, $H = 0$) penetration resistance was $\sim 2.5S_tV$.

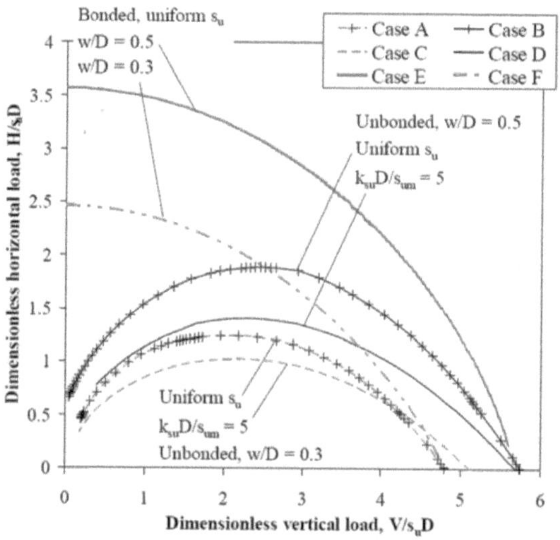


**Figure 33.** Example yield envelopes for bonded and unbonded pipes.

The full procedure for predicting the gradual embedment of the pipe during cyclic horizontal motion is described by Cheuk and White (2010). Fig. 35 shows good agreement of their predictions with the centrifuge data.

## 7.4 Large Displacement Finite Element Analysis of Pipeline Embedment

The most sophisticated approach to simulating pipeline embedment under cyclic motion is by means of large displacement finite element (LDFE) analysis. Simulation of the centrifuge model tests referred to above has been

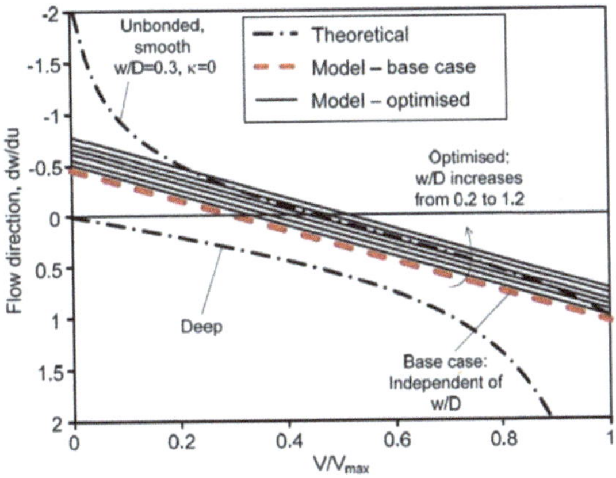


**Figure 34.** Flow direction derived from pipeline yield envelopes.

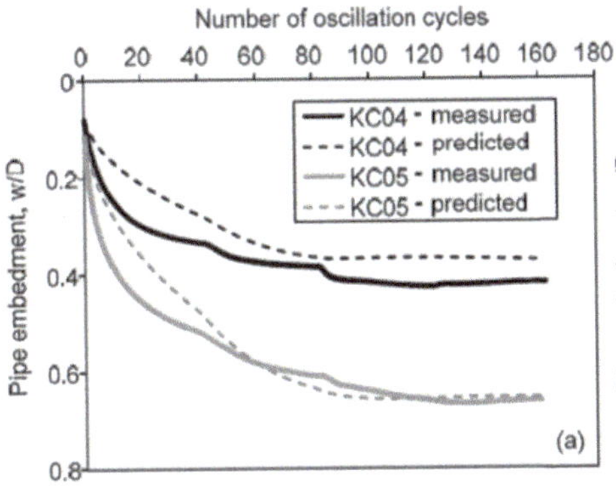


**Figure 35.** Comparison of calculated pipe embedment with measured data.

reported by Wang et al. (2009). The large displacement (or updated geometry) is essential in order to model the significant changes in geometry that occur, and to allow for the flow of softened soil around the pipeline. Softening and rate dependency is modelled in the same way as for the analysis of full-flow penetrometers described in the previous lecture. An example penetration history showing the magnitude of cyclic horizontal load for motion of $\pm 0.05D$, is shown in Fig. 36.

In order to match the model test data, it was necessary to adopt a soil sensitivity of $S_t = 4$, which is higher than the resistance sensitivity indicated from the T-bar tests (see earlier Fig. 30). Part of the difference is because the reduction in penetration resistance from a T-bar test will be less than the sensitivity of the soil, but in addition water entrainment is likely to have occurred during the cyclic tests (and also during a real pipe lay process). That would lead to increased water content of the soil, and increased loss of shear strength as it is remoulded. Comparisons of the pipe embedment with increasing number of cycles showed excellent agreement between the model test data and the LDFE results (Wang et al., 2009).

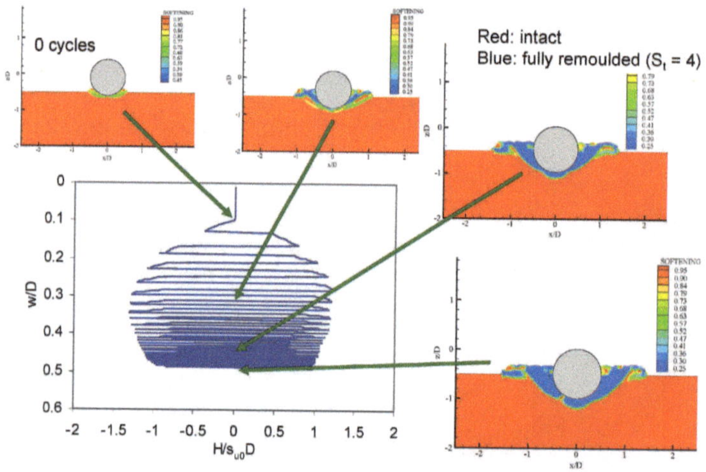


**Figure 36.** Gradual pipeline embedment and local softening of soil.

## 8 Pipeline Motion During Lateral Buckling

As a final illustration of the power of LDFE analysis for modelling soil-structure interaction while the soil is being remoulded, Fig. 37 shows a simulation of the lateral displacement of a pipeline, under fixed vertical load, such as might occur during thermally-induced lateral buckling. Comparison of the LDFE results (Wang et al., 2010) and data from centrifuge model tests (Dingle et al., 2008) shows very good agreement, with the only slight discrepancy being omission of the initial peak in lateral resistance from the centrifuge tests, which is attributable to suction on the trailing face of the pipe.

As the pipeline moves laterally, a berm of (partially remoulded) soil accumulates in front of it (see Fig. 38a, from LDFE analysis). The lateral resistance is affected by the berm, which may be modelled as equivalent to increased embedment, $w'$, as indicated in Fig. 38b (Wang et al., 2010). Assuming that the berm is square in elevation, and that the average degree of remoulding is about 50%, the equivalent embedment is given by

$$\frac{w'}{D} = \frac{w}{D} + \frac{h'_{\text{berm}}}{D} = \frac{w}{D} + \frac{\sqrt{A_{\text{berm}}}}{0.5 S_t D} \tag{13}$$

where $h'_{\text{berm}}$ represents the adjusted height of the lateral soil berm, and $A_{\text{berm}}$ its section area.

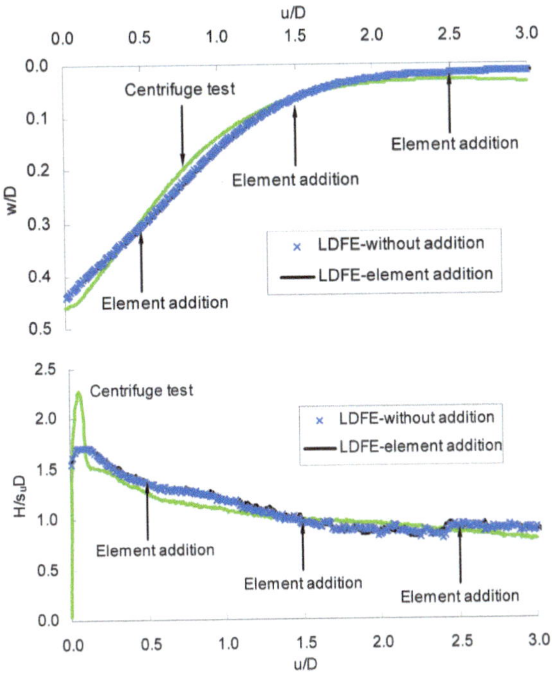


**Figure 37.** Pipeline response under lateral buckling.

It transpires that a common envelope of lateral resistance, $H/Ds_u$, as a function of equivalent embedment, $w'/D$, is followed, independent of the value of $V/V_{\max}$ (or the inverse, the overload ratio, $R$). This is illustrated in Fig. 39 (Wang et al., 2010).

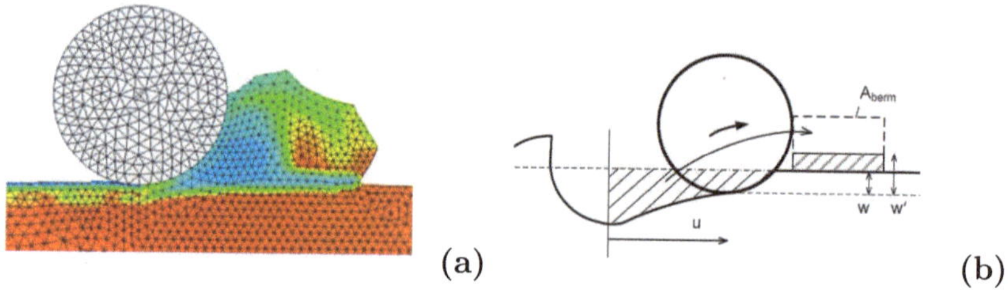


**Figure 38.** Development of soil berm and modelling as equivalent embedment: (a) partially softened soil berm from LDFE analysis; (b) modelling of berm as equivalent embedment

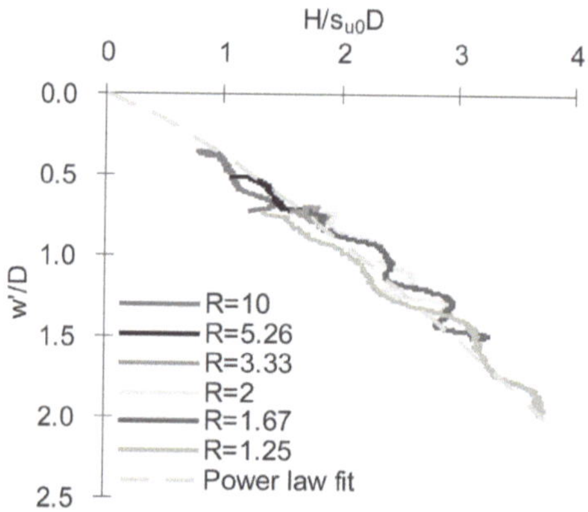


**Figure 39.** Envelope of pipeline lateral resistance with embedment.

## 9 Summary

Interpretation of field penetrometer tests has made significant advances with the development of full-flow penetrometers of simple cylindrical or spherical geometry. Strain rate dependence of shear strength, and partial softening during the flow of soil around the penetrometer, have been shown to have a significant effect, particularly since strain rates in field tests (both penetration and vane shear tests) are several orders of magnitude higher than a slow laboratory element test. In estimating the remoulded shear strength, consideration must therefore be given to the strain rates during the test, and also the strain rate relevant to the design application.

Cyclic penetration and extraction tests using full-flow penetrometers allow direct in situ measurement of the penetration resistance for remoulded conditions. Empirical correlations against remoulded shear strength data are often very scattered, but numerical modelling has allowed a theoretical framework to be established. A key finding is that resistance factors for remoulded conditions will be significantly higher than for initial (intact) conditions, with the difference increasing with increasing sensitivity of the soil.

Gradual remoulding of soil due to cumulative shear strain during cyclic loading is an important aspect of the interaction of pipelines and risers into the seabed. Site investigation practice for deep water pipelines now generally incorporates in situ cyclic T bar penetrometer testing, allowing the sensitivity of the seabed to be assessed, including any tendency to entrain water and thus soften further.

Applications considered in this chapter include (a) steel catenary risers, where the interaction with the seabed in the touchdown zone determines the fatigue life due to bending moment changes in that region; and (b) pipeline embedment during laying due to cyclic motion arising from wave-induced movement of the lay vessel.

Examples have been given of the various degrees of sophistication used in practice to model the softening of the soil due to (relatively large) cyclic strains. The phenomenological hyperbolic model for riser-seabed interaction might be regarded as excessively crude, but it is a significant advance on the widespread practice of representing the seabed as elastic springs. Future advances would endeavour to incorporate the more sophisticated models currently being developed for pipeline embedment. These are based on combining yield functions in horizontal-vertical load space, with a damage factor for the soil strength based on the cumulative normalised cyclic displacement.

Numerical analysis is also developing to the extent where good simulation of test data is being achieved. Large displacement formulations are necessary, modelling non-linear geometric changes as the pipeline penetrates the soil, or moves laterally by a significant distance, ploughing a berm of material ahead of it. A by-product of the large displacement approach (with remeshing and interpolation of stresses and other material parameters) is that it is straightforward to incorporate rate dependency and softening of the shear strength.

## 10 Acknowledgements

This chapter draws from the activities of the Centre for Offshore Foundation Systems (COFS), established under the Australian Research Council's

Research Centres Program and currently supported as a Centre of Excellence by the State of Western Australia and through grants FF0561473 and DP0665958 from the Australian Research Council. The author would like to acknowledge the significant contribution from colleagues at COFS and elsewhere in the world in the material presented.

## List of Symbols

| | | |
|---|---|---|
| $a$ | nondimensional coefficient | §6 |
| $A$ | $= \dfrac{P_u(z) - P_0}{P_u(z)}$ | §7.1 |
| $A_{\text{berm}}$ | section area of the lateral soil berm | §8 |
| $A_s$ | nominal cross-sectional area below the level of the seabed | §6 |
| $b$ | nondimensional coefficient | §6 |
| $d$ | penetrometer diameter in the FE computations | §4.1, §7.3 |
| $d_{\text{field}}$ | penetrometer diameter in field tests | §4.1 |
| $D$ | pipeline diameter | §6, §7, §8 |
| $f$ | interpolating factor | §3 |
| $f_b$ | buoyancy multiplier | §6 |
| $f_s$ | sleeve friction | §2 |
| $f_{\text{suc}}$ | $= \dfrac{P_{u,\text{suc}}}{P_u}$ | §7.1 |
| $h'_{\text{berm}}$ | adjusted height of the lateral soil berm | §8 |
| $H$ | horizontal force per unit length | §7, §8 |
| $k$ | seabed stiffness | §7.1, §7.3 |
| $k'$ | effective small displacement stiffness | §7.1 |
| $K_{\text{max}}$ | stiffness multiplier | §7.1 |
| $N_b$ | $= f_b \dfrac{A_s}{Dw}$ | §6 |
| $N_c$ | $= a \left(\dfrac{w}{D}\right)^b$ | §6 |
| $N_k$ | resistance factor | §3 |
| $N_{\text{rem}}$ | remoulded resistance factor | §3, §4 |
| $P$ | penetration/uplift force per unit length | §7.1 |
| $P_0$ | value of $P$ at the last reversal point | §7.1 |
| $P_u$ | limiting penetration force per unit length | §7.1 |
| $P_{u,\text{suc}}$ | limiting uplift force per unit length | §7.1 |
| $q$ | net bearing resistance | §3 |
| $q_{\text{in}}$ | penetration resistance | §3 |
| $q_{\text{out}}$ | extraction resistance | §3 |

| | | |
|---|---|---|
| $q_{\text{rem}}$ | penetration resistance in remoulded soil | §3 |
| $R$ | overload ratio | §8 |
| $q_t$ | total bearing resistance | §3 |
| $s_u$ | shear strength | §1, §3, §4, §6, §7 |
| $s_{u0}$ | original shear strength prior to any softening | §4.1 |
| $s_{\text{um}}$ | mudlines strength intercept | §6, §7.3 |
| $s_{u,\text{invert}}$ | shear strength at the pipeline invert | §6 |
| $s_{u,\text{rem}}$ | remoulded shear strength | §1 |
| $S_t$ | soil sensitivity | §1, §3, §4, §7, §8 |
| $t$ | time | §4.1 |
| $u$ | horizontal displacement | §7.2, §7.3 |
| $v_p$ | penetration rate in field | §4.1 |
| $V$ | vertical force per unit length | §6, §7, §8 |
| $V_{\text{max}}$ | maximum vertical force per unit length | §7.3, §8 |
| $w$ | nominal embedment | §6, §8 |
| $w'$ | embedment equivalent to the soil berm | §8 |
| $z$ | depth | §6, §7 |
| $\gamma'$ | effective unit weight of the soil | §6 |
| $\dot{\gamma}$ | shear strain rate | §4.1 |
| $\dot{\gamma}_{\text{max}}$ | maximum shear strain rate | §4.1 |
| $\dot{\gamma}_{\text{ref}}$ | reference shear strain rate | §4.1 |
| $\delta$ | displacement | §4.1 |
| $\delta_{\text{rem}}$ | strength ratio of soil between fully remoulded and intact state | §4.1, §7.3 |
| $\delta_{\text{str}}$ | strength ratio of soil between structure or cementation and intact state | §4.1 |
| $\varepsilon_1$ | major principal strain | §4.1 |
| $\varepsilon_3$ | minor principal strain | §4.1 |
| $\zeta$ | $= K_{\text{max}} \dfrac{z}{D}$ | §7.1 |
| $\zeta_0$ | value of $\zeta$ at the last reversal point | §7.1 |
| $\zeta^*$ | adjustment factor | §7.3 |
| $\lambda_{\text{rep}}$ | controlling parameter of delay | §7.1 |

| | | |
|---|---|---|
| $\lambda_{\text{suc}}$ | controlling parameter of suction resistance | §7.1 |
| $\mu$ | rate parameter | §4.1 |
| $\xi$ | plastic shear strain | §4.1 |
| $\xi_{95}$ | value of $\xi$ for the soil to undergo 95% remoulding | §4.1 |
| $\xi_k$ | average shear strain | §4.2 |
| $\rho$ | gradient of $s_u$ with depth | §6 |
| $\sigma_{\text{v}}$ | total overburden stress | §3 |
| $\tau_{\text{interface}}$ | shear stress mobilized along the lateral surface of the pipe | §6 |

## Bibliography

K. H. Andersen, T. Lunne, T. J. Kvalstad, and C. F. Forsberg. Deep water geotechnical engineering. In *Proc. 24th Nat. Conf. of Mexican Soc. of Soil Mechanics*, pages 1–57, Aguascalientes, Mexico, 2008.

C. Aubeny, C. Gaudin, and M. F. Randolph. Cyclic tests of model pipe in kaolin. In *Proc. Offshore Technology Conf.*, volume OTC 19494, Houston, Texas, 2008.

C. P. Aubeny, H. Shi, and J. D. Murff. Collapse loads for a cylinder embedded in trench in cohesive soil. *ASCE Int. J. Geomechanics*, 5(4):320–325, 2005.

G. Biscontin and J. M. Pestana. Influence of peripheral velocity on vane shear strength of an artificial clay. *ASTM Geotechnical Testing Journal*, 24(4):423–429, 2001.

C. D. Bridge and H. A. Howells. Observations and modelling of steel catenary riser trenches. In *Proc. 17th Int. Offshore and Polar Eng. Conf., ISOPE 07*, volume ISOPE-2007-468, Lisbon, Portugal, 2007.

C. D. Bridge, K. Laver, E. Clukey, and T. Evans. Steel catenary riser touchdown point vertical interaction models. In *Proc. Offshore Technology Conf.*, volume OTC 16628, Houston, Texas, 2004.

C. Y. Cheuk and D. J. White. Modelling the dynamic embedment of seabed pipelines. *Géotechnique*, in press:doi: 10.1680/geot.2010.60.00.1, 2010.

S. F. Chung and M. F. Randolph. Penetration resistance in soft clay for different shaped penetrometers. In *Proc. 2nd Int. Conf. on Site Characterisation*, volume 1, pages 671–678, Porto, Portugal, 2004.

H. R. C. Dingle, D. J. White, and C. Gaudin. Mechanisms of pipe embedment and lateral breakout on soft clay. *Canadian Geotechnical Journal*, 45:636–652, 2008.

I. Einav and M. F. Randolph. Combining upper bound and strain path methods for evaluating penetration resistance. *Int. J. Numer. Meth. Engng.*, 63(14):1991–2016, 2005.

S. Hansbo. A new approach to the determination of the shear strength of clays by the fall-cone test. In *Proc. Roy. SGI*, volume 14, pages 7–48, Stockholm, Sweden, 1957.

A. J. Hill and H. Jacob. In-situ measurement of pipe-soil interaction in deep water. In *Proc. Offshore Technology Conf.*, volume OTC 19528, Houston, Texas, 2008.

M. S. Hodder, D. J. White, and M. J. Cassidy. Centrifuge modelling of riser-soil stiffness degradation in the touchdown zone of an scr. In *Proc. 27th Int. Conf. on Offshore Mech. and Arctic Eng.*, volume OMAE2008-57302, pages 241–249, Estoril, Portugal, 2008.

Y. Hu and M. F. Randolph. A practical numerical approach for large deformation problems in soil. *Int. J. Numer. Anal. Methods Geomech.*, 22(5): 327–350, 1998.

D. Jayson, P. Delaporte, J. P. Albert, M. E. Prevost, D. A. S. Bruton, and F. Sinclair. Greater Plutonio Project - subsea flowline design and performance. In *Offshore Pipeline Technology Conf.*, Amsterdam, Netherlands, 2008.

R. Karlsson. Suggested improvements in the liquid limit test, with reference to flow properties of remoulded clays. In *Proc. 5th Int. Conf. Soil Mech. Found. Engng.*, volume 1, pages 171–184, Paris, France, 1961.

P. J. Kelleher and M. F. Randolph. Seabed geotechnical characterisation with the portable remotely operated drill. In *Proc. Int. Symp. on Frontiers in Offshore Geotechnics (ISFOG)*, pages 365–371, Perth, Australia, 2005. Taylor and Francis, London.

T. Koumoto and G. T. Houlsby. Theory and practice of the fall cone test. *Géotechnique*, 51(8):701–712, 2001.

H. E. Low, T. Lunne, K. H. Andersen, M. A. Sjursen, X. Li, and M. F. Randolph. Estimation of intact and remoulded undrained shear strength from penetration tests in soft clays. *Géotechnique*, 60(11):843–859, 2010.

T. Lunne and K. H. Andersen. Soft clay shear strength parameters for deepwater geotechnical design. In *Proc. 6th Int. Conf. Offshore Site Investigation and Geotechnics*, pages 151–176, London, U.K., 2007. Society for Underwater Technology.

C. M. Martin and M. F. Randolph. Upper-bound analysis of lateral pile capacity in cohesive soil. *Géotechnique*, 56(2):141–145, 2006.

R. Merifield, D. J. White, and M. F. Randolph. The ultimate undrained resistance of partially-embedded pipelines. *Géotechnique*, 58(6):461–470, 2008a.

R. Merifield, D. J. White, and M. F. Randolph. The effect of pipe-soil interface conditions on the undrained breakout resistance of partially-embedded pipelines. In *Proc. 12th Int. Conf. of Int. Assoc. for Comp. Methods and Advances in Geomechanics, IACMAG-08*, pages 4249–4256, Goa, India, 2008b. Society for Underwater Technology.

R. Merifield, D. J. White, and M. F. Randolph. The effect of surface heave on the response of partially-embedded pipelines on clay. *ASCE J. Geotech. and Geoenv. Eng.*, 135(6):819–829, 2009.

Orcina. *OrcaFlex User Manual.* http://www.orcina.com, 2008.

J. Peuchen, J. Adrichem, and P. A. Hefer. Practice notes on push-in penetrometers for offshore geotechnical investigation. In *Proc. Int. Symp. on Frontiers in Offshore Geotechnics (ISFOG)*, pages 973–979, Perth, Australia, 2005. Taylor and Francis, London.

G. W. Quiros and A. G. Young. Comparison of field vane, CPT, and laboratory strength data at Santa Barbara Channel site. In A. F. Richards, editor, *Vane Shear Strength Testing in Soils: Field and Laboratory Studies*, pages 306–317. ASTM STP 1014, 1988.

M. F. Randolph. Characterisation of soft sediments for offshore applications. In *Proc. 2nd Int. Conf. on Site Characterisation*, volume 1, pages 209–231, Porto, Portugal, 2004.

M. F. Randolph and P. Quiggin. Non-linear hysteretic seabed model for catenary pipeline contact. In *Proc. 28th Int. Conf. on Offshore Mech. and Arctic Eng.*, volume OMAE2009-79259, pages 145–154, Honolulu, USA, 2009.

M. F. Randolph and D. J. White. Upper bound yield envelopes for pipelines at shallow embedment in clay. *Géotechnique*, 58(4):297–301, 2008a.

M. F. Randolph and D. J. White. Pipeline embedment in deep water: processes and quantitative assessment. In *Proc. Offshore Technology Conf.*, volume OTC 19128, Houston, Texas, 2008b.

M. F. Randolph, P. A. Hefer, J. M. Geise, and P. G. Watson. Improved seabed strength profiling using T-bar penetrometer. In *Proc Int. Conf. Offshore Site Investigation and Foundation Behaviour 'New Frontiers'*, volume OSIFB-98-221, pages 221–235, London, U.K., 1998. Society for Underwater Technology.

M. F. Randolph, C. M. Martin, Y., and Hu. Limiting resistance of a spherical penetrometer in cohesive material. *Géotechnique*, 50(5):573–582, 2000.

M. F. Randolph, H. E. Low, and H. Zhou. In situ testing for design of pipeline and anchoring systems. In *Proc. 6th Int. Conf. Offshore Site Investigation and Geotechnics*, pages 251–262, London, U.K., 2007. Society for Underwater Technology.

M. F. Randolph, D. Wang, H. Zhou, M. S. Hossain, and Y. Hu. Large deformation finite element analysis for offshore applications. In *Proc. 12th Int. Conf. of Int. Assoc. for Comp. Methods and Advances in Geomechanics, IACMAG-08*, pages 3307–3318, Goa, India, 2008. Society for Underwater Technology.

R. S. Semple and J. P. Gemeinhardt. Stress history approach to analysis of soil resistance to pile driving. In *Proc. Offshore Technology Conf.*, volume OTC 3869, Houston, Texas, 1981.

D. P. Stewart and M. F. Randolph. T-bar penetration testing in soft clay. *ASCE J. Geotech. Eng. Div.*, 120(12):2230–2235, 1994.

D. Wang, D. J. White, and M. F. Randolph. Numerical simulation of pipeline dynamic laying process. In *Proc. 28th Int. Conf. on Offshore Mech. and Arctic Eng.*, volume OMAE2009-79199, pages 145–154, Honolulu, USA, 2009.

D. Wang, D. J. White, and M. F. Randolph. Large deformation finite element analysis of pipe penetration and large-amplitude lateral displacement. *Canadian Geotechnical Journal*, 47(10):842–856, 2010.

D. M. Wood. Some fall cone tests. *Géotechnique*, 35(1):64–68, 1985.

N. J. Yafrate and J. T. DeJong. Considerations in evaluating the remoulded undrained shear strength from full flow penetrometer cycling. In *Proc. Int. Symp. on Frontiers in Offshore Geotechnics (ISFOG)*, pages 991–997, Perth, Australia, 2005. Taylor and Francis, London.

N. J. Yafrate, J. DeJong, D. DeGroot, and M. F. Randolph. Assessment of remolded shear strength in soft clay with full flow penetrometers. *ASCE J. Geotech. and Geoenv. Eng.*, 135(9):1179–1189, 2009.

H. Zhou and M. F. Randolph. Computational techniques and shear band development for cylindrical and spherical penetrometers in strain-softening clay. *ASCE Int. J. Geomechanics*, 7(4):287–295, 2007.

H. Zhou and M. F. Randolph. Penetration resistance of cylindrical and spherical penetrometers in rate-dependent and strain-softening clay. *Géotechnique*, 59(2):79–86, 2009a.

H. Zhou and M. F. Randolph. Numerical investigations into cycling of full-flow penetrometers in soft clay. *Géotechnique*, 59(10):801–812, 2009b.

H. Zhou, D. J. White, and M. F. Randolph. Physical and numerical simulation of shallow penetration of a cylindrical object in soft clay. In *Proc. of ASCE GeoCongress 2008*, pages 108–117, New Orleans, Louisiana, 2008. ASCE Geotechnical Special Publication 179.

FSC
www.fsc.org
MIX
Papier | Fördert
gute Waldnutzung
FSC® C083411

Kolibri 360 GmbH
Ferdinand-Jühlke-Straße 7
99095 Erfurt, Deutschland
produktsicherheit@kolibri360.de